BIOLOGICAL INDICATORS OF FRESHWATER POLLUTION AND ENVIRONMENTAL MANAGEMENT

POLLUTION MONITORING SERIES

Advisory Editor: Professor Kenneth Mellanby

Monks Wood Experimental Station,
Abbots Ripton, Huntingdon

Previous titles include
Quantitative Aquatic Biological Indicators
by DAVID J. H. PHILLIPS

Biomonitoring Air Pollutants with Plants
by WILLIAM J. MANNING and WILLIAM A. FEDER

Effect of Heavy Metal Pollution on Plants
Volume 1—Effect of Trace Metals on Plant Function
Volume 2—Metals in the Environment
Edited by N. W. LEPP

Biological Monitoring of Heavy Metal Pollution
by M. H. MARTIN and P. J. COUGHTREY

BIOLOGICAL INDICATORS OF FRESHWATER POLLUTION AND ENVIRONMENTAL MANAGEMENT

J. M. HELLAWELL

*Monitoring Ecologist, Nature Conservancy Council,
Peterborough, UK*

*formerly Principal Scientist, Environmental Aspects,
Severn–Trent Water Authority, Birmingham, UK*

ELSEVIER APPLIED SCIENCE PUBLISHERS
LONDON and NEW YORK

ELSEVIER APPLIED SCIENCE PUBLISHERS LTD
Crown House, Linton Road, Barking, Essex IG11 8JU, England

Sole Distributor in the USA and Canada
ELSEVIER SCIENCE PUBLISHING CO., INC.
52 Vanderbilt Avenue, New York, NY 10017, USA

WITH 82 TABLES AND 89 ILLUSTRATIONS

© ELSEVIER APPLIED SCIENCE PUBLISHERS LTD 1986

British Library Cataloguing in Publication Data

Hellawell, J. M.
  Biological indicators of freshwater
  pollution and environmental management.
  —(Pollution monitoring series)
  1. Freshwater biology
  I. Title II. Series
  574.92′9   024628   QH96

Library of Congress Cataloging in Publication Data

Hellawell, J. M., 1941–
  Biological indicators of freshwater pollution and
environmental management.
  (Pollution monitoring series)
  Bibliography: p.
  Includes index.
  1. Water quality bioassay. 2. Indicators (Biology)
  3. Water quality management. 4. Freshwater ecology.
  5. Water—Pollution. I. Title. II. Series.
  QH96.8.B5H43 1986   628.1′61   86–4441

  ISBN 1-85166-001-1

Photoset in Malta by Interprint Limited
Printed in Great Britain by Page Bros. (Norwich) Limited

Can the rush grow up without mire?
Can the flag grow without water?

Job 8:11

Preface

The preface of a book often provides a convenient place in which the author can tender his apologies for any inadequacies and affords him the facility to excuse himself by reminding the reader that his art is long but life, or at least the portion of it in which he has the opportunity for writing books, is short.

I, too, am deeply conscious that I have undertaken a task which I could not hope to complete to my own satisfaction but I offer, in self-defence, the observation that, inadequate though it is, there is no other book extant, so far as I am aware, which provides the information contained herein within the covers of a single volume. Often during the last decade, in discharging my responsibilities for the environmental aspects of the water authority's operations and works, I should have been deeply grateful to have had access to a compendium such as this. The lack of a convenient source of data made me aware of the need which I have attempted to fill and in doing so I have drawn on my experiences of the kinds of problem which are presented to biologists in the water industry. The maxim 'half a loaf is better than none' seems particularly apt in this context.

The book owes much to the classic work by Professor H. B. N. Hynes, *The Biology of Polluted Waters*, published in 1960, which was essential reading during my undergraduate days when I was privileged to be taught freshwater biology by him. However, much has changed over the last quarter century. Former concern over foaming detergents has disappeared to be replaced by the problem of detergent phosphates contributing to eutrophication. Pesticides such as insecticides and herbicides attracted little attention in those days and acid rain existed only in jokes about bicycles which fizzed when left out in the wet in towns like Widnes or Runcorn! Since the late 1950s, organic pollution in Britain has greatly diminished and as a consequence other environmental stresses are now more evident. We have since learned that while some substances may enter the environment in very low concentrations they move insidiously through food chains and, by a gradual process of biological concentration, reach harmful levels in certain groups of species. Increased environmental awareness by society at large has inevitably modified the professional ecologist's perception of the pollution problem and this has also stimulated much more emphasis on the need for positive conservation of the environment rather than simply ameliorating existing pollution. The environmental impact assessment seems to be here to stay.

In compiling the material in Chapter 2, on basic biology and ecological principles, I have had regard to my experiences over several years in lecturing on elementary applied hydrobiology on courses intended for non-specialists given under the auspices of the training section of the National Water Council, now the Water Industry Training Association. It is hoped that by providing this, admittedly simple, account, those without formal training in biology will be able to appreciate better the ecological significance of pollutants or environmental manipulation. Elsewhere I have drawn upon my experience as an external examiner for a number of higher degree courses in applied hydrobiology and pollution control and trust that these sections will prove valuable to postgraduate students.

I am conscious that the treatment of each topic is in no way exhaustive and that some repetition has been inevitable since I have attempted to ensure that each section is relatively self-contained. Topics do not exist as isolated compartments but have relevance for each other. One must either accept some overlap or suffer an excessive number of cross-references. The latter are provided, however, where these are considered to be helpful. The literature reviewed has had to be selective and must be regarded as illustrative rather than comprehensive. Wherever possible, preference has been given to review papers or to publications which are fairly readily available in order that the reader may consult the originals should he wish to delve deeper.

Thanks are due to Drs Malcolm Elliott and John Wright of the Freshwater Biological Association, for assistance with Chapters 5 and 10 respectively, to my colleague Katharine Bryan for helpful discussions and comment, to Mary McCullough for turning my virtually illegible manuscript into typescript and to the publishers for their unfailing courtesy and considerate guidance in preparing copy.

J. M. HELLAWELL

Acknowledgements

Acknowledgement is gratefully made to authors and publishers of material which has been redrawn or reset in tables or reproduced directly or reproduced with minor modification. The exact locations can be derived from the references. For permission to reproduce copyright material thanks are due to Academic Press Inc. Ltd for Figs. 5.8, 5.11 and 5.12 from *Stress in Fish* Blackwell Scientific Publications Ltd for Fig. 5.4 from *Journal of Animal Ecology*, Fig. 6.14 from *The Scientific Management of Animal and Plant Communities for Conservation* and Fig. 9.5a from *Freshwater Biology*, the Canadian Department of Fisheries and Oceans for Fig. 3.7 from *Journal of the Fisheries Research Board of Canada*, Elsevier Science Publishers BV, Amsterdam, for Fig. 3.9 from *Aquatic Toxicology* and Fig. 3.6 from *The Science of the Total Environment*, the Director of the Freshwater Biological Association for Fig. 8.3 from the *FBA Annual Report* and Fig. 9.5b from the *Occasional Publications*, the Controller of Her Majesty's Stationery Office for Figs. 5.13, 6.3, 7.5 and 8.8 (Crown Copyright Reserved), the Editor of *Oikos* for Fig. 9.5c, Pergamon Press Ltd for Fig. 7.2 from *Water Research*, The Royal Society for Figs. 3.1, 3.2, 3.5 and Table 9.4 from the *Proceedings of the Royal Society of London, Series B*, Springer-Verlag, Heidelberg, for Fig. 5.7 from *Oecologia (Berlin)* and the Vanderbilt University Press for Fig. 5.6 from *The Biological Effects of Thermal Pollution*.

The assistance and kindness of my friends H. A. Hawkes, T. E. Tooby, and Drs J. W. G. Lund, FRS, J. M. Elliott and J. B. Leeming in providing ideas or material for certain figures is also gratefully acknowledged.

J. M. H.

Contents

CHAPTER 1

Water and Man

1.1 HISTORICAL SETTING

Water comprises up to 70% of the human body, while some soft tissues are about 85% water. Typical daily consumption for drinking in temperate climates is only about 1·6 litres, but in the United Kingdom the average daily per capita usage of water for all domestic purposes is about 110 litres (24 gallons). It is not altogether surprising, therefore, to find that many of the earliest great civilisations developed in the basins of large rivers, where there was ample water for large concentrated populations, for irrigation of their crops and where raw materials and manufactured products could readily be transported along the watercourses. Familiar examples include the empires of Sumeria, Assyria and Babylon which developed in Mesopotamia ('in the midst of rivers') between the Tigris and Euphrates, Egypt which flourished along the course of the Nile, and the great Indian civilisations on the Indus and Ganges.

Among less settled peoples, the availability of reliable and secure water supplies was essential; hence there are many records of disputes over wells in Biblical times (e.g. Genesis 26: 18–33). Ancient fortifications were largely useless without an unfailing water supply when under siege. When the Assyrian King, Sennacherib, came against Jerusalem in 701 BC, King Hezekiah had diverted a spring through a hastily rock-hewn conduit to provide a supply within the city while simultaneously depriving his enemy (Keller, 1956; Abrahams, 1978). Similarly, the protracted seige of Masada by Flavius Silva in AD 72 was largely the consequence of an ingenious previous provision for ample water supplies within the fortress by Herod's engineers and virtually no water in the surrounding desert area for the Roman army (Yadin, 1966).

1

In more recent times, the industrial revolution sprang from the advent of power from water driven wheels, and many industries relying on water for their processes or for transport originated along river valleys.

Concentrations of individuals within the cities which cradled the civilised world brought problems of effluent disposal and, in particular, the necessity for methods which protected the clean sources of potable and bathing water. It is a sad comment on human progress that many ancient civilisations had means of water supply and effluent disposal which were not matched in subsequent history until the nineteenth century! For example, many ancient cities in Mesopotamia had water supplies and water-borne sewage systems. The technology of water supply and waste disposal was equally well developed in contemporary civilisations such as the Indus Valley and on the island of Crete (Gray, 1940). Roman engineers constructed magnificent aqueducts to convey supplies of wholesome water and also drainage systems to remove surface and underground water, although foul drainage was poorly developed, except locally. By contrast, in medieval European cities the sanitary conditions were virtually indescribable and epidemics of water-borne diseases, such as cholera and typhoid, were common. Indeed it is a strong condemnation of our twentieth century that even now, in many parts of the world, people are entertained by transistor technology and yet do not have a convenient supply of clean water nor adequate means of waste disposal.

The importance of water and riparian facilities in the economy of civilised societies is reflected in the development of legislation to protect resources, maintain certain rights and apportion others. Ancient water law and Roman water law has been reviewed by Caponera (1976). In Britain, common law on water and riparian rights developed as an adjunct to the law on ownership of the land and possession of the bank confers riparian rights. Interest in the fishery rights is normally related to bank ownership but the fishery can be held quite separately by a third party. The fish, being wild creatures, are free to come and go and so are not the property of the fishery owner until they are lawfully caught in his water, but when those same fish are killed by accidental spillage of a poison, the owner of the fishery is entitled to compensation.

When such a vital commodity as water is shared by separate groups having different and often conflicting interests, it is inevitable that legislation which seeks to reduce or resolve these should itself be complicated and even conflicting. The law on watercourses in the United Kingdom is considered by Wisdom (1962–1979; 1981) and, more briefly, the early historical development in the United States by Warren (1971). Recent legislation on the provision of water supplies and effluent control in the USA is provided by Okun (1977). Much of the conflict stems from the fundamental and just premise that the riparian owner has a basic right to the water and its use but not so as to deprive other owners of their rights. Now in a river system, and to a lesser extent along a lake shoreline, the activities of one riparian owner naturally impinge on his neighbours. Taking, by way of illustration, the construction of a dam and abstraction of water to irrigate crops, it is evident that downstream less water will flow, especially during the dry season, and the upstream owners might be deprived of their fishery if a migratory species is unable to surmount the dam. An industrialist who abstracts clean water for his manufacturing process but discharges a polluting effluent interferes with the rights of other industrialists downstream and would himself be directly injured by a competitor who set up a similar business upstream. The matter becomes more complex when those without direct access to a watercourse such as the inhabitants of towns and cities produce wastes and effluents which will, sooner or later and by one means or another, find their way into streams and rivers.

Growing concern for public health and the deteriorating quality of the environment as large conurbations, with their developing associated trades and industries, sprang up during the nineteenth century was the springboard for the launch of legislation to control the quality of effluents and the waters which received them. The key to progress was the enactment of laws towards the end of the nineteenth century which recognised that the disposal and treatment of wastes was a collective national responsibility. From this point it was clearly recognised that quality standards had to be determined and means found for their attainment. Legislation in Britain was largely iterative (Klein, 1962; Hynes, 1960) but the setting up of the Royal Commission on Sewage Disposal in 1898 was a major landmark. The Royal Commission published a series of nine reports from 1901 to 1915 and the final report was also published in 1915. Perhaps the most significant of these, in that its proposals were to have a dominating effect on pollution control practice for many decades, was the eighth which proposed standards based on the biochemical oxygen demand (BOD) test and measurement of the suspended solids content of the effluent. An effluent having a BOD of 20 parts per million (oxygen demand) and a suspended solids concentration of 30 ppm met the Royal Commission's standard provided it was diluted eight-fold in the receiving watercourse, which itself had to meet certain assumed quality criteria. This effluent standard was adopted widely, although it was often misapplied, largely through failure to observe the dilution criteria. It is also of some interest to note that the Royal Commission proposed a classification of watercourses based on biochemical oxygen demand and their biological condition.

Legislation in Europe, North America and elsewhere has increased both in its scope and efficacy since the Second World War. A useful review has been provided by OECD (1957). As the grosser forms of pollution have been brought under control attention has tended to focus upon the subtler and more insidious effects of materials entering the environment by indirect or less evident routes. Perhaps the most emotive of these have been pesticides, notably the insecticides DDT, dieldrin, aldrin and endrin and some herbicides, of which 2,4,5-T is probably the most familiar and which has caused concern mainly because of the presence of impurities, in particular of dioxin. Fuller appreciation of the complexities of pollutant pathways and of ecological phenomena, and a greater environmental awareness on the part of the general public in developed countries, have resulted in political responses to increased aspirations for environmental protection and restoration. One may cite the introduction of Environmental Quality Directives as an ingredient of legislation in the European Economic Community and of the establishment of the Environmental Protection Agency in the United States as examples of the political expression of popular concern and, unfortunately, of sources of conflict (Freeman, 1978). But this anticipates issues to which we must return later, in Chapter 10.

1.2 MANAGEMENT OF THE WATER CYCLE

The water cycle was understood in ancient times (Ecclesiastes 1:7) but only in the last few decades has the significance of the need to manage the whole cycle been appreciated. In practice, we can only manipulate certain phases of the cycle but the relevance of these to the whole has now entered our thinking on management strategy. A simple system might involve a reservoir, fed by natural catchment run-off, from which water is fed into domestic supply, returns to the river as effluent and finally enters the sea. As demand outstrips the yield of such upland storage, or the costs of long-distance supply pipelines from remote catchments escalate, the stage is reached where rivers carrying effluent

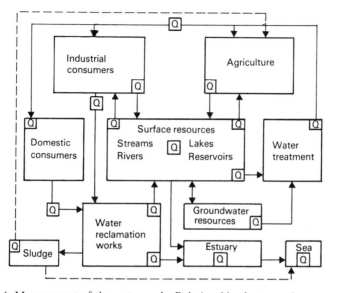

FIG. 1.1 Management of the water cycle. Relationships between the components and stages at which water or environmental quality (*Q*) is monitored. From Brewin and Hellawell (1980).

might have to be pressed into service as sources of raw water provided that the effluents were of a higher quality and more advanced treatment was applied to abstracted water, or both. Thus the water cycle acquires a much shorter radius since recycling of water now takes place within the watercourse, so one town's effluent becomes another's raw water supply and the change in terminology from 'sewage treatment' to 'water reclamation' ceases to be merely a euphemism (Brewin and Hellawell, 1980). In order to be able to manage the smaller cycle more effectively it is essential that the quality is monitored at key stages. This is illustrated in Fig. 1.1 where the inter-relationships between the various components of the cycle and their associated features are shown and the stages or points where water or environmental quality is monitored are indicated.

Differences in the approach to water management between the United States and Britain, and in particular the relevance of the mid-seventies re-organisation in England and Wales—which placed all functions within single or groups of river basins—for the North American administrative organisations are examined by Okun (1977).

1.2.1 The Biological Component of Water Cycle Management

The biological component of water cycle management is greater than is often appreciated and is to be found in almost all phases of the provision of water, subsequent treatment of effluents, and manipulation of water-courses. The terrestrial phases of the cycle and associated water management activities are portrayed in Fig. 1.2, where their biological components and ecological consequences, respectively, are identified. The ecological and environmental implications of water management are considered more fully in Chapter 4. The purpose of identifying them here is to emphasise that, while the provision of evident physical structures is important—for example, in sewage treatment or in slow sand-filtration of potable water—the actual processes are essentially biological. Operational problems are often biological phenomena; examples include tastes and odours in drinking water attributable to algal secretions, filter blockages caused by reservoir plankton, fly nuisances in the neighbourhood of treatment works, 'dirty' water or 'foreign bodies' emerging from the customer's taps (US 'faucets') and, perhaps the most significant of all, enteric infections. This last example also serves to emphasise that many activities in water management are undertaken in order to protect biological components, in this case the customer! But many others, such as setting and monitoring environmental quality objectives, largely through effluent control policies, are ultimately intended to protect human life, the wellbeing of resources such as livestock, crops or fisheries (both for food and recreation), wildlife and the aesthetic quality of our environment. The standards are not ends or goals in themselves but means to an end. Sadly, all too often great effort is expended in securing legislative powers, and resources are committed to facilitate the enforcement of such standards, but relatively little attempt is made to ensure that the standards are adequate or even relevant to the biota which they are, presumably, intended to protect. Equally unfortunate is the other extreme where, in order to be sure that the environment is safe, in the absence of adequate data, standards have been unrealistically stringent, even to a point beyond the limit of detection of practicable chemical analysis procedures.

Since the biological component of water management is so important it would seem appropriate that biological methods should play a large role in monitoring the effectiveness of management procedures and that biological data should contribute to formulation of management policy. The necessity for greater involvement of biological methods was recognised by Hynes (1960) with particular reference to pollution control

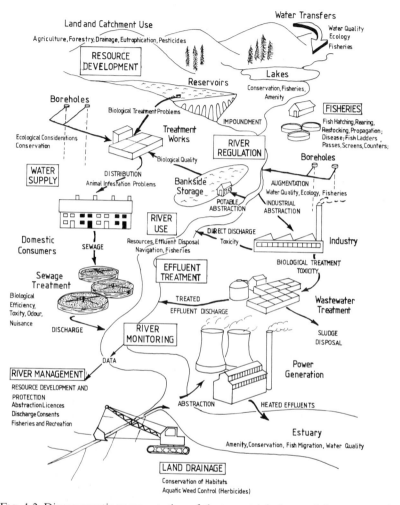

FIG. 1.2 Diagrammatic representation of the terrestrial phases of the water cycle showing the biological components and the environmental significance of water management activities.

and over the last couple of decades biological expertise has contributed to other aspects of water management.

1.2.2 The Contribution of Biological Methods

The use of biological systems as monitors or indicators of environmental health means that one avoids the necessity for abstract concepts of environmental wellbeing independent of the entity towards which our concern is directed. To assume that we can define realistic universal environmental parameters which, if met, will safeguard all the biota is missing the point. It would be equally pointless to overlook the fact that measuring a physical or chemical parameter is usually easier than undertaking a biological investigation. If one chooses a simple example, such as an upper temperature limit which will avoid significant biological damage, it is much easier to measure the maximum temperature experienced over a given period than to investigate whether the biota has suffered any heat damage. However, simply noting that the temperature did not exceed, say, 35°C would not, of itself, contribute a great deal to the protection of the biological system, unless one was certain that a sustained high temperature of 30–34°C would have no serious ecological consequences.

Other physico-chemical variables are less easily measured: *in situ* apparatus is available for some determinands but many others have to be estimated by conventional sampling and chemical analysis. Discrete samples taken at intervals have the inherent disadvantage that they only provide a transient picture, that is a 'snapshot' rather than a moving film; if repeated frequently enough they provide a better idea of what may be happening but place heavy demands on time and resources. Automatic sampling apparatus, which takes small water samples at regular intervals and bulks them, gives a better idea of the mean conditions but it is not then possible to measure extremes and these are of great interest in environmental protection. There is little doubt that progress will lead to better methods, not only in terms of selective problems or sampling techniques, but also in data logging and analysis. This latter aspect is particularly important in relation to continuous records: miles of charts or almost infinite columns of figures can be disconcerting and even counter-productive. Automatic data analysis to provide relevant statistical parameters is essential if the data are to make a contribution to environmental management. The problem is to determine which parameters are most informative.

In such a context, the use of biological monitors becomes most attractive since they are continuously exposed to all the variations in environmental quality and their response is an integrated one: they reflect the ecological impacts of peaks or troughs or sustained levels of quality because they are the ecological consequences of the environmental variability. All that is required is that we should be able to describe the biota adequately and comprehend what we perceive, that is, be able to detect and interpret differences between what we expect to exist or occur and what we observe to be the case from our sampling, analyses or experiments. We may then take steps to remedy these in terms of pollution control or environmental manipulation, and use the same methods to confirm the efficacy of our remedial action. This is the rationale for the material which is developed in subsequent chapters.

CHAPTER 2

Freshwater Ecosystems

2.1 INTRODUCTION

It would be presumptuous to pretend that one can give more than the barest of outlines of freshwater ecosystems within the compass of a single chapter. However, it is possible to provide guidance on some ecological principles which will help to prepare the ground for the issues which are to be raised in subsequent chapters. The emphasis here, then, will be placed on those matters which are important for a better understanding of the ways in which the consequences of particular activities, for example the use of pesticides or the discharge of certain effluents, may extend beyond the immediate areas of spatial or temporal influence.

Take, as a simple, hypothetical example, the introduction of a new herbicide which is suitable for the clearance of weeds from slow-flowing rivers or canals. Naturally, concern will be felt regarding its toxicity to other aquatic life and in particular to fishes. Toxicity tests will be conducted on adult fish and, perhaps, also on fry and eggs. Suppose that these reveal no acutely harmful effects at concentrations which are effective for weed control. One might conclude that the herbicide may be used without any deleterious effect on the fish populations. However, the herbicide will destroy plants on which the adult fish spawn and which shelter the fry from predators. In addition, many of the food organisms of both fry and adult fish depend on the plants for shelter or their food supply. Although the herbicide is likely to have little immediate effect upon the fishes, with repeated or long-term use the destruction of the weeds may cause serious consequential effects as the reproductive capacity and feeding of the fishes are impaired. In extreme cases their ultimate fate will be the same as that which a toxic herbicide would have caused. Examples of similarly extended consequences could be multiplied; this rather simple case serves to illustrate the type of issues which may be involved. In subsequent sections others will be encountered.

Readers who wish to explore ecological principles more fully are referred to an extensive literature. For a basic grounding in freshwater ecology books by Maitland (1978) and Whitton (1975) will be found helpful. Works which are oriented towards pollution ecology include Hynes (1960), Mason (1981), Warren (1971) and Welch (1980). For more detailed accounts the reviews of Hynes (1970) and Hutchinson (1957–1975) are recommended.

2.2 CLASSIFICATION OF ORGANISMS

To the uninitiated, the mysteries of biological taxonomy probably constitute the greatest primary obstacle to communication. Once the principles of classifying organisms are mastered, they are perceived as an aid to reducing the vast variety of biological complexity to some degree of order, rather than an underhand ruse whereby biologists try to keep chemists, engineers and other laymen in their place!

The scientific classification of organisms is hierarchical, each level embracing the characteristics which its components have in common but differing from all the other equivalent taxa in some important aspect. The basic unit is the *species* which, although difficult to define adequately in absolute terms (for there is a sense in which it is merely a figment of the taxonomist's imagination) it may, for most intents and purposes, be regarded as that group of individuals capable of breeding with each other so as to produce fertile offspring. Although this definition of a species is theoretically sound, it does not enable one to recognise the limits of species in nature without testing the breeding capability of all individuals, and clearly this would be impossible. In practice, a species is recognised by the close and fairly constant physical similarity of its members, together with the breeding facility, which is usually evident for at least a significant population and which apparently does not occur with other species.

Species are grouped into *genera*, each genus being equivalent to the surname of the two-part Latinised specific name whilst the specific epithet is comparable with a forename. For example, the scientific name of the lion is *Felis leo*, equivalent to Smith, John, whilst another member of this genus, the tiger, is *Felis tigris*, which, continuing with our analogy, is the counterpart of Smith, Mary. Sometimes species exhibit minor

TABLE 2.1

A SIMPLIFIED CLASSIFICATION OF ORGANISMS SHOWING THE TAXONOMIC RELATIONSHIPS OF THE PRINCIPAL GROUPS ENCOUNTERED IN FRESHWATER

PLANT KINGDOM

BACTERIA	*Sphaerotilus*
FUNGI	*Penicillium*
ALGAE	
Cyanophyta ('blue-greens')	*Anabaena, Microcystis*
	Nostoc, Oscillatoria, Rivularia
CHRYSOPHYTA	
Xanthophycae ('yellow-greens')	*Vaucheria*
Bacillariophycae (diatoms)	*Asterionella, Cocconeis,*
	Cyclotella, Diatoma, Fragilaria,
	Gomphonema, Melosira, Tabellaria
RHODOPHYTA ('reds')	*Batrachospermum, Lemanea*
CHAROPHYTA (stoneworts)	*Chara, Nitella*
BRYOPHYTA (non-vascular,	
non-flowering plants)	
HEPATICAE (liverworts)	*Riccia, Scapania*
MUSCI (mosses)	*Ambystegium, Fontinalis,*
	Rhynchostegium, Sphagnum
PTERIDOPHYTA (vascular,	
non-flowering plants)	
EQUISETALES (horsetails)	*Equisetum*
PTEROPSIDA (ferns)	*Azolla, Salvinia*
SPERMATOPHYTA (vascular, seed	
bearing (flowering) plants)	
MONOCOTYLEDONES	*Elodea, Lemna, Phalaris,*
	Phragmites, Potamogeton,
	Sparganium
DICOTYLEDONES	*Myriophyllum, Nymphea,*
	Polygonum, Ranunculus

ANIMAL KINGDOM

PROTOZOA (acellular, mainly	*Amoeba, Paramecium, Vorticella*
microscopic animals)	
PORIFERA (sponges)	*Euspongilla*
COELENTERATA (jelly fishes)	*Hydra*
PLATYHELMINTHES (flatworms)	
TURBELLARIA	*Polycelis*
ANNELIDA (worms, leeches)	
OLIGOCHAETA (worms)	*Limnodrilus, Tubifex*
HIRUDINEA (leeches)	*Glossiphonia, Haemopsis*
MOLLUSCA (shellfish)	
GASTROPODA (snails limpets)	*Ancylus, Campeloma, Lymnaca,*
	Physa
BIVALVIA (mussels, clams)	*Anodonta, Pisidium, Unio*
ARTHROPODA	
CRUSTACEA (crabs, lobsters,	
shrimps, etc)	

TABLE 2.1—*contd.*

ANIMAL KINGDOM

BRACHIOPODA	
Cladocera (water fleas)	*Daphnia*
OSTRACODA (beanshrimps)	*Cypris*
COPEPODA	*Cyclops*
MALACOSTRACA	
Isopoda (waterlice, sowbugs)	*Asellus, Lirceus*
Amphipoda (scuds)	*Gammarus, Hyalella*
Decapoda (crayfishes)	*Astacus, Orconectes,*
	Palaemonetes, Promcambarus
ARACHNIDA	
HYDRACARINA (water mites)	
ARANEAE (spiders)	
INSECTA (insects)	
PLECOPTERA (stoneflies)	*Acroneuria, Nemoura,*
	Pteronarcella, Pteronarcus
EPHEMEROPTERA (mayflies)	*Baetis, Cloeon, Ephemerella*
TRICHOPTERA (caddis flies)	*Brachycentrus, Hydropsyche*
COLEOPTERA (beetles)	*Dytiscus, Elmis, Gyrinus*
ODONATA (damsel-, dragon-flies)	*Ischnura*
MEGALOPTERA (alderflies)	*Sialis*
HEMIPTERA (bugs)	*Aphelocheirus*
DIPTERA (two-winged flies)	*Chaoborus, Culex, Simulium*
Chironomidae	*Chironomus, Tanypus, Tanytarsus*
CHORDATA	
CYCLOSTOMATA (lampreys)	*Lampetra, Petromyzon*
ACTINOPTERYGII (bony fishes)	
SALMONIDAE	*Salmo, Salvelinus, Oncorhynchus*
CYPRINIDAE	*Pimepheles, Rutilus*
CENTRARCHIDAE	*Lepomis, Micropterus*
ICTALURIDAE	*Ictalurus*
AMPHIBIA (frogs, toads, salamanders, newts)	
REPTILIA (snakes, lizards, turtles)	
AVES (birds)	
MAMMALIA (mammals)	

Exclusively marine forms, some parasites and certain minor or rare groups are omitted.

discrete variations and these may be recognised by allocating a third, sub-specific, name. By convention, scientific names are printed in italics and, therefore, usually underlined in manuscripts. The generic name always begins with a capital letter and the specific, or trivial, name is always entirely lower case, even when the name commemorates a famous scientist or location, e.g. *Salmo clarki* and *Elodea canadensis*.

Genera are grouped into Families, Families into Orders, Orders into Classes and Classes into Phyla, the last representing the major divisions of the animal and plant kingdoms. Since at each of these levels the groups are not all the same size, it may be necessary to institute intermediate categories, such as sub-families or supra-orders in order to provide manageable units.

A simplified classification of organisms is provided in Table 2.1. This should enable the reader to place most of the organisms mentioned in subsequent chapters into their correct taxonomic context. It must be emphasised, however, that this is by no means a complete classification of the plant and animal kingdoms.

2.3 FOOD CHAINS AND TROPHIC PYRAMIDS

The fundamental energy source in all ecosystems is sunlight. Light captured by chlorophyll in plants provides energy for a complex series of chemical reactions whereby carbon dioxide, water and simple inorganic nutrients are combined to form carbohydrates, fats and proteins which in turn are elaborated into tissues as the plant grows. Organisms such as plants (and some bacteria) which are able to photosynthesise their components in this way are said to be *autotrophic*, literally 'self-feeding', and since all other organisms ultimately depend upon this process they are also termed *primary-producers.*

All organisms other than these autotrophs are *heterotrophic*, that is they 'feed on others', either directly as their prey or on their decomposing remains. The smallest plants, the algae and the bacteria, provide food for the smallest animals, the protozoa. In their turn protozoa, along with algae and higher plants, form the food of other, larger invertebrate animals. These, or at least the larger ones, visible to the naked eye and forming the omnibus group of 'macroinvertebrates', are important as food for larger animals and in particular of fish. The last link in this food chain is forged when these fish are eaten by other larger fish, fish-eating birds (e.g. herons) or mammals (e.g. otters or man). Of course, this description is over-simplified in that there are many cross-links in the chain, so much so that it resembles chain-mail or, as it is often described, a food-web or food-net.

At each stage in the food chain there are, obviously, more organisms forming the prey than there are predators. The daily food requirement of a piscivorous bird or mammal might be of the order of one to ten fish, depending on their respective sizes. At the next level below, a fish may

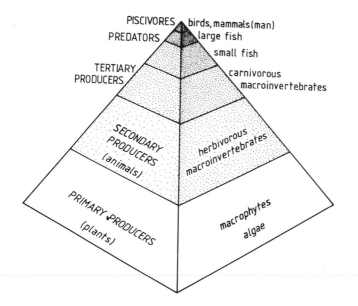

FIG. 2.1 Diagram of a 'trophic pyramid' indicating the relative biomass at each trophic level and the potential for bioaccumulation.

require ten to a hundred macroinvertebrates, and so on down the chain. In reality, then, we have a *trophic pyramid*, as indicated in Fig. 2.1, in which the amount of organisms, i.e. their *biomass*, which exists at each level is greater than that to which it contributes. A few moments thought will show that this must be so since food conversion is not 100% efficient and all the organisms must spend energy in feeding, respiring, reproduction and all other functions.

Trophic pyramids have great significance in pollution biology. Low environmental concentrations of pollutants, for example heavy metals or pesticides, may be taken up by the primary producers (or other stages in the pyramid) and although the amounts are not deleterious to that particular trophic level, as they pass upwards to the next and subsequent levels so they are effectively bioconcentrated or bioaccumulated. This *bioaccumulation*, which is enhanced in those substances which have greater affinity for material in living tissues (e.g. lipids) than for water, may cause death or damage to animals which occupy positions at the apex of the pyramid even though the actual environmental concen-

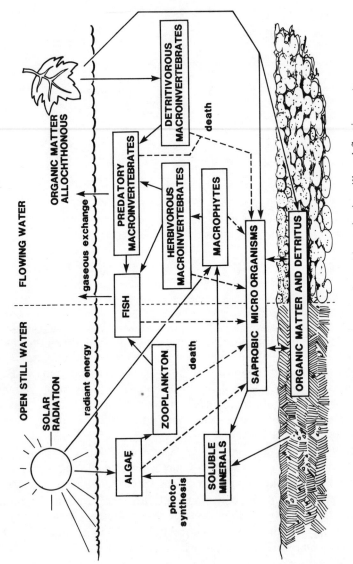

FIG. 2.2 Diagrammatic comparison of trophic cycles in still and flowing water.

trations of the substances may be relatively harmless. Examples of bioconcentration are to be found in Section 3.6.3.

So far we have only considered the upward movement of energy from prey to predator. No individual is immortal; when animals or plants die their decomposing tissues are consumed by carrion-eating animals and are further broken down by bacterial and fungal action. These remains, especially those of vegetable matter, form *detritus*—an extremely important food source for many invertebrates. The ultimate fate of all such material, and detritivores which feed on it, is complete mineralisation. These minerals are then available to photosynthesising plants and can be recycled once more. A simplified account of these relationships is provided in Fig. 2.2.

2.4 DISTRIBUTION OF SPECIES

Species are not ubiquitous nor uniformly distributed. This is largely because the environment in which they live is not homogeneous and there are areas where the prevailing conditions are outside the range of tolerance of some organisms. Most species have particular needs and can only flourish where these are found. Thus, the distribution of species within the environment at large reflects aspects of the variation in quality of that environment, and this feature is exploited in the concept of biological indicators (Chapter 3). This concept is not new, as the epigraph indicates, and is part of everyday experience: when walking across a meadow we are able to avoid crossing very wet or boggy patches because a characteristic flora, often of sedges or rushes, indicates the location of these areas long before we sink to our ankles or knees in mud!

The variation in distribution patterns is attributable to many factors; subsequent sections outline some of the important aspects of this topic.

2.4.1 Tolerance Limits of Environmental Factors
The principal factors which control the distribution of organisms in freshwater are current speed and the stability of water depth (spates and drought, in flowing water); light and temperature regimes; substrate condition and stability; dissolved oxygen concentration and water quality (acidity, hardness, nutrient concentrations and salinity).

The importance of some of these factors is immediately self-evident:

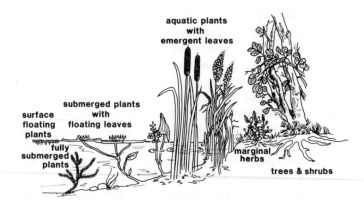

FIG. 2.3 Diagrammatic cross-section of the shore and margin of a productive (eutrophic) lake to illustrate the concept of 'zonation'.

flowing water the bed gradient will also relate to water velocity, although not in any simple way. These changes, and others, control the community of animals and plants which is able to establish itself. Some plants such as water lilies (*Nymphea* or *Nuphar*) require deep water. Others are restricted to shallow margins. All plants require light and so there are limits to the depths at which plants will grow in lakes. A typical cross-section of the shore and margin of a productive lake is shown in Fig. 2.3.

This variation in what we may call micro-habitat, and the local zonation of organisms reflected by it, can be changed when human interference such as dredging, channel modification or other engineering operations destroys or alters features on which the characteristic community depends. That which appears relatively small in engineering terms, for example raising or lowering water levels by less than 1 m, can have marked influences for the ecology of the site.

On a larger scale, one may also observe zonation along a river. This has been the subject of much study and many workers have attempted to provide classifications of river zones. The classical studies include the work of Thienemann (1925) who proposed six zones for continental European rivers; spring brook, trout zone, grayling zone, barbel zone, bream zone and brackishwater, each based on the presence of characteristic fish species. This elegant concept persisted in the systems devised by Carpenter (1928) and Huet (1949, 1954). Carpenter's classification divided watercourses into two major parts—highland brooks and lowland

TABLE 2.2

COMPARISON OF RIVER ZONE CLASSIFICATION SCHEMES PROPOSED BY SEVERAL EUROPEAN WORKERS, SHOWING THE CHARACTERISTIC FISH SPECIES IN EACH ZONE

SOURCE →

Thienemann (1925)	Carpenter (1928)	Huet (1954)	Illies (1961)
Spring	Head stream		Rhithron
Spring brook			
Trout region	Trout beck	Trout zone	
Trout (*Salmo trutta*)	Trout (*Salmo trutta*)	Trout (*Salmo trutta*)	
Grayling region	Minnow reach	Grayling zone	
Grayling (*Thymallus thymallus*)	Minnow (*Phoxinus phoxinus*)	Grayling (*Thymallus thymallus*)	
Barbel region	Upper reach	Barbel zone	
Barbel (*Barbus barbus*)	Gudgeon (*Gobio gobio*)	Barbel (*Barbus barbus*)	
	Chub (*Leuciscus cephalus*)		
	Dace (*Leuciscus leuciscus*)		
	Bleak (*Alburnus alburnus*)		
Bream region	Lower reach	Bream zone	Potamon
Bream (*Abramis brama*)	Bream (*Abramis brama*)	Bream (*Abramis brama*)	
	Carp (*Cyprinus carpio*)		
	Roach (*Rutilus rutilus*)		
	Tench (*Tinca tinca*)		
Brackish water region	Brackish estuary		
Ruffe (*Gymnocephalus cernua*)			
Flounder (*Platichthys flesus*)			

MOUTH

courses—of which the former had three zones—head streams, trout becks and minnow reaches (the grayling being less widely distributed in Britain than on the Continent)—while the lowland courses had upper and lower reaches. Characteristic invertebrate communities were included in this classification, although fish were still a key feature: the upper lowland reach was typically inhabited by gudgeon, chub, dace and bleak, while the lower lowland reach would contain bream, carp, roach and tench. Huet's four zones (trout, grayling, barbel and bream) were characterised by gradient and breadth and he also provided a 'slope graph' which enabled one to predict the fish zone from physical parameters. River valley shape could also be generally associated with typical fish communities. The approximate relationships between these classifications are set out in Table 2.2.

Studies of the distribution of freshwater invertebrates have revealed that many species are confined to certain sections of rivers and streams (Fig. 2.4). As one travels downstream, when species are first encountered they are relatively rare, then increase in abundance, later decline and finally are not found at all, or hardly ever. The distributions of individual species may overlap and this, coupled with the variations in abundance, ensures that while the composition of the invertebrate community alters spatially, the changes may appear to be transitional rather than discrete. Where the rates of change are greatest or the differences most pronounced, one might conclude that this marks the boundary of a 'zone'.

In an extensive review of river zonation and classification systems, Hawkes (1975) noted that, although the concept of fish faunal-zones had proved useful, all the sytems proposed had their limitations and these became more apparent the further one moved away from the geographical area in which they were originally conceived. The occurrence of longitudinal zonation is evidently a world-wide phenomenon, having been well-documented by workers in North America and observed in Asia, South America, Africa and Australasia. Hawkes (1975) has provided a very useful and comprehensive comparison of many systems of classification and appears to have achieved a measure of success in equating the corresponding zones of each system. This suggests that, in spite of the obvious dissatisfaction with the proposals of others and the keen desire on the part of each investigator to promulgate his own system, the evident degree of similarity is an indication that river zonation is a reality.

One proposal which seems to have considerable merit, in that it is based on abiological features and is, therefore, applicable world-wide

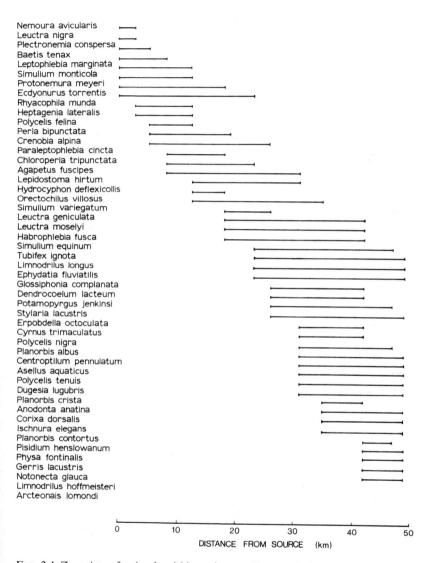

Fig. 2.4 Zonation of animals within a river as illustrated by the distribution of selected macroinvertebrates within the River Endrick, Scotland. Drawn from data provided in Maitland (1978).

without reference to the local fish-fauna (or other indicator group), is Illies' (1961) division of running waters into two main zones, the upper *rhithron* and the lower *potamon*. The rhithron is that part of the stream in which the annual range in mean monthly temperature does not exceed 20°C. This upper portion is characterised by high water velocity but low discharge and an eroding substratum. The potamon is the remaining downstream portion of the river, having a lower velocity and depositing substratum. The rhithron and potamon can be divided further, if required.

The significance of studies of zonation and the longitudinal distribution of organisms in the present context is that manipulation of river profiles and flow regimes by flood alleviation and land drainage schemes could destroy or modify certain zones with consequential effects on the characteristic community, or lead to misplaced or uncharacteristic river stretches with similar results. Recently, some very interesting work has been undertaken in order to try to classify relatively natural rivers on the basis of their physical, geological and water quality characteristics, and also on their macroinvertebrate fauna (Wright *et al.*, 1984; Furse *et al.*, 1984). It is hoped that it will prove possible to predict the typical community which should occur in a given river stretch, thereby enabling comparisons to be made between the expected and actual faunas and hence estimate the degree of change which may be attributable to pollution or other stress. More exciting still is the prospect that it should prove possible to predict the likely changes which would occur in the biota when modifications are proposed to existing river configurations or flow regimes in river management schemes.

2.4.4 Dispersal Mechanisms

The distribution of species is also a function of their dispersal mechanisms although this is less important in stable environments since one can reasonably assume that all the species capable of living in a given stable environment have already reached it and are now present. The question of dispersal becomes more important when we consider stressed environments, especially those which are subjected to intermittent influences. When some catastrophic event occurs, such as a very severe accidental pollution of a watercourse, the rate of recovery of the system depends a great deal upon the rate at which the habitat is recolonised from unaffected areas.

Birds are implicated in the transfer of microorganisms such as algae and protozoa between lakes or ponds (Atkinson, 1972) and the problem of passive dispersal of aquatic organisms between isolated bodies of water has been reviewed by Maguire (1963). Aquatic insects are able to disperse widely during their adult aerial phase and even poor fliers could be carried long distances in strong winds. However, oligochaete worms, molluscs and crustacea (as well as other minor groups) are clearly at a disadvantage in this respect although it has been suggested that wading birds may transport eggs in mud or other material adhering to their feet. Fish which are not exclusively lacustrine may colonise other lakes through river systems while ponds may be 'restocked' during periods of high discharge when rivers overflow their banks, inundate the flood plain and provide an aquatic 'bridge' from pool to pool.

2.4.4.1 Drift

Benthic invertebrates may be caught in nets placed above the bottom of streams and rivers as they move downstream in the current. This drifting is a well researched phenomenon (e.g. Waters, 1962; Elliott, 1969) and is found in many groups but especially amongst insects and crustacea; it occurs less commonly among species of caddis with heavy cases and molluscs. Marked diurnal variations in the density of drift have been observed, highest numbers being taken during the hours of darkness. The significance of this behaviour is debated. It may be a mechanism to prevent overcrowding and the drift fauna may represent the over-production within a given habitat (see Hynes, 1970, pp. 268 and 427). It will also ensure that no downstream habitat in which the organism could flourish is left unexploited. Recovery of the fauna following pollution, or other catastrophe, is greatly facilitated by organisms drifting down from unaffected upstream reaches. If drift does represent the surplus biomass it affords a most convenient way of estimating the biological productivity of a stretch of river (Hynes and Coleman, 1968; Waters, 1961).

Loss of 'surplus' individuals is not of itself significant but a net movement of animals downstream leads one to pose the question of how the community is able to maintain itself indefinitely. Some answers to this question are provided in the next section.

2.4.4.2 Upstream Colonisation

The maintenance of populations of organisms in flowing waters in spite of the tendency for downward drift is one of the more intriguing problems of freshwater ecology. When one contemplates, for example, the destructive potential of severe spates and floods upon sedentary forms, and on macrophytic vegetation in particular, it is difficult to understand how the community remains as stable as it clearly does. Leaving aside the fundamental question of how they arrived in the first place, it is self-evident that all species must have some means of maintaining their positions and of regaining them when displaced downstream. On a very long time scale rivers also change, with a general tendency towards an upstream migration of each zone. Perhaps the best and most readily understood example of this is a waterfall which, as rock erodes, moves upstream, albeit slowly.

Downstream propagation of macrophytes by means of fragments, fruits or winter buds is well known (Haslam, 1978) but it seems unlikely that these would be carried upstream even through the agency of birds or mammals. Plants having creeping underground stems, or with above-ground shoots but growing at the slower-flowing stream margins, could make slow headway against the direction of the prevailing current.

Aquatic insects having aerial phases, which is the vast majority, are able to fly upstream to lay their eggs and many groups have been observed to do so, sometimes over long distances. Even aquatic larvae and nymphs are able to move appreciable distances upstream. Crustacea and molluscs are also known to move upstream, often by several kilometres. The literature has been reviewed by Hynes (1970) who noted that some insects which emerge in cold seasons, at temperatures which may be so cold as to preclude flight, actually walk considerable distances overland in order to facilitate upstream re-colonisation! Even the most feeble of freshwater invertebrates could move upstream within the substrate where current velocities are low. In a study in which bacterial cells were labelled with radioactive phosphorus, Ball *et al.* (1963) observed a steady upstream progression of radioactivity in the biota. Progress of almost $0·5$ km was seen over a five week period.

2.5 ECOLOGICAL BALANCE

The 'balance of nature' is a widely held and understood concept, though in using the expression laymen often overlook the fact that it is a dynamic and not a static balance. As a consequence the fulcrum may not be fixed and the system will oscillate, different parts having different amplitudes.

2.5.1 Control of Numbers
Much work has been done on this aspect of ecology in an attempt to

understand the mechanisms which control animal numbers. The classical studies have concerned predator–prey systems, for example the lynx and snowshoe hare systems, which show cyclical changes in abundance with a period of about nine or ten years (MacLulich, 1937), or lemmings, the animal whose cyclical abundance and suicidal control mechanism is legendary, with a period of about three or four years (Elton, 1942). The cyclical abundance of locusts has been observed from antiquity. Although climatic factors resulting from cyclical sun-spot activity have been suggested as a mechanism for some cycles (however, Odum, 1971, has shown there is no correlation) it is likely that simple feedback mechanisms could operate in many cases.

In simple terms, taking a single predator–prey relationship as an example, although the basic principles would apply to others, one could imagine that a predator feeding on abundant prey would breed rapidly and increase its numbers. Greater numbers of predators would reduce the prey which, breeding at its maximum rate, would not be able to sustain population numbers. A scarcity of food would impair the reproductive success of the predator and its numbers would decline, perhaps quite quickly. Pressure would be lifted from the prey organism and if it had the capacity for rapid reproduction its numbers would increase dramatically. The predator would now have an abundant food supply but its utilisation would depend on the speed with which it could breed.

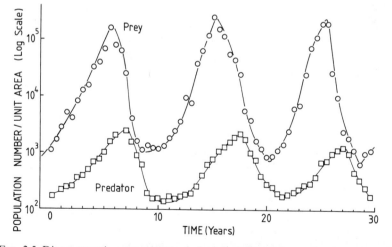

FIG. 2.5 Diagrammatic representation of cyclical fluctuations of a hypothetical predator and its principal prey species.

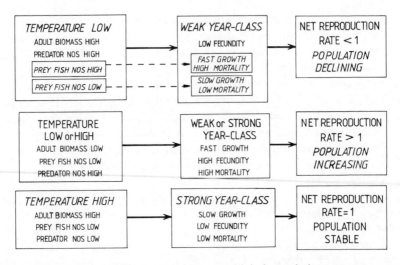

FIG. 2.6 Influence of environmental factors and inital population parameters on the reproductive performance of the perch (*Perca fluviatilis*) in Lake Windermere. After Craig (1980).

Eventually maximum predator numbers would be reached and the prey would decline again. This cyclical process is outlined in Fig. 2.5.

In freshwater environments there is little evidence of protracted cyclical events of this kind although within a given season there may be some cyclical changes, for example in algae and zooplankton (see Section 2.7) but more commonly the oscillations in numbers of such broad categories is, in reality, a succession of different species equivalent to the spatial zonation observed along a river but occurring temporally in one area of the lake.

More commonly, cyclical events in freshwater are triggered by stochastic climate events. These are well documented for fish. For example, warm dry summers appear to be conducive to enhanced fry survival and hence dominant year classes (Le Cren, 1955; Hellawell, 1971). Long-term studies of populations of perch (*Perca fluviatilis*) in Lake Windermere (Craig *et al.*, 1979) have enabled the sequences of events to be worked out in some detail and have led to construction of a simple model (Craig, 1980). This is illustrated in Fig. 2.6.

Finally, one may mention the possibility that there are inherent population-regulation mechanisms (Wynne-Edwards, 1962; 1965) in which organisms are able to sense the size of their population and so control reproduction directly or through behavioural 'stress' responses to population density.

2.5.2 Competition

As environmental resources are finite it is self-evident that organisms must compete with each other. This inter-specific competition ensures maximum utilisation of resources, usually through reproductive over-capacity. The classical example of this occurs in those fish, such as the cod (*Gadus morhua*), which lays many thousands of eggs but, fortunately, all do not survive. Similarly, many common garden birds raise several broods of young each season but the actual numbers nesting each year remains much the same. Each individual competes with its fellows for food, territory and often also for a mate and, according to the familiar maxim, we observe 'the survival of the fittest'.

Competition between species must also occur although it is difficult to assess and much of the evidence is indirect. It has been noted, for example, that although the total number of species of different freshwater groups is very similar between rivers, the actual species lists differ widely; the majority only occur sporadically and very few species appear to be virtually ubiquitous (Patrick, 1961, 1964). This leads to the conclusion that there are far more species than available niches, that is they compete severely for them or, species have very limited tolerances and very exacting requirements, and so can only occur where conditions precisely match their requirements. Species of variable distribution may occur in particular locations according to the presence or absence of competitors whose own distribution may depend on more exacting environmental conditions or be the result of zoogeographical accidents.

Many similar species display characteristics which appear to limit competition. First, these species may appear occasionally along a river or stream, as in the zonation described in Section 2.4.3. They may also choose different microhabitats so that although several species occur within a relatively small area the diversity of the habitat enables them to exploit different areas. Some, particularly fishes, are territorial and aggressively interact with both members of other species and their own. This ensures spatial separation which, although it does not limit competition for space, may avoid competition for food. Second, some very similar species exhibit different timing of their life cycles. This is quite evident in some insects which emerge successionally to mate and oviposit. They can thus coexist because at any given time they are at different stages in their life histories and/or differ in size. In this way they achieve temporal separation even though they may occupy the same micro-habitat. Third, some coexisting species have different dietary requirements or preferences. This is often seen in fish, for which extensive data on diets exist, where sympatric species may have identical lists of dietary items but the relative importance of these is quite different. However, such data may be accounted for in other ways. For example, where food is abundant and not the limiting factor, identical diets might be observed (Hynes, 1970). There may also be differences in feeding behaviour which effectively isolate two, or more, species. Finally, physiological differences such as oxygen requirements may cause two similar species to occupy different habitats or predominate at different seasons.

2.5.3 Causes of Imbalance
As indicated above, ecological balance is dynamic and may be changed by external influences such as pollution or environmental modification. A new, but less desirable balance may then be struck. As before, the loss of predators or competitors may have consequences beyond those of the immediate food web. Changes in water quality may prevent previously important species from maintaining their numbers and, in the absence of competition, others may quickly establish themselves. Organic enrich-ment or addition of nutrients are common sources of ecological imbal-ance where the ratios of producers to consumers may radically alter. A final example of the way in which the precarious dynamic balance can be upset is the introduction of alien species, such as the plants *Elodea canadensis* and *Eichhornia*, or fish such as carp (*Cyprinus*), rainbow trout (*Salmo gairdneri*) and brown trout (*S. trutta*).

2.6 COMMUNITY STRUCTURE

If a large sample of organisms is taken from a given natural habitat the number of individuals belonging to each species will be found to vary widely. Only a small number of species are likely to be very numerous while the majority are represented by several, a few or even single individuals (Fig. 2.7). The fundamental structure of plant or animal communities is amenable to mathematical description. Several models have been proposed (Pielou, 1969) of which the logarithmic series, lognormal and ordered random-interval or 'broken-stick' models may be cited as examples. In the lognormal model, species are grouped into

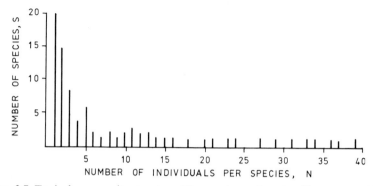

FIG. 2.7 Typical community structure. The numbers of species (S) represented by 1, 2, ... N individuals in a large sample of the community.

abundance categories (octaves, i.e. powers of 2) which then follow a normal curve (Fig. 2.8). In the 'broken-stick' model the community structure can be likened to a stick which is broken into segments at random. If the number of pieces is equal to the number of species then

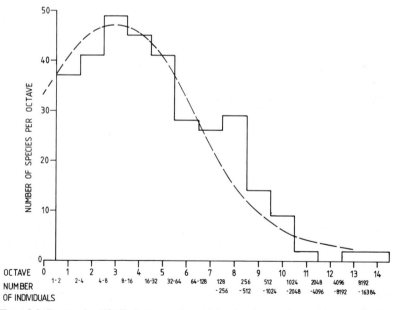

FIG. 2.8 Preston's (1948) lognormal model of community structure. Species abundances arranged by octaves (equivalent to log scale to base 2) in a sample from a community.

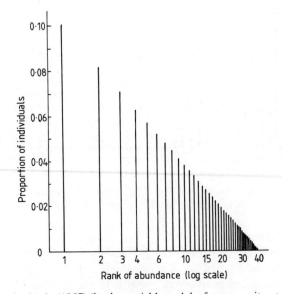

FIG. 2.9 MacArthur's (1957) 'broken stick' model of community structure. The proportion of individuals of each species is equivalent to the portion of a stick broken repeatedly at random.

the lengths will be proportional to the number of individuals in each species. When the pieces are arranged in decreasing length at logarithmic intervals the ends of the sticks will form a straight line (Fig. 2.9).

The value of such theoretical concepts of community structure lies not in any intrinsic merit but in the possibility which they offer for comparing the actual structure of any given community with the model and for measuring any deviations which might, for example, have resulted from some environmental stress or pollutant (see 9.4.1.1). Experience has shown that stressed communities differ from the ideal structure, usually in having fewer total species although those present might be represented by more individuals than is usual. This general principle may be illustrated by comparing the rich flora of meadow or natural grassland and the paucity of species in the town park, which is subject to mowing and trampling, or the bird communities of the open countryside and city centres. The bird fauna of the city is composed largely of many individuals of a few opportunist or exploiter species (e.g. sparrows, pigeons, etc.) while, by contrast, in the countryside many species are present but none is excessively dominant.

The structure of the community also reflects, to some degree, the structure of the habitat. The relationship between habitat and niche was considered in Section 2.4.2. In a diverse habitat one might expect that the wide variety of opportunities to be exploited would provide a wide range of niches and, hence, a multiplicity of species. When a habitat is monotonous and offers opportunities for only a few specialised species, then community diversity is likely to suffer.

Measures of community diversity, usually in the form of an index, have been used in order to provide an estimate of the intensity of environmental stress (Hellawell, 1978; and see Section 9.4.1.1). Diversity has two components which are derived from the number of species present (species richness) and the relative abundance of those species (equitability). High diversity obtains when numerous species are represented by equal numbers of individuals, the upper limit being infinite species each represented by a single individual. Lowest diversity would be an infinite number of organisms in a single species.

2.7 STILL AND FLOWING WATERS

The ecology characteristics of still (lentic) and flowing (lotic) waters are, as might be anticipated, quite different in many respects and, in the context of considering the consequences of environmental contamination, these differences have profound implications for the significance of pollutant inputs and their subsequent pathways. The principal factor is the rate of exchange of water through the system and hence the retention time of contaminants and the possibilities for their ultimate loss. Lakes and reservoirs are, in comparison with rivers, virtually closed systems although they do have inflows and outflows and a notional renewal time of many weeks, months or even years. For descriptive purposes a protracted retention time will be assumed.

When nutrients or other pollutants enter still waters they tend to be retained within the system although their biological significance may depend on their behaviour, for example the ease with which they are lost from the water column and enter sediments. This tendency for materials to accumulate within static waters, eventually reaching a threshold after which marked and perhaps irreversible changes occur, contrasts with the continual uni-directional renewal of water in flowing waters which provides a mechanism for dissipating pollutants or nutrients. Lakes and reservoirs thus have a 'history' or sequence of development (which is of great interest to palynologists who are able to study the evidence contained in benthic sediment layers) while rivers exhibit a dynamic steady-state condition reflecting the balance of gains and losses at any given time.

Similarly, when a pollutant ceases to enter a flowing water the loss of any residual material is rapid; in static water, residues may persist for considerable periods after pollution stops, since the exit pathways tend to be much more restricted. Reduced intensity of algal blooms in static waters may not be concomitant with lowered nutrient inputs although, clearly, it could help avoid future exacerbation of the problem. This apparent hysteresis is attributable to the former input of nutrients which have accumulated in the benthic sediments and which continue to be available through diffusion and circulation through the lake. The movement of nutrients between various components of the lake ecosystem, i.e. the water column, sediments and biota, is complex, being influenced by such factors as the dimensions and physical properties of the lake, including its propensity to stratify (see Section 2.7.1 below), its overall productivity and the chemical characteristics of the elements involved.

Primary productivity is intrinsic (autochthonous) within lakes and secondary production depends on the availability of an algal phytoplankton or, in shallow lakes, the contribution from benthic or marginal macrophytes. In flowing waters much secondary production is generated from allochthonous (i.e. imported) material, detritus being particularly important as a source of invertebrate food. This difference means that mineral nutrient input into lakes tends to stimulate primary productivity, with other production increasing consequentially, while in flowing water nutrient enrichment has far less effect. Plant growth in many flowing waters does not appear to be nutrient limited and enrichment rarely causes problems. Organic enrichment, however, is akin to the natural detrital input and often stimulates secondary production. Nutrient induced enhancement of primary productivity in static water bodies has become such a well-known phenomenon that it has passed into everyday environmental terminology as 'eutrophication' (see Section 2.72).

2.7.1 Stratification
Stratification may be observed in many kinds of lakes but the cyclical pattern of changes which occurs in deep lakes in temperate zones will be described here to illustrate the general principles.

During the warmer, summer, months the upper part of temperate lakes is heated by solar radiation. If water temperatures are measured at

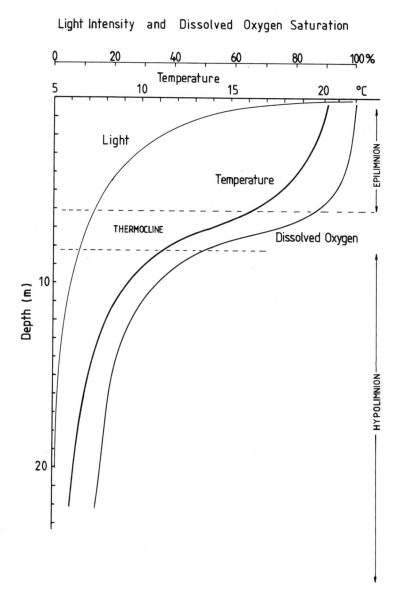

FIG. 2.10 Diagrammatic representation of temperature, light intensity and dissolved oxygen profiles in a northern temperate zone lake during summer stratification.

intervals of depth it will be found that a region exists in which the rate of temperature change is greatly increased (in excess of $1°C\,m^{-1}$); this 'layer' is called the thermocline (Fig. 2.10). The warmer surface water is less dense than the general mass of water beneath and the greater the difference in density the more stable will be the stratification since it will not be easily displaced by wind action across the lake surface. The depth at which the thermocline forms is not fixed, being determined by the degree of solar heating, the transparency of the water and the morphometry of the lake, but wherever it forms it effectively divides the water body into two layers, the upper, epilimnion (literally the 'upon-lake') and the lower, hypolimnion (literally the 'lake below') (Fig. 2.11). This division is not merely an interesting physical phenomenon but it has consequential effects on the biology of the lake. The epilimnion is well lit and oxygenated with sufficiently high temperatures to promote algal productivity and hence to support zooplankton and fish. When nutrients are in ample supply, algal growth is accelerated and excessive blooms may occur. By contrast the hypolimnion is cold, dark and becomes progressively de-oxygenated as the decaying remains of organisms rain down from the epilimnion. Conditions in the hypolimnion may become so extreme that anoxia ensues whereupon biological productivity becomes minimal. Under less extreme circumstances the epilimnetic material provides an energy source for benthic invertebrates. Anoxic sedi-

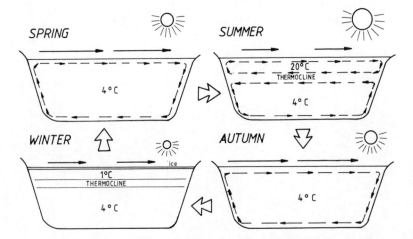

FIG. 2.11 Seasonal sequence of stratification and overturn in a typical temperate lake.

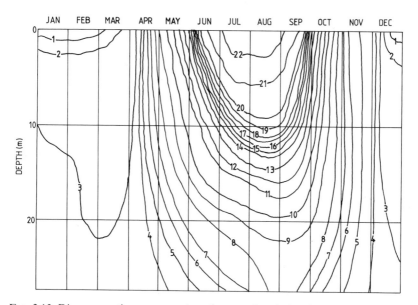

FIG. 2.12 Diagrammatic representation of seasonal variation in temperature with depth indicated by isotherms (°C) in a northern temperate lake showing summer stratification and autumn overturn.

ments may promote the release of certain elements, especially iron and manganese, which cause additional treatment problems when hypolimnetic water is used for public supply. The reduced biological productivity in this layer often makes it more attractive, especially when the epilimnion has excessive algal blooms with attendant problems of higher treatment costs and the risk of unpleasant tastes or odours. The sinking of dead algae and zooplankton through the thermocline not only contributes to the potential de-oxygenation of the hypolimnion but also prevents immediate recycling of nutrients. Nutrient depletion may occur to such an extent that algal growth, or at least that of certain groups of algae, is limited.

In autumn the solar radiation begins to decline and the temperature difference across the thermocline is reduced. Eventually the consequent density difference is inadequate to maintain stratification against strong surface winds and mixing or 'over-turn' occurs. This has several effects: oxygen is transported to the formerly depleted lower depths, nutrients are brought to the surface and the surface waters become cooler. The lower temperature and declining light intensity ensure that algal blooms, if they occur, are not so pronounced as in spring or summer even though more nutrients are now available in the former epilimnion. Winter gales ensure that the lake remains fully mixed. In spring, light and temperature increase, plant nutrients are high (both from redistribution by mixing and from external sources, primarily surface run-off via the lake feeder streams) and these factors contribute to a peak of biological activity as a prelude to summer stratification when the cycle here described is completed (Fig. 2.12).

Recently, reservoirs and lakes used for public supply have been artificially over-turned or destratified by the installation of jetting devices or other means of inducing circulation in order to prevent anoxic conditions and partly in the hope that algal blooms might be minimised (Tolland, 1977).

2.7.2 Eutrophication
The growth of algal populations in lakes and reservoirs is, as indicated above, largely dependent upon light and nutrients. Light is critical for photosynthesis and nutrients are needed for the anabolic activities of the algal cells. Nutrient limitation and its consequences are well illustrated by reference to the seasonal variation in abundance of diatoms and concentration of silicon in the water column of temperate lakes. Silicon is utilised by diatoms for the production of their frustules, the box and lid arrangement of their cellular skeleton, and during spring, when light and temperature increase, diatom numbers increase rapidly. However, while other nutrients are still available, and ambient light and temperature are still increasing, the diatom bloom ceases when the silicate silica is virtually exhausted (Lund, 1950, 1964). Later in the year, when diatoms have decayed thereby releasing some silicate, and remixing of the stratified layers has occurred, a short minor autumnal bloom of diatoms may occur (Fig. 2.13).

In lakes with natural scarcity of essential plant nutrients, phytoplankton densities are low, the water remains relatively clear and marginal macrophytes are also rare. Such lakes are said to be oligotrophic, that is 'underfed'. By contrast, lakes with ample nutrients supporting rich algal blooms and profuse growths of macrophytes are eutrophic, or 'well-fed'. Table 2.3 provides a comparison of the main classical features of oligotrophic and eutrophic lakes. Eutrophication is the name given to the process whereby lakes are enriched naturally or, more usually, as a result of human activity. This process was of considerable interest during the

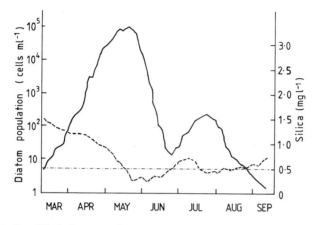

FIG. 2.13 Simplified diagram of the seasonal variation in diatom biomass and silica concentration in a temperate lake. After Lund (1950).

1970s when accelerated rates of change in some lakes and reservoirs became a cause for concern, both for aesthetic and economic reasons. Excessive algal blooms exerted high nocturnal respiratory oxygen demands, causing deaths of fish or, when blooms subsided, their decay increased BOD with similar results. Participation in water-based recreation was spoiled when winds caused algal blooms to accumulate at the edges or on the shores and their subsequent decay provided further sources of nuisance. Eutrophication of lakes or reservoirs which served as sources of potable water imposed increased costs in treatment or caused disagreeable tastes and odours, as algal metabolic by-products or break-down products passed into supply systems.

The key nutrients responsible for eutrophication are nitrogen and phosphorus but in most freshwater lakes algal growth is usually limited by the availability of phosphorus. Where nitrogen might otherwise be the limiting nutrient, in that phosphorus is adequate, some algae (Cyanophycae—blue-green algae) are able to fix dissolved nitrogen which diffuses readily from the atmosphere and so again the phosphorus levels become the critical factor. Should nutrients be very plentiful, algal growth is then likely to become light limited through self-shading effects: the density of algae near the surface absorbs sufficient light to prevent the indefinite survival of other algae at greater depths.

Excess nutrients come from a variety of sources although two, sewage

TABLE 2.3

COMPARISON OF MAIN CLASSICAL CHARACTERISTICS OF OLIGOTROPHIC AND EUTROPHIC LAKES

1. *Physical features*	*Oligotrophic lakes*	*Eutrophic lakes*
Altitude	Often high	Usually low
Proportions	Narrow and deep	Wide but shallow basin
Substratum	Inorganic silt and stones	Organically enriched mud
Light penetration	Good, water clear	Poor, transparency low
2. *Chemical water quality*		
Nutrients	Poor, low in nitrogen	High, especially nitrogen and phosphorus
Hardness	Soft	Hard
Suspended solids	Low	High
Dissolved oxygen	High, even on stratification	Variable, high in epilimnion but hypolimnion anoxic on stratification
3. *Biology*		
Phytoplankton	High diversity (few individuals but of many species), low biomass; Chlorophyceae typically present, including desmids (e.g. *Staurastrum, Staurodesmus*) and Chlorococcales (e.g. *Oocystis*); diatoms (e.g. *Cyclotella, Tabellaria*) and Chrysophyceae (e.g. *Dinobryon*) occur, as well as Dinoflagellates	Low diversity (few, abundantly represented species), high biomass; blue-green algae (Cyanophycae) typically present (e.g. *Anabaena, Aphanizomenon*); diatoms (e.g. *Asterionella, Fragilaria, Melosira* and *Stephanodiscus*) and Chlorophyceae (e.g. *Pediastrum* and *Scenedesmus*) also occur
Macrophytes	Few species and rarely abundant; rare on shore but may occur at depth	Many species, abundant in shallow margins and along shoreline
Benthic invertebrates	High diversity but low biomass	Low diversity, often very high biomass
Zooplankton	High diversity but low biomass	Low diversity but high biomass
Fishes	Few species, characteristically Salmonidae (e.g. *Salvelinus*) and especially *Coregonus* (whitefish); low biomass	Many species, typically Cyprinidae (e.g. *Abramis, Cyprinus, Tinca*), Percidae, Centrarchidae and Cichlidae

effluents and agricultural run-off, are particularly important. The development of water-borne sewage systems within lake drainage basins to transport wastes for treatment at specific works has provided a rich supply of nutrients which enter these relatively static waters, in due course. Efficient treatment of the sewage does not ameliorate the situation but may rather exacerbate it since the nutrients are made more readily available. The principal source of phosphorus in domestic sewage is synthetic detergents in which phosphates are used as 'builders'. Attempts to use alternatives have not met with much success.

In addition to the 'point sources' of nutrients from sewage-works outfalls there are more diffuse sources, and these are principally agricultural or forestry run-off. Artificial fertilisers are applied in intensive arable farming and also to newly established coniferous forests. These, together with leachates from the application to land of farm slurry (generated in intensive livestock rearing and especially from pig and poultry farming), account for much of the nutrient entering surface water in rural areas. In these inputs nitrogen predominates because it is more soluble and, therefore, tends to be more mobile in soil water. Phosphorus, being more readily precipated, is leached much more slowly. Some phosphorus is lost from agricultural land through erosion, in association with fine soil particles, as a result of normal cultivation practice and especially when soil is ploughed in autumn and subjected to high winter precipitation. This phosphorus may not be readily available as a nutrient initially but when deposited as mud in the substrate of lakes the anaerobic conditions may result in enhanced rates of release.

2.8 BIOLOGICAL PRODUCTIVITY

During the late 1960s much ecological research was concentrated upon measurements of the productivity of biological systems, stimulated, no doubt, by the International Biological Programme which was concerned with 'the biological basis of productivity and human welfare'. Earlier, attempts had been made to trace the patterns of energy flow through ecosystems (e.g. food webs, trophic pyramids) but, with a few notable exceptions (e.g. Ivlev, 1945; Odum, 1956, 1957) little effort had been made to quantify the processes involved. Even now, most effort has concentrated on parts of the system or, more accurately, on particular groups of organisms. Far greater progress has been made in estimating algal or fish productivity than, for example, macroinvertebrates, largely because fish production is of commercial interest, although their position at or near the summit of trophic pyramids means that they provide a useful summary or indicator of the productivity of the system, and algal productivity is often measured indirectly using well-developed chemical or biochemical techniques.

Productivity has several components. One, biomass (B) or standing crop, is already familiar (see Section 2.3), and is the amount of biological material, usually expressed in terms of the numbers of organisms, their weight or their calorific value per unit area or volume of their habitat. Biomass is a static measure, providing a 'snapshot' of the condition of the population or community at a given time, a point which the alternative term 'standing crop' emphasises. Production (P) is generally defined as the total mass of tissue elaborated within a stated interval and includes material which does not endure to the end of the period. It is, therefore, a measure of the dynamic state of the biomass, the rate at which material is being gained (or lost) within a given time, and thus it is usually expressed as the mass (or calorific value) per unit area per unit time (e.g. $g\,m^{-2}\,a^{-1}$). The relationship between biomass and production is analogous to financial capital invested and the rate of return which it yields or the turnover of a company and the profit which it makes. Although in general these factors are related, it is quite possible (indeed all too possible!) to have a company with a very large turnover but making very little profit or, more rarely, to obtain high interest on the investment of little capital. Similarly, in biological systems the presence of a large biomass may not necessarily imply that production is high. Indeed, the ratio of production to biomass (P/B) or 'turnover ratio' is a useful parameter with which to compare different systems. Production may be *gross*, that is, including all the energy assimilated, or *net*, that is, the amount which contributes to tissue growth, the difference being the energy which is used in the metabolic processes which maintain life, principally respiration. Algal production is usually expressed in gross terms (i.e. net photosynthetic activity added to respiratory activity) while fish production is usually net production, principally because this is what is of interest and is what is most readily measured. The ratio of production to respiration (P/R) is another useful measurement for comparative purposes.

Production by plants, and in particular planktonic algae, is estimated by measuring net oxygen production in photosynthesis and oxygen uptake by respiration under identical conditions, but in darkness. Assuming that the respiration rate is constant in light and in darkness,

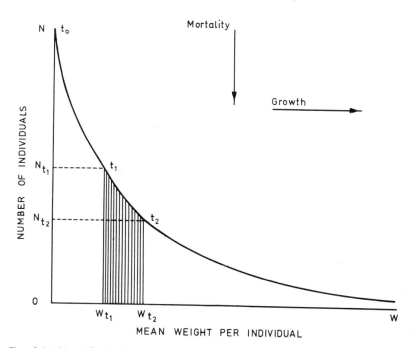

FIG. 2.14 Simplified 'Allen curve' for the graphical estimation of fish production. The number of individuals N in each cohort is plotted against the mean weight W. Production between times t_1 and t_2 is given by the shaded area under the curve.

the gross primary production may be calculated. Fish production is usually assessed more directly by estimating population numbers and the rate of growth in either of each cohort (age-class). Fishes, and some molluscs, have regular growth patterns in their hard tissues (e.g. opercular bones, scales, otoliths) which facilitate age determination and thus cohorts can be fairly easily identified.

Fish and invertebrate production may be calculated using the equation

$$P = G\bar{B}$$

where G is the instantaneous growth rate and B the mean biomass. In practice, G is determined from the difference between $\log_e$ weights and B is the mean biomass of the population over the period of the study. Production may also be derived graphically by plotting survival against growth and measuring the area beneath the curve (Fig. 2.14). Complications arise when weight is lost during winter, as stored products or tissue are metabolised, or in loss of gametes during reproduction. Strictly, these losses should be included since they are material elaborated by the population even though it did not survive to the end of the study period. Production of invertebrates is hampered by difficulties in identifying the cohorts, especially in multivoltine species with poorly synchronised life cycles.

CHAPTER 3

Biological Indicators

3.1 INTRODUCTION

The idea that organisms can provide an indication of the quality of their environment is widespread and basically sound but the terms 'indicator' or 'indicator species' may be used and understood in different ways. For example, some species are known to have particular requirements with regard to nutrients or levels of dissolved oxygen. Once these are defined, the presence of a particular species in a habitat indicates that the given determinand or parameter is within the tolerance limits of that species. It is in this sense that the term 'indicator' is used here. This concept of indicators may be extended beyond simply noting the presence or absence of species. Some indicators may continue to exist in a polluted environment but suffer physiological stress which is revealed in diminished rate of growth, impaired reproductive capacity or modified behaviour. This is essentially a 'bioassay' of the environmental contamination and, in allowing us to detect the change and, perhaps, estimate its intensity, the indicator has become a 'bio-sensor' for that pollutant or stressor.

Another concept associated with the term 'biological indicator' is that of an organism which accumulates substances in its tissues in a way so as to reflect environmental levels of those substances or the extent to which the organism has been exposed to them. When such indicators are collected and their tissues subjected to chemical analysis it is often possible to estimate prevailing environmental concentrations. Such organisms are 'bio-accumulators' of these substances and are often particularly useful when they concentrate very low environmental levels of substances, thereby facilitating detection and analysis (see Table 3.1).

In freshwater the former approach is most commonly used and it is

TABLE 3.1.

ACCUMULATION OF POLLUTANTS BY ORGANISMS: EXAMPLES OF THE TISSUE CONCENTRATIONS OF HEAVY METALS, PESTICIDES AND ORGANIC INDUSTRIAL COMPOUNDS AMONG A WIDE TAXONOMIC RANGE OF PLANTS AND ANIMALS

Substance	Organism	Tissue	Tissue concentration	Environmental concentration	Reference
Metals					
Aluminium (Al)	Brown trout, *Salmo trutta*	—	358 mg kg^{-1}	up to $3 \cdot 7$ mg litre^{-1}	Hunter *et al.* (1980)
Cadmium (Cd)	Bacteria	—	$6\,100$ mg kg^{-1}	1 mg litre^{-1}	Remacle (1981)
	Bacteria	—	$13\,000$ mg kg^{-1}	—	Remacle and Houba (1980)
	Alga, *Chlorella pyrenoidosa*	—	381 mg kg^{-1}	$38 \, \mu$g litre^{-1}	Gipps and Coller (1980)
	Alga, *Lemanea fluviatilis*	2 cm tips	$342 \, \mu$g g^{-1}	$0 \cdot 003\,9 \, \mu$g litre^{-1}	Harding and Whitton (1981)
	Liverwort, *Scapania undulata*	1 cm tips	$13 \cdot 3 \, \mu$g g^{-1}	$<0 \cdot 08$ mg litre^{-1}	Whitton *et al.* (1982)
	Moss, *Fontinalis antipyretica*	2 cm tips	$354 \, \mu$g g^{-1}	$0 \cdot 7 \, \mu$g litre^{-1}	Say *et al.* (1981)
	Moss, *Rhynchostegium riparioides*	2 cm tips	$433 \, \mu$g g^{-1}	$0 \cdot 7 \, \mu$g litre^{-1}	Say *et al.* (1981)
	Water hyacinth, *Eichornia crassipes*	leaves	510 mg kg^{-1}	10 mg litre^{-1}	Chigbo *et al.* (1982)
	Stonefly, *Pteronarcys badia*	whole body	$24 \, \mu$g g^{-1}	5 mg litre^{-1}	Clubb *et al.* (1975)
	Stoneflies, Pteronarcidae	whole body	$1 \cdot 01 \, \mu$g g^{-1}	$0 \cdot 5 \, \mu$g litre^{-1}	Van Hassel *et al.* (1980)
	Stoneflies, Perlidae	whole body	$0 \cdot 88 \, \mu$g g^{-1}	$0 \cdot 5 \, \mu$g litre^{-1}	Van Hassel *et al.* (1980)
	Craneflies, Tipulidae	whole body	$0 \cdot 97 \, \mu$g litre^{-1}	$0 \cdot 5 \, \mu$g litre^{-1}	Van Hassel *et al.* (1980)
	Amphipod, *Gammarus tigrinus*	whole body	$0 \cdot 77$ mg kg^{-1}	$2 \, \mu$g litre^{-1}	Zauke (1982)

Metal	Species	Tissue	Tissue concentration	Water concentration	Reference
	Mollusc, *Anodonta anatina*	soft tissues	5·9 µg g⁻¹	—	Manly and George (1977)
	Bluehead chub, *Nocomis leptocephalus*	whole body	0·42 µg g⁻¹	0·5 µg litre⁻¹	Van Hassel et al. (1980)
	Blacknose dace, *Rhinichthys atratulus*	whole body	0·76 µg g⁻¹	0·5 µg litre⁻¹	Van Hassel et al. (1980)
	Fantail darter, *Etheostoma flabellare*	whole body	0·60 µg g⁻¹	0·5 µg litre⁻¹	Van Hassel et al. (1980)
Chromium (Cr)	Rainbow trout, *Salmo gairdneri*	gills	31·7 µg g⁻¹*	2·0 mg litre⁻¹	Van der Putte et al. (1981)
Copper (Cu)	Alga, *Chlorella* sp.	—	4·77 mg g⁻¹	0·2 mg litre⁻¹	Klotz (1981)
	Alga, *Scenedesmus* sp.	—	2·90 mg g⁻¹	0·2 mg litre⁻¹	Klotz (1981)
	Moss, *Fontinalis antipyretica*	2 cm tips	103·0 µg g⁻¹	<0·008 mg litre⁻¹	Say et al. (1981)
	Moss, *Rhynchostegium riparioides*	2 cm tips	189·0 µg g⁻¹	<0·008 mg litre⁻¹	Say et al. (1981)
	Mollusc, *Anodonta anatina*	soft tissues	103·3 µg g⁻¹	—	Manly and George (1977)
	Rainbow trout, *Salmo gairdneri* (juv)	whole body	37·0 µg litre⁻¹ (8·3 µg/g*)	131 µg litre⁻¹	Dixon and Sprague (1981)
Lead (Pb)	Alga, *Lemanea fluviatilis*	2 cm tips	1 347 µg g⁻¹	0·13 µg litre⁻¹	Harding and Whitton (1981)
	Liverwort, *Scapania undulata*	1 cm shoot tips	5 910 µg g⁻¹	<1·82 mg litre⁻¹	Whitton et al. (1982)
	Moss, *Fontinalis squamosa*	2 cm tips	99 µg g⁻¹	0·005 mg litre⁻¹	Say et al. (1981)
	Moss, *Rhynchostegium riparioides*	2 cm tips	110 µg g⁻¹	0·004 mg litre⁻¹	Say et al. (1981)
	Water hyacinth, *Eichornia crassipes*	leaves	0·48 mg g⁻¹	10 mg litre⁻¹	Chigbo et al. (1982)
	Stoneflies, Pteronarcidae	whole body	20·4 µg g⁻¹	4 µg litre⁻¹	Van Hassel et al. (1980)
	Stoneflies, Perlidae	whole body	27·6 µg g⁻¹	4 µg litre⁻¹	Van Hassel et al. (1980)

TABLE 3.1—contd.

Substance	Organism	Tissue	Tissue concentration	Environmental concentration	Reference
	Craneflies, Tipulidae	whole body	$17\cdot3\ \mu g\ g^{-1}$	$4\ \mu g\ litre^{-1}$	Van Hassel et al. (1980)
	Snail, Lymnaea pulustris	flesh	$300\ mg\ kg^{-1}$	$1\ \mu g\ litre^{-1}$	Borgmann et al. (1978)
	Mollusc, Anodonta anatina	soft tissue	$42\cdot5\ \mu g\ g^{-1}$	–	Manly and George (1977)
	Rainbow trout, Salmo gairdneri	bone	$10\ mg\ kg^{-1}$*	$100\ \mu g\ litre^{-1}$	Hodson et al. (1982)
	Lake smelt, Osmerus mordax	–	100 to $250\ \mu g\ kg^{-1}$*	$<0\cdot8\ \mu g\ litre^{-1}$	Hodson et al. (1982)
	Fantail darter, Etheostoma flabellare	whole body	$19\cdot5\ \mu g\ g^{-1}$	$4\ \mu g\ litre^{-1}$	Van Hassel et al. (1980)
Tetramethyl lead	Rainbow trout, Salmo gairdneri	fat	$415\ \mu g\ g^{-1}$*	$51\ \mu g\ litre^{-1}$	Wong et al. (1981)
Mercury (Hg)	Water hyacinth, Eichornia crassipes	leaves	$0\cdot51\ mg\ g^{-1}$	$10\ mg\ litre^{-1}$	Chigbo et al. (1982)
	Amphipod, Gammarus pulex	–	$2\cdot52\ \mu g\ g^{-1}$	$0\cdot2\ \mu g\ g^{-1}$ (substrate)	Kristensen (1982)
	Mollusc, Anodonta anatina	soft tissues	$8\cdot2\ \mu g\ g^{-1}$	–	Manly and George (1977)
	Ricefish, Oryzias latipes	eggs	$56\ mg\ kg^{-1}$*	$30\ \mu g\ litre^{-1}$	Heisinger and Green (1975)
Methyl mercury	Pike, Esox lucius	whole body	$0\cdot09\ \mu g\ g^{-1}$*	(Food)	Phillips and Gregory (1979)
Nickel (Ni)	Stoneflies, Pteronarcidae	whole body	$3\cdot4\ \mu g\ g^{-1}$	$3\ \mu g\ litre^{-1}$	Van Hassel et al. (1980)
	Stoneflies, Perlidae	whole body	$11\cdot0\ \mu g\ g^{-1}$	$4\ \mu g\ litre^{-1}$	Van Hassel et al. (1980)
	Craneflies, Tipulidae	whole body	$6\cdot7\ \mu g\ g^{-1}$	$4\ \mu g\ litre^{-1}$	Van Hassel et al. (1980)

Organism	Tissue			Reference
Mollusc, Anodonta anatina	soft tissues	24·9 µg g^{-1}	—	Manly and George (1977)
Bluehead chub, Nocomis leptocephalus	whole body	2·5 µg g^{-1}	4 µg litre^{-1}	Van Hassel et al. (1980)
Blacknose dace, Rhinichthys atratulus	whole body	3·0 µg g^{-1}	4 µg litre^{-1}	Van Hassel et al. (1980)
Fantail darter, Etheostoma flabellare	whole body	5·9 µg g^{-1}	4 µg litre^{-1}	Van Hassel et al. (1980)
Zinc (Zn)				
Alga, Lemanea fluviatilis	2 cm tips	3 682 µg g^{-1}	2·02 µg litre^{-1}	Harding and Whitton (1981)
Liverwort, Scapania undulata	1 cm shoot tips	3 310 µg g^{-1}	1·86 mg litre^{-1}	Whitton et al. (1982)
Moss, Fontinalis squamata	2 cm shoot tips	5 430 µg g^{-1}	0·25 mg l^{-1}	Say et al. (1981)
Moss, Rhynchostegium riparioides	2 cm shoot tips	6 705 µg g^{-1}	0·25 mg litre^{-1}	Say et al. (1981)
Stoneflies, Pteronarcidae	whole body	254 µg g^{-1}	14 µg litre^{-1}	Van Hassel et al. (1980)
Stoneflies, Perlidae	whole body	235 µg g^{-1}	14 µg litre^{-1}	Van Hassel et al. (1980)
Craneflies, Tipulidae	whole body	106 µg litre^{-1}	20 µg litre^{-1}	Van Hassel et al. (1980)
Mollusc, Anodonta anatina	soft tissues	1 739 µg g^{-1}	—	Manly and George (1977)
Bluehead chub, Nocomis leptocephalus	whole body	98 µg litre^{-1}	20 µg litre^{-1}	Van Hassel et al. (1980)
Bluenose dace, Rhinichthys atratulus	whole body	331 µg g^{-1}	20 µg litre^{-1}	Van Hassel et al. (1980)
Fantail darter, Etheostoma flabellare	whole body	147 µg g^{-1}	20 µg litre^{-1}	Van Hassel et al. (1980)
Pesticides				
DDT				
Pike, Esox lucius	fat	up to 2 mg kg^{-1}*.	—	Boileau et al. (1979)

TABLE 3.1—contd.

Substance	Organism	Tissue	Tissue concentration	Environmental concentration	Reference
Dieldrin	Caddisfly larva, *Hydropsyche* sp.	—	23 mg kg^{-1}*	17 µg litre^{-1}	Wallace and Brady (1971)
	Caddisfly larva, *Cheumatopsyche* sp.	—	103 mg kg^{-1}*	17 µg litre^{-1}	Wallace and Brady (1971)
	Blackfly larva, *Simulium vittatum*	—	24 mg kg^{-1}*	17 µg litre^{-1}	Wallace and Brady (1971)
	Snail, *Physa* sp.	—	62 mg kg^{-1}*	17 µg litre^{-1}	Wallace and Brady (1971)
Endosulfan	Catfish, *Heteropneustes fossilis*	liver	4·7 µg g^{-1}*	0·5 µg litre^{-1}	Rao and Murty (1982)
Endrin	Stonefly, *Pteronarcys dorsata*		0·31 µg g^{-1}*	0·03 µg litre^{-1}	Anderson and DeFoe (1980)
	Channel catfish, *Ictalurus punctatus*		0·38 µg g^{-1}*	0·5 µg litre^{-1}	Argyle *et al.* (1973)
	Bullhead, *Ictalurus melas*		1·6 µg g^{-1}*	0·26 µg litre^{-1}	Anderson and Defoe (1980)
Fenitrothion	Moss, *Fontinalis antipyretica*	—	139 µg kg^{-1}*	18·8 µg litre^{-1}	Morrison and Wells (1981)
	Brown trout, *Salmo trutta*	muscle	126 µg kg^{-1}*	18·8 µg litre^{-1}	Morrison and Wells (1981)
	Crustacean, *Daphnia magna*		0·078 mg kg^{-1}*	0·15 µg litre^{-1}	Skaar *et al.* (1981)
	Bluegill, *Lepomis macrochirus*		1·68 mg kg^{-1}*	0·15 µg litre^{-1}	Skaar *et al.* (1981)
Lindane (HCH)	Crustacean, *Daphnia magna*		3·75 µg g^{-1}	10 µg litre^{-1}	Hansen (1980)

Methoxychlor	Stonefly, Pteronarcys dorsata		1.46 μg g⁻¹*	4.23 μg litre⁻¹	Anderson and DeFoe (1980)
	Snail, Physa integra		28.2 μg g⁻¹*	4.23 μg litre⁻¹	Anderson and DeFoe (1980)
Mirex	Crustacean, Daphnia magna		0.125 mg kg⁻¹	0.15 μg litre⁻¹	Skaar et al. (1981)
	Fathead minnow, Pimephales promelas	whole body	268 μg g⁻¹*	34 μg litre⁻¹	Buckler et al. (1981)
	Bluegill, Lepomis macrochirus	whole body	1.98 mg kg⁻¹*	0.15 μg litre⁻¹	Skaar et al. (1981)
Permethrin	Salmon, Salmo salar		1.21 μg g⁻¹*	0.022 mg litre⁻¹	Zitko et al. (1977)
Industrial organic chemicals					
PCB, Aroclor 1254	Amphipod, Gammarus pseudolimnaeus		552 μg g⁻¹*	5.1 μg litre⁻¹	Nebeker and Puglisi (1974)
	Coho salmon, Oncorhynchus kisutch		659 μg g⁻¹*	(dietary)	Mayer et al. (1977)
	Fathead minnow, Pimephales promelas		1 036 μg g⁻¹*	4.6 μg litre⁻¹	Nebeker et al. (1974)

Concentrations which are expressed in the literature as parts per million (or billion) have been converted in $mg\,kg^{-1}$ (or $\mu g\,kg^{-1}$). It should be noted that $mg\,kg^{-1}$ and $\mu g\,g^{-1}$ are equivalents. Data are based on dry weights: those which are indicated by an asterisk * are based on fresh (wet) weights: an asterisk in brackets indicates a probable fresh weight.

usual to employ the term 'indicator' in the sense of biological 'litmus-paper', although, of course, few examples are quite so definitive. Some workers have used tissue analysis as a means of assessing environmental levels of contaminants in freshwater, and this will be considered later in the chapter, but such an approach seems to have gained much more support in marine environments (Phillips, 1980).

3.2 SELECTION OF INDICATORS

In order to be able to choose indicators we must first determine the question of what is to be indicated. Far too often this is not done and, carrying over the litmus-paper analogy, one gains the impression that some organisms are regarded as 'universal indicators'. Almost any species can be an indicator of something but since our knowledge of the autecology of the majority of species is minimal and, even were that not the case, our resources are limited, we must select those organisms which are potentially most useful for the particular problem in hand. At present this is largely confined to monitoring water quality to appraise resources for potable supplies, manage or control effluent discharges and protect fisheries. More recently, biological surveillance has become more important in the United Kingdom, and no doubt elsewhere, as an adjunct to general nature conservation. In selecting indicators for environmental protection the following attributes may be particularly desirable.

Ideal indicators should, of course, unambiguously indicate by their presence very narrowly defined environmental parameters. This ideal is rarely realised, but good environmental indicators

(a) are *readily identified*—taxonomic uncertainties can confuse data interpretation;
(b) may be *sampled easily*, that is, without the need for several operators or expensive equipment, *and quantitatively*;
(c) have *cosmopolitan distribution*—the absence of species with very narrow ecological requirements and limited distribution may not be associated with pollution, etc.;
(d) are associated with *abundant autecological data*—this is of considerable assistance in analysing survey results and devising pollution, or biotic, indices;
(e) have *economic importance as a resource or nuisance or pest*: species

which are economically important (fish) or are nuisances (some algae) have intrinsic interest;

(f) *readily accumulate pollutants*, especially so as to reflect environmental levels since this facilitates understanding of their distribution in relation to pollutant levels;

(g) are *easily cultured in the laboratory*, which also assists in relating experimental studies of their responses to pollutants and field observations;

(h) have *low variability*, both genetic and in their role (niche) in the biological community.

The major taxonomic groups of organisms possess the above attributes to varying degrees. A tabular summary of the basic merits and difficulties of each of the major groups of organisms has been presented elsewhere (Hellawell, 1977a) but a fuller account is given here.

3.2.1 Bacteria

Certain bacteria have been studied for a considerable period because of their associations with public health and the spread of water-borne disease, and techniques for the isolation and examination of bacteria of faecal origin are, therefore, extremely well developed. For example, it is possible to detect one *Escherichia coli* organism in 100 ml (Evison, 1979). General bacteriological methods are also well developed and automatic methods for obtaining total counts of bacterial cells are available. Samples for bacterial analysis are easily obtained and manpower demands are low. The generation time of bacteria is very rapid and so their response times to organic enrichment or to toxic substances are likely to be quite rapid also. However, apart from their value for detecting contamination of potable water, or health risks to bathers or livestock, their use in general environmental monitoring is limited. In flowing water the origin of bacteria in a water sample is unknown and it is likely that the maximum numbers of cells will be recovered some distance downstream of the source of organic enrichment, and recovery from intermittent pollution is likely to be rapid. Other practical problems with bacterial indicators are the need for facilities for sterilising equipment, and incubating cultures.

One of the major disadvantages of bacterial methods is the delay, often of several days, in obtaining results from cultures. This may be compared with the immediate direct observation of other groups. The results may also prove difficult to interpret: the cells which grow on culture media are viable, but were they active in the sample when it was taken? A similar problem exists when one uses direct counting of bacterial cells: it is difficult, without special techniques, to distinguish between living and dead cells.

Finally, our understanding of the bacteriology of cleaner waters is not as far advanced as our knowledge of the significance of other groups.

3.2.2 Protozoa

Like bacteria, protozoa are relatively easy to sample and their responses to organic enrichment are well known. The documentation on the position of certain protozoa within the 'saprobien system' is extensive (Bick, 1968) and even species within individual genera, for example *Vorticella* and *Opercularia*, are claimed to exhibit different responses to organic enrichment (Sladecek, 1971, 1981). In order to utilise this group, taxonomic expertise of a high order is necessary although it has to be admitted that this applies to other groups also, including algae and some macroinvertebrate taxa. The difficulties of drift and the tremendous differences in the protozoan faunas of microhabitats detract from the value of this group, although the proposal by Cairns *et al.* (1973a,b) and Cairns and Yongue (1974) to use artificial substrates for colonisation by protozoa would help reduce this difficulty, although it might create others, particularly of interpretation of the results in an environmental context.

3.2.3 Algae

The particular value of using algae as indicators is in connection with eutrophication studies; indeed there is no alternative group for studies of nutrient enrichment in open water. The group is undemanding in sampling resources but quantitative sampling is difficult for attached forms, although the use of glass slides as artificial substrates for algal colonisation (see Section 9.2.5) helps to overcome this problem. Their organic pollution tolerances are well documented (e.g. Patrick, 1954; Fjerdingstad, 1964, 1965; Palmer, 1969) but they are not so directly useful for indicators of pesticide pollution or the majority of the heavy metals, although copper is an exception to this generalisation. As with bacteria, there may be difficulty in distinguishing between living and dead cells. Taxonomically the group is quite demanding and counting cells can be tedious but neither of these disadvantages is the exclusive prerogative of algae: some invertebrate groups are both difficult to identify and may occur in large numbers!

3.2.4 Macroinvertebrates

Macroinvertebrates constitute a heterogeneous assemblage of animal phyla and consequently it is probable that some members will respond to whatever stresses are placed upon them. Many are sedentary, which assists in detecting the precise location of pollutant sources, and some have relatively long life histories (measured in years) and this provides both a facility for examining temporal changes and also integrating the effects of prolonged exposure to intermittent discharges or variable concentrations of a pollutant. Extended life cycles, say of many months, also justifies the adoption of periodic sampling. Such a procedure would be invalid for groups having short life histories. Qualitative sampling of benthic macroinvertebrates is relatively easy, the methodology is well developed and equipment need not be elaborate. Taxonomic keys are available for most groups although certain 'difficult' taxa exist, notably the larvae of chironomid flies, some caddis larvae (Trichoptera) and oligochaete worms. Progress is being made in providing keys even for these groups and, on the whole, macroinvertebrates probably do not present any more taxonomic difficulties than algae or protozoa. Quantitative sampling of macroinvertebrates is difficult; their patchy distribution within the substrate often means that large numbers of samples are required in order to be able to make reasonable estimates of population densities. Again, this is a problem which is not confined to macroinvertebrates.

Perhaps the most useful feature of this group is that many methods of data analysis, including pollution indices and diversity indices, have been devised. The widespread, almost universal, use of these techniques testifies to their value in water quality management practice. The multiplicity of different techniques (see Section 9.4) suggests that not all are entirely satisfied with the results which they provide!

3.2.5 Macrophytes

Macrophytes have two important advantages as indicators: they are stationary and visible to the naked eye. However, in temperate climates they are subject to marked seasonal variations in biomass and for considerable periods (perhaps the greater part of the year) not readily seen. Until relatively recently their responses to pollutants were not well documented, but attempts have now been made to provide schemes for assessing environmental damage (e.g. Harding, 1981; Haslam, 1982b).

The demands on manpower for macrophyte surveys is low—another distinct advantage in using this group. Macrophyte communities appear to be governed by climate, geology and soil type and many waters, such as high altitude streams, have sparse macrophytic vegetation. In lowland waters, channel maintenance by mechanical means and by application of herbicides may greatly influence the macrophyte flora, which is a disadvantage when one is attempting to use this group for water quality assessment but a distinct advantage if one wishes to assess environmental quality. One hopes that future developments in techniques for using macrophytes will enable wider use to be made of a group with considerable potential for environmental impact assessments.

3.2.6 Fish

Fish have the distinction of being monitored by the public at large, either by interested parties such as anglers and commercial fishermen or when large numbers of fish corpses floating downstream or at the margins of lakes indicates that pollution has occurred.

As they occupy positions towards the apex of food pyramids, fish reflect effects of pollutants on other organisms as well as direct stresses on themselves. However, fish are mobile and are well able to avoid polluted water, returning when conditions are favourable. One might be misled, therefore, by periodic surveys into thinking that fish were always present (or absent) and that conditions were acceptable and stable (or always unacceptable). Identification of fish species in Europe is relatively easy but perhaps slightly more difficult in North America and elsewhere, where the fish fauna is considerably richer. Fortunately, the freshwater fish faunas of most regions of the world are better documented than, say, invertebrates.

The methodology for fisheries investigations is probably the most advanced of all the groups considered here and is covered in an extensive literature. In spite of this, and the other advantages listed, the use of the fish community in routine environmental surveillance is hampered by the necessity for extensive manpower and the difficulty in obtaining samples in deep, fast-flowing rivers.

3.2.7 Recommended Indicator Taxa

In an examination of the literature on indicators, Hellawell (1977a) found considerable variation in the extent to which different groups had been recommended for assessing water quality (Fig. 3.1). It is evident that two groups were recommended more often than others, but this is slightly misleading since all publications were given equal weight; workers who published extensively, writing many short, similar papers, could easily shift the balance. Whitton (1979) bemoaned the disparity between the ap-

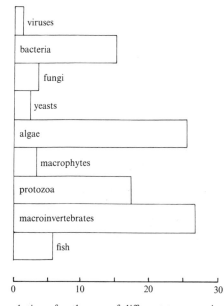

FIG. 3.1 Recommendations for the use of different taxonomic groups as biological indicators. Percentage distribution of literature citations up to 1970. From Hellawell (1977b).

parent virtual equality of algae and macroinvertebrates as recommended in the literature and the preponderance of the use of macroinvertebrates, by practising hydrobiologists, for assessing the quality of flowing waters. But biologists charged with environmental surveillance are pragmatic: pollution indices using algae have been rather limited (Watanabe, 1962; Palmer, 1969), and in an assessment of the Palmer index, Edwards *et al.* (1972) commented that this index tends to be self-compensating in that scores assigned to the few species which can survive in highly polluted waters are virtually equivalent to the total attained by the many, lower scoring species, present in less severe pollution. On the other hand, pollution indices which utilise macroinvertebrates are numerous (see Section 9.4.2) and effective. It is unlikely that the evident popularity of macroinvertebrates is merely self-perpetuating, but rather that this group is first choice for routine surveillance because the methodology is relatively simple yet well-developed and often requires only a single operator; the taxonomy is none too difficult and keys are available for most groups; the group comprises a wide range of organisms thereby offering the possibility of varied responses to different environmental stresses and, finally, the sedentary habits of most members of the group are conducive to spatial analyses while their relatively long life cycles permit temporal changes to be followed.

3.3 INDIVIDUALS AND POPULATIONS AS INDICATORS

Individual organisms or populations of a species might be thought to have great potential for environmental monitoring and yet, in freshwater at least, this potential has not been realised. Many factors associated with individuals or populations could be used to assess environmental quality. For individuals one could use growth rate, ultimate size or age, size or age at sexual maturity, fecundity, or the amount of fat or other stored material. Populations might be assessed by their density, age structure or sex ratio. Environmental conditions which impaired growth and development or reproductive success would become evident when populations were examined. Single organisms might be less useful since it would be difficult to be sure that significant departure from the norm was not simply the result of some genetic deficiency or disease.

There is a wealth of evidence from field studies and from laboratory investigations that many organisms show anatomical and physiological responses to environmental conditions. For example, it has long been known that the growth of fishes is correlated with water quality. The classical case of this phenomenon is the variation in size for age in brown trout (*Salmo trutta*) with water hardness or conductivity (see Frost and Brown, 1967). In nutrient-poor, soft acid upland streams trout often live for up to ten or more years but remain relatively small (less than 100 g) while in lowland, nutrient-rich, hard-water chalk streams growth is rapid, ultimate size is greater (more than 1 or 2 kg) but their life is shorter, perhaps only 3 to 5 years.

Some invertebrates are also known to vary according to environmental conditions. Probably the best-known example of this is cyclomorphosis in the Cladocera. These planktonic crustacea show marked variation in shape of their 'helmet', or variation in carapace or tail spine lengths (Hutchinson, 1976). Other examples of anatomical variations in organisms which appear to be related to environmental factors are reviewed by Hynes (1970). The majority of these are related to interspecific differences but also included is the relationship between shell shape and current speed in the freshwater 'limpet', *Ancylus fluviatilis*, the shell being higher and more steeply conical in faster water (Starmühlner, 1953).

However, most studies reported in the literature relate to variations in organisms which are exhibited in response to the intensity of some natural environmental phenomenon. There appears to be no intrinsic reason why organisms should not show morphological or biochemical modifications in response to pollution or other environmental stress but documentation on this is sparse.

Experimental studies of the effects of changes in environmental quality or of toxic substances upon the growth or reproduction of organisms have been extensively reported and are considered in more detail in Chapter 7. An example of the kind of change which is of potential value is provided by Winner (1981) who showed that the body length of primiparous female *Daphnia magna* was significantly reduced following exposure to copper and zinc.

The greater potential for the use of individuals or populations is almost certainly in the use of those groups of organisms, such as fish or some molluscs, which can be aged. This enables one to derive a size-age or age at maturity relationship against which the parameters for individuals or populations can be assessed. The basic techniques for aging many fish are well established but there are potential sources of error and care has to be exercised (Bagenal, 1974).

The major difficulty in using field surveys of individuals or populations is that of interpretation and this may be the explanation for the apparent lack of interest in this approach by environmentalists. It is one thing to establish that an individual or population differs from the norm and quite another to deduce the cause. Similarly, one can expose individuals or populations to known concentrations of pollutants or extreme environmental conditions under laboratory conditions and observe a resulting deviation from normality. But it is rarely possible to infer the reason why an individual or population appears to be abnormal.

3.4 COMMUNITY STRUCTURE AS AN INDICATOR

Few, if any, methods of routine surveillance in freshwater utilise simple presence or absence criteria for individual species. Most often the composition of the community of organisms, or more strictly of a particular taxonomic group such as macroinvertebrates, is used as an

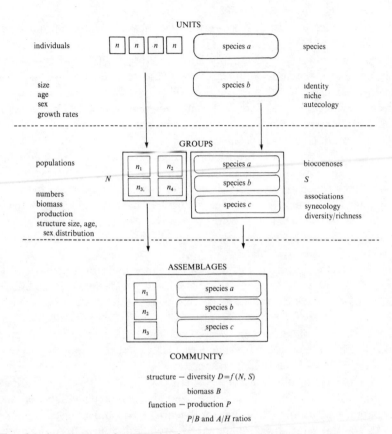

FIG. 3.2 A conceptual framework of community organisation and the structural or functional measures appropriate to each level. From Hellawell (1977b).

indicator of prevailing environmental quality or of changes in that quality. In order to understand the significance of community responses it is necessary to consider how the community is organised. A conceptual framework of community organisation is indicated in diagrammatic form in Fig. 3.2. A community consists of species, each of which is represented by a number of individuals.

The diversity which the community exhibits is a function of the numbers of individuals and species. Often the distribution of individuals within species follows a recognisable pattern which is, put simply and in purely descriptive form, that a few species will be represented by many individuals (i.e. they are common or abundant species) while many

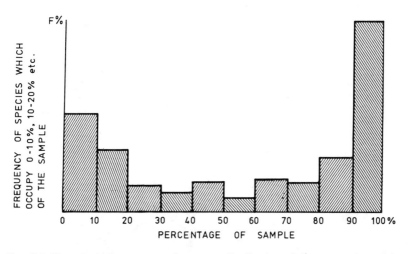

FIG. 3.3 Hypothetical percentage frequency distribution of the proportion of the sample occupied by species. After Raunkiaer (1934).

species are represented by fewer individuals and some species are quite rare (Fig. 3.3). This pattern has been observed in many natural communities (Williams, 1953) and a number of mathematical models have been proposed as descriptions of this structure (Raunkiaer, 1934; Fisher *et al.*, 1943; Preston, 1948; MacArthur, 1957; Pielou, 1969) and more recently May (1974) has claimed that they are all cases of a log-normal

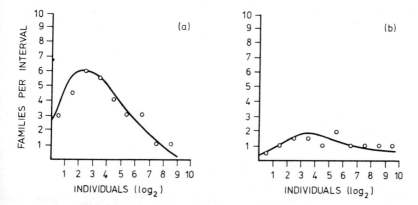

FIG. 3.4 Comparison of lognormal curves of community structure from (a) unstressed and (b) stressed habitats.

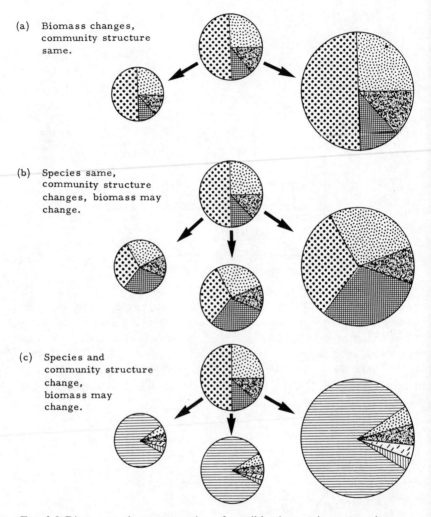

(a) Biomass changes,
community structure
same.

(b) Species same,
community structure
changes, biomass may
change.

(c) Species and
community structure
change,
biomass may
change.

FIG. 3.5 Diagrammatic representation of possible changes in community composition and biomass. From Hellawell (1977b).

distribution. Their value in biological surveillance is that they provide a comparative basis against which changes in the structure of communities in polluted or stressed environments may be compared (see Section 9.4.1.1).

This approach has been utilised by Patrick *et al.* (1954) in assessing changes in diatom communities and Brown (1971) has used it to study changes in macroinvertebrate communities. A typical form of the log-normal curve of community structure is shown in Fig. 3.4. When the community suffers stress the curve becomes flatter, indicating a reduction in the abundance of some of the rarer and more sensitive species, while the right-hand end of the curve is extended because there is an increase in the numbers of a few insensitive species which now dominate the community.

Possible options for change in community composition are indicated in diagrammatic form in Fig. 3.5. Here the community is represented by a circle in which the relative abundances of four species are indicated. The first option for change is merely one of size, enlarging or diminishing, but without any modification of the relative proportion of individual species. In the second option, the community may change in size but the composition shifts also—some species increase in relative importance, others decline. Finally, in addition to changes of the kind observed in the second option, new species join the community while others disappear. Besides these structural changes and variation in the total biomass there may be dynamic, functional changes, too, just as there were in Section 3.3 for individuals and populations.

3.5 FUNCTIONAL CHANGES IN COMMUNITIES

Functional changes are almost invariably accompanied by structural changes, although Hall et al. (1970) have described studies under experimental conditions in which nutrient enrichment increased the production of zooplankton communities without affecting their composition. Changes in the balance between autotrophs and heterotrophs are a common feature of organic or nutrient enrichment and indeed are utilised in the Assimilation–Zehrungstest or A/Z test (Knöpp, 1961) (see Section 8.2.1.2.3).

3.6 BIOACCUMULATIVE INDICATORS

The term 'bioaccumulative' indicator has been used here in order to emphasise the difference between ecological or pollution indicators, the presence or unexpected absence of which gives an indication of the habitat or environmental quality, and those organisms which accumulate polluting substances from their surroundings or food or both, sequestering them in their bodies, so that when tissues are analysed an indirect estimate of prevailing environmental concentrations of these substances may be made.

Extensive reviews have been made of the use of such indicators for metals (Phillips, 1977) and pesticides, especially organochlorine insecticides (Kerr and Vass, 1973; Johnson, 1973; Phillips, 1978).

3.6.1 Characteristics of Ideal Indicator

The ideal attributes of a bioaccumulative indicator differ somewhat from those set out in Section 3.2, although there is some overlap. Phillips (1978) has summarised the desirable attributes for an accumulative indicator as published in the literature and amended by his own experience. Although the requirements were principally drawn up for studies of metal pollution in marine habitats, they may easily be adapted for freshwater environments as follows:

(1) All individuals of the indicator species should exhibit the same simple correlation between their residue content and the average pollutant concentration in their surrounding water, or substrate deposits, or food, at all locations and under all conditions.

(2) The species should accumulate the pollutant without being killed or rendered incapable of sustained reproduction by the maximum levels encountered in the environments.

(3) The organism should be sedentary in order to be sure that the findings relate to the area of the study.

(4) The species should occur abundantly throughout the study area and should preferably have a widespread distribution in order to facilitate comparisons between areas.

(5) Long-lived species are desirable since they enable samples to be taken over several year-classes, if required. Long-lived species also suffer exposure over long periods and, therefore, provide evidence of long-term effects.

(6) The species should be of sufficient size to provide adequate tissues for analysis. This attribute is also of assistance in dissection when studies are to be made of accumulation in specific organs.

(7) The species should be easy to sample (i.e. collect) and robust, surviving well under laboratory conditions, allowing evacuation of gut contents before sampling or studies of pollutant uptake and the rate of loss of pollutants in clean water.

In reality, of course, organisms having all these desirable features are not known, but some meet one or more of the criteria. Having set them down it immediately becomes obvious that problems are encountered in the field. Taking the first criterion—simple correlations between environmental and tissue levels—this is rarely found although useful relationships have been observed, for example, between heavy metals residues in the growing points of bryophytes or macroalgae and concentrations in polluted waters (Harding and Whitton 1981; Say *et al.*, 1981; Whitton *et al.*, 1982).

3.6.2 Factors Which Affect Indicator Reliability
The 'reliability' of indicators, that is, their efficacy in presenting an estimate of prevailing environmental concentrations of substances, is governed by many factors. These are considered in other sections relating to the specific substance or indicator group but an outline of the principal factors, together with some examples, are given below. It should be remembered that some of the factors are not independent: physiological condition and age or size or sexual maturity are often related.

3.6.2.1 Accumulation and excretion rates
In fluctuating environmental concentrations of pollutants, differences between the rates of uptake and loss of contaminants will affect the net residue concentration at any given time. If, for example, a substance is accumulated rapidly but lost slowly then body burdens are likely to reflect peak environmental concentrations rather than mean levels. Often this property is valuable, since it enables the most critical conditions to be monitored. Substances which are lost rapidly are likely to be found at high concentrations only immediately after exposure: when environmental levels are reduced the material is lost. In a study of the application of fenitrothion to forest Morrison and Wells (1981) found that the concentration–time profile in the fish followed closely the profile in the water (Fig. 3.6).

3.6.2.2 Physiological condition of indicator species
Rates of accumulation and more particularly equilibrium concentrations of some substances depend much on the condition of the accumulator. A good example of this factor is the relationship between lipid content and whole-body levels of organochlorine pesticide residues (Anderson and Everhart, 1966; Anderson and Fenderon, 1970).

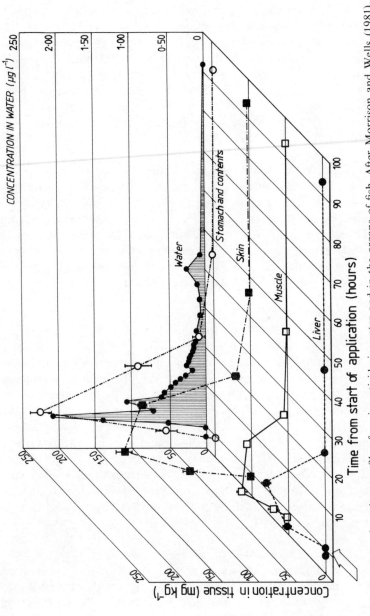

Fig. 3.6 Concentration–time profile of an insecticide in water and in the organs of fish. After Morrison and Wells (1981).

Physiological condition is often a seasonally related factor: changes in lipid storage, feeding activity or reproduction are commonly associated with season. Seasonal variations in total DDT and PCB residues were studied in *Gammarus pulex* by Södergren *et al.* (1972). They noted that different seasonal profiles were obtained according to whether the results were expressed on the bases of wet weight or on the basis of lipid weight. But however expressed, these data, and others from other invertebrates, indicate seasonal shifts in organochlorine residues.

Having reviewed examples from the literature on seasonal variations in organochlorine residues in fish, Phillips (1978) notes that the form of the seasonal concentration profiles together with the timing of maximum concentrations may be related to the timing of the spawning period, fish migration, fish age or size, the location of the study area or simply whether the results are expressed on the basis of wet weight or lipid weight. These sources of variation can affect the validity of the interpretation of survey data. He emphasises the need to avoid studies in which only a single sample is taken at a given time at each sampling station since the populations occurring at each location may be at a different phase in seasonal cycles, and advocates the collection of time-series samples at each site.

3.6.2.3 Age or size of indicators

The age or size of organisms is frequently associated with variations in residue levels. For example, the levels of DDD in catfish (*Ictalurus catus*) and largemouth bass (*Micropteris salmoides*) have been related to age (Hunt and Bischoff, 1960). Similar age-related increases in DDT and DDE concentrations have been observed in the Atlantic salmon, *Salmo salar* (Anderson and Everhart, 1966). Other examples include a tenfold increase in total DDT concentrations in salmonids from Lake Michigan (Macek and Korn, 1970), increases in total DDT concentrations with size in the mosquito fish, *Gambusia affinis* (Murphy, 1971) and significant correlations between both length and age and total DDT residues in lake trout *Salvelinus namaycush* by Youngs *et al.* (1972).

Phillips (1978) noted that, while there was an age-related variation in organochlorine residues, which could not always be accounted for by commensurate changes in lipid content and must be truly age or size dependent, no generalisations could be made about size as a cause of indicator variability. He concludes that the age of the organisms sampled should be considered as 'a potential interfering factor' in indicator surveys, especially when using fishes.

3.6.2.4 Interference between substances
Studies of the acute toxicity of mixtures of poisons have shown that, with some exceptions, their effects are additive, that is, the relative toxicities of each substance when summed give an estimate of the toxicity of the combination (see Section 7.5). A few combinations have much higher than predicted toxicities ('synergism' or supra-additive interaction) while others may have less toxicity when combined and, in extreme instances, may actually be less toxic in combination than when present alone at the same concentration ('antagonism'). Similar effects have been noted in the accumulation of tissue burdens of some pollutants although the data are relatively sparse (EIFAC, 1980). Interactions in the uptake of metals have been observed; for example copper and cadmium are both accumulated in fish gills more readily when both are present (Hewitt and Anderson, 1978) while cadmium inhibited zinc uptake although zinc had no effect on cadmium accumulation (Eisler and Gardner, 1973). However, in long-term tests with sub-lethal mixtures of cadmium and zinc, neither metal appeared to influence the uptake of the other (Spehar *et al.*, 1978).

Interactions between pesticides are only slightly better documented. The mutual effect of organochlorine pesticides appears to be either neutral or inhibitory, although there are exceptions; for example, dieldrin appears to enhance the uptake and storage of DDT (Macek *et al.*, 1970; Mayer *et al.*, 1970). In studies of the interaction of three organochlorine insecticides when taken up by rainbow trout, *Salmo gairdneri*, in food, Mayer *et al.* (1970) found that methoxychlor tended to reduce DDT storage, and DDT in combination with dieldrin reduced methoxychlor storage, depending on the dosage of the latter. Both DDT and methoxychlor decreased dieldrin storage but methoxychlor had a more pronounced effect. Thus clear interactions between pairs of organochlorines, and sometimes between all three, were evident. It was suggested that pesticide-metabolising enzymes had been induced in the liver by DDT and possibly methoxychlor. The interactions of labelled DDT and dieldrin uptake in food by rainbow trout were also studied by Macek *et al.* (1970). The interaction was found to occur in the pyloric caecae, where DDT was found to decrease the rate of dieldrin uptake while, as noted above, dieldrin enhanced DDT accumulation.

No interactive affects on the uptake of the organochlorine substances dieldrin and polychlorinated biphenyls by white suckers, *Catostomus commersoni*, were observed by Frederick (1975) when they were exposed to these compounds separately and in combination. Similarly, Addison *et al.* (1976) observed no interactions in the excretion of intermuscular injections of labelled aldrin and DDT by fry of Atlantic salmon (*Salmon salar*).

The factors which may account for the ways in which different toxicants can influence the uptake of other xenobiotic substances have been summarised in EIFAC (1980) as follows:

(1) Detoxifying enzymes, which can influence the uptake or loss of xenobiotic substances, may be inhibited.
(2) Permeability of membranes may be changed thereby affecting the rate of uptake of other substances.
(3) Internal physiological changes, such as increased rate of blood flow through the gills of fish, may increase the accumulation rate of substances.
(4) There may also be external interaction of toxicants, perhaps in the formation of complexes, which may influence uptake.

There is, clearly, no inevitable interaction between compounds but sufficient evidence exists to indicate that in using some organisms as bioaccumulators it is possible that the body burdens of a particular contaminant may be influenced by the presence of other substances. It may prove necessary, therefore, to study a broader band of residues than those for which a particular study or monitoring programme has been instigated in order more confidently to interpret the results. This aspect becomes even more significant when one considers how rarely the environment is likely to be influenced by a *single* pollutant!

3.6.2.5 Environmental variables
Included under this heading are factors such as temperature, turbidity, water hardness and salinity. Temperature not only affects the solubility of many substances but also appears to influence the rate of uptake of toxicants. For example, Murphy and Murphy (1971) found that mosquito fish accumulated more DDT at 20 °C than at 5 °C while Veith *et al.* (1979) found that fathead minnows, green sunfish (*Lepomis cyanellus*) and rainbow trout showed increased bioconcentration factors of polychlorinated biphenyl at higher temperatures. Reinert *et al.* (1974) studied the rates of uptake of DDT by rainbow trout (*Salmo gairdneri*) at 5, 10 and 15 °C and found that uptake was correlated with temperature (Fig. 3.7).

Turbidity and high organic content of water reduce the availability of organochlorine residues and so residues in oligotrophic waters are likely to remain available for accumulation for much longer than in eutrophic

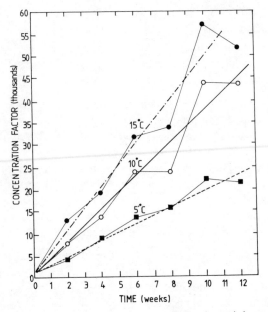

FIG. 3.7 Variation in the rate of uptake of DDT by rainbow trout (*Salmo gairdneri*) with temperature. After Reinhart *et al.* (1974).

lakes (Johnson, 1973). Some metals, such as copper, readily form complexes with organic matter and most metals are less toxic in hard water (Alabaster and Lloyd, 1980). Some studies of pesticide toxicity have also shown that many are less toxic in hard water, yet others are not influenced by hardness while, rarely, some are more toxic in softer water (Holden, 1973). Salinity appears to be an important factor in the net uptake of organochlorines in estuarine species but at present the effects are poorly defined (Phillips, 1978).

3.6.2.6 Trophic level

Another factor which is held to affect the body burden of residues, especially those which could accumulate mainly through food, is the relative position of the indicator in the trophic scale. Primary producers, such as algae, absorb substances directly from the surrounding water but creatures which occupy higher positions on the trophic 'ladder', such as macroinvertebrates and fish, are able to absorb material which has been accumulated on the lower 'rungs' by their food organisms, in addition to direct uptake from their surroundings. One would expect, therefore, that as one ascends the trophic scale so the potential for increased accumulation ('biomagnification' or 'bioconcentration') is greater.

Although some support is available for this hypothesis, especially in terrestrial food chains which culminate in birds of prey or mammals (Edwards, 1973a), in aquatic environments the evidence is conflicting (Johnson, 1973; Phillips, 1980). Not all substances are accumulated in food; some reach similar concentrations at all trophic levels and when differences do occur they cannot always be explained solely by position in the trophic scale. This last factor is often true in field studies and has not been entirely eliminated in laboratory studies or by using relatively simple or 'model' ecosystems (Metcalf *et al.*, 1971; Sanborn and Yu, 1973; Hamelink *et al.*, 1971; Hamelink and Waybrant, 1976). Other factors such as longevity, size, lipid content and the excretion efficiencies of organisms, some of which also correlate with trophic level, could equally be responsible for the apparent differences.

Most of the work which has a bearing on this aspect relates to pesticide residues. Associations between metal levels and trophic position have also been described. For example, the amplification of concentrations of methyl mercury in food chains appears to be well established but Phillips (1980) speculates that this may also be related primarily to increasing lipid content or, perhaps, merely by long environmental persistence.

The important conclusion which one may draw from the possibility that trophic level may influence bioaccumulation is that in selecting indicators, tissue levels based on, say, herbivorous invertebrates or fish in one habitat cannot be compared with those derived from carnivores or detritivores in the same habitat or elsewhere.

3.6.3 Examples of the Use of Bioaccumulative Indicators

Two of the principal ways in which bioaccumulation indicators have been employed for environmental monitoring in freshwater are the use of macrophytes for studies of heavy metal pollution and the analysis of fish for organochlorine pesticide residues. These are selected as examples in order to illustrate the technique (see also Sections 8.2.1.8 and 8.2.1.9).

3.6.3.1 Macrophytes as indicators of metals

The distribution of certain macrophytic algae and bryophytes in streams receiving minewaters, and the ease with which they may be sampled and analysed, were, no doubt, important contributory factors in the re-alisation that here was a potentially valuable and effective group of organisms with which to study heavy metal pollution. An extensive literature on this topic now exists (McLean and Jones, 1975; Empain, 1976a, b; Benson-Evans and Williams, 1976; Say *et al.*, 1977, 1981; Harding and Whitton, 1981; Whitton *et al.*, 1982).

It has long been recognised that the macrophytic red alga *Lemanea* is among the few plants which appear to be tolerant of heavy metals in the outfalls of mining activity (Carpenter, 1924). Its usefulness as an accumulator of heavy metals has been investigated in detail in a variety of habitats encountered in a range of rivers throughout Europe by Harding and Whitton (1981). *Lemanea* was found to occur at concentrations of zinc up to $1 \cdot 16$ mg litre $^{-1}$. No evidence was found of genetic differences in the ability of different populations to tolerate high levels of zinc, cadmium and lead. The alga was sampled by taking 20 mm lengths from the tips of filaments from 100 mm^2 of rock surface and pooling these for analysis. Using four samples from each of 59 sites, significant correlations were found between the logarithm of concentration of zinc, cadmium and lead in the *Lemanea* tips and the logarithm of the mean concentration of these metals in both filtered ('dissolved' metal) and unfiltered ('total') samples. The enrichment ratios (plant to water) decreased as the environmental concentration of metals increased and, although the metal concentrations are interrelated, differences in the ratios of the metals were reflected in the plant tissue composition. The results indicated that uptake of metal was less in soft waters.

When samples of *Lemanea* were transplanted, the levels of zinc in the transplants changed and stabilised at concentrations close to those observed in the native filaments. On the basis of these studies these workers suggest that *Lemanea* would be useful for monitoring heavy metal concentrations in rivers.

Aquatic mosses were used by Say *et al.* (1981) to monitor heavy metal contamination of a river polluted by industrial discharges and treated sewage effluents. Three tributaries contributed pollutants including zinc while one tributary was thought to be contaminated by chromium. In addition to making metal analyses of the mosses, samples of water and sediment were taken for analysis. Significant correlations were observed between the concentration of zinc in the river water and in the terminal 20 mm section of shoots of *Fontinalis squamata* and *Rhynchostegium riparioides* (Fig. 3.8). Although chromium could not be detected in the water samples, appreciable concentrations were found in the mosses, which suggests that intermittent discharges of this metal did occur.

The aquatic liverwort *Scapania undulata* has also been considered as a potential bioaccumulating monitor for the heavy metals zinc, cadmium

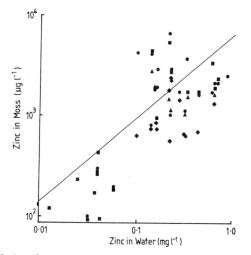

FIG. 3.8 Correlation between concentration of zinc in river water and the terminal sections of shoots of four mosses ($r = 0.67$, $p = <0.001$). ▲ *Amblystegium riparium,* ◆ *Fontinalis antipyretica,* ■ *Fontinalis squamosa,* ● *Rhynchostegium riparioides.* After Say *et al.* (1981).

and lead, by Whitton *et al.* (1982), following studies of the plant in a range of European streams and rivers, including one in which zinc was present at 7 mg litre^{-1}, cadmium at 0·028 mg litre^{-1} and lead at 1·74 mg litre^{-1}. As with *Lemanea*, highly significant correlations were observed between the logarithm of the environmental concentration of the metals and the logarithm of the concentrations in pooled samples of 10 mm tips of shoots. The data suggested that at high pH or high calcium concentrations (i.e. 'hard' water) there was increased accumulation of zinc and cadmium, and possibly lead also.

The potential of *Scapania* for use as an indicator is limited by two factors; it does not occur in nutrient-rich waters, rendering it unsuitable for use in rivers which receive sewage effluents or many kinds of industrial wastes, and also its rate of increase in metal accumulation is low in relation to increased water concentrations. However, its advantages as an indicator are that it is one of the larger liverworts and it is relatively easy to recognise. It has a widespread distribution and can tolerate high metal concentrations, even as much as 13·4 mg Zn litre^{-1} in intermittent doses (Say and Whitton, 1981). Until it is possible to relate observed tissue metal levels to the duration of exposure to given level of metal, Whitton *et al.* (1982) were unwilling to recommend its use for routine surveillance but did suggest that it might be useful when transplanted in the manner advocated by Benson-Evans and Williams (1976).

Employing macrophytes as monitors of metals is obviously potentially very useful but the adoption of this approach by statutory pollution-control authorities appears to be limited. Although the technique is in its infancy and much more experience is necessary before it becomes possible to use it as a regulatory monitor, nonetheless one can confidently expect that the method will be adopted widely, especially for alerting authorities to intermittent discharges of metals.

3.6.3.2 Fish as bioaccumulators

The literature relating to the uptake by fish of metals, pesticides and other pollutants is extensive (Phillips, 1980). From the tremendous range of possible studies, two have been selected to illustrate the ways in which analysis of fish tissues may be used to investigate environmental contaminants.

The bioaccumulation of copper by juvenile rainbow trout was studied by Dixon and Sprague (1981) as part of an investigation of the mechanisms whereby this fish acclimates to copper. Young rainbow trout, *Salmo gairdneri*, were exposed for up to three weeks to a range of copper concentrations (30 to 194 µg Cu per litre in approximately 30 µg steps) which represented approximately 0·1 to 0·6 of the incipient lethal level (ILL) for this species (330 µg litre^{-1}) in hard water (374 mg litre^{-1} as $CaCO_3$). Small amounts of copper were also present in the diet.

It was found that the whole body copper content of the trout increased significantly at all concentrations as the exposure continued and the amounts accumulated at a given time increased with the exposure concentration. Maximum whole body copper concentrations were about 38 µg g^{-1} dry weight (or 8·3 µg g^{-1} wet weight).

The relationship between whole body copper concentration, exposure level and duration of exposure were found to be well represented by a three-dimensional response surface (Fig. 3.9) and those authors also provided an equation for calculating the whole body copper concentration given the exposure level and duration which accounted for about 98% of the variation in response.

When exposed fish were transferred to clean water, copper residues diminished rapidly and after three weeks had fallen to 6·9 µg g^{-1}, which may be compared with the level of 3·5 µg g^{-1} before exposure and accumulation. Interestingly, the increased tolerance to copper of pre-

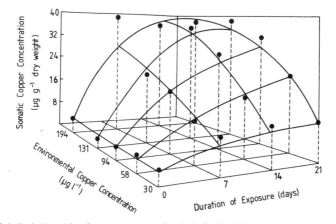

FIG. 3.9 Relationship between somatic (whole body) copper concentration, exposure level and duration of exposure, in young rainbow trout. From Dixon and Sprague (1981).

viously exposed fish was reduced and the ILL became similar to that of controls, thus indicating a relationship between loss of acquired tolerance and body burden reduction.

Those studies illustrate the ideal response of an indicator to an environmental pollutant: the tissue concentration is a definite function of the environmental level and duration of exposure yet when contaminant levels are lowered the body burden is also reduced.

Bioaccumulation of organic chemicals by the fathead minnow (*Pimephales promelas*) and prediction of bioconcentration factors from the *n*-octanol/water partition coefficients of those chemicals was reported by Veith *et al.* (1979). These workers explored a number of factors affecting uptake including age and source of the test fish, temperature and differences between species.

Using hexachlorobenzene as the test substance, Veith *et al.* (1979) studied uptake by four age groups (newly hatched larvae, juvenile minnows 30 and 90 days old, and adults) and found remarkably similar bioconcentration factors over the test period of 120 days. Uptake was rapid at first but stabilised after about ten days and then increased only slightly over the next hundred or so days. The source of test fish (from laboratory culture and from a farm pond) was not found to have any effect on the bioconcentration factors of polychlorinated biphenyl; differences were as great between batches from the same source as between those from different sources.

TABLE 3.2

BIOCONCENTRATION FACTOR (BCF) AND BIOCONCENTRATION POTENTIAL, CALCULATED RELATIVE TO THAT OBSERVED FOR p,p'-DDE, FOR SEVERAL ORGANIC SUBSTANCES, DETERMINED FOR THE FATHEAD MINNOW, *Pimephales promelas*, AFTER 32 DAYS OF EXPOSURE
(from Veith *et al.*, 1979)

Substance	Mean concentration during exposure (μg litre^{-1})	Biocon-centration factor (BCF)	Biocon-centration potential (p,p'-DDE $= 100$)
Lindane	3·4	180	0·4
Pentachlorophenol	11·1	770	1·5
1,2,4-trichlorobenzene	1·6	2 800	5·5
Methoxychlor	3·5	8 300	16
Heptachlor	3·1	9 500	19
Mirex	1·2	18 100	35
Hexachlorobenzene	2·6	18 500	36
p,p'-DDT	6·5	29 400	58
op'-DDT	5·1	37 000	72
Chlordane	5·9	37 800	74
Aroclor 1016	8·7	42 500	83
p,p'-DDE	7·3	51 000	100
Aroclor 1248	4·0	70 500	138
Aroclor 1254	4·3	100 000	196
Aroclor 1260	1·0	194 000	300

Bioaccumulation rates of polychlorinated biphenyl varied with temperature and differed between species. Fathead minnows and green sunfish (*Lepomis cyanellus*) showed very similar bioconcentration factors at 10 °C and above and for both species the greatest increase in the factor occurred between 5 and 10 °C. Virtually no change in accumulated rate for rainbow trout (*Salmo gairdneri*) was observed between 5 and 10 °C but it increased steadily above 10 °C. Rainbow trout were not tested at 25 °C. At 20 °C fathead minnows and green sunfish accumulated three times more PCB residues than rainbow trout and similar differences in uptake were seen in tests using hexachlorobenzene and 1,2,4-trichlorobenzene at 15 °C. These results provide ample evidence for the need to ensure that comparisons are based on the same species and following exposure at a similar temperature. The bioconcentration factors of two substances, p, p'-DDE and heptachlor epoxide, were much the same whether presented singly or in combination to fathead minnows. This finding is at variance with the conclusions of Mayer *et al.* (1970) and Macek *et al.* (1970) but those studies were concerned with uptake in food. Veith *et al.* (1979) concluded that it might be possible to perform simultaneous tests of the bioconcentration of many compounds, thereby saving considerable time and expense. They also proposed that p,p'-DDE should be used as a reference substance and the bioconcentration factor should be expressed relative to that, thereby enabling meaningful comparisons to be made between different laboratories and different species. A selection of bioconcentration factors and bioconcentration potentials, that is relative to p,p'-DDE, is given in Table 3.2. The value of the data provided by Veith and his co-workers is that they relate to one species measured under standard conditions and they are, therefore, comparable.

CHAPTER 4

Environmental Stress

4.1 NATURAL ENVIRONMENTAL STRESSES

Natural environments cause stress to the organisms which inhabit them. For example, on a rocky shore, tidal movements cover and uncover the denizens twice daily, to a greater or lesser extent, depending upon the phase of the cycle of spring and neap tides. During exposure, perhaps for several hours according to their vertical position on the shore, organisms may be subjected to the withering heat of a summer noonday sun or the more subtle stress of freshwater from a rainstorm. With each returning tide the surge of the waves grinds or batters the animals and plants clinging to the exposed rock surfaces while with each retreat shear forces would rapidly wrench them away were they not so pliant, so well cemented or equipped with powerful suckers. One has only to walk the high-tide mark after a violent storm to see how precarious is continued survival on a rocky shore. In terrestrial environments stresses are generated by high winds, low and high rainfall, frost action, snowfall and intense sunshine. Even the constant pull of gravity generates stress, as is evident when climbing plants fail to find support or fully-laden branches of fruit trees give way under the strain.

Aquatic environments are not exempt from environmental stresses. Extremes of temperature; discharge, with related changes in depth or velocity; turbidity, with effects on light penetration and deposition of materials; low or high dissolved oxygen concentrations; and movements of unstable substrata, all contribute to the environmental stresses of life in freshwater. Most, if not all, organisms in freshwater are at risk from disturbed saltwater balance and must expend energy in order to maintain it within viable limits. The classic example of this is the maintenance of body fluid constancy in freshwater fishes. The blood of fishes has a greater concentration of dissolved substances, and particularly salts, than that of freshwater and so water passes, by osmosis, through the semi-permeable membranes of the gills and gut into the body. Were this to continue indefinitely, the body fluids would be diluted to such an extent that death would ensue, if it did not first occur through physical damage to bloated tissues. Balance is maintained by the osmo-regulatory activity of the kidneys which excrete the excess water, but at the price of energy expenditure. Low dissolved oxygen concentrations may mean that organisms may have to indulge in greater respiratory activity, either in increased blood flow or more pronounced movement of gills or ancillary organs, the price of which is also increased energy expenditure.

Most freshwater organisms are well able to cope with the stresses which their environment places upon them, being well adapted both physiologically and morphologically. This chapter is concerned with the effects of stresses which exceed those normally encountered and which change the character of the habitat and the type of community to be found there.

4.2 IMPOSED ENVIRONMENTAL STRESSES

In addition to the extensive range of natural stresses encountered by organisms in their habitats, human activities and environmental manipulation may generate others which are similar to natural ones in kind, but not in degree, or they may be of a type completely alien to the normal environment. Examples of the latter category would include the addition of toxic organic chemicals, such as herbicides, or alteration of flows by large amounts or by abrupt changes such as might occur below impoundments or in hydroelectric power generation. Many forms of human interference are, of course, analogous to natural phenomena. The discharge of heated effluents is the counterpart of thermal springs; toxic effluents containing heavy metals may be intrinsically similar in their effects to percolation from metal-bearing rock strata while organic sewage effluents are an extreme form of the enrichment which occurs in streams flowing through deciduous woodland during leaf-fall. This similarity in nature, if not in intensity, between natural phenomena and what is covered by the omnibus term 'pollution' raises questions of how we define our terms, since we need to understand what pollution is before we can usefully consider the different kinds and their significance.

TABLE 4.1

ENVIRONMENTAL IMPLICATIONS OF THE DISCHARGE OF

Factor	Principal environmental effect	Potential ecological consequences
Organic enrichment		
1. High biochemical oxygen demand (BOD) caused by bacterial breakdown of organic matter	Reduction in dissolved oxygen concentration	Elimination of sensitive oxygen-dependent species, increase in some tolerant species; change in community structure
2. Partial biodegradation of proteins and other nitrogenous material	Elevated ammonia concentrations; increased nitrite levels; increased nitrate levels	Elimination of intolerant species since ammonia is toxic; reduction in sensitive species; potential for increased plant growth in nutrient-poor waters
3. Release of suspended solid matter	Increased turbidity and reduction of light penetration	Reduced photosynthetic activity of submerged plants; abrasion of gills or interference with normal feeding behaviour (see inert solids below)
4. Deposition of organic sludges in slower water	Release of methane and hydrogen as sulphide matter decomposes anoxically	Elimination of normal benthic community
	Modification of substratum by blanket of sludge	Loss of interstitial species; increase in species able to exploit increased food source
Toxic wastes		
1. Presence of poisonous substances	Change in water quality	Water directly and acutely toxic to some organisms, causing change in community composition consequential effects on prey–predator relations; sub-lethal effects on some species (impaired reproductive capacity, changes in behaviour, etc.)
Inert solids		
1. Particles in suspension	Increased turbidity Possibly increased abrasive action	Reduced photosynthesis of plants Impaired feeding ability through reduced vision or interference with collecting mechanisms of filter feeders (including abrasion or reduction in nutritive value of collected material)
2. Deposition of material	Blanketing of substratum, filling of interstices and/or substrate instability	Change in benthic community, loss of interstitial species, reduction in diversity, increased number of a few adventitious species; substrate is unstable

The effects of the three major categories of effluents, namely degradable organic matter, toxic substances and inert solid particles, are considered separately. Many effluents are composed of more than one type and the proportions of these vary according to the source.

Probable severity	Remedial or ameliorative action	Comments
Dependent upon degree of de-oxygenation, often very severe	Pretreatment of effluent; ensure adequate dilution	BOD can be reduced substantially by adequate treatment of effluent before discharge
Variable, locally severe Mild Mild/moderate	Provision of improved treatment to ensure complete nitrifica-tion; nutrient stripping possible but expensive	As above, adequate treatment is best solution to this problem; de-nitrification is the ultimate solution to nitrate problems
Moderate, usually local	Provide improved settlement, ensure adequate dilution	
Variable, may be severe Variable	Discharge where velocity adequate to prevent deposition	This tends to be a locally restricted phenomenon
Highly variable, depending upon substance and its concentration	Little can be done except to provide increased dilution	Toxic effluents cover a very wide range of substances and it is, therefore, difficult to generalise
Variable, often moderate	Provide improved settlement facility	Inert solids may cause greater change than organic wastes since, although they change the character of the substrate and are unstable, they provide no additional nutrition
Variable, often severe	Discharge where velocity adequate to ensure dispersion	

TABLE 4.2

ENVIRONMENTAL IMPLICATION OF THE ADDITION OF HEAT FROM INDUSTRIAL COOLING WATER OR ELECTRICAL POWER GENERATION

Factor	Physical or chemical environmental effect	Potential ecological consequences	Probable severity	Remedial or ameliorative action	Comments
1. Abstraction of cooling water	Local increased water velocity	Attraction of and damage to fish species (and invertebrates)	Variable, usually not serious	Good design of screens ensure low velocity as possible	
2. Anti-fouling protection of pipes, etc.	Possibility of escape and loss of biocides, especially chlorine	Toxic effects on normal river biota; death of organisms passing through system	Low	Careful design and operation of system	
3. Return of heated water	1. Elevation of water temperatures	Heat stress or death of sensitive species; acceleration of growth, shifts in timing of life cycles, increased rates of feeding of fish, invertebrates, etc.	Low	Minimise temperature differences	Heat stress may render some species more prone to disease.
		Enhanced microorganism respiration, especially in organically enriched waters, reducing dissolved oxygen levels	Moderate		Cooling towers often provide net benefit through aeration of organically enriched water
		Increased toxicity of many poisons	Variable		Critical temperature for repulsion of many temperature species is 30°C
		Attraction of mobile species (esp. fish) to thermal plume, repulsion of others	Variable		
		Survival of exotic species, accidentally or deliberately introduced, to compete with indigenous species	Usually not serious	Complete removal rarely possible	Populations may adapt to lower temperatures by selection, and spread further
	2. Lowering of dissolved gas solubility				
	(i) Oxygen	Effects on sensitive biota, especially important when biochemical oxygen demand is elevated	Moderate	Minimise temperature differences; provide aeration or mixing facilities	
	(ii) Nitrogen	'Gas bubble disease' in fish and invertebrates	Low		

4.2.1 Pollution Defined

Several definitions of pollution have been proferred, their differences almost certainly reflecting different facets of the problem and serving to emphasise that we are not dealing with a single entity but with a concept which changes with one's viewpoint. The legal definition given by Wisdom (1956), namely that pollution arises by 'the addition of something to water which changes its natural qualities so that the riparian owner does not get the natural water of the stream transmitted to him' suffers from heavy dependence on foreknowledge of what are the 'natural qualities' of 'natural water' (Hynes, 1960; Hawkes, 1962b). The definition hinges on value judgments (made by 'reasonable' men, no doubt, who are, in turn, undefined products of legal minds), as does the succinct definition of a pollutant as 'something that is present in the wrong place, at the wrong time, in the wrong quantity' (Holdgate, 1971). This must be true of all pollutants but who decides whether place, time or quantity are wrong? A more pragmatic definition (Edwards, 1972) is that pollution is 'the release of substances or energy into the environment by man in quantities that damage either his health or resources'. This definition stresses the importance of a perceived and measurable detrimental effect on human health or resources, the degree of pollution then depending upon the loss which is sustained. Such an anthropocentric view of the problem is inevitable: pollution arises as a product of the activities of one group of individuals which then manifests itself as a disbenefit to another.

4.2.2 Kinds of Pollution

Pollutants may be classified into a number of broad categories on the basis of their origins (domestic, agricultural, industrial, mining, etc.) or their principal components (organic, metallic, saline, heated, etc.) or their properties and effects (putrescible, toxic, inert, colloidal, etc.). Many effluents, both from discrete and diffuse sources, contain more than one kind of pollutant and materials from quite different sources may cause very similar effects. The major kinds of pollution are outlined below and their general environmental effects, potential ecological consequences, probable severity and the possible remedial or ameliorative action which might be taken in order to mitigate these are given in Tables 4.1 and 4.2.

4.2.2.1 Sewage and organic wastes

By far the largest volume of effluent entering watercourses in developed countries is sewage. Its condition ranges from virtually 'raw' or 'crude' sewage to highly treated effluents from which even the bulk of mineralised nutrients have been removed. Much of the material is of domestic origin but it may also contain industrial wastes which have been discharged to sewer; often these wastes are more readily treated when combined with domestic sewage. The principal characteristic of sewage effluents is their organic content although wastes from some industries such as intensive agriculture, food processing or paper manufacturing have similar properties. The primary consequences of the discharges of these effluents is depletion of dissolved oxygen as the material is degraded by microbial activity, and the suspension or deposition of the more refractory materials. These effects are discussed more fully in Sections 6.3 and 6.4.

4.2.2.2 Toxic wastes

Some toxic substances, for example ammonia or hydrogen sulphide, occur naturally through the decay of allochthonous organic matter and toxic heavy metals may leach from natural deposits, but many industrial processes produce wastes which contain substances which rarely, if ever, occur naturally and certainly not at such high concentrations. Unlike organic wastes, which are to some extent 'self-destructive' in that they support the microbial activity which ultimately reduces their polluting effects, toxic wastes are not readily biodegradable. Some, for example mercury, may be rendered more harmful by microorganisms, while others may enter the environment at very low concentrations but become significant toxicants as they are concentrated in food chains. Many pesticides attain environmental significance in this way.

As toxic substances differ widely in their effects, few generalisations can be applied to this group of pollutants. Fuller treatment is given in Chapter 7.

4.2.2.3 Inert solids

Inert solids, such as material from coal washeries or china-clay production, become serious pollutants when they are present in such large quantities that they change the physical characteristics of the habitat by blanketing the river bed or their instability precludes the development of benthic communities. Unlike organic solids, which are available as a nutritional source for a number of 'opportunist' species, the effective sterility of inert solids means that where large quantities settle out the river bed becomes virtually devoid of any life. The ecological consequences of pollution by inert solids are considered in Section 5.1.

4.2.2.4 Heat

Many industrial effluents and most sewage effluents have temperatures above ambient, especially in winter, and so convey heat to the receiving streams, but by far the greatest source of additional heat energy comes from the return of water used for cooling purposes in electrical power generation. Raising the temperature of river water may exceed the limits of tolerance of cool-water species, deplete oxygen directly or by enhancing the rate of microbial respiration in organically enriched waters, and may increase the rates of growth or development of certain species (Table 4.2). In general, addition of heat has not proved a particularly difficult kind of pollution, except locally and where inputs have caused large increases in temperature. Indeed, the euphemism 'thermal enrichment' has been coined to emphasise the more beneficial aspects of heat pollution.

4.3 ENVIRONMENTAL MANIPULATION

In addition to the typical stresses which arise from the discharge of abnormal kinds of quantities of substances, as briefly considered in the preceding sections, biological communities may suffer when their habitat is subjected to environmental manipulation. This last term is a convenient description for the growing tendency to modify river channels, impound water, regulate flows or transfer water between catchments. These activities hardly qualify as pollution, but their consequences can be just as serious. A selection of the main kinds of environmental manipulation and their ecological implications is reviewed below.

4.3.1 Channel Modification

River channel modification is usually undertaken to enlarge its carrying capacity and thereby reduce the frequency of flooding, or the channel may be re-aligned or constrained in order to prevent erosion or other natural, but often inconvenient, phenomena. Engineering works may destroy the existing habitat simply by removing it and using the materials of which it is composed for the construction of flood banks or spreading on adjacent land. The enlarged channel tends to have a different substrate and shallower water during normal flows while the tendency to provide a straightened watercourse with trapezoidal cross-sections reduces the natural diversity of the habitat. In effect, channel

Factor	Physical or chemical environmental effect	Potential ecological consequences
1. Enlargement of channel to provide increased flow capacity	Change in physical dimensions of habitat to give (i) reduced depth under dry weather flow (ii) greater channel width (iii) change in water velocity for given discharge	Removal of biota from existing channel by reconstruction work; high, temporary, turbidity reducing plant photosynthesis, blanketing substrate downstream and affecting macroinvertebrates and possibly feeding of fish
2. Modification of channel shape, both in profile, plan and cross-section	Reduction in habitat diversity (i) smooth profile removes variation in depth (pool:riffle configuration) with tendency towards uniform substrate material	Loss of many microhabitats and their associated flora and fauna; reduction in overall community diversity
	(ii) trapezoidal cross-section destroys habitat diversity, especially shallower margins	Loss of habitat diversity; marginal plants unable to establish foothol loss of some macroinvertebrates
	(iii) straight channel removes meanders having deep fast water on outside and shallow on inside of bends; modifies velocity and suspended solid carrying capacity	As above; channel length reduced and even with increased width, habitat area may be lost
3. Bank modifications	(i) removal of trees to provide access (mainly for machines) and reduce obstruction of flood plain	Loss of shading so that increased light reaching water encourages algal and macrophyte growth; los of detrital input during leaf-fall and aerial insects for fish food
	(ii) removal of bankside vegetation by mechanical means or herbicides	As above Herbicide spray drift may affect other plants
	(iii) construction of raised or flood-banks	Increased carrying capacity of channel may modify habitat
4. Maintenance of channels	(i) removal of substrate and vegetation by mechanical means	Habitat modification, removal of benthic fauna, loss of flora and fauna associated with it
	(ii) removal or control of vegetation by means of herbicides	Loss of plants and associated animals; invasion of more resista species leading to community ch

Probable severity	Remedial or ameliorative action	Comments
Often very variable, but of short duration	Reinstatement by reintroduction; working from downstream to upstream helps; little can be done to avoid this but working short distances at any one time helps to reduce severity	Recolonisation occurs by drift from upstream; high turbidity and suspended solids are 'natural' phenomena associated with floods but duration may be longer during engineering operations
Severe	Dig out deeper pools below level of designed profile	Fortunately, tendency for channel to return to natural configuration unless constrained by massive structures (e.g. concrete channel
Severe	Construct with an irregular channel cross-section and especially with	or piling etc.)
Severe	marginal ledges ('berms') to allow reinstatement of marginal plants	
	If flood channel must be straight, encourage dry weather channel to meander within its confines	Not an ideal solution but will encourage some habitat diversity
Variable, usually moderate	Remove only from one (preferably north) bank. Plant trees in rows parallel with flow to reduce risk of flood loss	Tree loss is slow to recover but other vegetation may return within one or two seasons
Moderate	Restrict control of position of bank each season	
Probably insignificant		
Variable, moderate to severe	Restrict to partial treatment in any one season	
Variable, usually mild to moderate		

TABLE

ENVIRONMENTAL IMPLICATION OF RESERVOIR CONSTRUCTION

Factor	Principal environmental physical or chemical effect	Potential ecological consequences
1. Construction of reservoir	Barrier to normal river flow	Prevention of upstream migration of fishes and certain invertebrates. Effective barrier to upstream reproductive flights of some insects, and downstream drift of many inverts; may interfere with normal downstream movement of fi
	Physical changes in impounded river (i) depth increase	Loss of normal riverine community, survival and sometimes explosive increase of a few species
	(ii) flow changes (a) upstream (b) downstream	Reduction in flow changes habitat from lotic to lentic. Modified flow pattern ('compensatic flows'); habitat modification
	(iii) inundation of adjacent land	Complete destruction of normal habitat; may contribute to increase aquatic productivity in early years
	Disturbances caused by civil engineering operations (i) turbidity	Blanketing of plants and river bed downstream, change in substrate condition; exacerbated by loss of normal flushing action of spates
	(ii) increased pollution potential	Loss of sensitive species following pollution incidents. Recolonisation by natural drift is impeded when dam is completed.
2. Reservoir operation	Release of stored water for power generation[a] (i) increased discharge, velocity	Modification of substrate, tendency for organisms to be swept away ar or stranded when flows are reduce
	(ii) sudden fluctuations in discharge; 'drawdown' of reservoir as stored water is used	Exposure of shore and erratic movement of shore-line limits the diversity of the biological community

[a] See Table 4.5 where effects of releases are covered in greater detail.

DIRECT WATER SUPPLY OR POWER GENERATION

Possible severity	Remedial or ameliorative action	Comments
Very severe Moderate	Provision of effective fish passes, ladders or lifts	Downstream effects less severe where other tributaries enter river short distance below dam, but headwaters above dam may suffer badly
Severe Severe Moderate Severe	None None Provide adequate, variable flows of surface 'compensation' water None	These are permanent changes which create a new habitat (artificial 'lake') upstream and a modified river downstream
Variable Variable	Little can be done about riparian construction but interception of other run-off, with settlement, helps Provision of adequate treatment facilities, oil bunds, etc.; exercise of care	This is a temporary factor; careful siting of ancillary works (e.g. stores, workshops, offices and living accommodation) can reduce risks
Moderate Variable, often severe	Ensure all changes in discharge are gradual and mimic normal spates Provision of offshore bunds will maintain some shallows during severe drawdown	Fluctuations are more devastating at lower discharge stages when area of wetted river bed varies most

modification produces another kind of river, and an unnatural one at that, and so it is only to be expected that the changes have ecological consequences (Table 4.3). Fortunately, the natural forces which moulded the original configuration continue to operate and the river often reverts in time to its former state, or something very similar. Only when the channel is confined within a concrete or steel piling 'strait-jacket' is it truly compliant to the imposition of the engineer's will, and even then it may seem to retaliate at the extremities of its confinement.

4.3.2 River Impoundment

From the start of the Industrial Revolution, rivers have been impounded to provide water supplies, generate power or to facilitate transport, but the scale of these enterprises has increased quite dramatically since about the turn of the century. The construction of an impounding dam or weir causes changes both upstream and downstream. These are additional to those, usually only of a temporary nature, which occur during the actual civil engineering works (Table 4.4). The greatest change is, of course, the effective disappearance of the original river upstream of the dam and its replacement by an artificial lake. The presence of the newly created static water body causes significant changes in the river downstream, especially if this is maintained largely by 'compensation' flows or subjected to rapid fluctuations according to the demands for power.

4.3.3 River Discharge Regulation

Many rivers have been required to support increasing abstractions for potable supply, industry or irrigation. During periods of low natural flows, augmentation may be provided through discharges from impounding reservoirs. These reservoirs are also of value in flood control since they are usually managed so as to provide storage of peak natural discharge. Such river management means that the river downstream is more rarely subjected to very high discharge and, during periods of natural drought, does not experience very low flows. Thus flows in regulated rivers tend to be more equable than those of unregulated rivers. At first sight it might seem that, although the hydrological regime is different, the loss of extremes might be generally beneficial to the ecology of the river. However, it must be recognised that regulation may be a mixed blessing; benefits certainly accrue, but the communities present before regulation have been selected and moulded by the natural regime. Changes in that pattern are bound to influence the community. Examples of the changes which may occur when rivers are regulated by

TABLE 4.5

IMPLICATIONS OF RIVER REGULATION BY MEANS OF DISCHARGE FROM IMPOUNDING OR PUMP-STORAGE RESERVOIRS

Factor	Physical or chemical effect	Potential ecological consequences	Probable severity	Remedial or ameliorative action	Comments
1. Increased discharge (flow augmentation)	Change in normal hydrograph to give higher summer flows; Difference increases with severity of drought				In pump-storage reservoirs higher (winter) discharges may have reduced peaks as reservoir is replenished
	(i) Increased water velocity and depth	Changes in plant growth patterns and distribution of invertebrates	Variable	Little can be done	Resuspension of accumulated silt and detritus occurs if regulating discharge is sudden
	(ii) Disturbance of substrate, and removal of detritus	Changes in invertebrate community though habitat changes and loss of food supply of detritivores	Moderate (most severe at start)	Increase flows gradually and reduce slowly	
	(iii) Increased turbidity	Suppression of plant growth	Slight (important only at start)		
2. Physical and chemical properties of stored water	(i) Lowered temperature in summer (bottom water)	Retarded growth and development; reduced primary and secondary productivity	Moderate	Use surface or mid-water if variable draw-off available	
	(ii) Reduced dissolved oxygen concentrations (bottom water)	If severe and close to outlet, asphyxiation of sensitive species	Slight	Provide aeration cascades at outlet	
	(iii) Fe/Mn and turbidity	Deposits on plants suppressing growth or on bed, modifying habitat	Negligible	Provide destratification facilities	Mainly a water treatment problem
3. Biological character of stored water	Surface water supports algal and zooplankton communities which				Section of river where benefit is evident will depend on balance between overall environmental deterioration and enhanced food input
	(i) are unlikely to be maintained under conditions prevailing in river	Increase in population of some filter feeding and drift-consuming organisms	Variable but generally moderate	Probably a beneficial effect which will increase overall productivity	
	(ii) add to 'drift' in river and form additional food source for invertebrates and fish	Potential for increased fish production			

Factor	Physical or chemical effect	Potential ecological consequences
1. Increased discharge (flow augmentation)	Change in normal discharge pattern, i.e. increase in summer flows especially when otherwise low	
	(i) Increase in depth and velocity	Effects on plant growth, invertebrate distribution and community structure
	(ii) Movement of accumulated detritus (food source)	Growth of invertebrates (and fish may be retarded; loss of some detritivores
2. Physical properties of groundwater	Temperature constant at about 10 °C	
	(i) Suppression of diurnal variation	Reduced photosynthesis (lower risk of supersaturation) and reduced nocturnal respiration (smaller oxygen deficit)
	(ii) Lowering of mean summer temperatures, elevation of autumn (and spring) temperatures	Retardation of growth and development of biota
3. Chemical quality of groundwater	Modification of quality of receiving watercourse	
	(i) Lowered dissolved oxygen concentration	If severe, asphyxiation of fish and invertebrates; otherwise risk of reduction or loss of sensitive species
	(ii) Increased dissolved nitrogen concentration	'Gas-bubble disease' in fish (and probably also invertebrates)
	(iii) Increased iron concentration	Precipitation changes character of river bed, smothers plants, prevents survival of many specie
	(iv) Increased nitrate concentration	Eutrophication and possible increase in plant growth (if nitr levels very high in potable wate human health could be affected methaemoglobinemia)
4. Intermittent use of pumps	Discharge of sand particles accumulated from erosion of boreholes during periods of disuse	Blanketing substratum and possib changing character thereby affe macroinvertebrates present

Severity	Remedy	Comments
Moderate	Collection of contributions from several boreholes to discharge flows in catchment into at least equivalent flows	Discharge at lowest convenient point reduces length of river affected by groundwater May not exceed natural long-term mean flows at appropriate season Unlikely to be critical
Moderate	Collect contributions to discharge into adequate diluting flows	Less river exposed to effects Reduced plant growth beneficial (less cutting, oxygen super-saturation and nocturnal deficit)
Moderate	Include cascades or other aeration structures and discharge into adequate dilution volume	Severe oxygen deficit quickly made up on exposure to atmosphere but full saturation may not occur for some distance downstream
Slight	As for oxygen deficit	
	Blend with other groundwater to reduce	Saline intrusion may give high chloride levels; blend to reduce or
Negligible	—	ensure adequate dilution
Slight	Provide sand-traps	

TABLE

ENVIRONMENTAL IMPLICATIONS OF WATER TRANSF

Factor	Principal environmental effect	Potential ecological consequences
1. Abstraction of water from donor catchment		
(a) Construction of regulating impoundment	(i) Reduced (compensation) flow in river below dam	Habitat modification by prese of dam; downstream habita changes in silt, detritus, deposition; loss of flushing action of floods
	(ii) Exposure of shore by drawdown	Loss of marginal flora and fa reduced fish production
(b) River abstraction	(i) Reduction in normal river flows	Habitat modifications (as abo
	(ii) Diversion of water	Loss of fish stocks, especially young fish and salmonid smolts
		Confusion of returning migra species, especially salmonid imprinted by natal stream pheremones
2. Civil engineering construction and works for water transfer		
(a) Channel capacity enlargement of donor and recipient rivers		
(b) Operation of connecting tunnels and pipes	Intermittent releases of stored water, possibly low in dissolved oxygen and high in sulphides etc., when scheme is operated infrequently	Death by asphyxiation of sensitive fish and invertebrates
3. Release of water to recipient catchment	1. Change in normal hydrological regime (usually approaching an inversion)	Modification of normal biol communities through char in bed stability, deposition detritus (food for many invertebrates); potential consequences for reproduc success of fish species
	2. Changes in water velocity	
	(i) when flows augmented	Fish (especially fry), invertebrates and plants displaced
	(ii) when support ceases	Less active species stranded

BETWEEN CATCHMENTS

Probable severity	Remedial or ameliorative action	Comments
Variable but may be very severe	Allocate adequate proportion of stored water for variable compensation, including flushing	See Table 4.4
Moderate	Provide bunds to hold water and maintain shallow margins	
Moderate		
Unknown, potentially serious		Headwaters of receiving river may have no (or inadequate) spawning grounds
		See Table 4.3; donor stream enlargement applies if flows are augmented by regulation
Severity depends on BOD etc. and duration of storage	Injection of oxygen or ensure pipes and tunnels are emptied when transfer ceases or continue some water movement at all times	Likely to be serious only in large schemes or when use is infrequent
Variable but potentially very severe	Ensure augmentation is not excessive	Aseasonal high and low flows asynchronous with normal ecological cycles
Potentially severe	Ensure rates of change of flow do not exceed those in natural floods	Accurate and fully variable control mechanisms required

TABLE 4.7—*contd.*

Factor	Principal environmental effect	Potential ecological consequences
	3. Modification of water quality	
	(i) chemical composi-tion (e.g. hardness)	Loss or addition of species, depending upon tolerances
	(ii) temperature differences	Modification of growth rates and timing of life cycles
	4. Transfer of biological material	
	(i) organisms	Transfer of phytoplankton, potential food source for invertebrates; facilitates dispersion of species, including unwanted 'nuisance' species, 'trash fish', unwanted predators, fish diseases, parasites and their intermediate hosts
	(ii) fish pheromones	Confusion of migratory salmonids returning to spawn in natal streams

impoundments are sudden increases in discharge, especially at low flows, in response to demand, alterations in the physical and chemical characteristics of the river and addition of biological material in the form of plankton from the reservoir (Table 4.5). Far less frequently, rivers have been regulated by means of groundwater. In this technique the aquifer is recharged during wet years and is pumped to maintain river flows during drought. The ecological consequences of such schemes are considered in Table 4.6.

Probable severity	Remedial or ameliorative action	Comments
Mild/moderate	Little can be done	Largely depends on geology of donor and recipient catchments
Mild	Little can be done	Depends on whether augmentation is from impounding reservoir and lower draw-off points used
Variable, potentially very serious	Provision of screens, filters, use of biocides which quickly inactivate	
Potentially very serious	Little can be done	Politically a very sensitive issue; owners of fisheries in donor streams where smolts raised, naturally concerned when adults return to recipient stream; population at risk if no suitable spawning grounds in recipient catchment

4.3.4 Inter-catchment Transfers

The transfer of water between catchments is usually resorted to in order to avoid the high cost of direct pipelines. Much depends on the topographical configuration of the watercourses but schemes have been developed which utilise appropriate segments of rivers to convey water from a large impoundment, or an under-utilised river source, to a convenient abstraction point. Many of the ecological implications of transfer schemes are similar to those of river regulation but others are peculiar to this particular kind of environmental manipulation (Table 4.7). For example, the possible transfer of 'nuisance' species, or the diseases and parasites of commercially important ones, between river basins is a hazard which is almost exclusive although it is possible that other environmental modifications could create conditions which supported unwanted species or provided reservoirs of disease organisms and their vectors.

4.4 COMBINED STRESSES

Before leaving this brief review of the kinds of natural and imposed stresses on individual organisms, and hence on populations and communities, it is worth reflecting on the dangers of classifying entities. Although it is convenient to place each type of pollutant or category of environmental manipulation into its own compartment and then catalogue its characteristic properties and effects, reality is rarely so simplistic. Human influences on watercourses are often superimposed, and thus mixed effluents may enter regulated rivers, the channels of which have been enlarged to improve land drainage or impounded to facilitate navigation. In utilising the indicator potential of biological systems to identify or measure environmental stress it is usually difficult, and often impossible, to apportion the impact of the individual sorts of stress which may occur at any site. Much is yet to be learned in simple situations and few opportunities exist in which one can change the various components at will in order to observe their individual and combined effects. The cost of experimental manipulation on the scale required would be prohibitive and probably unacceptable on social or ethical grounds. This means that it is all the more imperative that any opportunities to study the effects of schemes, and especially those which include elements which are changed from time to time, should be grasped fully. The value of such studies would extend beyond the immediate area of environmental impact; if conducted carefully they should also help to elucidate fundamental ecological processes.

CHAPTER 5

Effects of Physical Disturbances

Physical disturbance of the aquatic environment is considered here to include those factors which change the character of the water or the habitat or those which, when they settle, will modify the character of the substratum. But these changes only involve the physical properties of the environment and do not modify the chemical composition, for example by adding toxic quantities of heavy metals.

Similarly, the addition or removal of heat by means of heated effluents or cold regulating reservoir discharges will exert their influence through physical principles although some effects, for example changes in the quantity of dissolved gases, may be considered to be chemical rather than physical.

Changes in pH are within the province of physical chemistry but are conveniently considered with physical phenomena since they are, in practice, closely associated with precipitation of dissolved matter which then causes the changes mentioned above.

Finally, one must include here the consequences of, literally, physically disturbing the environment through engineering works such as are encountered in land drainage activities (channel widening, deepening, regrading, realignment), drainage improvement (weed cutting or killing) and general river management. The environmental consequences of these activities tend to be large and obvious; their main interest in the context of biological indicators is how they may be used to determine the degree and rate of recovery of the communities present and whether there have been any permanent or, at least, long-term shifts.

5.1 EFFECTS OF SUSPENDED SOLIDS

5.1.1 Introduction

Almost all inland waters carry some suspended solids. In flowing waters even under natural conditions, such as during spates, levels of 4000 mg litre^{-1} may be exceeded (McCarthy and Keighton, 1964) and even 300 000 mg litre^{-1} may be reached (Brown, 1960). Ponds and lakes, although relatively static, may also become laden with suspended material through surface run-off or through turbulence caused by wind and wave action on the shore. The communities of organisms present in these habitats are able to survive these intermittent high suspended solid loads since they are dissipated by the high flows which generate them or, in the case of static waters, soon settle. Suspended solids only pose a threat to freshwater communities when they are present either at abnormally high levels or for unusually long periods and thereby change the character of the natural habitat.

Changes caused by suspended solids may result from one or more of the following factors

(i)   Light penetration may be reduced as a consequence of increased turbidity and the photosynthetic rates of algae and submerged macrophytes may be affected. This is important not only for the maintenance of the plant community but also for the animals which depend upon them for food, shelter and support. Turbidity also affects the transmission and radiation of heat (Ellis, 1936).

(ii)   Suspended solids may exert direct mechanical effects on organisms by increasing abrasion, clogging the respiratory surfaces of gills and/or interfering with feeding through inadvertent collection on feeding appendages or nets of filter-feeding invertebrates. The accumulation of considerable amounts of inert matter by filter-feeding organisms must reduce the nutritional value of their food, perhaps to the point of starvation.

(iii) Most often, suspended solids modify the nature of the habitat by changing the character of the substratum when they collect by settlement as flows are reduced, both generally, after spates, and locally as a stream or river returns to a normal pattern of riffles and pools. This factor is especially important when abnormally high suspended solid loads enter the eroding zones of rivers since the increased quantities of finer materials fill the interstices of gravel or silt and change the structure of the bed. Many invertebrates and some fishes depend upon the permeability of the gravel for respiration or feeding and may suffer when this is reduced by the accretion of a coating of finer material on pebbles and within the bed. The accumulation of this material may also physically prevent the normal movement of animals within the substratum and, in extreme cases, there may be no interstices for any indigenous animals to occupy. Furthermore, the replacement substrate of fine sand or clay tends to be less stable and some animals and plants may be unable to maintain their position at a given site.

(iv) Even in depositing zones, increased deposition of predominantly inert mineral solids can change the character of the bed. The proportions of inert and organic components may change so that the burrowing, filter-feeding or surface detritivores are overwhelmed by inedible material and the habitat is virtually sterilised. Higher loads of organic matter increase food supplies for those species which are capable of exploiting them but often the increased quantity of putrescible matter generates a hostile anoxic environment which further limits the range of organisms which are able to survive there.

The ultimate effect on the biota of suspended solids is clearly dependent upon the nature of the material held in suspension, including its size, density, potential for bacterial decomposition, nutritive value and toxicity, and upon the increased quantity present or the prolonged duration of its presence.

5.1.2 Sources of Suspended Solids

In addition to the normal burden of suspended material, which is derived from natural weathering and soil erosion by run-off, the solids content of surface waters can be augmented from several different sources. First, agricultural activities, for example ploughing, may enhance soil erosion and civil engineering operations, which cause large areas of soil and subsoil to be disturbed, may lead to increased loads of silt and soil particles entering streams and rivers. Similar conditions are created by forestry operations, including planting, thinning and felling. Secondly, some suspended solids enter surface waters by direct discharge. Among these are sewage and sewage-treatment effluents; industrial wastes of many kinds, including pulp and paper manufacture; mining wastes from the coal or china-clay industries, quarrying, sand or gravel extraction.

TABLE 5.1

SUSPENDED SOLIDS CONTENT OF INDUSTRIAL AND OTHER EFFLUENTS AND RIVERS
INTO WHICH THEY ARE DISCHARGED

Substance/effluent	Typical range or mean value (mg litre^{-1})	Reference
China clay (rivers)	28–91 268	Nuttall and Bielby (1973)
	500–100 000	Herbert *et al.* (1961)
Paper manufacture	200–3 000	Shell (1976)
	1 180	Nemerow (1978)
Leather (tannery)	3 000–8 000	Shell (1976)
	2 000–3 000	Gurnham (1965)
Petroleum production	441	Nemerow (1978)
Metal finishing and plating	200–1 000	Shell (1976)
Cotton textiles	30–300	Gurnham (1965)
Laundry	2 000–5 000	Shell (1976)
Meat packing	900–3 200	Gurnham (1965)
	1 400	Nemerow (1978)
	500–5 000	Shell (1976)
Cannery	200–2 500	Shell (1976)
	1 350	Nemerow (1978)
Fruit canning	100–750	Gurnham (1965)
Vegetable canning	30–2 220	Gurnham (1965)
Sugar beet	800–4 300	Gurnham (1965)
Poultry	200–2 500	Shell (1976)
Primary sewage treatment	252	Nemerow (1978)
Sewer storm overflows	2 968	Klein (1962)

Tipping of industrial or domestic wastes can also lead to increased amounts of solid matter entering watercourses. In some of the examples just listed, and in particular the industrial wastes, it is not always easy to distinguish between the effects caused by the physical properties of the materials involved and those which arise from other attributes, including their toxicity. Typical levels of suspended solids in effluents from a number of sources are shown in Table 5.1.

5.1.3 Effects of Suspended Solids on Invertebrates

Early work on the influence of inorganic sediment on aquatic life has been reviewed by Cordone and Kelley (1961). The effects of inert suspended solids from a number of industries including china-clay mining, sand and gravel extraction, coal mining and civil engineering (motorway construction) with reference to macroinvertebrates are described below.

5.1.3.1 China-clay mining
Early work on the effects of china-clay extraction in Cornwall, UK, centred on their importance for salmonid fisheries but indirect evidence of effects on food organisms was obtained by Herbert *et al.* (1961). More detailed investigations by Nuttall and Bielby (1973) showed that where concentrations exceeded 2000 mg litre^{-1}, or where mineral particles blanketed the substrate, rooted vegetation was absent and macroinvertebrate communities were sparse and poor in species. Of those present, *Baetis rhodani*, *Perlodes microcephala* and the burrowing Tubificidae, Naididae and Chironomidae were more abundant ($p < 0.01$) in china-clay polluted reaches than in similar, unpolluted stretches. There were no macroinvertebrate species which occurred in significant numbers only in the clay-polluted sections. The reduced abundance of plants and animals was attributed to the deposition of fine, inert, solids rather than to increase turbidity or abrasion by particles in suspension.

The presence of increased numbers of *Baetis* and *Perlodes* was explained by their essentially carnivorous diets: the absence of algae and other vegetation need not, therefore, be a limiting factor. In addition, *Baetis* tends to live on the top of stones and might thus be expected to be less affected by the accumulation of interstitial clay particles. The observation of increased densities of Tubificidae and Naididae in areas where china-clay deposits accumulated is corroborated by the findings of Learner *et al.* (1971) with reference to naidids in fine coal deposits and Hynes (1960) who noted their association with depositing reaches of rivers. Other important differences between clay-polluted and unpolluted streams noted by Nuttall and Bielby (1973) were the absence of *Simulium* larvae, the net-spinning caddis, *Hydropsyche instabilis*, and the flatworm *Polycelis felina*. The absence of *Simulium* was thought to have been caused by the inability to find firm attachment in unstable clay deposits. Interference with the feeding mechanisms of the net-spinning caddis *Hydropsyche* and the mucus-string traps of *Polycelis* was probably the explanation for their absence. It might also have been assumed that inert clay particles would interfere with filter feeding in *Simulium* although field experiments by Wu (1931) suggested that substrate and not turbidity is the main factor in the distribution of this genus.

5.1.3.2 Sand extraction or discharge
Luedtke and Brusven (1976) investigated the effects of heavy sand accumulations resulting from mining for garnets in Idaho, USA, on insect drift, colonisation and upstream migration and concluded that the

combination of increased exposure to currents and the instability of a bed composed of sand grains was responsible for the restriction of upstream movement by insects. Only the heavy-cased caddisfly, *Dicosmoecus*, was able to move upstream over sand substrates at water velocities as low as $0.12 \, \text{m s}^{-1}$. Long sandy stretches did not impede downstream drift by insects.

Changes caused by large quantities of medium to coarse sand (0.3–0.6 mm) which were carried into the River Camel, Cornwall, UK, have been described by Nuttall (1972). The sand resulted from erosion following diversion of the river channel and was carried mainly during spates to be deposited at low flows. Although macrophytes were not abundant in the catchment they were absent from stations affected by sand deposition as were filamentous algae, although these were frequently found at stations upstream of the affected reaches. The macroinvertebrate community was also seriously depleted to about 15% of the upstream species and this was attributed to the unstable shifting nature of the sand deposits rather than any increased turbidity or abrasion. Nuttall provided a useful list of the reactions of different macroinvertebrates, showing those which were enhanced in numbers, eliminated or unaffected by sand deposition (Table 5.2).

Chutter (1969), in discussing the effects of sand and silt deposition on

TABLE 5.2

EFFECTS OF SAND DEPOSITION ON MACROINVERTEBRATES
(After Nuttall, 1972[1] and Hamilton, 1961[2])

1. Species eliminated by sand deposition:
   Amphinemura sulcicollis[1], *Baetis pumilis*[1], *Centroptilum luteolum*[2], *Ephemera danica*[1], *Gammarus pulex*[1], *Leuctra geniculata*[1], *L. hippopus*[2], *L. nigra*[1], *Perlodes microcephala*[2], *Polycelis felina*[1], *Sericostoma personatum*[1].

2. Species which increased in numbers following sand deposition:
   Baetis rhodani[1], *Ephemerella ignita*[2], *Nais* sp.[2], *Rhithrogena semicolorata*[1], *Stylaria*[2], Tubificidae[2].

3. Species largely unaffected by sand deposits:
   Amphinemura sulcicollis[2], *Ancylus fluviatilis*[1,2], *Atherix* sp.[2], *Caenis rivulorum*[1,2], Chironomidae[2], *Chloroperla torrentium*[2], *Ecdyonurus venosus*[2], *Eiseniella tetrahedra*[2], Enchytraeidae[2], *Gammarus pulex*[2], *Gyrinus* sp.[1], *Helesus* sp.[2], Hydracarina[2], *Hydroptila* sp.[2], *Hydropsyche instabilis*[1], *Leuctra fusca*[1,2], Lumbriculidae[2], Naididae[1], *Paraleptophlebia* sp.[2], *Protonemura meyeri*[1,2], *Rhithrogena semicolorata*[2], *Rhyacophila dorsalis*[2], *Sericostoma* sp.[2], *Silo pallipes*[2], *Simulium* sp.[2].

invertebrates in South African streams, noted that when the bed was completely blanketed it was quite understandable why the fauna virtually disappeared, but when biotopes were not smothered the effects were less obvious. He observes that in gastropods the eggs may be adversely affected (Harrison and Farina, 1965) or in bivalve molluscs, feeding and respiration may suffer interference (Ellis, 1936). From his own investigations he concluded that increased sand and silt in a stony biotope (riffle) reduced the diversity of macroinvertebrates and the abundance of certain groups but did not necessarily change the overall faunal density.

Some studies of the effects of discharges of sand have shown that the qualitative composition of the bottom fauna was unaltered by the presence of sand and silt in suspension, but direct deposition did have effects. For example, Hamilton (1961) reported his observations on the effects of sand-pit washings on Fruin Water in Scotland and found that some twelve species, six genera and three families were equally abundant above and below the point of discharge. These are given in Table 5.2 where they may be compared with those of Nuttall (1972).

5.1.3.3 *Civil engineering works*
The effects of the construction of the M11 motorway in Essex, UK, were studied by Extence (1978). A tributary carrying the run-off from the area which was disturbed by the civil engineering operations carried up to 336 mg litre^{-1} suspended solids (mean, 86 mg litre^{-1}) while the river into which it discharged had up to 130 mg litre^{-1} (mean 32 mg litre^{-1}) solids in suspension. Other determinands differed only very slightly between the two waters, except for iron which was approximately doubled in the 'effluent'. The bed of the river upstream of the confluence was stony (particle size 10–150 mm) while that below the discharge of motorway construction run-off, although originally similar, acquired a covering of sand (mostly less than 1 mm) and sand banks began to build up.

The macroinvertebrate communities above and below the entry of motorway run-off became progressively dissimilar over the study period. Certain groups, such as stoneflies, mayflies and cased caddis were largely absent at the outset: this was attributed to the discharge of sewage effluents further upstream. At an early stage some species such as the filter-feeding caddis *Hydropsyche augustipennis* and the mollusc *Lymnaea peregra* were present at reduced densities below the discharge. The leeches *Erpobdella octoculata* and *Glossiphonia complanata* also became much less abundant (less than 1% of their density upstream), a change which was ascribed to the loss of a suitable substratum. This was also thought to be the probable cause of the complete disappearance of *Ancylus fluviatilis*. Although *Asellus aquaticus* became more abundant upstream during the period of the study it was unable to establish itself below the discharge. Even *Caenis moesta*, a mayfly generally recognised as tolerant of silty conditions, was less abundant in the affected section. The bivalve mollusc *Pisidium*, also typical of depositing regions of rivers and associated with silt and mud, was found more frequently below the discharge, although densities were low, perhaps because the suspended matter was inert and lacking an organic component which might have provided a food source.

Only tubificid worms and chironomid fly larvae appeared to be almost unaffected by the changes: as noted above, both are burrowing or tube-building forms and are commonly associated with soft, depositing substrates. Even these groups, however, tended to be present at lower densities than upstream which suggests that the instability of the sand substrate and its lack of organic matter both contributed to the reduction in numbers. The filamentous alga *Cladophora* was recorded above the outfall throughout the study but after a reduced summer peak disappeared from the stream below the motorway run-off.

These studies show that the high suspended solids carried by run-off during civil engineering operations can have a marked effect on the ecology of the received stream. Their long-term effects could, however, probably prove to be small since, once the works are completed and winter spates have carried the bulk of the material away, recolonisation can occur from upstream. This view finds support in the studies of Barton (1977) who noted that the reduced fish populations (24 to 10 kg ha^{-1}) immediately below the site of highway construction returned to the original levels after the work had been completed. Invertebrate densities were unaffected even though suspended solid levels were as high as 1390 mg litre^{-1} and sediment deposition increased ten-fold. The composition of the benthic community did change, however; net-spinning caddis and mayflies were reduced while cased-caddis and chironomids increased. No common species was eliminated and it was surmised that the change in community composition did not result directly from siltation.

5.1.3.4 Channel modification and maintenance

Dredging is commonly used for the maintenance of drainage channels. In digging out accumulated silt there is considerable physical disturbance of the habitat since the roots and rhizomes of submerged and marginal macrophytes are also removed. In addition large amounts of bed material are released into suspension. Advantage may be taken during these works of modifying the course or configuration of the channel. Sometimes the appearance of the channel is considerably changed and one instinctively expects that this has very severe consequences for the bottom fauna and fishes.

A two-year quantitative study of the recovery of the benthic fauna of a slow flowing mill-stream following draining and dredging was undertaken by Crisp and Gledhill (1970). The initial benthic faunal density after and during dredging was about 4000 individuals per m^2 (about 5 g dry weight) and the Chironomidae, Oligochaeta and Mollusca together formed about 85% of the benthos. During the winter following dredging the biomass increased to $8 \, g \, m^{-2}$ with a faunal density of about 8000 individuals per m^2 but decreased the following spring to the levels of the previous summer.

However, during the summer numbers increased five- or six-fold (20 000–30 000 individuals per m^2 or $30–40 \, g \, m^{-2}$). During the spring and summer of the second year densities were lower (spring minimum of 8000 individuals per m^2) and by the following autumn the maximum was 20 000 to 30 000 individuals per m^2. Unfortunately, an accidental pollution incident precluded further studies of the benthos. It was concluded that recovery was virtually complete about one year after dredging and the long-term effects on the benthos were negligible.

The effects of modifications of six streams in Pennsylvania, made in order to increase their flood carrying capacity, were studied by Duvel *et al.* (1976). It was found that 'channelisation' had a direct deleterious effect on trout populations which was attributed to habitat modification rather than any change in trophic status. Large trout were denied suitable hiding places as channels were modified and natural features (holes, undercut banks, vegetation) were removed. No long-term deleterious effects were observed for forage fish species, attached algae or benthic macroinvertebrates.

Again in North America, in the Delmarva peninsula, Delaware, Whitaker *et al.* (1979) studied the effects of channel modification on macroinvertebrate community diversity in small streams (approximately 3·6 m, 12 ft wide). These workers concluded that, on the basis of indices of diversity (see Section 9.4.1.1), the macroinvertebrate community stabilised shortly after channel modification and in less than a year were indistinguishable from similar, nearby, streams.

5.1.3.5 Coal mining wastes
Several suspended solids problems are associated with coal mining. In this section solid wastes (coal or minerals particles) from coal washing or spoil disposal will be considered. The other principal suspended solid waste associated with mining, iron hydroxide, is considered separately below.

Hynes (1960) described the typical results of the discharge of colliery wash water. The receiving stream, Ditton Brook in Lancashire, was quite opaque and the stream bed consisted of a thick layer of black grit and mud mixed with vegetable debris from adjacent trees. The fauna was dense and consisted mainly of chironomid larvae, and tubificid and naidid worms. The unpolluted headwaters of the stream supported patches of *Ranunculus* and dominant animals included *Gammarus*, *Baetis* with Trichoptera, platyhelminths, Hirudinea and molluscs present. Chironomid larvae, Naididae and Tubificidae were unimportant components of the upstream fauna. The large densities of those organisms below the coal washing was attributed to an additional pollutant; pig farm effluent entered the stream above the coal washery. This feature is most significant: as explained below, it is quite probable that without the organic enrichment the coal-washery residues would be relatively 'sterile' and support few organisms.

In studies of rivers within the coalfields of South Wales, Edwards *et al.* (1972), Learner *et al.* (1971) and Scullion and Edwards (1980a) also found that oligochaete worms and chironomid larvae were the dominant components of the fauna in sections affected by the deposition of coal particles. In the Taff river system (Edwards *et al.*, 1972) most of the chironomids were of the sub-family Orthocladiinae, generally considered to be herbivorous, while the sub-family Chironominae was not well represented. This subfamily contains many filter-feeders and detritivores and their absence may, perhaps, be explained by the presence of large quantities of coal particles. The chironomids were the most abundant insects but oligochaete worms, principally naidids and tubificids, were the most numerous macroinvertebrates, constituting over 75% of the fauna at most sampling stations. Although the rivers in the Taff system are essentially characterised by eroding substrates, the presence of coal-particle deposits filling the interstices of the stony river bed provides a niche which may be exploited by these burrowing worms, more commonly associated with soft depositing substrates. Naidid worms were represented by some ten species, *Nais elinguis* and *Chaetogaster* spp. being most widely distributed. Where deposits of coal particles were deeper and enriched by organic deposits, such as in stretches of the lower parts of the system and especially upstream of weirs, tubificid worms assumed greater importance.

More detailed studies of one tributary in the Taff system, the River Cynon, were made by Learner *et al.* (1971). These workers noted a dramatic change in the fauna downstream of the point where coal particles in washery effluent were introduced via a tributary. Further introductions of coal particles (and other effluents) at points downstream maintained or heightened the change in faunal composition. In this tributary chironomid larvae usually outnumbered oligochaetes, except where highest suspended solids loads were combined with high organic enrichment. Naidid worms attained a density of 8000 m^{-2} and the thirteen species were almost identical to those recorded in the River Kinzig, a polluted German trout stream (Besch *et al.*, 1967). *Nais barbata* proved to be the most abundant species. Eleven species of tubificids were identified but most were numerically important in the lower reaches where organic pollution was associated with coal particle deposition. Chironomid larvae attained densities of 18 000 m^{-2} and, as in the studies of Edwards *et al.* (1972), the Orthocladiinae were most abundant in the middle and lower reaches influenced by coal wastes, especially where the alga *Cladophora* was well-developed. The dominant species were *Cricotopus bicinctus, Orthocladius rubicundus, Syncricotopus rufiventris* and *Rheocricotopus foveatus.*

Other faunal features associated with suspended coal particles included the disappearance of the only simuliid present in unaffected stations, *Odagmia ornata*, perhaps as a result of overloading of its filter-feeding mechanism or difficulty in settling on unstable sediments, and the reduction in both numbers and size of the net-spinning caddis, *Hydropsyche pellucidula.* It was noted that molluscs, especially *Lymnaea peregra, Physa fontinalis* and *Ancylus fluviatilis*, were associated with organic enrichment and also tolerated the high concentrations of fine coal particles.

Scullion and Edwards (1980), in a comprehensive study of the effects of coal industry pollutants on the fauna of the Taff Bargoed, another tributary of the R. Taff system, concluded that the effect of deposition of coal particles was to reduce abundance of animals and change the proportional composition of the fauna rather than to eliminate any common species or encourage the establishment of previously absent species. In general, the Ephemeroptera, Plecoptera and Trichoptera were sensitive to siltation although exceptions were evident, especially *Baetis rhodani, Amphinemura sulcicollis, Hydropsyche pellucidula* and *Rhyacophila dorsalis.* Apart from Empididae and Dolichopodidae, which increased in density, dipteran larvae and oligochaetes were largely unaffected. The mollusc *Potamopyrgus jenkinsi* and the crustacean *Gammarus pulex* were present, but in low numbers. Effects directly attributable to coal particles were difficult to isolate: the intermittent discharge of sewage and consequent organic enrichment increased the abundance of chironomid larvae and oligochaetes as in the other tributaries, described above.

5.1.3.6 Power generation wastes
A source of suspended solids which is of ecological significance on the north American continent, but apparently not in Europe, is power generation by means of coal. Disposal of the ash is a major problem although recently its use in construction, for example in the production of insulating building blocks or as a constituent of concrete, has turned a previous 'waste' to good effect. Some ash disposal is by tipping or by marine dumping but a common technique is to form a slurry, for ease of handling, pump this to settling lagoons and then recycle the supernatant liquid, to conserve water and prevent pollution. Supernatant may be discharged where quantities are small in relation to the dilution afforded by the receiving watercourse. Reduced efficiency of the settling lagoons, storm overflows or breaches of the lagoons may cause discharge of ash to nearby streams.

In a study of the effects of coal ash effluent in South Carolina, Cherry *et al.* (1979a) noted the effects on macroinvertebrates and fish in the receiving stream. While the settling basin was filling some macroinvertebrates were able to survive, although invertebrate density was clearly affected by the effluent. The most tolerant species were dragonflies (*Libellula* and *Enallagma*), crayfish (*Procambarus*), *Gammarus*, the mollusc *Physa* and chironomid midge larvae. When the basin overflowed the Crustacea and molluscs disappeared within three months and all invertebrates had disappeared within six months. Although heavy metals were present in the ash, these workers concluded that the increased turbidity and the smothering effects of the ash (some 15 cm deep) were the most important factor affecting the biota. Later studies (Cherry *et al.*, 1979b) showed that, when efficient primary and secondary settling facilities were installed, most invertebrate groups were able to recover to a level of abundance equal to or greater than that which existed four years earlier. Only mosquito fish (*Gambusia affinis*) populations failed to recover and densities fell to less than 20% of their levels during the period of high suspended-solids discharge.

In addition to the effects of direct discharge of coal ash, percolation of water from the lagoons through the embankments into neighbouring watercourses may, depending on the nature of the ash, cause secondary suspended-solids problems. Studies of ash disposal at Bull Run Steam Plant, which discharges supernatant from the settling pond into the Clinch River, Tennessee, showed that water percolating through the pond embankment contained 44 times more iron than the pond outfall (Coutant *et al.*, 1978). This iron precipitated as ferric hydroxide and formed a typical ochreous blanket (see next section). No significant benthic fauna could be found in the area affected by the discharge.

5.1.3.7 Iron compounds
Subterranean waters often contain high concentrations of dissolved iron and when these are discharged at the surface, for example as a result of mining operations, the iron precipitates as the hydroxide and ochreous deposits form on the bed of the receiving watercourse. Similar effects may be observed when rainfall percolates through coal mine spoil heaps and enters rivers. Coal ash lagoons may also give rise to iron deposits if leachate percolates to nearby watercourses (Coutant *et al.*, 1978).

Bacterial activity is thought to play a significant role in those processes. The associated acid, caused by oxidation of iron pyrites in the coal-bearing rock to produce ferrous sulphate and sulphuric acid, is considered in the section dealing with the effects of low pH (Section 5.34).

Ochreous deposits affect the stream environment in much the same way as other suspended solids but the effluents may have more serious direct consequences if the iron precipitates on the gills or other respiratory surfaces of fish or invertebrates. Precipitation of iron deposits on the leaves of macrophytes or the surfaces of algae may inhibit photosynthesis and, if this is severe, ultimately lead to the disappearance of the flora.

The close association of ferruginous deposits and acid water often makes for difficulty in separating the effects to be attributed to each factor. However, in a study of acid mine-drainage in western Pennsylvania, an area, along with the Appalachian region of the USA, much affected by acid mine-drainage, Letterman and Mitsch (1978) reported that, due to an alkaline discharge and significant upstream alkalinity, the pH below mine discharges remained between 6·5 and 8. Thus, the effects of low pH could be eliminated and the major factor affecting benthic communities and fish populations, the ferric hydroxide deposits, could be investigated. The iron deposits blanketed several kilometres of stream bed and macroinvertebrate biomass decreased to 1–11% of its former value. Invertebrate diversity was not as sensitive an indicator as biomass. Only the trichopteran *Hydropsyche* and chironomid larvae appeared to be able to tolerate these conditions although the plecopteran *Isogenus* was evident at a station a little way below the outfall. The cooler mine discharges apparently reduced water temperatures to those favoured by Plecoptera and partially counteracted the effects of ferric hydroxide precipitate. The disappearance and recovery of selected indicator invertebrates below mine water discharge is shown in Fig. 5.1.

The effects of drainage from coal mine spoils was investigated by Greenfield and Ireland (1978) in tributaries of the Ribble catchment of

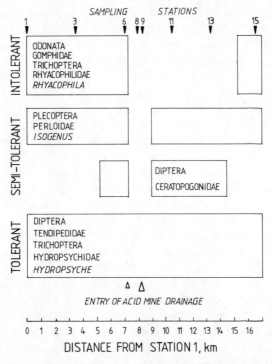

FIG. 5.1 Disappearance and recovery of selected indicator macroinvertebrates below a minewater discharge in a North American stream. After Letterman and Mitsch (1978).

Lancashire, UK. They observed that the macroinvertebrate community was diverse above the site of pollution, but was considerably reduced, as a result of the blanketing effects of iron compounds, when it consisted exclusively of oligochaetes and chironomid larvae. Similar reductions in invertebrate community diversity have been observed below the site of a ferruginous discharge on the River North Esk (Balloch *et al.*, 1976).

Scullion and Edwards (1980) observed that ferric hydroxide deposition promoted a fauna similar to that observed in areas of coal particle siltation. The fauna was dominated by chironomids (8000 m^{-2}) and oligochaetes (60 000 m^{-2}) but *Baetis rhodani, Bezzia, Chelifera flavella, Limnius volckmari, Gammarus pulex, Polycelis* and *Trocheta subviridis* were fairly abundant.

5.1.3.8 Organic solids

The major source of organic solids is sewage treatment. In addition to the blanketing effects associated with inert solids these substances are oxidisable by microorganisms and therefore exert an oxygen demand both in suspension and when settled. Their significance is related more to the nature of the receiving water, especially its hydraulic characteristics such as flow, depth and turbulence, than the biological oxygen demand of the waste. Clearly, the impact of these wastes is likely to be greater in deep, slowly-flowing waters where the rates of gaseous exchange are less than in shallow, turbulent rivers. This difference in significance is heightened by the increased probability of settlement of solids in slow, deep stretches of rivers so that, in addition to blanketing effects, there is also a risk of forming a highly anaerobic substrate. It is not possible, therefore, to determine acceptable concentrations of oxidisable solids for the maintenance of aquatic life.

The problem is further complicated by the strong association of other toxic compounds in sewage effluents. In water reclamation works with poorly nitrified effluents, ammonia may contribute to the toxicity of the discharges. Trade effluents are commonly mixed with domestic sewage to improve their treatment and thus effluents from reclamation works which receive both industrial and domestic wastes may discharge suspended matter in which toxic metals are closely bound up with organic solids. Although results from investigations of the effects of combined heavy metal and organic matter suspension are limited, it appears that the toxicity of some metals, and other pollutants, is reduced in the presence of suspended organic matter. Up to 50% of toxic metals (copper, zinc, nickel, cadmium and chromium) may be associated with suspended solids from reclamation works and a similar proportion has been found in the suspended solids of the River Trent (Alabaster, 1972).

The valuable reduction in the toxicity of metals, which the presence of organic matter provides, may be offset when the metal-laden suspended matter settles to form a metal-enriched mud. Further decomposition of the organic component and the tendency towards anoxic conditions in the muddy or silty substrate may release the metal, thereby creating a toxic habitat for benthic organisms. When the substrate is resuspended at higher flows the metal may be transferred to the water phase, enter solution and increase the general level of toxic substances present.

Another, recently highlighted, aspect of the interaction of heavy metals and organic sediments is the influence of bacteria and other microorganisms. The conversion of inorganic mercury to the much more dangerous methyl-mercury by microorganisms in sediments (D'Itri, 1973) may result in their accumulation in food chains and ultimately prove a hazard to human health ('Minamata disease').

The effects of organic solids and organic pollution in general are inextricably linked and fuller treatment of the consequences of deposition of organic solids is covered in Section 6.4.6.

5.1.3.9 Agriculture and forestry

The principal aspects of agriculture and forestry which contribute to high suspended solids are through physical disturbance of soil in ploughing or other similar cultivation (Saunders and Smith, 1965) and in the construction of roads for logging. Tebo (1955) has described the effects of siltation on the bottom fauna of a small trout stream in the southern Appalachian mountains which was subjected to unsupervised logging. The logs were dragged along steep slopes usually parallel and adjacent to the stream channel. During storms the drainage from the logged area contained 1200–1400 mg litre^{-1} suspended solids compared with 25–67 mg litre^{-1} in the unaffected stream, and was attributed to enhanced soil erosion from the roads and skid trails. Increased siltation below the logged area was associated with reduced benthic biomass. Some of the reduction is probably attributable to the instability of the silted river bed which was vulnerable to the frequent winter floods. After floods had removed silt the stream tended to be colonised by larger numbers of ephemeropteran nymphs (mayflies) than the unaffected reaches. This was attributed to reduced competition in an improved habitat.

The effects of logging on stream macroinvertebrates was studied in California by Newbold *et al.* (1980) by comparing the communities in a range of streams subjected to negligible upstream disturbance ('controls'), to logging without stream protection (unbuffered), and to logging with a 3–60 m strip of unharvested or lightly harvested timber adjacent to the stream ('buffered'). Significant ($p < 0.05$) differences in community structure were noted between buffered and unbuffered streams and diversity in unprotected streams was lower ($p < 0.01$) than in controls. The density of total macroinvertebrates was greater, however, in unprotected streams ($p < 0.05$); this was mainly caused by increased populations of Chironomidae, *Baetis* and *Nemoura*. These workers noted that there was a tendency for the width of the buffer strip to correlate with the degree of change from the control condition. For example, streams with wide buffer strips (30 m or wider) were similar to controls but different from unbuffered streams. The higher density of certain macroinvertebrates following logging was attributed to increased sunlight, temperature and nutrients enhancing algal growth. No evidence of increased sediment deposition was noted in logged streams but sampling sites were restricted to riffles on high gradient reaches.

5.1.3.10 Summary of effects of solids on invertebrates

Certain recurrent features will have been noted above with respect to the effects of deposition of solids on the benthos. These include the lowering of benthic community diversity through the disappearance or marked reduction in biomass and numbers of certain sensitive species, often those requiring 'open' eroding substrata for attachment or feeding (especially filter feeders). The replacement fauna consists of burrowing forms, typical of soft, depositing substrates, provided that the deposits are neither excessive in their rates of accumulation nor completely sterile and devoid of nutriment. Almost invariably the dominant organisms are chironomid larvae and oligochaete worms and whenever the deposits are also organically enriched then these groups may become extremely abundant.

The reasons why these taxa can exploit the changed habitat do not appear to have been formally investigated but a number of probable factors may be identified. Both groups are widely distributed in still and flowing waters and are present, although in moderate numbers, in appropriate microhabitats even in eroding zones of rivers. They occur deep in the substrate (hyporheal) and in the marginal sand and silt (psammon). When the substratum is blanketed by solids their microhabitat is effectively enlarged, while predation and competition from other organisms are, presumably, reduced. Such conditions may not be enjoyed permanently since winter spates are almost certain to resuspend the deposits. In order fully to utilise the opportunities presented by such unstable ephemeral deposits they must be exploited quickly.

Oligochaetes are rarely encountered in invertebrate drift so that rapid colonisation is unlikely to be attained by drifting organisms settling and remaining wherever they find conditions favourable. Chironomids are, however, commonly found to drift and could easily colonise new habitats in this way. In addition, it is thought that many species are multivoltine with short life cycles so that numbers could increase rapidly through the summer. Oligochaetes also have effective reproduction and may produce young over a long period (Ladle, 1971). Some, for example Naididae, reproduce asexually by budding and fragmentation, a feature which enables them to increase their populations quite rapidly. Oligochaetes and many chironomid larvae are detritivores and some feed directly on bacterial growths (Ivlev, 1945; Brinkhurst and Chua, 1969, Brinkhurst *et al.*, 1972), a food supply which may still remain abundant in certain, otherwise inert, suspended solids (Cherry *et al.*, 1979).

Chironomidae may have acquired a reputation as ubiquitous exploiters simply because the family consists of several hundred species, few of which are readily separated as larvae. One might reasonably expect that some of these would be well adapted to take advantage of almost any change in environmental conditions.

5.1.4 Effect of Inert Suspended Solids on Fish

The relationship between fisheries and suspended solids was the first to be considered by the European Inland Fisheries Advisory Commission in their Technical Paper series (EIFAC, 1964) and it has since been reviewed by Alabaster (1972) and Alabaster and Lloyd (1980).

It is important when considering the status of a fishery and the prevailing levels of suspended solids to distinguish between direct effects of suspended matter on the fish and indirect effects, such as paucity of food organisms or unsuitable spawning areas. The work of Herbert *et al.* (1961) illustrates the indirect effects: differences in densities of trout populations in china-clay polluted and unpolluted Cornish rivers could not be fully explained as direct effects on the fish since the few fish present in polluted stretches were in good condition (Alabaster, 1972). However, experimental studies (Herbert and Merkens, 1961) have shown that trout survival is reduced in suspensions of china-clay of 270 mg litre^{-1} (and perhaps even 90 mg litre^{-1}). It is also of interest to note that the effects of diatomaceous earth (a silicious deposit of diatom frustules) were similar to those of china-clay (kaolin, hydrated aluminium silicate) under experimental conditions and suggests that the direct effect is a physical one which depends more on the form of the suspended particles rather than any chemical properties which they may have. One may conclude that, in field situations, inert suspended solids such as china-clay act both directly and indirectly on fish populations.

The effects of waste solids from a coal washery, smaller than 64 μm, on the survival of rainbow trout was investigated by Herbert and Richards (1963). The experiment lasted for about 9 months and suspensions of 50, 100 and 200 mg litre^{-1} were used. None of the control fish nor any of those kept in the three concentrations of coal solids died during the experimental period and all appeared to be healthy at the end of the investigation. It would seem, therefore, that 200 mg litre^{-1} of coal washery solids has no detrimental effect on rainbow trout survival or health. An interesting observation was that growth of the experimental fish was depressed, compared with the controls, and was poorer with increasing suspended solids load. These results suggest that the suspended matter may interfere with feeding efficiency, perhaps by making food harder to find, or impose an additional physiological 'stress' burden on energy expenditure, possibly through a reduction in respiratory efficiency or even at the simple level of increased mucus secretion in response to the presence of inert solids.

Trout populations in stream sections affected by high suspended solids (mean 110 mg litre^{-1}, occasionally exceeding 2000 mg litre^{-1}) from mine water discharges had lower densities than in unaffected stretches and also exhibited lower condition factors (Scullion and Edwards, 1980b). The diet of these fish was almost exclusively terrestrial in origin. These findings support the earlier laboratory findings of Herbert and Richards (1963).

The status of fisheries and levels of inert solids was examined in a literature survey by Herbert and Richards (1963). Care was taken to exclude any data relating to rivers in which it was suspected that pollutants other than solid materials were involved. Similarly, wastes from paper manufacture were excluded since these invariably also involve serious oxygen depletion. Although these constraints reduced the range of observations relating to rivers in which fish populations were adversely affected or absent, there appeared to be a clear separation between rivers with normal fish populations and those with impaired fisheries or which were fishless at suspended solid concentrations of about 600 mg litre^{-1} (Fig. 5.2).

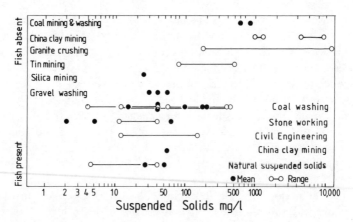

FIG. 5.2 Distribution of normal and impaired fish populations in relation to concentrations of inert solids in rivers. After Herbert and Richards (1963).

The EIFAC (1964) criteria have now been adopted in the EEC Directive on the quality of fresh waters needing protection or improvement in order to support fish life (EEC, 1978), where the guideline value for mean suspended solids concentrations of harmless substances in salmonid and cyprinid waters is not more than $25 \, \text{mg litre}^{-1}$. Derogations from this value are permissible under exceptional weather or for other natural phenomena such as floods or spates.

5.1.5 Effects of Other Suspended Matter on Fishes
Other kinds of suspended matter include wood fibres, solids precipitated from acid mine drainage (principally iron hydroxide) and organic solids from sewage treatment or food processing.

5.1.5.1 Wood fibres
Wood fibres, mainly from paper pulp manufacturer, have an oxygen demand but this is usually at a lower rate than organic solids from food processing or sewage treatment. The effects of wood fibre (ground spruce pulp) on rainbow trout was investigated by Herbert and Richards (1963). A population of these fish kept in suspensions of $200 \, \text{mg litre}^{-1}$ showed more or less steady mortality rates over eight months by which time approximately 80% had died. Rainbow trout kept for the same period at concentrations of 50 and $100 \, \text{mg litre}^{-1}$, however, showed no mortality although those in the latter concentration developed slight fin rot.

The mechanical action of suspended wood fibres was found to have no effect upon egg survival in brown trout (*Salmo trutta*), rainbow trout (*S. gairdneri*) (Kramer and Smith, 1965) and walleyes (*Stizostedion vitreum vitreum*) (Kramer and Smith, 1966) at concentrations up to 250 mg litre^{-1} but survival, respiration and growth of trout alevins kept in wood fibre suspensions were depressed. Fibres were found to clog the buccal and gill cavities and it was concluded that, although the eggs survived unharmed in suspensions, in natural waters the blanketing effects of settled wood fibres would have a deleterious effect. (Simultaneous experiments with the mercuric bactericide 'Slimicide' added to the fibre at the papermill indicated that the observed effects could be attributed to the presence of the fibre alone.)

In streams the deposition of wood fibre leads to increased mortality of walleye eggs by encouraging the growth of *Sphaerotilus natans* (Smith and Kramer, 1963).

Survival of fathead minnows (*Pimephales promelas*) and young walleyes (*Stizostedion vitreum vitreum*) exposed to wood fibres was studied by Smith et al. (1965). Walleye fingerlings were more susceptible than the minnows and ground woodpulps were more lethal than chemically produced pulps. Conifer ground woodpulp was the most lethal fibre tested. Lowered oxygen levels were found to increase mortality rates of walleyes in both ground wood and chemically derived fibres.

5.1.5.2 Iron compounds

Ferric iron in suspension appears to exert little direct effect upon fish even at high concentrations (Wallen et al., 1957). A week of exposure to 3·5 mg Fe per litre by salmon caused little mortality (Sprague, 1966) but slight reduction in growth of brook trout (*Salvelinus fontinalis*) was observed at 12 mg Fe per litre and marked retardation of growth at 50 mg litre^{-1} although much of this was precipitated hydroxide. Reports of fish deaths or absence of fish in water polluted by ochre (Larsen and Olsen 1950; Nichols and Bulow, 1973; Coutant et al., 1978) were probably caused by a combination of enhanced mucus production, induced by low pH, compounded by precipitation of iron on the gills and gill filaments. The presence of iron compounds does not, of itself, seem to be deleterious to fish, except perhaps indirectly by destroying benthic food resources. This view finds support in the study by Letterman and Mitsch (1978) of the effects of mine drainage in which ferric hydroxide deposits were not accompanied by significant lowering of pH. Although the biomass and numbers decreased by over 95%, much of the reduction was in three bottom-dwelling species, the mottled sculpin (*Cottus bairdi*), the white sucker (*Catostomus commersonii*) and the hog sucker (*Hypentelium nigricans*). These fish are largely sedentary, feeding mainly on benthic invertebrates and detritus, and thus would be most susceptible to the effects of blanketing deposits of hydroxide which also destroyed the benthos. Brown trout (*Salmo trutta*) were not affected to the same degree, perhaps because of their tendency to feed on drifting organisms. One species, the creek chub (*Semotilus atromaculatus*) actually increased, both in biomass and numbers, below the mine drainage; this was explained by the well-known adaptability and tenacity of this species.

Additional support for the view that ferruginous deposits affect fishes indirectly is to be found in the studies of Scullion and Edwards (1980b) of the effects of ferruginous drainage at neutral pH from an abandoned mine. Lower densities of trout (a reduction to 17% of the upstream value) was associated with reduced benthic invertebrate density and diversity (Scullion and Edwards, 1980a). The remaining fauna consisted largely of burrowing oligochaetes and chironomids which were probably largely inaccessible to the trout. The mortality of trout eggs was high (greater than 80%) in areas affected by ferruginous deposits. In experimental studies of the effect of lime-neutralised iron hydroxide suspensions on juvenile brook trout (*Salvelinus fontinalis*), increasing concentration was associated with greater suppression of growth (Sykora *et al.*, 1972). For example, the fish exposed to 50 mg Fe per litre had a mean weight only 16% of that of the controls after 35 weeks. It was postulated that impaired visibility due to the high turbidity of the test solutions prevented the fish from feeding successfully.

5.2 EFFECTS OF THE ADDITION OF HEAT

5.2.1 Introduction

The thermal regime of any habitat has a considerable influence in determining the community which is to be found there. This factor is recognised in the classification of freshwaters and may be seen, for example, in the recognition of river zones which are determined not only by hydrological characteristics, such as dimensions, slope, discharge and the nature of the substratum, but also by thermal characteristics. The widely accepted primary division of flowing waters into rhithron and potamon (Illies, 1961) contains an important element regarding temperature.

The *rhithron* is the uppermost section of the stream, from the source to the point where the annual range of monthly mean temperatures does not exceed 20 °C. Typically the velocity of the current is high but the discharge is small. The stream bed may be rock, boulders, gravel or sand but mud is only rarely found, in pools and slacks.

The *potamon* is the remaining lowermost stretch of the river in which the annual range of monthly mean temperatures exceeds 20 °C. In the tropics the summer maximum of monthly mean temperatures may exceed 25 °C. The velocity is lower than in the rhithron and sand or mud is deposited on the river bed.

As a consequence of the temperature criterion, the proportion of rhithron to potamon varies with latitude and the altitude of the stream source.

Hawkes (1975), in a thorough review of river zonation and classification, has stressed the significance of the range of temperatures experienced rather than some critical temperature which differentiates between rhithron and potamon.

Much of the original concern with regard to controlling thermal pollution (or 'thermal enrichment', as some of those who are responsible euphemistically prefer to call it) arose from the effects of discharges of cooling water from power stations. But thermal regimes may also be modified by the removal of heat when water colder than the receiving stream is discharged. This may occur when water is released from the hypolimnion of reservoirs, either for power generation in hydro-electric schemes or for river regulation, or by the release of groundwater for the latter purpose. Some temperature reduction may also occur during the discharge of cool minewaters.

5.2.2 Natural Temperature Regimes

The natural temperature regimes of surface waters in temperate climates tend to follow closely those of air, being likewise dependent on seasonal variations in the intensity of solar radiation, heat flux and losses through the latent heat of evaporation, although, usually, water temperatures do not exhibit the same extremes as air temperatures and may lag slightly behind them in the seasonal cycle.

In addition to the overall seasonal pattern there is, superimposed, a daily pattern of temperature fluctuations (see Fig. 5.3), the magnitude of which is controlled by season, by immediate climatic conditions and, for flowing waters, by the hydrological conditions (e.g. stage of discharge).

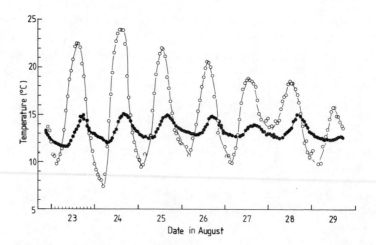

FIG. 5.3 Daily variation in air and water temperatures in a lowland English stream during summer.

Greatest daily range in water temperatures and often the highest temperatures are observed in the uppermost reaches of rivers since the lowest volumes of discharge are to be found there. Many hydrobiologists (e.g. Reid, 1961; Macan, 1974) consider the temperature regime to be the most important physical factor controlling the ecology of aquatic communities. Temperature may act directly, that is acutely, whenever the lethal limits of some species are exceeded, or indirectly, that is chronically, by affecting the rates of physiological processes. All organisms associated with freshwater, with the exception of birds and mammals, are poikilothermic, that is, they are unable to control their body temperatures and are, therefore, highly dependent on ambient temperatures. Variations in temperature affect all physiological processes and this is revealed in changes in growth rate, attainment of sexual maturity, reproduction and behaviour. Departures from the normal temperature pattern, particularly sudden or aseasonal changes, could disturb these processes and cause accelerated, or retarded, growth and abnormal timing of life cycles.

5.2.3 Sources of Additional Heat
The major additional heat contribution to freshwaters is from industrial cooling water, especially that of electrical power generation. It is estimated that electricity generation accounts for 90% of heated discharges

in Britain and 80% in the United States (O'Sullivan and Collinson, 1976). The cooling water is needed to increase turbine efficiency. In the direct (or 'open') system of cooling, river water is passed to condensers and returned, warmed, to the river. If the normal dry weather flows are insufficient, cooling towers must be utilised and losses caused by evaporation are made up from the river. Sometimes this indirect (or 'closed') system is combined with the direct system, depending on the adequacy of river flows. In the combined or 'mixed' system, the water may pass either through cooling towers or be taken from the river when supplies are adequate. Recirculating the water through cooling towers dissipates heat to the atmosphere but the temperature of the reduced volume of effluent may be higher. The degree of recirculation governs the temperature of the discharged effluent. Discharges of heated effluents from direct or mixed systems are not constant, but depend on generating load which is itself subject to seasonal and diurnal fluctuations. Peak demand occurs in daytime and during winter months. Summer peak demand coincides with highest river temperatures and lower efficiency of cooling towers. Variations in heated effluent discharge are smallest in stations with a high load factor, that is, those which operate almost continuously to supply the base load (the amount of electricity which is always required) while those which are used to meet peak demands have quite large variations in heated effluent output.

The difference in temperature between heated discharges and receiving waters ranges from 6·3 to 10·4 °C but in mixed cooling systems with recirculation of the water the difference may reach 13 °C (Alabaster, 1964). Figure 5.4 shows the seasonal variation in the difference in temperature between a river without thermal discharges and a point some 5 km (3 miles) below a power station of 600 MW using a direct cooling system.

An important factor which determines the extent to which heated effluents increase the temperature of receiving water in contact with the biota, and hence their biological significance, is the degree of mixing. Since hot water is less dense than cold water it may persist as a surface layer for some distance downstream from the point of discharge. This tends to occur more readily when only a fraction of the river flow is used for cooling or is used intermittently. The difference between the surface and the river bed may typically approach 6 °C.

5.2.4 The Significance of Heated Effluents or 'Thermal Enrichment'

There are three important effects of the introduction of heated effluents

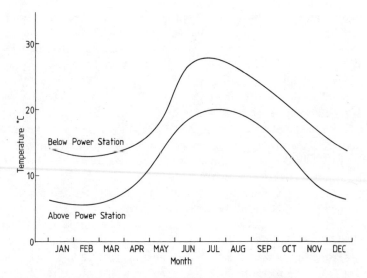

FIG. 5.4 Seasonal variation in the mid-day river temperature upstream and downstream of the 600 MW power station at Castle Donington on the River Trent. Smoothed curves derived from data presented in Sadler (1980).

into rivers and lakes. The first is to raise the ambient temperature. The significance of this for the biota depends upon its range of temperature tolerance, especially the upper lethal limit. The second is an indirect physico-chemical phenomenon concerning the solubility of gases, especially oxygen, which changes the nature of the environment in that the quantity of dissolved gas varies inversely with temperature (Fig. 5.5). The third effect relates to the rate of biochemical processes which, within limits, varies directly with temperature. A generally held relationship is that the biochemical rate approximately doubles with each ten degree Centigrade rise in temperature (i.e. $Q_{10} = 2$). These factors tend, therefore, to be synergistic in that as temperature increases, dissolved oxygen is reduced while at the same time the rate of metabolic processes (such as the bacterial breakdown of organic matter) is increased. The net effect may be much lower oxygen levels or even an oxygen deficit: in this way the effects of 'thermal enrichment' may be as serious as the addition of putrescible organic matter or certain industrial wastes and the community of organisms present in heated waters may be restricted to those which are able both to tolerate the prevailing higher temperatures and the reduced oxygen levels.

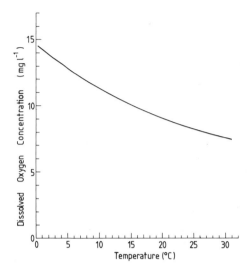

FIG. 5.5 Variation with temperature in the solubility of oxygen in water in equilibrium with air at normal atmospheric pressure.

The effect of increased temperature on the 'self-purification' capacity of a receiving water depends, in part, on the variation with temperature of the rate-constants of deoxygenation (K_1) and oxygenation (K_2). Both constants increase with temperature but the effect of deoxygenation (K_1) increases at a greater rate than re-aeration (K_2) and the latter is negated by the lower dissolved oxygen at higher temperatures (Krenkel and Parker, 1969). This last mentioned phenomenon is exacerbated by the lower turbulence which is encountered in the lower reaches of rivers or when otherwise turbulent rivers are impounded by weirs. An example of the changes induced by these factors is illustrated in Fig. 5.6 where the oxygen 'sag' in a river with a given additional organic load is shown under the influence of two re-aeration coefficients.

5.2.5 Effects of Temperature Variations on the Benthos
In a study of the effects of a cooling water effluent from two power stations on the River Severn, Langford and Daffern (1975) concluded that temperature increases did not have a significant effect on the total numbers and overall emergence period of Trichoptera, Ephemeroptera and Megaloptera. The species composition of catches from above and below the power stations was also very similar. Coutant (1968), however,

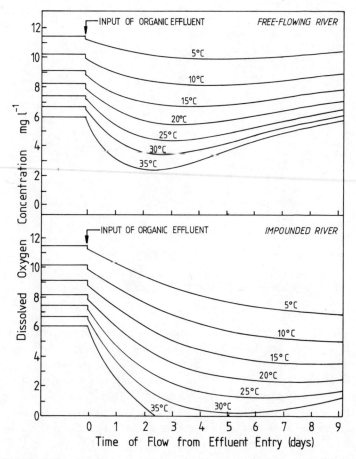

FIG. 5.6 Variation in oxygen 'sag' curves with temperature in a river receiving heated effluents following an additional organic load with constant deoxygenation coefficients at high (free flow) and low (impounded) re-aeration coefficients. After Krenkel and Parker (1969).

observed that a two-week advance in the emergence rate of Trichoptera in the Columbia River was promoted by a 1 °C temperature rise while Gledhill (1960), on the other hand, concluded that temperature was not responsible for initiation of emergence by insects in an English Lake District stream.

Galepp (1976), in an experimental study of the effects of temperature and daylight on the periodicity of feeding in the North American trichopteran, *Brachycentrus occidentalis*, concluded that temperature could, in the absence of daylight fluctuations, act as a cue for diel feeding periodicity. In studies of the feeding activity of the leech, *Erpobdella octoculata*, Greene (1974) found that temperature did not play much part between 7 and 12 °C and feeding continued at temperatures as low as 2 °C. Similarly, Aston and Brown (1975) observed that the seasonal pattern of reproduction and growth was largely unaffected by condenser effluents from power stations on the River Trent, UK, although the emergence of young leeches was advanced by up to one month, there was more rapid growth, earlier maturity and leeches were larger in the stretches below outfalls.

Aston (1973) found an increase in the abundance of sexually mature tubificid worms (*Limnodrilus hoffmeisteri*) below heated effluents from a power station. In experimental studies it was found that *L. hoffmeisteri* increased its rate of egg production up to 25 °C while another species, *Tubifex tubifex*, was unaffected over the range 10–25 °C. In laboratory and experimental field studies of the effect of temperature on the time of hatching in the mayfly, *Baetis rhodani*, Elliott (1972) showed that the period between oviposition and the start of hatching was reduced from 17 weeks at 3 °C to about 1 week at 22 °C and that the duration of hatching was reduced from about 34 days at 3 °C to 3 days at 22 °C (Fig. 5.7).

Another laboratory study, this time of the effects of thermal shocks on the behaviour of drifting aquatic insects using an ephemeropteran, *Isonychia* (Baetidae), and the caddis *Hydropsyche* showed that no differences in mortality were observed between experimental and control groups until the shock temperature approached the upper lethal limit for these species (Sherberger *et al.*, 1977). Behavioural studies showed no differences with respect to rheotaxis, phototaxis and substrate orientation nor to susceptibility to predation. The moulting frequency of *Isonychia* was unaffected but the pattern of moulting changed.

Teckelmann (1974) studied the effects of temperature, in field and laboratory, on the amphipod (*Gammarus fossarum*). Growth was found to be dependent upon temperature over the range at which the species survived (0 to 25 °C) but reproduction was successful only over the range 3 to 18 °C. Optimum temperature for feeding and growth was about 15 °C. Differences in the composition of the bodies of these animals (protein, lipid and polysaccharides) occurred with changes in temperature.

Studies of the effects of thermal discharges on the benthos of lakes

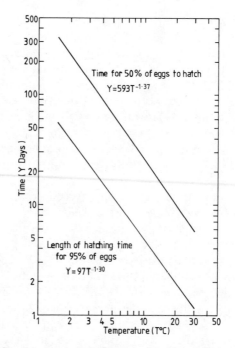

FIG. 5.7 Effect of temperature on the time to hatching and duration of hatching of eggs of the mayfly *Baetis rhodani*. After Elliott (1972).

have been less frequently undertaken than equivalent studies in rivers. This may, of course, simply be a reflection of the greater use of rivers as a source of cooling water. Weiderholm (1971) investigated the bottom fauna of Lake Malaren in Sweden near Vasteras where the lake receives heated effluent from two fossil fuel power stations in addition to organic enrichment from the town. It proved difficult to separate the effects of thermal discharges from the organic pollution which appeared to contribute to the high biomass. However, the biomass of benthos was even higher under the influence of thermal enrichment, mainly contributed by increased numbers of oligochaetes, and was supposed to result from enhanced microbial activity in the mud which could be exploited by these animals. Elsewhere the fauna was composed of some 41 species of chironomid larvae, of which *Chironomus thummii* was dominant in the severely polluted zones and in less severe environments Tanypodinae and *Cryptochironomus* spp. were important. It is quite probable that the preponderance of oligochaetes in the warm, organically enriched areas was an effect of very low oxygen levels. Other components of the benthic fauna included *Chaoborus* larvae and a sparse population of the lamellibranch mollusc *Pisidium* (3 species). Two other interesting effects which may be attributed to heated effluents were the presence of the exotic warm water oligochaete *Branchiura sowerbyi* and three species of Chironomidae which were new to Sweden, having previously had a much more southerly distribution.

In a study initiated to determine the effects of a large thermal discharge into Lake Erie, which had already been subjected to pollution by organic and toxic materials, Cole and Kelly (1978) concluded that macrobenthos abundance in the lake was unaffected by the 4 km long thermal plume. The fauna of the discharge channel and the shallow bay adjacent to it was, however, severely depleted. Chironomid larvae tended to disappear while oligochaetes (especially *Limnodrilus hoffmeisteri*) remained relatively abundant. This change was attributed to preferential selective grazing of the chironomid larvae by the fishes which frequented the discharge channel. The maximum discharge temperature (34 °C) was considered to be relatively low and was thought to account for the presence of fish predators. In addition, it was noted that macroinvertebrate production rates were directly related to temperature (Johnson and Brinkhurst, 1971) and that the increased production may have augmented the food of fish in the vicinity of the power station. The warm water oligochaete *Branchiura sowerbyi* was found in the bay associated with the thermal effluent channel before the power station was operative and, surprisingly, it failed to become more abundant when the temperatures were raised.

5.2.6 Effects of Temperature Variations on Fish

5.2.6.1 Introduction
Considerable attention has been given to the effects of temperature on the biology of fish in general and the significance of heated effluents in particular. The Slavonic literature on temperature relations of fish was reviewed by EIFAC (1968b) and supplemented by a further list of papers taken from two important reviews (Kennedy and Mihursky, 1967 and Reney and Menzel, 1967) by EIFAC (1969). These EIFAC reviews have been updated to include recent Slavonic work and are now more readily available in Alabaster and Lloyd (1980) but no attempt was made to incorporate more recent non-Slavonic literature. Early work on the

thermal requirements of fish was reviewed by Brett (1956, 1960) and general reviews on the significance of heated effluents have been provided by Alabaster (1964, 1969a), Hawkes (1968) and Sylvester (1972b). Recently Elliott (1981) has provided a comprehensive review of thermal stress in freshwater fish.

5.2.6.2 Thermal requirements

Fish are poikilothermic, that is, unable to regulate their temperature so as to maintain optimal temperatures for metabolic activity. However, within quite wide limits they are able to remain quite active and investigations have revealed that they are able to achieve a degree of independence of temperature through regulation of enzymes. For example, a reduction of ambient temperature for a fish accustomed to warm water means that its enzymes would be inefficient and perhaps virtually inactive. But, after a period of exposure to reduced temperatures, iso-zymes (iso-enzymes) appear which are able to function at these lower temperatures. The presence of pairs of enzymes with quite different temperature optima, preferably towards the extremes of the ambient range, enables the fish to adjust or maintain its metabolic activity as temperature changes (Hochachka and Somero, 1971). Adjustments to changing temperature, or acclimation, occur faster in most fishes when

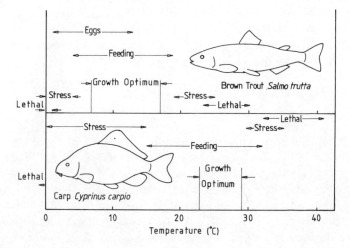

FIG. 5.8 Optimal temperature ranges for a warm water species (carp, *Cyprinus carpio*) and a cold water species (trout, *Salmo trutta*). After Elliott (1980).

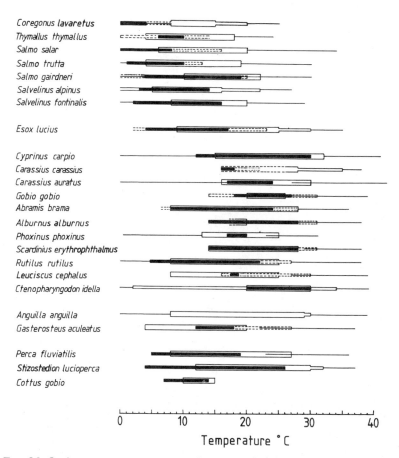

FIG. 5.9 Optimum temperature ranges for several fish, based on published data. Solid black line indicates spawning temperature range and pecked black the range for eggs. Double width open line indicates optimum temperature range; narrow open line indicates sub-optimal and single line indicates critical temperature. Postulated critical temperatures are indicated by the pecked line.

temperatures are raised than when they are reduced. The significance of these responses is considered below.

The thermal requirements differ between species and between different stages in the life history of a single species. Upper thermal limits do not seem to be associated with enzyme inactivation alone. Most enzymes are only inactivated by temperatures in excess of 35 °C (Hawkes, 1968) so

TABLE 5.3

TEMPERATURE TOLERANCE RANGES AND OPTIMAL TEMPERATURES FOR A SELECTION OF NORTH AMERICAN FRESHWATER FISH SPECIES

(Data mainly from Hawkes, 1968, supplemented by Elliott, 1980, 1981)

(a) *Warm water species*

Active at 25–30°C	Active at 30–35°C	Active at 35–40°C	Survival at 35–40°C	Unable to survive at temperatures above 38°C
Ambloplites rupestris (rock bass)				*Ambloplites rupestris*
Ictalurus punctatus (channel catfish)	*Ictalurus nebulosus* (brown bullhead)			*Ictalurus nebulosus*
	Micropterus salmoides (large mouth bass)		*Micropterus salmoides*	
Micropterus dolomieu (small mouth bass)	*Micropterus dolomieu*	*Lepomis macrochirus* (bluegill sunfish)	(*Lepomis macrochirus*)	*Erimyzon sucetta* (chub sucker)
		Lepomis gibbosus (pumpkinseed sunfish)	(*Lepomis gibbosus*)	*Notomigonus crysteleucas* (golden shiner)
Alosa pseudoharengus (alewife)				*Notropis heterodon* (black shin shiner)
				Notropis heterolepis (blacknose shiner)
	Anguilla rostrata (american eel)	*Lepomis auritus* (yellow belly sunfish)	*Lepomis cyanellus* (green sunfish)	*Etheostoma exile* (Iowa darter)
	Etheostoma nigrum (johnny darter)	*Etheostoma lepidum* (green throat darter)	*Lepomis megalotus* (longear sunfish)	*Ictalurus notalis* (yellow bullhead)
				Noturus miurus (brindled madtom)
	Carpoides cyprinus (quillback carpsucker)			*Noturus gyrinus* (tadpole madtom)
	Proxomis nigromaculatus (black crappie)	*Fundulus diaphanus* (banded killifish)		*Pimephales notatus* (bluntnose minnow)

(b) *Cold water species*

Habitat temperature always below 10 °C	Species requiring temperatures below 10 °C for spawning	Active between 0–10 °C	Active between 10–15 °C	Active between 15–20 °C
Salvelinus namaycush (lake trout)	*Salvelinus fontinalis* (brook trout)	*Salvelinus fontinalis*	*Salvelinus namaycush*	*Salvelinus fontinalis*
	Oncorhynchus tshawytscha (chinook salmon)		*Salvelinus fontinalis*	*Oncorhynchus tshawytscha*
			Oncorhynchus tshawytscha	*Oncorhynchus gorbuscha*
Pomolobus pseudoharengus			*Oncorhynchus gorbuscha* (pink salmon)	*Oncorhynchus keta*
Osmerus mordax (American smelt)			*Oncorhynchus keta* (chum salmon)	*Oncorhynchus nerka*
Leucichthys artedi			*Oncorhynchus nerka* (sockeye salmon)	*Oncorhynchus kisutch*
			Oncorhynchus kisutch (coho salmon)	
		Salmo gairdneri (rainbow trout)	*Salmo gairdneri*	*Salmo gairdneri*
			Salmo salar (Atlantic salmon)	*Salmo salar*
	Esox lucius (northern pike)		*Esox lucius*	
	Stizostedion vitreum (walleye)		*Stizostedion vitreum*	
			Coregonus clupeaformis (Lake white fish)	
			Catostomus commersoni (common sucker)	
			Catostomus catostomus (longnose sucker)	

that the observed lower thermal limits (about 20–25 °C) for some species must involve other factors, perhaps the rates of metabolic processes such as oxygen transport or membrane permeability or changes in lipids in cell membranes.

The differences in thermal requirements between species and the different optimal ranges associated with different life stages or functions are illustrated in Fig. 5.8 for a warm-water species, the carp (*Cyprinus carpio*), and a cold-water species, the trout (*Salmo trutta*). The optimum temperature range for a number of European freshwater fishes is indicated in Fig. 5.9 while data on selected North American species are to be found in Table 5.3. Temperatures found to disturb fishes and which ultimately proved lethal to lake dwelling European species (Horoszewicz, 1973) are given in Table 5.4.

TABLE 5.4

TEMPERATURES FOUND TO BE DISTURBING AND LETHAL IN EUROPEAN FISHES
(data from Horoszewicz, 1973)

Species	Disturbing temp (°C)	Lethal temp (°C)
Rudd, *Scardinius erythrophthalmus*	30–33·6	38·2
Roach, *Rutilus rutilus*	29·2–33·0	36·4
Chub, *Leuciscus cephalus*	33·6–34·0	38·0
Ide, *Leuciscus idus*	–	37·9
Bleak, *Alburnus alburnus*	–	37·7
Gudgeon, *Gobio gobio*	30·8–30·9	36·7
Carp, *Cyprinus carpio*	32·3–32·5	40·6
Crucian Carp, *Carassius carassius*	34·6–36·0	38·5
Tench, *Tinca tinca*	32·1–33·7	39·3
Bitterling, *Rhodeus amarus*	30·5–31·8	36·5
Perch, *Perca fluviatilis*	30·5–32·0	? 35·5
Pike-perch, *Stizostedion lucioperca*	33·0–33·3	37·0
Ruffe, *Gymnocephalus cernua*	29·4	34·5
Stickleback, *Gasterosteus aculeatus*	30·5–31·8	36·5

5.2.6.3 Temperature adaptation

When fish are exposed to increasing ambient temperatures it is found that their upper lethal limit is raised. For example, Cocking (1959) noted that when the acclimating temperature of roach (*Rutilus rutilus*) is raised by 3 °C the lethal temperature rises by 1 °C. This occurs over much of the thermal range and continues until a maximum lethal temperature is

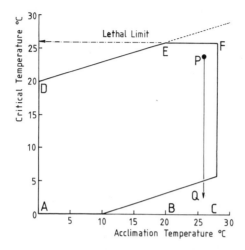

FIG. 5.10 Temperature tolerance polygon for a fish (roach, *Rutilus rutilus*). After Cocking (1959).

reached above which no further adaptation is possible. One may thus construct a temperature tolerance polygon (Fig. 5.10) which is an envelope of limiting temperatures which vary according to the history of previous exposure.

Raising the acclimation temperature over the range A to B leads to an increased upper tolerance limit (D to E) but continuing to raise the acclimation temperature fails to secure any increase in the tolerance of high temperatures and thus the upper lethal limit is reached (E to F). Similarly, for fish acclimated to high temperatures an equivalent reduction in ambient temperature may bring the fish to a point outside its tolerance polygon. For example, if a fish acclimated at high temperature (P) is suddenly exposed to a low temperature (say to 4 °C at point Q), perhaps as a result of reduced discharge of heated turbine effluents on a snow-melt river or the release of low-level draw-off regulating discharges from a reservoir in midsummer, then the temperature may now be below the tolerance range and death may ensue. This is probably a problem of enzyme adaptation since iso-enzymes (isozymes) must be induced in order to be able to function at low temperatures. It is well known that most fish can adapt more readily to higher temperatures than lower temperatures. Adaptation to higher temperatures is a rapid process, often taking less than one day, and is lost only slowly.

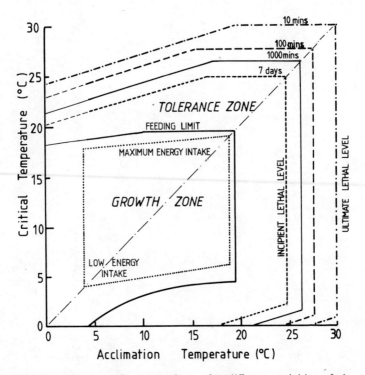

FIG. 5.11 Temperature tolerance polygon for different activities of the trout (*Salmo trutta*). From Elliott (1980).

The significance of temperature adaptation is rendered more complex by the existence of different limits for the different aspects of fish biology (cf. Fig. 5.8). Thus, a temperature range which may be tolerated may lie outside the limit for normal feeding and so death may ensue from starvation (see Fig. 5.11). The form of the tolerance polygon will vary between species (Fig. 5.12) and also according to the duration of exposure (Fig. 5.11).

5.2.6.4 Behavioural responses
Since fish are highly mobile they are able to respond to changing temperatures by movement. They can avoid high and low temperatures, especially from discharges, and where a gradient exists they can select an optimal temperature. The actual choice may depend on previous experience (acclimation). In a gradient there is a tendency to select higher

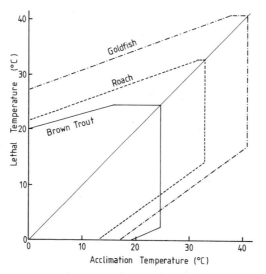

FIG. 5.12 Temperature tolerance polygons for three species of fish: goldfish (*Carassius auratus*), roach (*Rutilus rutilus*) and brown trout (*Salmo trutta*). From Elliott (1980).

and higher temperatures (Fig. 5.13) as fish become acclimated during movement within the gradient. The greater rate of acclimation to higher temperatures compared with acclimation to lower temperatures is probably responsible for this phenomenon (Reynolds, 1978). The final preferenda of several species are indicated in Table 5.5.

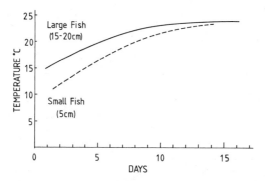

FIG. 5.13 Behavioural response of fish to a temperature gradient. Selection of preferred temperature in a gradient by roach (*Rutilus rutilus*). After Alabaster (1964).

TABLE 5.5

FINAL PREFERENDA OF FISH SPECIES IN A TEMPERATURE GRADIENT

Species	Final preferendum (°C)	Reference
Roach, *Rutilus rutilus*	23	Alabaster, 1964
Carp, *Cyprinus carpio*	32	Pitt *et al.*, 1956
Tench, *Tinca tinca*	20·3	Schmeing-Engberding, 1953
Salmon, *Salmo salar* (alevins)	14–15	Alabaster and Lloyd, 1980
Brown Trout, *Salmo trutta*	12·4–17·6	Ferguson, 1958
Rainbow Trout, *Salmo gairdneri*	13·6	Garside and Tait, 1958
Orfe, *Leuciscus idus*	17*	Alabaster and Lloyd, 1980
Bream, *Abramis brama*	19*	Alabaster and Lloyd, 1980
Crucian Carp, *Carassius carassius*	27*	Alabaster and Lloyd, 1980

*Estimated.

Fish are able to detect small temperature differences: in conditioning experiments differences as low as 0·03 °C and commonly between 0·05 and 0·1 °C have been detected (Murray, 1971). In channel experiments, Alabaster (1964) has investigated the effects of temperature gradients on bream, roach and perch. These species were found to move when exposed to a change in ambient temperature of between 0·5 and 7·6 °C but tended to select a temperature close to that initially enjoyed, though usually slightly higher (perhaps as a result of acclimation). Studies of fish densities in the River Trent were found to be correlated with temperature; highest densities were found in the range 22–26 °C which was intermediate between the temperatures of unheated and heated stretches. Similarly, studies of fish distribution near a power station on the River Trent by Sadler (1980) showed that populations of bleak (*Alburnus alburnus*), gudgeon (*Gobio gobio*) and roach (*Rutilus rutilus*) were denser downstream of the power station.

5.2.6.5 Effects of heated discharges on fish

The effects of heated discharges, principally from industrial cooling water and in particular from electrical power generation, may affect fish in several ways. First, fish acclimated to normal temperatures may be affected by sudden and large changes in temperature near outfalls. Small fish seem to be especially vulnerable (Alabaster, 1964). More often, however, heated effluents tend to form a warm layer at the water surface by virtue of their lower density. Fish are thus able to maintain themselves in the cooler water beneath and, where mixing occurs, unless temperatures exceed 30 °C many species will be able to survive or avoid the affected stretches. Evidence of attraction to warmer water (Alabaster, 1964; Sadler, 1980) and lack of evidence of serious mortality below heated outfalls suggests that the discharge of hot water is unlikely to affect fish populations directly. Even migratory salmonids can survive quite high temperatures for short periods so that their migratory passage is unlikely to be seriously hampered. Indirect effects of heated effluents are apparently not well understood.

5.2.6.6 *Effects of low temperatures on fish*
Low winter temperatures can cause freezing of streams at the surface. Once this has occurred further loss of heat is reduced and the water beneath the ice remains near freezing point. Under severe conditions shallow streams may freeze completely and fish can then survive only by moving to deeper stretches or pools. More serious, from the point of survival of fish, is the formation of underwater ice. This occurs by radiation of heat from the substrate in water at 0 °C and takes two forms; frazil or slush and anchor ice (Hynes, 1970). Frazil results from the formation of ice crystals in supercooled water, or by entrained snow flakes, and can occur in deep water to considerable depth. Anchor ice usually forms in shallow water as a thin dense layer on the substrate.

In lakes the peculiar density–temperature relationships of water work to the advantage of organisms. As the lakewater cools it reaches maximum density at 4°C so that freezing occurs at the surface only when the whole water column has reached 4 °C. Further surface cooling lowers the density and so water cooler than 4 °C remains at the surface until it freezes. The ice diminishes further heat loss so that most of the water remains at 4 °C.

The danger of freshwater fish freezing is small since the freezing point of water is about 0·5 °C above the freezing point of their body fluids (Devries, 1971). Only in climates where lakes freeze completely is there any risk and here fish remain inactive and burrow into the mud at the bottom of the lake.

5.2.6.7 *Effects of cold water discharges*
Discharges of cold water from regulating impoundments (which can be as low as 4 °C) or even cessation of heated discharges could have potentially serious effects (as outlined in Section 5.2.6.3). However,

evidence is lacking, possibly because of other simultaneous factors, of any clear effects directly attributable to cold water discharges.

5.3 EFFECTS OF CHANGES IN pH

5.3.1 Introduction

The pH of a water is not the same as its acidity or alkalinity, although lower pH values are said to be acid and higher pH values alkaline. It is also important to recognise that changes above and below pH 7, the neutral value, are logarithmic and that each unit of change (e.g. 7 to 6) is a change of one order of magnitude. Acidity and alkalinity are measures of the buffering capacity of a water, that is their ability to neutralise the effects of the addition of acids or bases and thus fluctuations in pH value. Normally the alkalinity (or acidity) is expressed in carbonate equivalents, the quantity of calcium carbonate which would have to be present to require the same addition of acid (or alkali) to effect a change of pH to 4·2 to 5·4 (total or methyl orange alkalinity, so named because of the indicator used) or to pH 8·3 (phenolphthalein alkalinity).

The natural variation in pH is extensive, ranging from 3·5 or 4 for acid moorland streams (with little buffering capacity) to 8 or so in chalk streams. The intense photosynthetic activity of algae and higher plants in bright weather may increase pH values above 8 to 9 or 10. This illustrates the interdependence of pH with the carbonate system, since the reduction in carbon dioxide levels as a result of photosynthesis may cause marked rapid changes in pH.

Some difficulty may be experienced in measuring pH near the extremes of the range, especially in nutrient-poor acid waters where pH is susceptible to rapid fluctuations as a result of carbon dioxide exchange.

5.3.2 Sources of Change in pH

Changes in pH may be brought about naturally, for example by drainage from peat moorlands which may overlie inert sandstones or granites where the contribution of humic acids gives rise to acid streams with very low pH (since there is virtually no buffering capacity because of the generally low mineral content). The situation is sometimes compounded by acid rainfall, arising from industries located nearby because of the availability of soft water from moorland drainage. Higher pH values are associated with run-off from calcareous rocks (chalk or limestone) and these may be increased by photosynthesis, as explained above.

Changes in pH which are attributable to human activity include the discharge of mineral acids or alkalis from industrial processes or, quite commonly, as run-off from mining spoil tips or discharge of minewater. Recently, concern has been expressed over the acidification of lakes and ponds in Scandinavia and eastern areas of Canada and the USA (e.g. Beamish, 1974) as a result of acid precipitation caused by sulphur dioxide discharges into the atmosphere from industrial activities, particularly the burning of fossil fuels. The problem has been particularly acute in areas where lake waters are naturally low in dissolved substances and hence have a low buffering capacity.

When considering the effects of changes in pH, especially extreme values, it is often difficult to dissociate the direct effects of hydrogen or hydroxyl (or hydroxonium) ions from related effects such as a toxic anion (e.g. acetic, tannic or other acids) or when the alkali, such as ammonia, is toxic as an undissociated molecule. Further complications arise when the change in pH is associated with other changes as, for example, in many acid mine drainage waters which are also ferruginous, and precipitation of iron hydroxide blankets the bed of the receiving water (see Section 5.3.4).

5.3.3 Effects of Extreme pH Values on Organisms
The direct effects of extreme pH are not easily demonstrated but it is believed that the main consequence is interference with normal physiological functions, especially those associated with respiration. It is thought that pH may affect the permeability of membranes. Fish placed in water of low pH show increased loss of sodium, and other ions, through gill membranes (Packer and Dunson, 1970; Fromm, 1980). Salt balance in invertebrates is likewise influenced by pH (Potts and Fryer, 1979).

Low pH may also affect the quantity of free carbon dioxide present, causing acidaemia. Some early work led to the conclusion that precipitation of mucus on the gill surface or precipitation of proteins within the gill epithelium caused death by suffocation (see EIFAC, 1969) although Lloyd and Jordan (1964) found no evidence for this in studies of the effects of low pH (4·5) on rainbow trout (*Salmo gairdneri*). Analysis of blood from the fish showed that increase in free carbon dioxide in the blood was unlikely to be important but the fish were clearly unable to maintain the bicarbonate alkalinity of the blood and death through acidaemia followed. Reduced oxygen consumption resulting from decreased oxygen transfer and lower oxygen-carrying capacity of the blood was observed in brook trout (*Salvelinus fontinalis*) exposed to acutely lethal low pH (Packer, 1979). Exposure to chronic acid stress causes haematological changes which presumably assist in restoring respiratory efficacy and allow fish to acclimatise to lower pH (Neville, 1979). Macan (1974) reviewed early work on the effects of pH including protozoan physiology (movement, osmoregulation etc.) and concluded that some protozoa are probably the only animals affected directly by pH. Certain indirect effects of pH are more readily identified. The toxicity of ammonia, for example, varies with pH, being more toxic at higher pH. Conversely, cyanide toxicity varies inversely with pH, as also does sulphide toxicity. Some heavy metals such as zinc are more toxic at higher pH values (Mount, 1966) while others, for example aluminium, appear to be more toxic at low pH (Haines, 1981).

5.3.4 Field Studies of the Effects of Low pH
Studies of the effects of pH in the field can conveniently be considered on the basis of the taxonomic category which is principally involved, namely microorganisms, macroinvertebrates and fish.

5.3.4.1 Microorganisms
Most studies of the effects of low pH in the field have been associated with acid mine drainage. The acidity is caused by the oxidation of pyrite assisted, it is believed, by bacterial action, in a series of chemical reactions which may be summarised as

$$2 \, FeS_2 + 7 \, O_2 \; + 2 \, H_2O \rightarrow 2 \, FeSO_4 + 2 \, H_2SO_4$$
$$\text{(pyrite + oxygen + water} \rightarrow \text{ferrous + sulphuric acid)}$$
$$\text{sulphate}$$

The ferrous iron may become oxidised and ferric hydroxide may be precipitated (see Section 5.1.3.7). The stages of pyrite oxidation and ferric iron precipitation are complex and for fuller details of the chemistry the reader is referred to Coutant *et al.* (1978).

The low pH of acid mine streams tends to reduce the viability of heterotrophic bacteria and, in particular, the survival of sanitary indicator bacteria such as *Escherishia coli* and *E. arogenes*, but some, for example *Streptococcus faecalis*, appear to be able to persist longer (Hackney and Bissonnette, 1978).

Iron bacteria are presumably active in most acid mine drainage streams. Although *Ferrobacillus ferrooxidans, Thiobacillus ferrooxidans* and *Metallogenium* are associated with pyrite degradation (Coutant *et al.*, 1978), the dependence of *T. ferrooxidans* on the creation of acidic conditions by *Metallogenium* (Walsh and Mitchell, 1972) for its own growth and the participation of these bacteria in the oxidation of ferrous iron suggests that they are an important component of the bacterial flora in acid streams immediately below the discharge of mine waters, especially where pH values are in the range 3·5 to 5.

The flora of a very acid stream studied by Harrison (1958), in which pH fell below 3, was dominated by rich growths of the moss *Sphagnum truncatum* in the pools and by the alga *Frustulia rhomboides* in the riffles.

Acid precipitation has been observed to cause increased growth of periphyton in streams (Hendrey, 1976), and in experimental studies using artificial acidification, Muller (1980) noted increased biomass (but not production) in littoral periphyton and a change in community composition in which diatoms were replaced by filamentous green algae. The genus *Mougeotia* was most tolerant of high acidity and tended to dominate the flora, which was characterised by fewer species and lower diversity. Reduction in the number of algal species was noted by Warner (1971) in streams receiving acid mine drainage. The reduction was greater where ferruginous deposits coated the river bed and greatest close to the mine discharge. The most tolerant species and most numerous in the reaches affected by acid were *Ulothrix tenerrima, Pinnularia termitina, Eunotia exigua* and *Euglena mutabilis*. Six other species, *Microthamnion strictissimum, Microspora pachyderma, Closterium acerosum, Chlamydomonas* sp., *Frustulia rhomboides* and *Surirella ovata*, were also tolerant of acid conditions. *Ulothrix tenerrima* formed dense mats below mine adits draining into a tributary of Turtle Creek, Pennsylvania, where pH values around 2·6 were experienced (Koryak *et al.*, 1972).

5.3.4.2 Macroinvertebrates

Very acid streams (pH sometimes below 3) were studied by Harrison (1958) in South Africa. These received sulphuric acid pollution from coal and gold mining and, in the reaches studied, were not affected by ferric hydroxide precipitation. The fauna was extremely restricted, consisting largely of the orbateid mite *Hydrozetes* and the chironomids *Pentapedilum anale* and *Chironomus linearis*. Other organisms present included *Argyrobothrus* (a hydroptilid caddis) and the chironomids *Lymnophyes spinosa* and *Tanytarsus pallidulus*.

In a study of streams affected by a variety of pollutants, including acid mine-drainage, Koryak *et al.* (1972) were able to distinguish the effects of low pH from those caused by organic enrichment and increased suspended solids load. They noted that, at sites of low pH (mean 2·6), the benthos was composed predominantly of midge larvae (*Tendipes gr. riparius* ≡ *Chironomus riparius*) and a few tipulid larvae while at pH 3·0 a few neuropteran larvae and Coleoptera were also present. The *Tendipes* larvae were not found in nearby streams with higher BOD loads although normally they are associated with heavy organic pollution. It may well be that this commonly observed association is not specifically related to the organic enrichment but rather to the ability of this species to exploit almost any stressed environment when there are virtually no competitors. In the stream studied by Koryak *et al.* (1972) it was extremely abundant in the dense mats of the filamentous alga *Ulothrix* which grew below the mine adit and peak larval biomass was 14·2 gm^{-2} during July. At other stations on Turtle Creek the precipitation of iron hydroxide and its deposition on the bed precluded any conclusions regarding possible recovery of the benthic community with increasing pH. In general, as the amounts of iron diminished and pH increased so the community increased, although it never reached the biomass of stations unaffected by acid mine drainage.

Studies of an acid mine drainage polluted stream, Roaring Creek in West Virginia, also revealed extremely high densities (16 675 m^{-2}) of the chironomid larvae, *Tendipes* (*Chironomus*) *plumosus*, at median pH values of 2·8 (Warner, 1971). Alderfly larvae (*Sialis* sp.), dytiscid beetles and other chironomids were also present, while in stiller water the caddis, *Ptilostomis* sp., could be found. Warner noted that above a median pH value of 4·5 the Simuliidae, Ephemeroptera, Plecoptera and Trichoptera were well represented. This observation agrees with the findings of Cherry *et al.* (1979) who studied the recovery of benthic communities following discharge of acid water from a fly-ash settling lagoon.

The effects of both constant and intermittent acid mine drainage on the insect fauna of some western Pennsylvania streams were studied by Roback and Richardson (1969). These studies showed that under conditions of constant acid mine drainage the Odonata, Ephemeroptera and Plecoptera were eliminated and Trichoptera, Megaloptera and Diptera were represented by fewer species. Species tolerant of these conditions included the caddis fly *Psilostomis*, *Sialis* and *Chironomus attenuatus*. Certain Hemiptera and Coleoptera were present in large numbers. In a stream affected by intermittent acid mine drainage the insect fauna

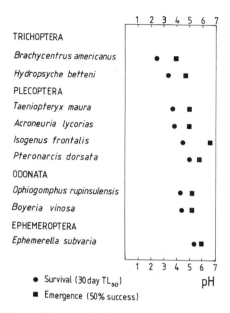

FIG. 5.14 Effect of low pH on survival and emergence of nine North American aquatic insects. Data from Bell (1971).

differed little from similar unpolluted streams except for the absence of some sensitive Ephemeroptera and Diptera.

Experimental studies of the effect of low pH and the survival and emergence of North American aquatic insects has been undertaken by Bell (1971). All the nine test species survived low pH during the 30 day tests although they were more sensitive to pH during emergence. The most tolerant species was *Brachycentrus americanus* and the most sensitive *Ephemerella subvaria* (Fig. 5.14). In general, caddis flies were very tolerant of low pH, stoneflies and dragonflies moderately tolerant and mayflies quite sensitive. Bell noted that these laboratory-determined tolerances were corroborated by field observations in that the mayfly *Ephemerella* was not collected in waters with a pH lower than 5·5 and that the dragonfly *Ophiogomphus* was present at a pH of 4·2 but was eliminated when the pH dropped to 3·4 (Parsons, 1968).

The fauna of the Nant Gyrawd, a tributary of the Taff Bargoed in South Wales, was subject to acid drainage from coal stockpiles which reduced the pH values below 3·5 (Scullion and Edwards, 1980a). Only the mayfly *Baetis rhodani* and the chironomid *Conchapelopia pallidula* were able to survive these conditions. Scullion and Edwards (1980a) have provided a list of acid tolerant European species which includes *Baetis rhodani, Amphinemura sulcicollis, Leuctra hippopus, Plectrocnemia conspersa, Rhyacophila dorsalis, Hydropsyche pellucidula*, Ceratopogonidae (*Bezzia/Palpomyia* group), Tipulidae, *Simulium ornatum, Tabanus* sp., *Conchapelopia pallidula, Microspectra* sp., *Tanytarsus* sp., *Rheocricotopus foveatus, Brillia modesta*, Orthocladiinae including *Prodiamesa olivacea, Limnius volckmari* and *Hygrobates fluviatilis*.

North American invertebrate species which are tolerant of low pH are given in Table 5.6.

TABLE 5.6
NORTH AMERICAN INVERTEBRATES TOLERANT OF LOW pH
(Data mainly from Hart and Fuller, 1974, with additions)

Species recorded at pH values:

(1) less than 3·5	(2) between 3·5 and 4·5 (inclusive)	(3) between 4·6 and 5·5
Hemiptera	Spongillidae	Spongillidae
Hesperocorixa	*Anheteromeyenia*	*Corvomeyenia everetti*
Megaloptera	*argyrosperma*	*Eunapius mackayi*
Sialis sp.	*A. ryderi*	*Spongilla lacustrio*
Coleoptera	*Eunapius fragilis*	*Trochospongilla horrida*
Acilius spp.	*Heteromeyenia*	Tricladida
Agabus spp.	*baileyi*	*Dugesia dorotocephala*
Agaporus spp.	Astacidae	*D. tigrina*
Hydroporus spp.	*Cambarus*	*Phagocata velata*
Ilybius spp.	*longulus*	Hirudinea
Laccophilus spp.	Gastropoda	*Dina fervida*
Helophorus spp.	*Campeloma*	*Erpobdella punctata*
Hydrobius spp.	*decisum*	*Glossiponia complanata*
Trichoptera	Hemiptera	*Helobdella stagnalis*
Ptilostomis sp.	*Callicorina audeni*	Odonata
Diptera	*Gernis marginatus*	*Ophiogomphus rupinsulensis*
Chironomus	Megaloptera	*Boyeria vinosa*
attenuatus	*Nigronia* sp.	*Plathemis lydia*
Chironomus	Coleoptera	Ephemeroptera
riparius	*Bidessus* spp.	*Isonychia* sp.
Polypedilum	*Enochrus* spp.	*Heptagenia* sp.
illinoense	Trichoptera	
Polypedilum	*Brachycentrus*	
scalaeaum	*americanus*	

TABLE 5.6 —*contd.*

Species recorded at pH values:

(1) less than 3·5	(2) between 3·5 and 4·5 (inclusive)	(3) between 4·6 and 5·5
	Diptera	Stenonema spp.
	Tipula abdominalis	*Paraleptophlebia* sp.
	Procladius bellus	*Caenis* spp.
	Cardiocladiues	Plecoptera
	obscurus	*Pteronarcys* spp.
		Pteronarcys dorsata
		Perlesta placida
		Acroneuria abnormis
		Phasganophora capitata
		Hemiptera
		Gerris canaliculalus
		Trepobates spp.
		Megaloptera
		Corydalis cornutus
		Coleoptera
		Haliplus spp.
		Peltodytes spp.
		Copelatus spp.
		Dineutes spp.
		Gyrinus spp.
		Tropisternus spp.
		Helichus fastigiatus
		Ancyronyx variegatus
		Machronychus glabratus
		Dubiraphia quadrinotata
		Stenelmis crenota
		Lepidoptera
		Paragyractis spp.
		Trichoptera
		Neureclipsis crepuscularis
		Polycentropus crassicornis
		Hydropsyche betteni
		Diptera
		Clinotanypus pinguis
		Ablabesmyia monilis
		Psectrocladius elatus
		Rheorthocladius sp.
		Harnischia pseudotener
		Rheotanytarsus exiguus
		Calopsectra guerla
		Chrysops spp.

5.3.4.3 *Fish*

More than a decade ago, the European Inland Fisheries Advisory Commission Working Party on Water Quality Criteria for European Freshwater Fish published a comprehensive critical review of published and unpublished data on the direct and indirect effects of extreme pH values on fish (EIFAC, 1969) in order to determine the range of pH values within which one may reasonably expect to be able to develop an acceptable fishery. This review established that one could not define a pH range in which a fishery is unharmed and outside which it is damaged, but it was clear that a gradual deterioration set in the further pH values were removed from those obtaining in most natural freshwaters. The range which was not directly lethal to fish was found to be from pH 5 to pH 9, though since the toxicity of certain common pollutants (such as ammonia and cyanide) is modified by changes in pH it is quite possible that changes in pH within this range could have significant indirect effects (Fig. 5.15).

Addition of acid may release sufficient carbon dioxide from the bicarbonate in the water to kill the fish even though the pH level itself would not be directly lethal. Lloyd and Jordan (1964) noted that one of the important conclusions to be reached from their investigations of factors which affected the resistance of rainbow trout to acid waters was the effect of sub-lethal concentrations of free carbon dioxide. They considered that this alone could account for the considerable variation in the lethal pH values which were quoted in the existing literature and observed that their own work (which had determined the 24 hour LC_{50} pH value to be 3·6 when little free carbon dioxide was present and the 15 day LC_{50} pH value to be 5·6 in the presence of 50 ppm free CO_2) covered most of the pH values said to be toxic by other authors.

EIFAC (1969) considered that although some species may become acclimated to pH values as low as 3·7, in general, whenever the pH falls below 5·0, some mortalities may be expected. Since, however, general productivity in acid environments is low, the yield from any fishery in such environments is also likely to be low.

Studies of the effects of acid waters on the hatching of salmonid eggs have shown that for both salmon (*Salmo salar*) and trout (*S. trutta*) a pH of 3·5 was lethal within 10 days but at higher pH (4·5 and above) no effect of acidity on hatching could be detected (Carrick, 1979). At pH 4·0–5·5 Peterson *et al.* (1980) found that hatching of 'eyed' salmon eggs was prevented or delayed but if the eggs were transferred to water of higher pH (6·6–6·8) then hatching occurred.

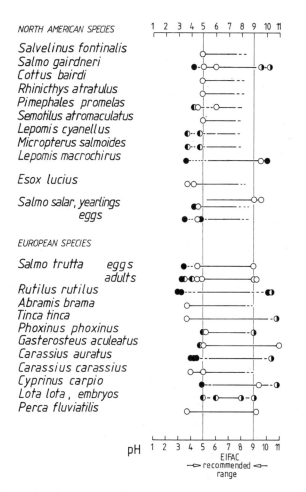

NORTH AMERICAN SPECIES

Salvelinus fontinalis
Salmo gairdneri
Cottus bairdi
Rhinicthys atratulus
Pimephales promelas
Semotilus atromaculatus
Lepomis cyanellus
Micropterus salmoides
Lepomis macrochirus

Esox lucius

Salmo salar, yearlings
eggs

EUROPEAN SPECIES

Salmo trutta   eggs
adults
Rutilus rutilus
Abramis brama
Tinca tinca
Phoxinus phoxinus
Gasterosteus aculeatus
Carassius auratus
Carassius carassius
Cyprinus carpio
Lota lota , embryos
Perca fluviatilis

pH 1 2 3 4 5 6 7 8 9 10 11
EIFAC
recommended range

FIG. 5.15 Variation in tolerance of extreme pH values by freshwater fish. Based on data in the literature. ● = death (acute response), ◐ = survival for a limited period (chronic response), ○ = indefinite survival.

5.3.5 Acid Precipitation

5.3.5.1 Introduction

Acid precipitation, or 'acid rain' as it is more usually called (Fig. 5.16), has come to prominence recently although the phenomenon was re-cognised in the middle of the nineteenth century (Smith, 1872; Cowling,

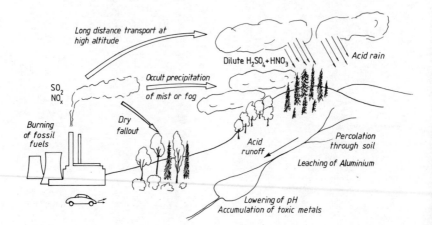

FIG. 5.16 Diagrammatic representation of the factors which are thought to contribute to 'acid rain', and its ecological consequences.

1982). During the late nineteenth century and the first half of the twentieth century, the effects of air-borne pollution were evident locally. The 'pea-souper' London fog (or more properly 'smog', i.e. smoky-fog) was an inevitable, though unpleasant, part of normal life in the English capital and similar conditions prevailed in many other large cities. In the 1950s a 'clean air' campaign reduced domestic smoke emissions and those industries which burned fossil fuels, and in particular the electric power generators, were required to construct tall chimney stacks to ensure effective high-level dispersion and much effort was expended in cleaning waste gases before their discharge. These measures greatly reduced the local effects; soot and grime were less evident and the pollution of city air by sulphur dioxide, a factor which contributed to the high incidence of bronchitis and related diseases in major conurbations, fell to insignificant levels. It now appears that the problem may simply have been transported elsewhere.

5.3.5.2 Geochemical contributory factors
Work by Goreham (1958a, 1958b) showed that rain which fell in the English Lake District contained higher than normal amounts of sulphate and nitrate when the prevailing wind passed over industrialised regions. The rain was usually more acidic than pH 4·5, which may be compared with that of rainfall in areas distant from industrial pollutants which usually has a pH value above 5·0. It is to be expected that natural rainfall

will be acidic since the approximate pH of distilled water in equilibrium with an atmospheric concentration of carbon dioxide is 5·6. 'Acid rain' is an appropriate description, therefore, of precipitation which has a pH below 5·6 although the significance of the extent to which it falls below this value will depend on many other factors, as explained below. Much of Europe and North America receives precipitation with a pH below 4·7 (Haines, 1981). Several trends have been observed in the chemistry of surface waters in some regions over recent decades; pH has fallen along with bicarbonate ion concentrations while aluminium and sulphate ion concentrations have increased. Another interesting paradoxical change has been the increasing nutrient content of rain (nitrogen and sulphate) coupled with the tendency to oligotrophic conditions induced by increasing acidity (Falkenmark, 1980).

The effects of acid rain are most evident in regions where streams drain igneous, metamorphic or hard sedimentary rocks and the water is relatively pure, that is, it contains very small amounts of dissolved substances such as calcium and magnesium which would provide 'buffering' capacity against the effects of acid precipitation. Such waters are often naturally acidic since the soils they drain are thin and poor or the location and climate is conducive to the accumulation of peat. Humic acids from the peat may contribute to the natural acidity. In these situations relatively small changes in the acidity of precipitation can cause marked changes in the pH of streams and highly oligotrophic lakes. Waters with an alkalinity of 20 mg litre^{-1} as $CaCO_3$, or less, are likely to be vulnerable to the effects of acid precipitation while those below 10 mg litre^{-1} can be regarded as particularly sensitive and almost certain to be affected, especially where the natural pH is low.

5.3.5.3 Biological contributory factors

Vegetation also influences the extent to which acid precipitation exerts significant effects, both directly and indirectly. Dry deposition of acidic substances on vegetation together with the interception of fog or mist by trees, especially by conifers—the so-called 'occult' precipitation—contributes to the total amount of acidic fall-out (Dollard *et al.*, 1983). This interception by forests plays an important role in intensifying the importance of acid precipitation. For example, the acidity of coniferous stem flow (that is, the water flowing down the trunks of trees) may increase the acidity of the rain eight-fold (Last and Nicholson, 1982). Two indirect effects of trees tend to exacerbate the impact of acid precipitation: diminution of run-off which occurs as a result of evapotranspiration from trees in afforested areas (Newson, 1979) reduces the amount of water available to dilute the acid deposits, while the uptake of nutrients such as calcium and magnesium by growing trees tends to reduce the buffering capacity of the soil. The above factors may help to explain why the deleterious effects of acid rain have been particularly evident in areas of coniferous forests, in Scandinavia, central Europe and the north-eastern areas of North America. It could well be that those features which contribute to the development of forests (high rainfall, poor shallow soil on slowly weathering rocks, steep gradients, high altitude, all of which are detrimental to agricultural development) are also likely to heighten the impact of acid precipitation.

5.3.5.4 Ecological consequences

The ecological significance of acid precipitation is not merely the reduction in pH of the receiving streams, although in extreme circumstances pH could be lowered to the point where it injured fish and other organisms directly. More serious is the change in ionic composition which occurs as acid precipitation percolates through the soil. Cations, principally calcium, magnesium and aluminium, are released in exchange for hydrogen ions. Where soils are poor in base cations such as calcium or magnesium, large amounts of aluminium may be released when acid rainfall comes into contact with alumino-silicate clay minerals. The increased concentration of aluminium in water entering streams or lakes is thought to be the prime factor in causing fish kills in acid lakes. The input of contaminants varies seasonally and in some areas appears to become particularly acute following spring snowmelt. Other areas experience peaks of acidity in autumn, probably as the onset of rains washes material deposited during summer. Many areas are subjected to highly variable acidity depending on whether the storm track includes industrial or urban areas.

The ecological effects of acid precipitation entering surface waters are varied and have been extensively reviewed by Haines (1981). Acidification appears to depress the normal decomposition of organic matter, probably by eliminating some kinds of bacteria or fungi, and one consequence of this has been an increase in algal biomass in lakes. Macrophyte communities are also modified, usually to become dominated by the moss *Sphagnum* which, it is thought, may actually contribute towards increasing the acidity. In lakes, zooplankton diversity is usually reduced and may become largely restricted to the genera *Bosmina* or *Diaptomus* and a few acid-tolerant rotifers. Benthic macroinvertebrate communities are also greatly diminished and, as might be anticipated, the mollusc fauna is particularly impoverished. Crustacea also tend to be absent from acidic waters but the responses of various groups of insects varies widely. For example, Ephemeroptera and, to a lesser extent, Plecoptera become less numerous as pH declines while Corixidae, Megaloptera and Coleoptera may be far more abundant in acid waters. The causes of these shifts are probably complex: some may suffer direct physiological damage or be affected by increases in toxic metals while, in others, the effect may be an indirect one through altered food resources or predator–prey relationships. The absence of fish from any highly acid waters could be responsible for at least some of the observed ecological changes. Fish mortalities have been observed at times of rapid change in pH, such as autumnal storms or at snowmelt, but the cause of death is most probably simultaneously elevated concentrations of aluminium. Low pH may be responsible for chronic toxicity by affecting the normal homeostatic mechanisms for ion regulation or, when pH is very low (less than pH 3·5), by causing respiratory failure. The long-term survival of fish populations in waters affected by acid precipitation probably depends most on their reproductive success since it is now evident that some life history stages are more susceptible than adult fish to acid stress, or to exposure to increased concentration of toxic metals.

5.3.5.5 *Economics of remedial action*
Remedial action to solve the problem of acid rain is likely to be prohibitively expensive since the economies of industrialised nations are far from buoyant and energy costs are an important factor in their survival. Economic losses sustained by areas which receive acid precipitation may be high but the cost of emission control could be much higher, although firm figures are not readily obtainable. It has been estimated that a 50% reduction in European emissions would cost 5 billion dollars a year but would only increase the pH of Scandinavian lakes by 0·2 units (Brown and Sadler, 1981). Palliative measures have been used, with varying success, on the receiving streams and lakes. These include attempts at neutralising the acid by applications of limestone or hydrated lime to catchments or waters, introductions of acid-resistant species or strains of fish, or repeated stocking.

The problem of acid precipitation is not a simple one and although it is currently an area of active research, much evidence is conflicting and many links in the processes involved are tenuous and very imperfectly understood. Research workers supported by organisations which may be responsible, and which certainly have been blamed, for acid precipitation have shown a healthy scepticism towards some of the work which purports to demonstrate their guilt. One cannot help speculating that much apparently well-founded science might be capable of alternative explanation were it to be associated with highly sensitive political and economic issues.

CHAPTER 6

The Effects of Organic Enrichment

6.1 INTRODUCTION

The discharge of organic matter or nutrients is probably the commonest, the most fully documented and the best understood form of pollution. Indeed, organic enrichment is the classical form of pollution and there is a common tendency to speak of 'pollution' when organic enrichment is intended.

The main source of organic enrichment is the treatment of domestic sewage but industries such as food processing (brewing, dairies and milk processing, vegetable canning, etc.) also make a significant contribution.

Depletion of dissolved oxygen is the principal effect of discharge of organic matter and results from heterotrophic utilisation by microorganisms. The 'strength' of organic effluents is commonly measured by means of the BOD (biochemical oxygen demand) test. Usually the oxygen concentration is measured before and after the sample has been incubated at 20 °C for 5 days in the dark (to avoid any autotrophic metabolism). The result of this test is expressed as the quantity of oxygen used in the degradation of the organic matter. Where necessary, the sample may be diluted with well-oxygenated water before incubation to avoid complete depletion of the oxygen before the test ends. Interference from the oxidation of other substances may occur. The commonest interfering compound associated with organic effluents is ammonia: this may be suppressed by allyl thiourea (ATU). Examples of the BOD values of common effluents are given in Table 6.1. It should be noted that biodegradation may continue beyond 5 days, perhaps up to 20 days, depending upon the ease with which substances may be assimilated. It may be, therefore, that the actual influence of an organic pollutant may differ from that implied by a BOD test.

TABLE 6.1
APPROXIMATE BIOCHEMICAL OXYGEN DEMAND (BOD) OF TYPICAL EFFLUENTS AND FRESHWATERS

		Range of BOD (mg litre^{-1})	Reference
Natural water,	upland streams	0·5–2·0	
	lowland streams	2·0–5·0	
	large lowland rivers	3·0–7·0	
Sewage effluents,	crude sewage	200–800	Klein (1959)
	treated	3–50	
Farm wastes,	pig	27 000–33 000	Weller and Willetts (1977)
	poultry	24 000–67 000	Weller and Willetts (1977)
Silage liquor		60 000	Weller and Willetts (1977)
Abbatoir		650–2 200	Azad (1976)
Meat packaging and processing		200–3 000	Azad (1976, Callely *et al.* (1977)
Fruit canning		635–2 100	Azad (1976, Callely *et al.* (1977)
Vegetable processing		480–4 400	Callely *et al.* (1977)
Sugar beet		3 800–4 200	Azad (1976)
Sugar refining		210–1 700	Callely *et al.* (1977)
Dairies, milk		300–2 000	Callely *et al.* (1977)
cheese		1 800–2 000	Callely *et al.* (1977)
Breweries		500–1 300	Callely *et al.* (1977)
Distilleries		over 5 000	Callely *et al.* (1977)
Tannery		250–5 000	Callely *et al.* (1977)
Textile waste		50–1 000	Callely *et al.* (1977)
Paper making		100–400	Callely *et al.* (1977)
Petrochemicals		200–8 000	Azad (1976)

6.2 GENERAL EFFECTS OF ORGANIC DISCHARGES

The main ecological effects of organic effluents in rivers are to change the biological community composition and the abundance of organisms below the point of discharge through drastic lowering of the dissolved oxygen content, modifications of the substrate (for example, by deposition of organic sludges) and by addition of nutritive material which may favour the dramatic increase in numbers of certain organisms. The relative importance of these factors depends, among other things, on the physical characteristics of the watercourse. Typical changes have been described by Bartsch (1948) and these are shown diagrammatically in Fig. 6.1 where it will be seen that successional (temporal) changes occur which manifest themselves as spatial changes along the river.

6.3 PHYSICAL AND CHEMICAL CHANGES

In outline, the physical and chemical changes which result from organic discharges are largely predictable and are attenuated by mixing and dilution. For example, the high suspended solids load is dissipated and settles while continued surface re-aeration through turbulence ensures that oxygen levels return to those present above the outfall. When these changes are plotted as a graph with dissolved oxygen levels on the ordinate and distance downstream on the abscissa, the resulting line is an 'oxygen-sag' curve. Settled organic matter imposes little biochemical oxygen demand on the water column and attainment of normal dissolved oxygen levels is accelerated once the demand imposed by microbial metabolism is reduced by the exhaustion of the organic substrate. Similarly, oxidation of ammonia to nitrate can proceed as oxygen levels are restored. Many of the changes shown in Fig. 6.1 are reciprocal effects. For example, high BOD is accompanied by low dissolved oxygen while a fall in the levels of suspended solids is associated with a build-up of sludge deposits. Some 'conservative' substances, for example inorganic salts, principally sodium chloride, are reduced in concentration simply by dilution. Phosphates, largely from detergents which are associated with organic enrichment from sewage wastes, appear to behave similarly. Although necessary, along with nitrates, as plant nutrients, these substances are commonly found at concentrations which far exceed the assimilative capacities of the flora as their growth is limited by other factors. Aquatic plants are rarely nutrient limited (Westlake, 1968; Ladle

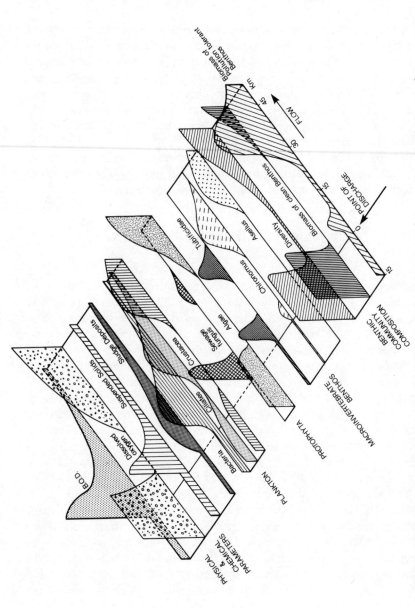

Fig. 6.1 Spatial variation of physical, chemical and biological consequences of the continuous discharge of a severe organic load into flowing water. After Bartsch (1948).

and Casey, 1971) and, therefore, nutrient losses by photosynthetic assimilation are likely to be too small to effect reductions in concentrations in river water although this may occur in small streams which support dense macrophyte growths (Westlake, 1973).

6.4 BIOLOGICAL CHANGES

Many of the biological phenomena occur in two phases, largely associated with the rapid reduction in dissolved oxygen which occurs below the discharge point and subsequent recovery as putrescible matter is oxidised and re-aeration occurs. One may identify a series of successional changes in the composition of the biological community which follow these phases. Several workers have described these successional changes and have also used them for assessing the pollutional status of a given location (Fig. 6.2). The classical description is the 'saprobien system' (Kolkwitz and Marsson, 1908, 1909). Four stages are recognised: *polysaprobic*, the condition immediately below the organic discharge in which large quantities of rapidly decomposing matter (albumens, polypeptides and carbohydrates) reduce oxygen levels and may lead to anaerobic production of hydrogen sulphide and, hence, unpleasant odour with accumulation of black sludge deposits, bacteria and protozoa abound; α- and β-*mesosaprobic*, two stages in the recovery process, to be described in more detail; and *oligosaprobic*, the condition of full recovery and which approaches the upstream condition of the river. The α-mesosaprobic zone is the first stage of recovery in which amino-acids are abundant, bottom muds are no longer black and odour nuisance ceases. Bacterial counts are still high and the fauna is restricted. The β-mesosaprobic is a further stage in the oxidation and mineralisation of the added material and is characterised by ammoniacal compounds of fatty acids. Bacterial counts decline and the diversity of both flora and fauna increases, although this is less than in oligotrophic zones. This last zone is the zone in which oxidation and mineralisation has been completed. The additional mineral content may encourage luxuriant growth of plants. In large rivers this will take the form of enhanced phytoplankton while in others some benthic algae and macrophytes may be stimulated. Of these, the alga *Cladophora* is quite characteristic.

Other classification schemes, which describe the zones in terms of the prevailing conditions such as 'degradation', 'active decomposition', 'septic', 'contaminated' or 'recovering', have been compared by Warren (1971) and are shown in Fig. 6.2.

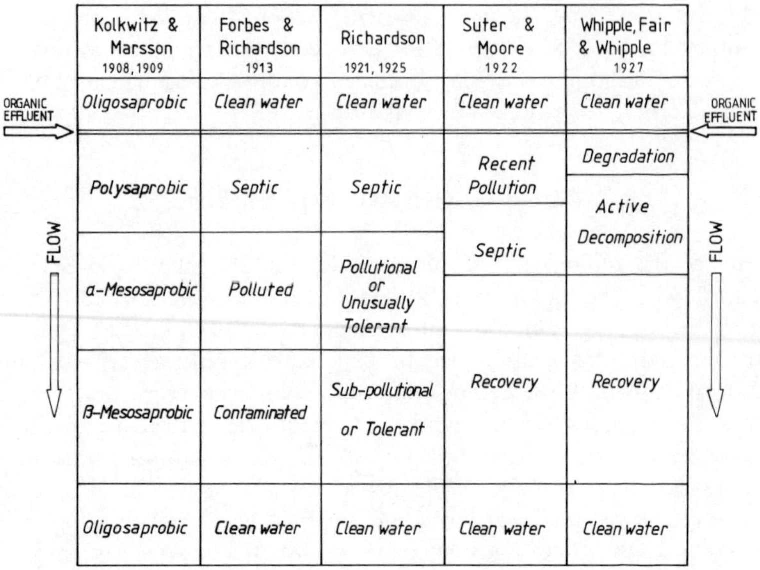

	Kolkwitz & Marsson 1908, 1909	Forbes & Richardson 1913	Richardson 1921, 1925	Suter & Moore 1922	Whipple, Fair & Whipple 1927	
ORGANIC EFFLUENT →	Oligosaprobic	Clean water	Clean water	Clean water	Clean water	ORGANIC EFFLUENT
FLOW	Polysaprobic	Septic	Septic	Recent Pollution	Degradation	FLOW
				Septic	Active Decomposition	
	a-Mesosaprobic	Polluted	Pollutional or Unusually Tolerant			
	β-Mesosaprobic	Contaminated	Sub-pollutional or Tolerant	Recovery	Recovery	
	Oligosaprobic	Clean water	Clean water	Clean water	Clean water	

FIG. 6.2 Comparative chart of several early systems for classifying the stages of recovery from the effects of an organic effluent in flowing water. After Warren (1971).

Successions in the flora and fauna outlined in Fig. 6.1, and described in more detail below, concern the appearance of abundant bacterial and other saprophytic populations (collectively called 'sewage fungus'), the development of large numbers of protozoa (feeding on the bacteria), changes in the composition of macroinvertebrate fauna and the development of algae and macrophytes as mineralisation proceeds.

6.4.1 'Sewage Fungus' and the Associated Organisms

The development of profuse growths of 'sewage fungus' is a characteristic indication of organic discharges. 'Sewage fungus' is not a single organism but a community of heterotrophic microorganisms which forms macroscopic, slimy, furry growths (sometimes not unlike plumes of cotton wool) on the river bed, banks, tree roots and any other solid surface. An account of studies of the composition and physiology of 'sewage fungus' has been provided by Hawkes (1962b) and Curtis (1969), the latter providing useful illustrations of important and common constituent organisms.

In a study of the occurrence of 'sewage fungus' in rivers in the United Kingdom, Curtis and Harrington (1971) noted that most outbreaks (75%) extended for less than 1 km downstream from the point of discharge, although 1% extended for more than 8 km. They also found that most frequently the effluents associated with outbreaks of 'sewage fungus' were from domestic sewage, although industrial effluents, especially those from distilleries, malting, fruit and vegetable canning, paper manufacture, textile and dyeing processes, were also important. Less common were agricultural wastes (e.g. piggery discharges) and refuse tip discharges. In attempting to relate the intensity of 'sewage fungus' growth to the quality of the effluent, it was found that heavy growths were commonly associated with BOD values in the range 5–30 mg litre^{-1} and with soluble organic carbon levels of 6–30 mg litre^{-1}. Outbreaks occurring where effluents are below these values are often associated with the presence of rapidly biodegradable materials (for example, sugars). Industrial effluents which support 'sewage fungus' frequently include readily degradable food sources. Food processing yields waste liquors which contain sugars, organic and amino acids; paper manufacture provides starch and some soluble wood polysaccharides; and textile processing releases starch, organic acids and esters. Below sewage works, the presence of 'sewage fungus' indicates inadequate treatment, either through overloading of the works or lack of full secondary biological treatment. Provision of the latter would favour the removal of readily biodegradable substances and reduce the probability that the effluent would contain sufficient concentration to support growths of 'sewage fungus'.

The 'sewage fungus' community comprises many species and varies in its composition: in an analysis of slime-forming organisms at 178 sites Curtis and Curds (1971) found 7 species of bacteria, 6 species of fungi, 21 species of algae and about 82 species of protozoa. There are, however, only a few species which are almost always present (see Table 6.2) and of these the bacterium *Sphaerotilus natans* Kutz. together with zoogleal bacteria (commonly *Zooglea ramigera*, *Pseudomonas* spp. and zoogleal forms of *Sphaerotilus*) are regarded as the most frequent organisms to form 'sewage fungus'. Other important components are the bacteria *Beggiatoa alba* (Vauch) and *Flavobacterium* sp.; the fungi *Geotrichum candidum* Link., *Leptomitis lacteus* Agardh and *Fusarium aquaeductum* (Radl & Rab.) Sacc.; the alga *Stigeoclonium tenue* Kutz. and the diatoms *Navicula*, *Fragilaria* and *Synedra*; and the ciliated protozoan *Carchesium polypinum* Linn. (Curtis and Harrington, 1971).

TABLE 6.2

LIST OF TYPICAL SEWAGE FUNGUS ORGANISMS

According to Curtis and Curds (1971)	According to Butcher (1932)
Bacteria	
[a]*Sphaerotilus natans*	[b] { *Sphaerotilus natans* / *Cladothrix dichotoma*
[a]*Zoogleal bacteria*	*Zoogloea ramigera*
[a]*Beggiatoa alba*	*Beggiatoa alba*
[a]*Flavobacterium* sp.	
Fungi	
[a]*Geotrichum candidum*	
[a]*Leptomitus lacteus*	*Leptomitus lacteus*
	Fusarium aquaeductum
	Penicillium fluitans
Ciliates	
Colpidium colpoda	
Colpidium campylum	
Chilodonella cucullulus	
Chilodonella uncinata	
Cinetochilum margaritaceum	
Trachelophyllum pusillum	
Paramecium trichium	
Paramecium caudatum	
Uronema nigricans	
Hemiophrys fusidens	
Glaucoma scintillans	
[a]*Carchesium polypinum*	*Carchesium lachmanni*
	Carchasium spectabile
Algae	
[a]*Stigeoclonium tenue*	
Navicula spp.	
Fragilaria spp.	
Synedra spp.	

[a] Indicates a common 'sewage fungus' organism.
[b] *Cladothrix dichotoma* is now generally regarded as a form of *S. natans*.

In a detailed study of the composition of the slime community, Curtis and Curds (1971) used cluster analysis to elucidate the species interrelationships and noted the high degree of association between *Sphaerotilus* and zoogleal bacteria. Other associations were attributed to predator–prey relationships or similar habitat requirements. Figure 6.3 shows the community relationships in the form of a minimum-spanning

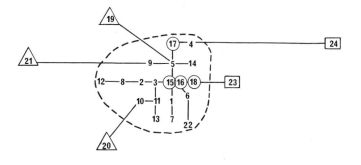

ALL ARE CILIATES EXCEPT

BACTERIA   ◯

FUNGI   △

ALGAE   ▢

1	*Trachelophyllum pusillum*	13	*Aspidisca lynceus*
2	*Chilodonella cucullulus*	14	*Glaucoma scintillans*
3	*C. uncinata*	15	*Sphaerotilus natans*
4	*Paramecium caudatum*	16	*Zoogloea*
5	*Colpidium colpoda*	17	*Beggiatoa alba*
6	*Hemiophrys fusidens*	18	*Flavobacterium* sp.X
7	*Cinetochilum margaritaceum*	19	*Leptomitus lacteus*
8	*Paramecium trichium*	20	*Geotrichum candidum*
9	*Colpidium campylum*	21	*Fusarium aquaeductum*
10	*Uronema nigricans*	22	Sessile ciliates
11	*Tachysoma pellionella*	23	*Stigeoclonium tenue*
12	*Litonotus fasciola*	24	*Cladophora glomerata*

FIG. 6.3 Minimum spanning tree of the composition of the 'sewage fungus' slime community. The distances between components indicate their closeness of association and the dotted line marks the boundary of the main community. After Curtis and Curds (1971).

tree (Gower and Ross, 1969) in which the distance between neighbouring species indicates the closeness of their association. The bacteria and ciliates form a homogeneous group (within the dotted line) centred around *Sphaerotilus* and the zoogleal bacteria. Fungi and algae are not associated with this group and Curtis and Curds (1971) concluded that the algae are probably not true 'sewage fungus' components but are associated with it and more typically represent the flora of a zone downstream of that of 'sewage fungus'. *Stigeoclonium tenue*, a filamentous alga, known to be tolerant of organic and inorganic pollution and occasionally a dominant organism in 'sewage fungus', may be an exception. The autecological characteristics of the many organisms which comprise part of the 'sewage fungus' community or are associated with it are considered below according to their taxonomic groupings.

6.4.1.1 Bacteria

(a) *Sphaerotilus natans*

Sphaerotilus natans is a Gram-negative, non-sporing sheathed bacterium (Chlamydobacteriaceae) which usually forms filaments of rod-shaped cells with rounded ends. The sheath may contain deposits of inorganic material and in this, and related species, the deposits may be of ferric hydroxide which has given rise to the genetic term 'iron bacteria' for the group. Some bacteria (e.g. *Cladothrix dichotoma* and *Leptothrix ochracea*) are now known to be forms of *Sphaerotilus natans* (Curtis, 1969). On reviewing the taxonomy of these bacteria Pringsheim (1949a, b) concluded that most of the group were forms of *S. natans* and the diverse morphological characters were a reflection of differing environmental conditions. Five forms of *S. natans* were distinguished: *eutrophica*, the normal form, rarely branching; *dichotoma* in less enriched water, with regular false branching; *ochracea*, heavy yellow or brown sheath with few cells; *sideropus*, iron or manganese encrusted short filaments and *patens* with short, club-like branches.

Sphaerotilus is characteristically an organism of flowing water and studies of the relationship between flow (velocity) and nutrient status have shown that dense growths occur at low nutrient levels in flowing water. It appears that slime growth is a function, not of the concentration of nutrients, but of the quantity (i.e. load) of nutritive material passing over the surfaces of the slime. Velocities of 0·1 to 0·6 m s^{-1} are associated with good growths under river conditions.

At high velocities, for example during river spates, *Sphaerotilus* is removed but may become re-established further downstream. In large rivers *Sphaerotilus* may not be attached to solid surfaces but can form a benthic 'plankton' carried along near the bottom.

Although laboratory studies suggest that optimum temperatures for *Sphaerotilus* growth are high (around 30 °C), 'sewage fungus' is typically a winter phenomenon in temperate rivers. This discrepancy may be explained by synecological factors: competition from other organisms may suppress *Sphaerotilus* in summer while in winter its ability to grow at low temperature (even below 4 °C) may give it an advantage. Effluents tend to deteriorate in winter, when lower temperatures make for less efficient treatment, and the additional food sources which are thus provided may also contribute to increased winter *Sphaerotilus* 'blooms'.

Sphaerotilus growth is favoured by alkaline conditions and it is known that calcium is necessary for sheath development. Optimum pH ranges from around 6·8 or 7·0 to about 8·0 and perhaps even 9·0.

Under field conditions *Sphaerotilus*-dominated slimes are usually found at medium to high dissolved oxygen levels (3–11 mg litre^{-1}) and apparently never below 2·5 mg litre^{-1}, but this may be simply a correlation with velocity. This observation finds support in laboratory studies, by means of recirculation channels, of the growth of *Sphaerotilus* under different physical conditions and nutritional status which revealed that initially dissolved oxygen concentrations below 5 or 6 mg litre^{-1} were necessary for the development of slimes (Curtis *et al.*, 1971). Once initiated, the *Sphaerotilus* slimes could be maintained at higher dissolved oxygen levels.

Organic nitrogen compounds are thought to be necessary for luxuriant *Sphaerotilus* growth although the only specific requirement appears to be trace amounts of vitamin B_{12}. It is now known that *Sphaerotilus* will grow when only inorganic nitrogen is available (i.e. ammonium, nitrite and nitrate ions). Very low concentrations of phosphorus (less than 0·01 mg P per litre) were not found to promote growth, as previously thought. The biomass of the slime produced was proportional to the concentration of organic carbon in the influent (whether the source was glucose or acetate) and the growth limiting level of nutrient was about 1 mg C per litre. The rate of accumulation of slime was greater (about 1·8 g m^{-2}, dry weight, per day) when acetate was the substrate than when glucose was used (1·2 g m^{-2} per day).

(b) *Zoogloea species*

Zoogloea ramigera (Itzigshon) is a Gram-negative, non-sporing bacterium in which the cells are embedded in a gelatinous mass instead of in filaments. However, since many bacteria (including *Sphaerotilus*) may manifest this form, the taxonomy is confused and isolation of representative material is virtually impossible; there is likely to be some difficulty in establishing the true identity of the 'causative organism' of zoogleal growths. It seems preferable, therefore, to use the term 'zoogloea' to describe the growth form without implying that the identity of the genus is known.

(c) *Beggiatoa alba*

This bacterium forms unbranched filaments of rod-shaped cells, not unlike *Sphaerotilus*, but from which it may be distinguished by the absence of a sheath and motility of the filaments. It is typically associated with sulphur-containing effluents since it is able to oxidise hydrogen sulphide and stores sulphur in granules which are evident within the filaments. Effluents from breweries, dairies and wood pulp processing may contain sulphides from the sulphites utilised in those industries and promote *Beggiatoa* growth. It is also associated with slow flowing water with high loads of faecal material.

(d) *Flavobacterium*

Curtis (1969) reported that an unidentified *Flavobacterium* was also commonly found in 'sewage fungus' where it formed pink or yellow growths, the pigment being an insoluble carotenoid. The filaments are of long, thin, Gram-negative, non-sporing cells which characteristically lie parallel but separated by a capsule.

6.4.1.2 Fungi

(a) *Geotrichum candidum*

This species, not commonly described as a 'sewage fungus', was recently found to be the most important fungal component of slimes, especially in heavy growths (Curtis, 1969; Curtis and Curds, 1971). It forms branching septate hyphae (10–15 µm diameter) and can oxidise lactic acid and so is often commonly associated with wastes from the manufacture of dairy products, silage making and pickling. It occurs widely in soils, water and also in activated sludge.

(b) *Leptomitus lacteus*

Leptomitus lacteus (*Apodya lactea*) is similar to *Sphaerotilus* but is not mucilaginous. The non-septate hyphae branch and may contain plugs of cellulin (Fig. 6.4) giving the impression that septate are present. This fungus requires high oxygen levels and prefers acid conditions (pH optimum about 2.9 to 5.4). Organic nitrogen is needed and sugars are not metabolised. It appears to be a common constituent of sewage fungus in larger continental European rivers.

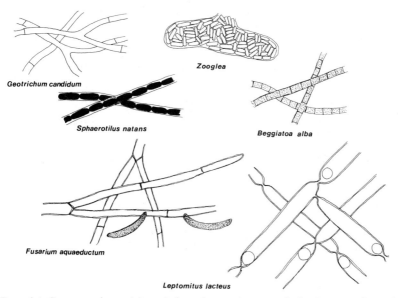

Fig. 6.4 Common bacterial and fungal constituents of the 'sewage fungus' community.

(c) *Fusarium aquaeductum*

Fusarium has branching septate hyphae and may be distinguished by its boat-shaped macroconidia (spores). In bulk, it may colour the slime pink or red. It appears to be relatively unimportant in sewage fungus outbreaks, having a preference for acid waters (pH 4–9) with high oxygen content.

6.4.1.3 Algae

Algae are not major constituents of 'sewage fungus' but one alga, *Stigeoclonium tenue*, tolerates severe organic pollution and may be present, especially where organic matter has been largely assimilated and inorganic nutrient levels are high. This alga has opposite branched tapering filaments which arise as a tuft from an attached prostrate portion of the thallus (see Fig. 6.5).

Other filamentous algae present in 'sewage fungus' include *Cladophora*, *Spirogyra*, *Ulothrix* and the blue-green algae (Cyanophyceae) *Oscillatoria* and *Lyngbya*. Several *Oscillatoria* species (*tenuis, angusta, chlorina, limosa* and *putrida*) are associated with *Sphaerotilus* (Fjerdingstad, 1964).

Diatoms are also often associated with 'sewage fungus' and the genera

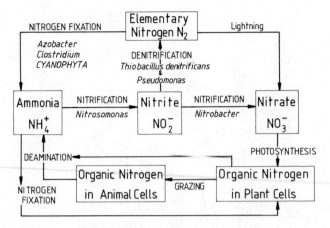

FIG. 6.5 The nitrogen 'cycle': principal pathways and sinks of nitrogen and its compounds in the aquatic environment.

Fragilaria, Synedra and *Navicula* are commonly present along with *Meridion, Melosira, Diatoma* and *Surirella.*

6.4.1.4 Protozoa

Protozoa are important constituents of the sewage fungus community though, with few exceptions, they do not form slimes and tend to be associated with the complex where they feed on bacteria, algae and other protozoa. The majority are ciliates, of which the attached, sedentary or sessile forms may be dominant members of the slime community.

Carchesium polypinum, a sessile colonial peritrich (see Fig. 6.9), may contribute substantial amounts of filamentous material in its stalks but the majority of ciliates present are holotrichs, species which swim or glide amongst the filaments of the 'fungus', including *Colpidium colpoda* and *Chilodonella cullculus.* Curtis and Curds (1971) noted that the 77 ciliate species which they recorded in 'sewage fungus' represented a much more diverse community than the list of species noted in sewage treatment works by Curds and Cockburn (1970).

6.4.2 Bacteria Other than 'Sewage Fungus'

Bacteria other than those associated with macroscopic growths, i.e. 'sewage fungus', will be considered under this heading. It will be noted that in Fig. 6.1 the large increase in bacterial growth which is associated with a drastic reduction in dissolved oxygen levels occurs after the development of 'sewage fungus'. This is because the generalised sequence displayed in this figure is based on the assumption that the load of organic matter is so great that the available oxygen is depleted and cannot be replenished rapidly enough by aeration. Under these conditions anaerobic fermentation will occur, especially in the sludge deposits which accumulate in slack water where gas production (chiefly methane and hydrogen sulphide) may cause bubbles of gas to be evident. These processes are complex and the details are by no means fully understood. Two stages are often described although single step and multiple step methanogenesis has also been proposed (Zehnder, 1978). The two phases involve acid-producing (non-methanogenic) and methane-producing (methanogenic) bacteria.

Acid-producing bacteria metabolise organic matter and produce lower fatty acids, along with some lower aldehydes and ketones. For example, the amino acid cysteine degrades to form acetic acid, carbon dioxide, ammonia, hydrogen sulphide and hydrogen:

$$4C_3H_7O_2NS + 8H_2O \rightarrow 4CH_3COOH + 4CO_2 + 4NH_3 + 4H_2S + 8H$$

      cysteine   acetic acid

The methanogenic bacteria degrade the acetic acid to methane, carbon dioxide and water:

$$4CH_3COOH + 8H \rightarrow 5CH_4 + 3CO_2 + 2H_2O$$

Useful reviews of the role of anaerobic bacteria and methanogenic bacteria will be found in Crowther and Harkness (1975) and Zehnder (1978).

Aerobic bacteria predominate in the region immediately below the bacteria will be found in Crowther and Harkness (1975) and Zehnder below the region dominated by the anaerobic bacteria responsible for continued degradation of the wastes, under the prevailing anoxic conditions.

Bacteria are introduced by the discharge of many organic effluents. For example, untreated sewage will introduce large numbers of bacteria in faecal matter since faeces have a bacterial flora in which obligate anaerobes such as *Bacteroides* spp. predominate (Pike, 1975) but which also contain coliforms, *Escherichia coli*, enterococci, *Clostridium welchii*, streptococci, lactobacilli, and staphylococci together with certain pathogenic organisms (see Table 6.3).

Although many sewage bacteria are removed in sewage treatment

TABLE 6.3

PATHOGENIC ORGANISMS KNOWN TO OCCUR IN SEWAGE EFFLUENTS OR ASSOCIATED
WITH SEWAGE TREATMENT

(After Hawkes, 1974, with additions and modifications)

Organism	Disease or condition	Comments
Viruses		
Polio virus	Poliomyelitis	Found in effluents, but not proven to be water-borne transmission
Infectious hepatitis virus	Infectious hepatitis	Only virus for which water route has been proven, epidemiologically
Bacteria		
Salmonella typhi	Typhoid fever ⎫	Common in sewage
Salmonella paratyphi	Paratyphoid fever ⎭	and effluents in epidemics
Shigella spp.	Bacterial dysentery	Source of infection, mainly polluted water
Bacillus anthracis	Anthrax	Spores resistant
Brucellosis spp.	Contagious abortion in livestock, undulant or Malta fever in man	Infection normally from contact or infected milk but sewage suspected also
Mycobacterium tuberculosis	Tuberculosis	Isolated from sewage? Possible mode of transmission
Vibrio cholerae	Cholera	Transmission by polluted water
Leptospira icterohaemorrhagiae	Leptospirosis, Leptospiral jaundice (Weil's Disease)	Carried by rats in sewers
Protozoa		
Entamoeba hystolytica	Amoebic dysentry	Contaminated water, tropical countries
Metazoa		
Schistosoma spp.	Bilharzia ⎫	
Taenia spp.	Tape worms ⎬	Spread by application
Ascaris spp.	Nematode worms ⎭	of sludge as agricultural fertiliser

plants (water reclamation works), discharges of sewage effluents may contain *Pseudomonas, Proteus, Bacillus subtilis, B. cereus, Aerobacter cloace, Zoogloea ramigera* along with *Escherichia coli, Aerobacter aerogenes* and *Streptococcus faecalis.*

Sterile organic effluents or effluents with bactericidal components may not begin to decompose until a sufficient inoculum has been introduced by mixing or until sufficient dilution has occurred to reduce toxicity, respectively. Self-purification in the river causes a change in the composition of the bacterial flora: a gradual reduction in proteolytic bacteria is accompanied by an increase in the number of cellulose decomposers.

Elevated levels of ammonia are associated with organic effluents, either as a result of putrefaction (anaerobic degradation) as described above or in treated sewage effluents. Ammonia is oxidised to nitrite by *Nitrosomonas* sp. and nitrite is further oxidised to nitrate by *Nitrobacter* sp. as shown in the following reactions:

$$NH_4^- + 1\tfrac{1}{2} O_2 \rightarrow NO_2^- + H_2O + 2H^+ (+76\ kcal)$$

$$NO_2^- + \tfrac{1}{2} O_2 \rightarrow NO_3^- (+24\ kcal)$$

These bacteria are strict autotrophs with specific requirements for these energy sources. Nitrite oxidising bacteria are temperature sensitive with a threshold around 10–15 °C and are, therefore, active only in summer (Rheinheimer, 1974) in temperate climates.

Denitrification is the reverse of the above processes, except that it ends in the release of free nitrogen, and occurs when facultative anaerobic bacteria use nitrate or nitrite as an oxygen source under anoxic conditions. The inter-relationships of these changes in nitrogen are summarised in Fig. 6.5. A very comprehensive review of inorganic nitrogen metabolism has been provided by Painter (1970).

6.4.3 Algae

Fjerdingstad (1964) made a comprehensive review of benthic phytomicroorganisms as indicators of the degree of pollution in streams. He noted that many workers previously had proposed that the fauna should be used since this was thought to be more sensitive but Fjerdingstad considered that benthic plants were equally appropriate. Planktonic algae were rejected on the grounds that they were carried by the current and could exist in conditions which were not typical of the point at which they were sampled. Macrophytes were also rejected since they are usually absent from heavily polluted waters. Fjerdingstad provides lists of algae which are classified into four categories: (i) saprobiotic species, that is, those which occur in large numbers only in heavily polluted waters; (ii) saprophilous species, i.e. those which occur generally in polluted waters but are found elsewhere; (iii) saproxenous species, i.e. those which generally occur in unpolluted water but may be found in

TABLE 6.4

CHARACTERISTIC INDICATOR GENERA OF MICROPHYTES IN DIFFERENT WATER
QUALITIES
(Mainly after Fjerdingstad, 1964)

	Polysaprobic	α-Mesosaprobic	β-Mesosaprobic	Oligosaprobic
Bacteria	*Spirillum* *Streptococcus* *Beggiatoa* *Sphaerotilus*			
Fungi	*Fusarium*			
Algae	*Bodo* *Cerobodo* *Euglena* *Oscillatoria* *Phormidium*	*Ulothrix* *Oscillatoria* *Stigeoclonium*	*Cladophora* *Phormidium* *Scenedesmus* *Pediastrum* *Ulothrix* *Vaucheria*	*Meridion* *Lemanea* *Batracho-* *spermum*

mildly polluted conditions; and (iv) saprophobous species, those which
do not thrive in polluted waters. Besides the autecological classification,
Fjerdingstad provides synecological information in which species associations or communities to be found under nine segments of the saprobic
spectrum are listed. These stages or zones are extensions and subdivisions of the classical four-part saprobien system and within them
communities are described as alternatives (depending, for example, on
the nature of the pollutant) or according to differences in degree of
intensity of conditions within the saprobic zone. A simplified account is
provided in Table 6.4 where characteristic genera which dominate the
communities found in different saprobic zones are listed. Some of the
algae are illustrated in Fig. 6.6.

The effects of organic enrichment on the algal flora which colonised
immersed microscope slides were studied extensively by Butcher (1947).
These investigations revealed that there were changes in the abundance
and also in the composition of the algal community with increasing
distance below the outfall. Hynes (1960) attributed the increase in
numbers to the release of inorganic nutrients from the decomposition of
the organic matter and thought that the differences in the location of
peak numbers were caused by the nature of the pollutant. For example,
readily decomposable milk wastes in the Bristol Avon caused a bloom
about 5 km downstream while in the River Tame the peak algal counts
were observed 50 km downstream (Fig. 6.7). Two peaks were observed in

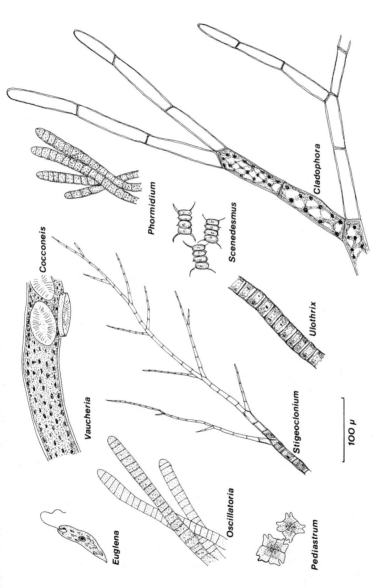

FIG. 6.6 Illustrations of some of the algae listed in Table 6.5 as characteristic of various degrees of organic enrichment.

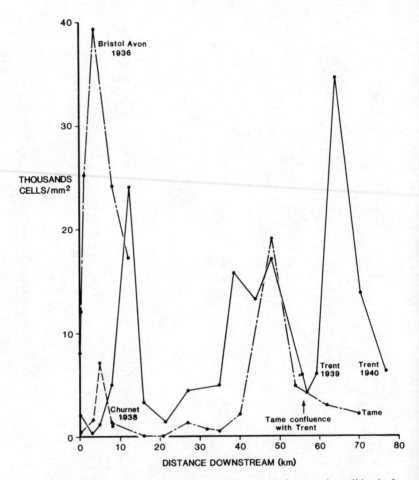

FIG. 6.7 Spatial changes in the algal populations growing on glass slides in four organically polluted British rivers. From data in Butcher (1947).

the River Trent at 15 and 40–50 km. Hynes (1960) ascribed these differences to the complexity of the composition of the discharges. This may well be correct, but one should not underestimate the influence of toxic substances (especially heavy metals, such as copper) which could suppress algal growth until sufficiently diluted.

A fairly well established pattern of succession in the dominant components of the algal community could be discerned (Butcher, 1947). The

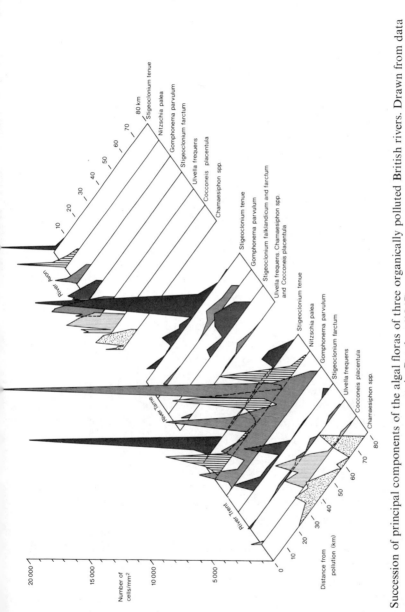

FIG. 6.8 Succession of principal components of the algal floras of three organically polluted British rivers. Drawn from data in Butcher (1947).

Stigeoclonium, Nitschia and *Gomphonema* community is the first to return when the algal flora recovers. Genera present upstream of the discharges (*Ulvella, Chamaesiphon* and *Cocconeis*) returned later and then other species typical of clean waters (Fig. 6.8).

In addition to the floral changes which are evident from microscope slides there may be growths of macro-algae, and of *Stigeoclonium tenue* and *Cladophora glomerata* in particular. *Stigeoclonium tenue* has been long recognised as tolerating high levels of pollution by organic matter (Zuelcr, 1908; Butcher, 1947) or at least α-meso saprobic conditions Fjerdingstad, 1964). McLean and Benson-Evans (1974) examined the distribution of this species in South Wales rivers and found that it occurred in a wide range of environments from organically polluted to 'fairly clean'. They suggest that, in order to avoid errors in attributing its presence to organic effluents, one should examine the abundance of the organism in spring. Only when it is highly abundant (that is, dominant) at this season can one reasonably diagnose organic pollution. These authors suggest that Fjerdingstad's (1964) description of this species as 'saprophilous' is most apposite, i.e. it is generally to be found in organically polluted waters but occurs elsewhere and is therefore largely indifferent to pollution. The autecology of this species of *Cladophora*, typical of waters recovering from organic enrichment and which may reach lengths of several metres, has been reviewed by Whitton (1970a). The precise reasons why this species can become so prolific as to constitute a nuisance have not yet been fully identified, although a number of factors appear to be involved. It is said to prefer cooler water yet may occur on cooling tower structures where temperatures are about 40 °C (Sladeckova, 1969) or in heated rivers (Whitton, 1967). High light intensity, high pH and good water flows are also thought to be conducive to good growth of this species. *Cladophora* growths are apparently promoted by increased nutrient levels, especially phosphates (Hawkes, 1963) and probably by a high phosphate–inorganic nitrogen ratio. This conclusion is supported by experimental studies by Bolas and Lund (1974) of the effect of enhanced nutrients on the growth of *Cladophora* in tanks supplied with river water containing a nitrate level of 3·5 mg litre^{-1} as N and a phosphate level of 2·0 mg litre^{-1} as P_2O_5 and which were regarded as exceeding limiting values. However, when the nutrients in the river water were augmented by different combinations of extra nitrate and phosphate it was found that *Cladophora* growths were largely unaffected by increased nitrogen but the response to increased phosphate was generally proportional to the increase in phosphorus.

Although considerable evidence has accumulated which suggests that *Cladophora glomerata* is very sensitive to the presence of heavy metals, some contradictory data are also available (Whitton, 1970a). It is interesting to note that in a study of the nutritional requirements of *Cladophora*, and phosphorus in particular, Pitcairn and Hawkes (1973) found that this species seems to require higher than usual levels of iron.

Studies of the epiphytes of *Cladophora* have shown that these are very similar in a wide geographical range of habitats (Whitton, 1970a) and the list of common genera includes several, such as *Chamaesiphon*, *Cocconeis* and *Gomphonema*, noted above as tolerant of conditions prevailing after recovery from the gross effects of organic enrichment. Another species commonly found in similar habitats is *Stigeoclonium tenue* (see below) but Whitton (1970b) has noted that *Cladophora* is often absent from stations in which *Stigeoclonium* is abundant, but where *Cladophora* is abundant *Stigeoclonium* is rarely absent. He attributed this to the level of heavy metals present, of which *Cladophora* is intolerant. Hawkes (1963) noted that in rivers receiving sewage effluents *Stigeoclonium* often appears downstream of the discharge before *Cladophora*. Whitton (1970c) suggests that this may be due to a poorer ability on the part of *Cladophora* to compete with sewage fungus or its greater sensitivity to heavy metals, especially zinc, which often occur in sewage effluents from industrial areas.

One may conclude that dense growths of *Cladophora* which become a nuisance by interfering with river gauging, angling or boating, causing fish deaths through deoxygenation by night-time respiration or high pH with associated increases in ammonia toxicity, are stimulated by discharges of sewage effluents containing substantial amounts of phosphate but that other, interrelated factors are also involved. Whitton (1970c) has suggested that some river management procedures might also encourage *Cladophora* growths. For example, impoundments may tend to reduce turbidity, lower peak flood flows and reduce temperature extremes. Increase in hardness by inter-river transfers might also encourage the growth of *Cladophora*.

6.4.4 Protozoa

The response of protozoa to organic enrichment is well documented and forms a mainstay of the saprobien system (Liebmann, 1951; Sladecek, 1964; Bick, 1968). The saprobic sequence within a single genus, *Vorticella*, comprising some 32 species and sub-species has been documented by Sladecek (1971). However, the published classifications or accounts of the

TABLE 6.5

LIST OF COMMON CILIATES WHICH ARE INDICATIVE OF DIFFERENT DEGREES OF
ORGANIC POLLUTION

1. Polysaprobic—grossly polluted
   Colpidium campylum (Stokes)   *Paramecium caudatum* Ehrenberg
   Colpidium colpoda Stein   *Paramecium putrinum* Clap. and Lachmann
   Colpoda steinii Maupas   *Paramecium trichium* Stokes
   Glaucoma scintillans Ehrenberg   *Vorticella microstoma* Ehrenberg
   Trimyema compressum Lackey

2. α-mesosaprobic—polluted
   Chilodonella cucullulus (Muller)   *Urocentrum turbo* (Muller)
   Chilodonella uncinata Ehrenberg   *Urotricha farcara* Clap. and Lachmann
   Colpoda cucullus Muller   *Stylonychia mytilus* Ehrenberg
   Cyclidium citrullus Cohn   *Stylonychia putrina* Stokes
   Spirostomum ambiguum Muller-   *Carchesium polypinum* Linnaeus
      Ehrenberg

3. β-mesosaprobic—mildly polluted
   Aspidisca costata (Dujardin)   *Litonotus lamella* Schewiakoff
   Cinetochilum margaritaceum Perty   *Paramecium bursaria* (Ehrenberg)
   Coleps hirtus Nitzsch   *Spirostomum teres* Clap. and Lachmann
   Euplotes affinis Dujardin
   Euplotes patella (Muller)
   Halteria grandinella (Muller)

autecological requirements of protozoan species or genera are by no
means unanimous. General trends have been established and typical
indicators of the stages of recovery from organic pollution (polysaprobic
to β-meso saprobic) are given in Table 6.5. In reaches where organic
pollution is severe and accompanied by putrifaction, only bacteria and
the flagellate *Bodo* will be found. The protozoa associated with the
development of *Sphaerotilus* are described above (Section 6.4.1.4).
Typical protozoa of the polysaprobic zone are *Colpidium*, *Paramecium*
and *Glaucoma* which feed on the large numbers of bacteria. As organic
material is further degraded these genera are supplemented by *Vorticella*,
Spirostomum and *Carchesium* and in the final stage of recovery genera
such as *Coleps*, *Euplotes* and *Litonotus* along with many others will be
found (Fig. 6.9). The trend is generally towards increasing diversity as
recovery continues.

6.4.5 Macrophytes

Macrophytes are severely depleted or eliminated by the high turbidity
which often accompanies organic discharges and by the smothering of

settled particulate matter in much the same way as with inert suspended solids (see Section 5.1.1). In addition, dense growths of sewage fungus on their surfaces may also quickly eliminate them. Some macrophytes are tougher, however, and may be able to maintain themselves even when conditions are quite severe. *Potamogeton pectinatus* is a species with long strap-like leaves and is able to withstand a considerable amount of silting, perhaps because of the elongated leaf form. Where velocities are high, for example in riffles, the bryophyte *Fontinalis antipyretica* seems to be able to tolerate considerable organic enrichment although this habitat is probably less severe since the high velocity keeps the plant clear of particulate matter and the relatively shallow water allows light penetration which supports photosynthesis.

In deeper, slower water marginal emergent plants, such as *Glyceria maxima*, or submerged macrophytes with floating leaves, for example *Potamogeton natans*, may survive moderate organic enrichment.

A list of the most tolerant and moderately tolerant species (Haslam, 1978) is provided in Table 6.6 together with the tolerances of a representative selection of species, used to derive a 'plant score' (Harding, 1981).

6.4.6 Macroinvertebrates

The responses of macroinvertebrates to organic pollution are the best documented and most thoroughly understood of all groups in freshwater and form the basis of a large number of systems or indices for classifying stages or degrees of this (and other) forms of pollution (see Section 9.4.2). Examples of macroinvertebrates associated with different degrees of organic pollution are illustrated in Fig. 6.10.

As with many other environmental stresses two interrelated trends are readily discerned when increasing organic enrichment occurs. The first is a reduction in the diverse macroinvertebrate community characteristic of clean water, in which many species are represented by relatively few individuals, towards a condition, typical of many stresses, in which, under the influence of severe pollution, a few species are represented by very large numbers of individuals. These species are those which are able to take advantage of the changes which the pollutant induces and to exploit the increased food supply which is provided. The second change is the progressive disappearance of particular indicator species until very few remain and their place is taken by species not previously present or, at least, not abundantly present.

In the most extreme conditions new species, only found under such

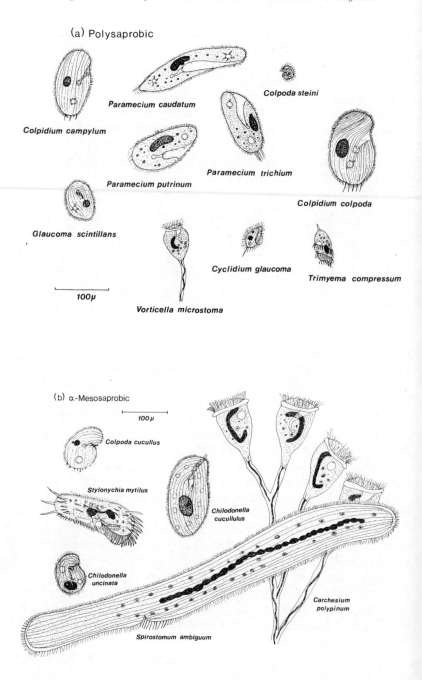

(a) Polysaprobic

Colpidium campylum

Paramecium caudatum

Colpoda steini

Paramecium putrinum

Paramecium trichium

Colpidium colpoda

Glaucoma scintillans

Cyclidium glaucoma

Trimyema compressum

100μ

Vorticella microstoma

(b) α-Mesosaprobic

100μ

Colpoda cucullus

Stylonychia mytilus

Chilodonella cucullulus

Chilodonella uncinata

Carchesium polypinum

Spirostomum ambiguum

(c) ß-Mesosaprobic

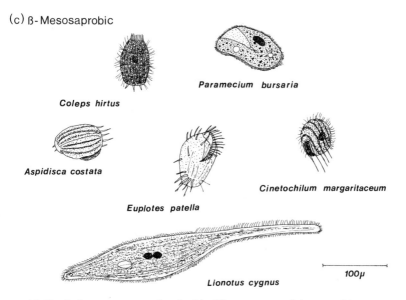

Coleps hirtus

Paramecium bursaria

Aspidisca costata

Euplotes patella

Cinetochilum margaritaceum

Lionotus cygnus

100μ

FIG. 6.9 Typical protozoa associated with different zones of the saprobiensystem. (a) Polysaprobic; (b) alpha-mesosaprobic; (c) beta-mesosaprobic.

conditions, may appear. Examples of these exploiters of severe organic enrichment are the larvae of the flies *Eristalis* and *Psychoda*. The latter species is a common constituent of the fauna of the percolating filters (bacteria beds) of sewage treatment works, while the former occurs in humus or sludge tanks (Learner, 1975). *Eristalis* is equipped with a telescopic air tube through which it breathes while it feeds in the anoxic organically enriched mud. It is restricted to shallow waters, not exceeding about 0·1 m, and is normally found in pond margins or shallow water enriched by cattle dung. *Psychoda* is essentially an inhabitant of mud flats but is probably more commonly associated with heavily loaded sewage-works where it may become so abundant as to cause nuisance (Learner, 1975).

Below strong organic discharges which reduce oxygen levels almost to zero the benthic macroinvertebrate fauna consists almost exclusively of tubificid worms, principally the genera *Tubifex* and *Limnodrilus*. These worms are able to live in the accumulated organic deposits where they feed on bacteria-enriched material (Brinkhurst and Cook, 1974; Brinkhurst *et al.*, 1972). They are responsible, by their burrowing acti-

TABLE 6.6
POLLUTION TOLERANCE RATINGS OF BRITISH MACROPHYTES

(a) According to Haslam (1978)

Most tolerant

Potamogeton pectinatus	*Sparganium emersum*
Potamogeton crispus	*Spargarium erectum*
Schoenoplectus lacustris	*Mimulus guttatus*

Moderate tolerance

Fontinalis antipyretica	*Lemna minor*
Agrostis stolonifera	*Nuphar lutea*
Butomus umbellatus	*Rorippa amphibia*
Glyceria maxima	*Ranunculus* spp.

(b) According to 'Plant Score' of Harding (1981)

Most tolerant[a]

Agrostis stolonifera	*Potamogeton berchtoldii*
Amblystegium riparium	*Potamogeton pectinatus*
Phalaris arundinacea	*Sparganium erectum*

Moderate tolerance[a]

Alisma plantago-aquatica	*Juncus effusus*
Apium nodiflorum	*Lemna* spp.
Elodea nutallii	*Mimulus* spp.
Epilobium hirsutum	*Oenanthe crocata*
Equisetum fluviatile	*Petasites hybridus*
Equisetum palustre	*Polygonum amphibium*
Fontinalis antipyretica	*Potamogeton crispus*
Glyceria declinata	*Potamogeton natans*
Glyceria fluitans	*Rorippa amphibia*
Glyceria maxima	*Sagittaria sagittifolia*
Glyceria plicata	*Sparganium emersum*
Iris pseudacorus	*Typha latifolia*

[a]'Most tolerant' is equivalent to scores of 1 or 2, 'moderate tolerance' is equivalent to scores of 3 or 4.

vities, for considerable overturn on the sediments. In the heavily polluted stretches their density can be high—up to $420\,000\,\text{m}^{-2}$—which may be attributed not only to the abundant food supply but also the absence of leech predators (Brinkhurst, 1963).

As conditions improve the tubificid worms are joined by chironomid larvae of which the 'bloodworm' *Chironomus* is predominant both numerically and in appearance, being quite large (about 25 mm long) and bright red in colour. The commonest species in Europe (and North America—Hynes, 1960) is *Chironomus thummi* (*riparius*). Learner and

Edwards (1966) have provided a useful summary of the biology of this species. Adult swarms, which may occupy a volume of some 1400 m³, consist principally of males. Females are attracted to these and mating occurs on nearby vegetation after which eggs are laid in a gelatinous mass on objects beneath the water surface. Most females lay only one batch of about 600 eggs though a few may lay two batches. The eggs hatch within four or five days at 18 °C. The larvae are detritus feeders consuming the organic material added in effluents and their associated bacteria (Baker and Bradnam, 1976). Larval densities may reach 100 000 m⁻² (Edwards, 1957) and penetrate the mud in which they build their tubes to a depth of about 10 cm. This means that they exert considerable influence on the rate of oxidation of the mud since the larvae draw water through their tubes by body movements. Although *Chironomus* larvae tolerate low dissolved oxygen levels they are less resistant than tubificids and cannot endure prolonged periods of anoxia. Learner and Edwards (1966) observed seven peaks of emergence during the year while Gower and Buckland (1978) suggest that there are at least five generations during the year, although the picture is somewhat obscured by continuous recruitment of first instar larvae which results in considerable overlap. These workers further postulate that *Chironomus riparius* may be quick to take advantage of organic enrichment because of the ovipositing behaviour of females and the multi-voltine life history thereby being particularly valuable as an early indicator of pollution.

The next major shift in the composition of the benthos as degradation continues is an increase in the abundance of the *Asellus aquaticus*. This species is a widely recognised indicator in Europe but does not appear to have any North American equivalent. Hynes (1960) noted that the transatlantic species, *Asellus communis*, was mentioned by Bartsch (1948) but could find no other evidence that this occupied a similar position in North America. The related European species, *Asellus meridianus*, is not found in polluted waters. Vast numbers of *A. aquaticus* are often found in dense stands of the macro-alga *Cladophora*.

Asellus aquaticus is the most characteristic species of the· zone of recovery following that in which *Chironomus* dominates, but other macroinvertebrates, for example certain tolerant leeches and molluscs, may also be plentiful. These include *Erpobdella testacea*, usually limited to anaerobic ponds and polluted waters (Elliott and Mann, 1979) and which appears to be well adapted to maintaining its metabolism in lower oxygen tensions (Mann, 1961); *Trocheta subviridis; Glossiphonia complanata*—an almost ubiquitous species, not only associated with

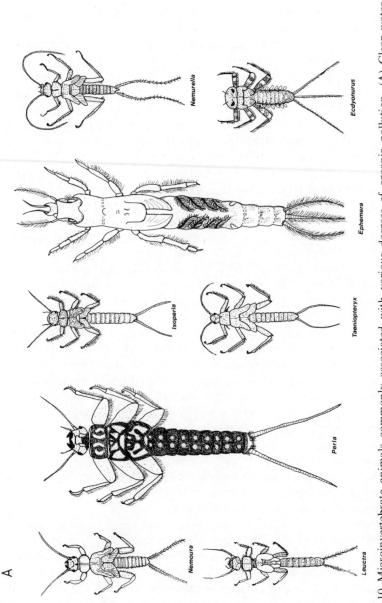

FIG. 6.10 Microinvertebrate animals commonly associated with various degrees of organic pollution. (A) Clean-water (sensitive) forms; (B) mild-pollution (less sensitive) forms; (C) moderate-pollution (relatively tolerant) forms; (D) severe-pollution (tolerant) forms.

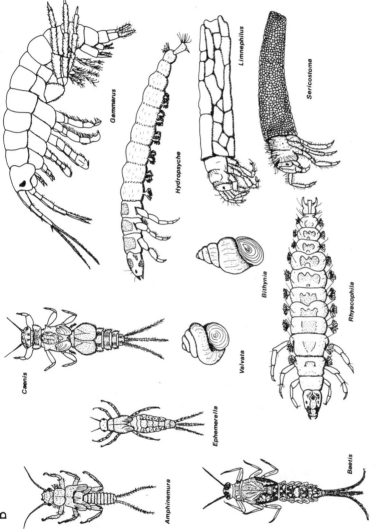

Fig. 6.10—*contd.*

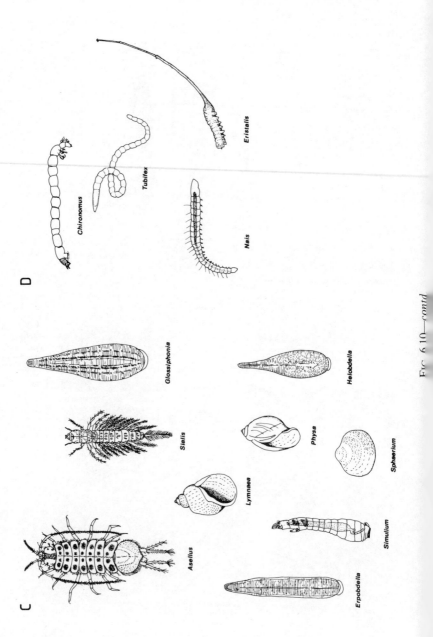

Fig. 6.10—*contd*

polluted waters; and *Helobdella stagnalis*. Tolerant molluscs include *Lymnaea peregra*, *Physa fontinalis* and *Sphaerium* spp.

Beyond the *Asellus*-zone, with further recovery, is the *Gammarus*-zone which may be regarded as the first stage in the return of a normal cleanwater fauna. Hawkes and Davies (1971) have observed that *Gammarus* is particularly sensitive to short periods of low oxygen (5 hr LC_{50} was $1 \, mg \, O_2$ litre^{-1} at 20 °C), particularly at night when plant respiration may depress dissolved oxygen levels, and so may not become established even though oxygen levels, as indicated by daytime measurements, appear to be adequate. The *Gammarus* zone is characterised by an even greater diversity of macroinvertebrates than the *Asellus*-zone. The most important group along with *Gammarus* are the Trichoptera (caddis flies), followed, as the zone merges with the next, by Ephemeroptera (mayflies). Some species of Ephemeroptera are well adapted to coping with silty conditions (e.g. *Ephemera* and *Caenis*) and so can also survive the deposition of materials in organic effluents provided that oxygen levels are not too depressed.

The extent to which other groups, for example Plecoptera and some Coleoptera, will return depends on the type of river or stream. In upland eroding substrates these groups are part of the normal community but are less likely to occur in lowland streams and rivers, even when they are unaffected by pollution.

The responses of European invertebrates to degrees of organic enrichment, determined largely in terms of the Saprobien system, are indicated in Table 6.7. Similar tables, based on North American species (Table 6.8) and on the limited published data for other regions (Table 6.9) have also been compiled. It should be remembered that the limitations of 'indicator species', discussed in Section 3.2, apply to these compilations.

6.4.7 Fish

As with most of the other organisms considered above, the existence of fish populations below discharges of organic matter depends to a large extent on dissolved oxygen levels. However, the presence of ammonia in many effluents, especially sewage, may have a greater influence on the survival of viable fish populations since most fish are extremely sensitive to undissociated ammonia. When oxygen is adequate, coarse fish populations are able to survive even in undiluted sewage effluent (Allan *et al.*, 1958; Alabaster, 1959). Other substances present in sewage effluents which contain industrial and domestic components, for example phenols,

TABLE 6.7
THE POLLUTION TOLERANCES OF A SELECTION OF COMMON EUROPEAN MACRO-
INVERTEBRATES, DERIVED LARGELY FROM THEIR POSITION IN PUBLISHED TABLES
BASED ON THE 'SAPROBIENSYSTEM'

(1) *Species largely insensitive to severe organic pollution* (*polysaprobic*)

Chironomidae	Oligochaeta
Camptochironomus tentans	*Limnodrilus claparedeanus*
Chironomus plumosus	*Limnodrilus hoffmeisteri*
Chironomus thummi	*Rhyacodrilus coccineus*
	Tubifex tubifex

(2) *Species tolerant of severe organic pollution* (*alpha-mesosaprobic*)

Chironomidae	Oligochaeta
Anatopyria plumipes	*Limnodrilus udekemianus*
Apsectrotanypus trifascipennis	*Lumbriculus variegatus*
Psectrotanypus varius	*Nais barbata*
Prodiamesa olivacea	*Nais communis*
Megaloptera	*Nais elinguis*
Sialis fulginosa	*Peloscolex ferox*
Sialis lutaria	*Psammoryctes barbatus*
Trichoptera	Hirudinea
Anabolia laevis	*Erpobdella octoculata*
Hydropsyche angustipennis	*Glossiphonia heteroclita*
Hydropsyche ornatula	*Haemopsis sanguisuga*
Hydropsyche pellucidula	*Helobdella stagnalis*
Coleoptera	*Hirudo medicinalis*
Haliplus lineatocollis	Crustacea
Odonata	*Asellus aquaticus*
Coenagrion pulchellum	Platyhelminthes
Ephemeroptera	*Dugesia tigrina*
Baetis vernus	*Planaria torva*
Caenis horaria	
Heptagenia longicauda	
Siphlonurus lacustris	

(3) *Species tolerant of moderate organic enrichment* (*beta-mesosaprobic*)

Chironomidae	Oligochaeta
Ablabesmyia monilis	*Aelosoma niveum*
Corynoneura celtica	*Aelosoma quaternarium*
Corynoneura scutellata	*Aelosoma tenebrarum*
Cricotopus bicinctus	*Amphichaeta leydigii*
Cricotopus triannulatus	*Branchiura sowerbyi*
Macropelopia nebulosa	*Chaetogaster cristallinus*
Metriocnemus fuscipes	*Chaetogaster palustris*
Metriocnemus hygropetricus	*Criodrilus lacuum*
Microtendipes cloris	*Euclyodrilus hammoniensis*
Polypedilum laetum	*Nais bretscheri*

TABLE 6.7—*contd.*

Ephemeroptera
  Baetis rhodani
  Brachycercus harisella
  Caenis moesta
  Cloeon dipterum
  Cloeon rufulum
  Ephemera danica
  Ephemerella ignita
  Heptagenia sulphurea
  Leptophlebia marginata
Plecoptera
  Isoperla grammatica
Heteroptera
  Brychius elevatus
  Nepa rubra
  Notonecta glauca
Odonata
  Ischnura elegans
Trichoptera
  Anabolia nervosa
  Athripsodes annulicornis
  Athripsodes cinereus
  Crunoecia irrorata
  Geora pilosa
  Hydroptila tineoides
  Limnephilus sericeus
  Limnephilus sparsus
  Polycentropus flavomaculatus
  Potamophylax stellatus
  Psychomia pusilla

  Nais variabilis
  Psammoryctes abicolus
  Stylaria lacustris
  Stylodrilus heringianus
Hirudinea
  Glossiphonia complanata
  Hemiclepsis marginata
  Theromyzon tessulatum
Mollusca
  Anodonta cygnaea
  Bithynia tentaculata
  Lymnaea stagnalis
  Physa fontinalis
  Planorbis planorbis
  Radix auricularia
  Sphaerium corneum
  Theodoxus fluviatilis
  Unio pictorum
Platyhelminthes
  Dendrocoelum lacteum
  Dugesia lugubris
  Polycelis nigra
  Polycelis tenuis
Coleoptera
  Dryops ernesti
  Dryops luridus
  Haliplus fluviatilis
  Helichus substriatus
  Hydrobius fuscipes
  Potamonectes assimilis

(4) *Species largely intolerant of organic pollution (oligosaprobic)*
Chironomidae
  Brillia longifurca
  Brillia modesta
  Cladotanytarsus mancus
  Eukiefferiella clavescens
  Eukiefferiella longicalcar
  Heterotrissocladius marcidus
  Microcricotopus bicolor
  Micropsectra atrofasciata
  Microtendipes britteni
  Parametriocnemus stylatus
  Psectrocladius dilatatus
  Pseudodiamesa branickii

Oligochaeta
  Haplotaxis gordioides
  Nais pseudobtusa
Mollusca
  Ancylus fluviatilis
  Bythinella austriaca
  Pisidium supinum
  Pisidium personatum
  Sphaerium solidum
Platyhelminthes
  Crenobia alpina
  Polycelis felina

TABLE 6.7—*contd.*

Plecoptera
  Amphinemeura sulcicollis
  Brachyptera risi
  Capnia bifrons
  Chloroperla torrentium
  Dinocras cephalotes
  Leuctra nigra
  Leuctra moselyi
  Nemoura cambrica
  Nemurella picteti
  Perla bipunctata
  Perlodes dispar
  Protonemura meyeri
  Taeniopteryx nebulosa
Ephemeroptera
  Ameletus inopinatus
  Baetis alpinus
  Ecdyonurus dispar
  Ecdyonurus venosus
  Rhithrogena semicolorata
  Paraleptophlebia submarginata

Trichoptera
  Agapetus fuscipes
  Agraylea multipunctata
  Apatania fimbriata
  Brachycentrus montanus
  Hydropsyche siltali
  Ithytrichia lamellaris
  Lasiocephala basalis
  Leptocerus tineiformis
  Philopotamus montanus
  Ptilocolepus granulatus
  Rhyacophila philopotamoides
  Rhyacophila tristis
  Silo nigricornis
  Silo pallipes
  Trichostegia minor

cyanide and certain heavy metals, may preclude the indefinite existence of populations of fish (Herbert *et al.*, 1965; Brown *et al.*, 1970; Alabaster *et al.*, 1972).

The effects of organic wastes on fish may be conveniently considered with reference to the three major relevant components of effluents, namely degradable carbonaceous matter (depressing dissolved oxygen levels), unionised ammonia and suspended solids. The last component is covered in Section 5.1.4 and need not be further considered here except to note that its effects on fish are likely to be indirect, through blanketing of the gravels used by certain species for spawning.

6.4.7.1 *Effects of low oxygen*

Low dissolved oxygen which results from the microbial degradation of organic wastes may affect fish differently during various phases of their life history. Species may survive protracted periods of low oxygen levels as adults which would be deleterious to eggs or fry. Similarly, different minimum oxygen levels may be necessary to support various activities such as reproduction, feeding, swimming performance or normal behaviour. A detailed review of these has been provided by Alabaster and

TABLE 6.8

POLLUTION TOLERANCES OF COMMON NORTH AMERICAN MACROINVERTEBRATES, BASED ON A NUMBER OF PUBLISHED STUDIES

(a) *Species which are largely indifferent to organic pollution*

(i) Oligochaeta
   Branchiura sowerbyi
   Limnodrilus cervix
   Limnodrilus claparedeanus
   Limnodrilus hoffmeisteri
   Limnodrilus maumeensis
   Limnodrilus udekemianus
   Ophidonais sp.
   Tubifex tubifex

(ii) Hirudinea
   Dina microstoma
   Dina parva
   Erpobdella punctata
   Glossiphonia complanata
   Helobdella nepheloidea
   Helobdella stagnalis
   Macrobdella sp.
   Mooresbdella microstoma
   Placobdella montifera

(iii) Mollusca
   Galba catascopium
   Goniobasis virginica
   Lymnaea ovata
   Musculium transversum
   Physa anatina
   Physa cubensis
   Physa helei
   Physa heterostropha
   Physa integra
   Physa pumila
   Pisidium abditum
   Pisidium idahoensis
   Planorbis panus
   Planorbis trivolvis
   Segmentina armigera
   Sphaerium notatum
   Unio complanata

(iv) Crustacea
   Cambarus diogenes
   Cambarus fodiens
   Cambarus striatus
   Hyallela knickerbockeri
   Palaemonetes exilipes
   Procambarus acutus
   Procambarus fallax

Procambarus troglodytes

(v) Ephemeroptera
   Caenis diminuta
   Callibaetis floridans

(vi) Odonata
   Ischneura verticalis

(vi) Hemiptera
   Belostoma sp.
   Corixa sp.
   Hesperocorixa sp.
   Hydrometra martini

(vii) Coleoptera
   Berosus sp.
   Dineutes americanus
   Gyrinus floridanus
   Laccophilus maculosus
   Stenhelmis decorata
   Tropisternus lateralis
   Tropisternus natator

(viii) Chironomidae
   Ablabesmyia illinoense
   Calopsectra gregarius
   Chironomus attenuatus
   Chironomus carus
   Chironomus crassicaudatus
   Chironomus fulvipilus
   Chironomus plumosus
   Chironomus riparius
   Chironomus stigmaterus
   Cryptochironomus fulvus
   Dicrotendipes incurvus
   Glyptotendipes barbipes
   Glyptotendipes lobiferus
   Glyptotendipes paripes
   Kiefferullus dux
   Pentaneura flavifrons
   Pentaneura melanops
   Polypedilum scalaenum
   Procladius culiciformis
   Procladius denticulatus
   Psectrotanypus dyari
   Psilotanypus bellus
   Rheotanytarsus exiguus
   Tanypus stellatus

TABLE 6.8—*contd.*

(ix) Other Diptera
    Bezzia glabra
    Brachydeutera argentata
    Culex pipiens
    Eristalis aenaus
    Eristalis bastardi
    Eristalis brousi
    Psychoda alternata

Psychoda schizura
Stilobezzia antenalis
Stratiomys discalis
Stratiomys meigeni
Syrphus americanus
Tabanus atratus
Tabanus benedictus
Tabanus lineola
Telmatoscopus albipunctatus

(b) *Species which tolerate moderate levels of organic pollution*

(i) Oligochaeta
    Dero sp.
    Helodrilus chlorotica
    Nais sp.
    Stylaria sp.

(ii) Hirudinea
    Dina sp.
    Piscicola punctata
    Placobdella rugosa

(iii) Mollusca
    Alasmondonta costata
    Amblema undulata
    Anadonta grandis
    Anadonta imbecillis
    Aplexa hypnorum
    Bulimus tentaculatus
    Campeloma contectus
    Campeloma fasciatus
    Campeloma integrum
    Campeloma rufum
    Campeloma subsolidum
    Ferrissia fusca
    Ferrissia tarda
    Goniobasis livescens
    Gyraulus arcticus
    Helisoma anceps
    Helisoma trivolvis
    Lampsilus alata
    Lampsilus anadontoides
    Lampsilus gracilis
    Lampsilus luteola
    Lasmigona complanata
    Lymnaea auricularia
    Lymnaea caperata

(Mollusca contd.)
Lymnaea humilis
Lymnaea obrussa
Lymnaea peregra
Musculium securis
Mytilopsis leucophaetus
Physa acuta
Physa fontinalis
Physa gyrina
Pisidum fallax
Pisidium henslorvanum
Pisidium subtruncatum
Pleurocerca acuta
Pleurocerca elevatum
Pseudosuccinea columella
Quadrula lachrymosa
Quadrula plicata
Quadrula pustulosa
Quadrula rubiginosa
Quadrula undulata
Rangia cuneata
Sphaerium corneum
Sphaerium moenanum
Sphaerium rhomboideum
Sphaerium stamineum
Sphaerium striatinum
Sphaerium sulcatum
Sphaerium vivicolum
Symphynota costata
Symphynota edentulus
Tritigonia tuberculata
Unio gibbosus
Unio tumidus
Valvata piscinalis
Valvata tricarinata

TABLE 6.8—contd.

(iv) Crustacea
      Asellus intermedius
      Cambarus floridanus
      Cambarus latimanus
      Crangonyx pseudogracilis
      Faxonella clypeata
      Gammarus sp.
      Hyallela azteca
      Lirceus sp.
      Orconectes erichsonianus
      Orconectes propinquus
      Orconectes rusticus
      Palaemonetes paludosus
      Procambarus angustatus
      Procambarus barbatus
      Procambarus chacei
      Procambarus enoplosternum
      Procambarus epicyrtus
      Procambarus howellae
      Procambarus litosternum
      Procambarus lunzi
      Procambarus paeninsulanus
      Procambarus pubescens

(v)   Plecoptera
      Acroneuria abnormis
      Allocapnia vivipaR
      Taeniopteryx maura

(vi) Ephemeroptera
      Hexagenia bilineata
      Stenonema ares
      Stenonema femoratum
      Stenonema heterotarsale
      Stenonema integrum
      Stenonema pulchellum
      Stenonema scitulum

(vi) Odonata
      Agrion maculatum
      Argia apicalis
      Argia translata
      Enallagma antennatum
      Gomphus pallidus
      Gomphus spinceps
      Gomphus vastus
      Libellula lydia
      Neurocordulia moesta

(vii) Megaloptera
      Corydalis cornutus

(viii) Coleoptera
      Agabus stagninus
      Dubiraphia sp.
      Gonielmis dietrichi
      Optioservus sp.

(ix) Trichoptera
      Cheumatopsyche sp.
      Hydropsyche bifida
      Hydropsyche bronta
      Hydropsyche orris
      Potamya flava

(x)   Chironomidae
      Ablabesmyia janta
      Ablabesmyia peleensis
      Ablabesmyia rhamphe
      Chironomus decorus
      Chironomus equisitus
      Chironomus flavus
      Chironomus staegeri
      Coelotanypus concinnus
      Coelotanypus scapularis
      Cricotopus exilis
      Cricotopus trifasciatus
      Cryptochironomus nais
      Cryptotendipes emorsus
      Dicrotendipes modestus
      Glyptotendipes amplus
      Glyptotendipes merdionalis
      Harnischia collator
      Parachironomus pectinatellae
      Polypedilum fallax
      Polypedilum illinoense
      Polypedilum tritum
      Tanypus carinatus
      Tanypus punctipennis

(xi) Other Diptera
      Odontomyia cincta
      Palomyia tibialis
      Simulium vittatum

(xii) Other phyla
      Cordylophora lacustris
      Cristatella mucedo
      Paludicella ehrenbergi
      Plumatella repens
      Spongilla fragilis
      Urnatella gracilis

TABLE 6.8—*contd.*

(c) *Species intolerant of organic pollution*

(i)   Mollusca
    Amnicola emerginata
    Amnicola limnosa
    Anodonta mutabilis
    Campeloma decisum
    Ferrissia rivularis
    Lampsilis parvus
    Leptodea fragilis
    Obliquaria reflexa
    Planorbis carinatus
    Pisidium fossarinum
    Proptera alata
    Somatogyrus subglobosus
    Sphaerium solidum
    Truncilla donaciformis
    Truncilla elegans
    Unio batavus
    Unio pictorum
    Valvata bicarinata
    Valvata tricarinata
    Viviparus contectoides
    Viviparus subpurpurea

(ii)   Crustacea
    Cambarus acuminatus
    Cambarus asperimanus
    Cambarus extraneus
    Cambarus hiswassensis
    Cambarus conasaugaensis
    Orconectes juvenilis
    Procambarus raneyi
    Procambarus spiculifer
    Procambarus versutus

(iii) Plecoptera
    Acroneuria arida
    Acroneuria evoluta
    Isoperla bilineata
    Taeniopteryx nivalis

(iv)   Ephemeroptera
    Baetis vagans
    Ephemera simulans
    Hexagenia limbata
    Pentagenia vittgera
    Stenonema exiguum
    Stenonema fuscum
    Stenonema interpunctatum

(iv)   Ephemeroptera (contd.)
    Stenomema proximum
    Stenonema rubromaculatum
    Stenonema smithae
    Stenonema terminatum
    Stenonema tripunctatum

(v)   Odonata
    Anax junius
    Gomphus externus
    Gomphus plagiatus
    Hataerina titia

(vi)   Neuroptera
    Climacia areolaris

(vii) Megaloptera
    Sialis infumata

(viii) Coleoptera
    Anacyronyx variegatus
    Helicus lithophilus
    Macronychus glabratus
    Microcylloepus pusillus
    Stenelmis crenata
    Tropisternus dorsalis

(ix)   Trichoptera
    Chimarra obscura
    Chimarra perigua
    Hydroptila waubesiana
    Hydropsyche frisoni
    Hydropsyche simulans
    Macronemum carolina
    Neureclipsis crepuseularis

(x)   Chironomidae
    Ablabesmyia aspera
    Ablabesmyia auriensis
    Ablabesmyia ornata
    Brillia par
    Chaetolabis atroviridis
    Chaetolabis ochreatus
    Chironomus anthracinus
    Chironomus paganus
    Chironomus tentans
    Clinotanypus caliginosus
    Corynoneura scutellata
    Corynoneura taris
    Cricotopus absurdus

TABLE 6.8—contd.

(x) Chironomidae (contd.)
Cricotopus bicinctus
Cricotopus politus
Cryptochironomus psittacinus
Diamesa nivoriunda
Dicrotendipes fumidus
Glyptotendipes senilis
Harnischia tenicaudata
Labrundinia floridana
Labrundinia pilosella
Labrundinia virescens
Micropsectra deflecta
Micropsectra nigripula
Microtendipes pedullus
Orthocladius obumbratus
Paratendipes albimanus
Pentaneura americana
Polypedilum nubeculosum
Polypedilum vibex
Prodiamesa olivacea
Stenochironomus hilaris
Strictochironomus devinctus
Strictochironomus varius

Tanytarsus dissimilis
Tanytarsus gracilentus
Thienemanniella xena
Tribelos fuscicornis
Tribelos jucundus
Trichocladius robacki
Xenochironomus scopula
Xenochironomus xenolabis

(xi) Other Diptera
Anophales punctipennis
Cnephia pecuarum
Prosimulium johannseni
Pseudolimnophila luteipennis
Simulium venustrum
Tabanus giganteus
Tabanus variegatus
Tipula abdominalis
Tipula caloptera

(xii) Other phyla
Lophopodella carteri
Pectinatella magnifica
Plumatella polymorpha

Lloyd (1980) but some general points may be noted. First, there is evidence that several species can acclimate to lower than usual levels of dissolved oxygen so that intermittent exposure or the ability of fish to swim into and out of areas affected by lowered oxygen may facilitate the development of tolerant populations in organically polluted waters. Second, some species are able to detect and avoid low oxygen levels, especially when these are accompanied by raised levels of carbon dioxide, a behaviour pattern which also should contribute to survival. The natural distribution of fish species in freshwater is, in part, a function of differing abilities to survive various levels of dissolved oxygen. The longitudinal zonation of fish in rivers (Carpenter, 1928; Huet, 1954) recently reviewed by Hawkes (1975) is well-correlated with oxygen requirements. Species characteristic of upland zones (Table 2.2) in which oxygen levels are at, or close to, air saturation include salmonids while in lower reaches species able to tolerate lower oxygen levels predominate. As a general guide, Alabaster and Lloyd (1980) suggest that salmonid species require a 50-percentile oxygen level of about 9 mg litre^{-1} while more tolerant 'coarse' fishes need a 50-percentile oxygen level of about

TABLE 6.9
COMPILATION OF POLLUTION RESPONSES OF MACROINVERTEBRATES IN
FRESHWATERS OTHER THAN EUROPE AND NORTH AMERICA
(South African data based on Chutter, 1972, and Allanson, 1961; Australian
data from Jolly and Chapman, 1966 and New Zealand data from Hirsch, 1958)

(i)   Species indifferent to pollution

South Africa

Oligochaeta	*Limnodrilus* sp.; *Nais* sp.; *Tubifex* sp.
Hirudinea	*Glossiphonia disjuncta*
Crustacea	*Ilyocryptus sordidus*; *Macrothrix*
	spinosa; *Moina dubia*; *Platycyclops*
	poppei
Diptera	*Chironomus calipterus*; *C.formasipennis*;
	C. pilosimanus; *Eristalis* sp.; *Psychoda*
	alternata
Hemiptera	*Sphaerodema nepoides*
Odonata	*Aeschna miniscula*; *Enallagma glaucum*;
	Paragomphus hageni; *Pseudoagrion*
	salisburyense

Australia

Oligochaeta	*Lumbriculus* sp.; *Tubifex* sp.
Mollusca	*Pettancylus* sp.; *Physastra*
	gibbosa
Diptera	Orthocladinae; Tabanidae;
	Tanypodinae

New Zealand

Oligochaeta	Naididae; Tubificidae
Mollusca	*Grundlachia neozelanica*; *Planorbis*
	corinna; *Potamopyrgus*
	antipodum; *P. badia*; *P. corolla*;
	Physastra variabilis
Diptera	*Chironomus pavidus*; *C. zealandicus*
Trichoptera	*Hydrobiosis parumbripennis*;
	H. umbripennis

(ii)   Pollution-tolerant species

South Africa

Coelenterata	*Hydra attenuata*
Platyhelminthes	*Phaenocora Foliacea*
Nemertini	*Prostoma* sp.
Oligochaeta	*Aulophorus* sp.; *Branchiura* sp.;
	Chaetogaster sp.; *Dero* sp.; *Pristina* sp.

TABLE 6.9—*contd.*

Crustacea	*Alona diaphana; Cypridopsis* sp.; *Daphnia similis; Leydigia propinqua Moina rectirostris; Potamon perlatus; Scapholebris Kingi; Simocephalus exspinosus; S. vetulus; Stenocypris* sp.
Ephemeroptera	*Austrocaenis* sp.; *Austrocloeon africanum; A. virgiliae; Baetis bellus; B. harrisoni; Centroptilum excisum; Choroterpes (Euthraulus)* sp.
Hemiptera	*Laccotrephes brachialis; Laccocoris limigenus; Micronecta quewale; M. scutellaris; Nychia marshalli; Plea pullula*
Trichoptera	*Amphipsyche scottae; Cheumatopsyche afra; C. thomasseti; Macronema capense*
Coleoptera	*Amarodytes peringueyi; Copelatus capensis; Guignotus capensis; Herophydrus oscillator; Hyphydrus aethiopicus*
Diptera	*Bezzia* sp.; *Cladotanytarsus reductus; Dichrotendipes perinquayanus; Ephydra* sp.; *Polypedilum natalense; P. vittatum; Simulium adersi; S. nigritarsis; S. ruficorne; Tabanus* sp.; *Tanypus guttatipennis; Trimicra* sp.
Mollusca	*Bulinus tropicus; Burnupia* sp.

Australia

Crustacea	*Boeckella* sp.; *Ceriodaphnia quadriangula; Daphnia carinata; Eucyclops agilis; Paracyclops fimbriatus*
Mollusca	*Lymnaea tomentosa*
Ephemeroptera	Baetidae; Caenidae; Leptophlebiidae; Potamanthidae
Plecoptera	*Dinotoperla* sp. *Leptoperla* sp.; *Trinotoperla* sp.
Hemiptera	Vellidae
Trichoptera	Beraeidae; Calamoceratidae; Hydrophilidae; Hydropsychidae; Leptoceridae; Odontoceridae; Philopotamidae; Polycentropidae;

TABLE 6.9—*contd.*

	Rhyacophilidae; Sericostomatidae
Coleoptera	Dytiscidae; Psephenidae
Diptera	*Antocha* sp.; Ceratopogonidae; Emphididae
Lepidoptera	Nymphulinae
Megaloptera	*Archichanliodes guttiferus*

New Zealand

Platyhelminthes	*Curtisia stagnalis; Spathula fontinalis*
Crustacea	*Paracalliope fluviatilis*
Mollusca	*Gundlachia lucasi; Pisidium novaezelandiae*
Trichoptera	*Hydropsyche* sp.; *Hudsonema amabilis; Olinga feredayi; Oxyethira abiceps; Pycnocentrodes* sp.; *Pynocentria* sp.; *Triplectides obsoleta*
Coleoptera	*Hydora* sp.
Neuroptera	*Archichauliodes dubitatus*

(iii) Pollution-intolerant species

South Africa

Crustacea	*Bosminia longirostris; Chydorus globosus; C. sphaericus; Ilyocypris* sp.; *Megalocypris tuberculata; Paracypretta* sp.; *Pionocypris* sp.; *Simocephalus serrulatus*
Ephemeroptera	*Adenophlebia* sp.; *Afronurus harrisoni; Baetis latus; Centroptilum medium; C. pulchrum; C. sudafricanum; Cloeon* sp.; *Euthraulus elegans; Neurocaenis* sp.; *Pseudocloeon maculosum*
Hemiptera	*Ctenipocoris africana; Enithares sobria; Micronecta dimidiata; Ranatra vicina*
Trichoptera	*Chimarra ambulans; Ecnomus* sp.; *Hydroptila capensis; Macronema* sp.
Plecoptera	*Neoperla spio*
Coleoptera	*Aulonogyrus abdominalis; A. alternata; Dineutus aereus*
Diptera	*Atrichopogon* sp.; *Chironomus leucochlorus; Simulium medusaeforme*
Odonata	*Aeshna* sp.; *Lestes* sp.; *Trithemis risi*

TABLE 6.9—*contd.*

Mollusca	*Anisus natalensis; Gyraulus corniculum; Lymanea columella; L. natalensis; Physopsis africana; Pisidium* sp.; *Unio* sp.
	Australia
Malacostrara	*Paratya australiensis*
	New Zealand
Ephemeroptera	*Atalophlebia* sp.; *Coloboriscus humeralis; Deleatidium* sp.; *Nesameletus* sp.

5 mg litre^{-1}. It should be noted that BOD has only an indirect bearing on the survival of fish: where waters are shallow and turbulent it may be that dissolved oxygen levels are maintained at a sufficient level to permit fish survival. Conversely, even under natural conditions without extra oxygen demand from organic wastes, lethally low oxygen-levels may prevail, for example at night when dense macrophyte growth is imposing a respiratory demand. When coupled with high temperatures such conditions may prove particularly devastating: severe mortality of adult salmon in the River Wye (Wales) occurred in 1976 when dissolved oxygen concentrations fell to 1 mg litre^{-1} and temperatures were around 25–28 °C (Brooker *et al.*, 1977).

6.4.7.2 Effects of elevated ammonia
The toxic effects of elevated ammonia levels in organic effluents are attributable to the proportion of un-ionised ammonia which in turn is related to pH and temperature (Fig. 6.11). Although different fish species vary in their acute sensitivity to ammonia, chronic sensitivity is similar for salmonids and coarse fishes (Table 6.10). The maximum level of un-ionised ammonia for the indefinite maintenance of fish populations is 0·025 mg litre^{-1} (Alabaster and Lloyd, 1980). At lower temperatures (below about 5 °C) the toxicity of the un-ionised ammonia may increase. This factor is particularly significant for polluted water because sewage effluents in temperate zones are likely to have higher winter ammonia levels since the sewage treatment processes are also temperature dependent. Low dissolved oxygen and high free carbon dioxide levels, both often associated with organic effluents, are other factors which influence

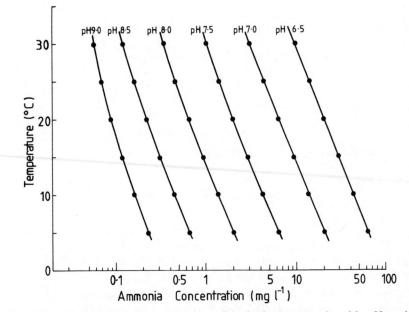

FIG. 6.11 Variation of the proportion of un-ionised ammonia with pH and temperature. Concentration of total ammonia (mg litre^{-1}) which contains an un-ionised concentration of 0·025 mg litre^{-1} under the specified conditions of temperature and pH. Drawn from data in Alabaster and Lloyd (1980).

TABLE 6.10

VARIATIONS IN SENSITIVITY TO AMMONIA BY DIFFERENT FISH SPECIES, BASED ON DATA IN THE LITERATURE

Species	Acute toxicity (mg NH$_3$ per litre)	Chronic toxicity (mg NH$_3$ per litre)
Salmo salar, smolts	0·28 (24 hr LC$_{50}$)	
Salmo trutta, fry	3·60 (10 hr LC$_{50}$)	
Salmo gairdneri, fry	0·07 (24 hr LC$_{50}$)	0·06–0·02
fingerlings	0·44 (96 hr LC$_{50}$)	0·2
adults	0·1 (24 hr LC$_{50}$)	0·49–1·8
	0·5	
Rutilus rutilus		0·42
Scardinius erythrophthalmus		0·24–0·44
Abramis brama		0·50
Perca fluviatilis		0·35–0·6
Cyprinus carpio	1·5–2·0	
Tinca tinca	2·0	

the toxicity of ammonia, although their effects are complex. Lower dissolved oxygen increases the susceptibility of fish to ammonia (Lloyd, 1961) but in field situations, and especially under the influence of organic effluents, lowering of dissolved oxygen is likely to be associated with increased free carbon dioxide and lower pH which tend to offset the higher ammonia toxicity. Exposure to sub-lethal ammonia concentrations increases the resistance of fish to lethal conditions so that populations accustomed to organic discharges may be able to withstand otherwise lethal levels, especially if these are encountered intermittently.

Studies of the field distribution of fish species have shown that coarse fish populations are maintained when the annual median concentration of un-ionised ammonia is between 0·01 and 0·04 mg litre^{-1}.

6.4.7.3 Fish as indicators

The responses of fish populations to eutrophication and other forms of organic enrichment have been reviewed by Larkin and Northcote (1969). The value of fish as indicators of pollution, especially organic enrichment, has been questioned in that they do not necessarily indicate the prevailing quality at a given point since their mobility enables them to avoid unfavourable conditions and return when water quality improves (Price, 1977). Thus, the presence of fish at any given time does not necessarily indicate that conditions are always favourable. Against this may be set the evidence of a restricted home range in many species, for example salmonids (Kalleberg, 1956; Northcote, 1967) and coarse fish (Stott, 1967). This behavioural trait has been utilised by Stott and Cross (1973) who devised an experimental system in which the degree of displacement from the home range may be used as a measure of the intensity of sub-lethal effects of perceived pollutants. The last factor is especially important: positive fish movements away from polluting discharges will only occur if the fish are able to detect the pollutant and respond by moving away. There is evidence that for many pollutants fish are unable to detect sub-lethal concentrations and even if they are able to do so may not be repelled by them. For example, green sunfish (*Lepomis cyanellus*) were not repelled by distressing concentrations of ammonia (Summerfelt and Lewis, 1967). Fish species community diversity has been used as an indicator of pollution in a marine context by Bechtel and Copeland (1970). In freshwater habitats, the numbers of fish species is often small and it may prove to be an insensitive technique. Fish condition factors and community diversity were compared with physico-chemical data and a biotic index by Laurent and Clavet (1977) who concluded that the results were similar and suggested that data from fish surveys could be utilised by less highly specialised staff than would be required for other methods. Certainly, it must be agreed that fish species are generally easily identified but the manpower resources required to collect samples of fish would often outweigh this gain and adequate information on water quality may be acquired from relatively simple analyses of invertebrate data (see Section 9.4).

6.5 FIELD STUDIES OF ORGANIC POLLUTION

In reviewing the many field studies of the effects of organic pollution Hynes (1960) concluded that results had been broadly similar whatever the nature of the organic effluent and whether the work had been conducted in Europe or North America. These results have also been reviewed by Hawkes (1962b) who selected examples of surveys conducted in North America, Great Britain and South Africa. Other studies have been conducted since these reviews were compiled; it is not possible, therefore, to provide a comprehensive account of the literature. However, a few examples from the literature are considered below in order to furnish typical results which illustrate some of the principles outlined in the sections above. It should be noted that they are not necessarily the best examples but in selecting them a conscious effort has been made to avoid those in which the pollution was complicated by the presence of toxic industrial effluents or inorganic wastes with a high content of inert suspended matter. These criteria have precluded some excellent work (for example. Learner *et al.*, 1971) but those included have the advantage that they were investigations of rivers with a single large organic input in their upper reaches, the effects of which could be followed relatively simply.

6.5.1 Langley Brook (Hawkes, 1964)
Langley Brook is a tributary stream of the River Tame which arises in a rural area north-east of Birmingham and received effluent from a sewage treatment works at Langley Mill situated about one-quarter of the distance from the source to the confluence. The effluent prior to 1949 was satisfactory since the works coped adequately with the flow from dispersed dwellings but the erection of a large housing development in the area resulted in progressive deterioration as the works became increasingly

overloaded. A new works was built in 1955 and an improved effluent was discharged in summer although the quality of the effluent in winter was less satisfactory.

The condition of the benthic riffle communities at four stations (above the outfall, 90 m, 0·6 km and 2·4 km below the outfall) was examined over a nine year period which included the decline and improvement of the effluent (Fig. 6.12). Several trends were discernible and three kinds of response were exhibited by different groups of organisms.

The typically dominant or common riffle community organisms declined numerically as the organic load increased: this is seen in the response of species such as *Gammarus pulex* and *Baetis rhodani* at stations 2 and 3. The stoneflies (Plecoptera) and other ephemeropteran (mayfly) nymphs (principally *Ecdyonurus*) and the mollusc *Ancylus fluviatilis* showed a similar response at station 4. Other members of the riffle community such as the leeches (Hirudinea), *Simulium* larvae and the caddis *Hydropsyche* appeared first to benefit from the organic enrichment with enhanced numbers but, as the organic loading increased their decline followed, according to their individual sensitivities.

A third group, not normally present in the riffle community or if present not forming a significant component of it, was able to exploit the conditions created by increasing organic loading and became the dominant organisms. Examples of this group include the crustacean *Asellus aquaticus* and the midge larvae *Chironomus riparius*. These animals were eliminated by the grossest organic pollution.

Changes were also evident in the flora growing on the stones. The dominant alga *Cladophora* was almost eliminated during the period of greatest organic loading while another alga, *Stigeoclonium*, uncommon before the pollution began, increased to become the dominant filamentous alga as the organic loading increased. The 'sewage fungus' community also increased and was commonly principally *Sphaerotilus* although the colonial protozoan *Carchesium* made up the bulk of the 'sewage fungus' during the period following the inauguration of the enlarged works.

Examination of the communities at each of the four stations demonstrated the general principles of self-purification of organic pollution in streams. The brook at station 1 (above the outfall) was about 1 metre wide and here the community was always dominated by *Gammarus pulex* while *Baetis rhodani* was common. Other components of the fauna of this station included *Simulium* larvae, *Hydropsyche* larvae, the leeches *Glossiphonia* and *Erpobdella*, chironomid larvae, the molluscs *Lymnaea*

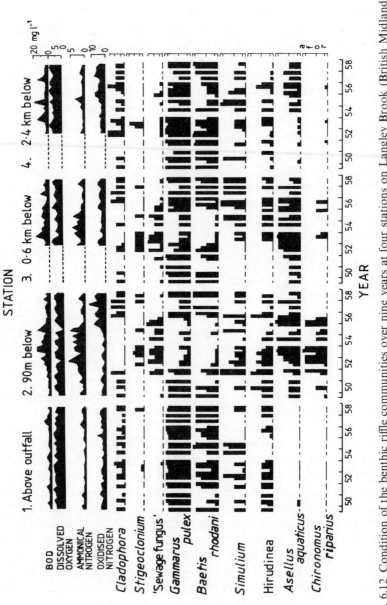

FIG. 6.12 Condition of the benthic riffle communities over nine years at four stations on Langley Brook (British Midlands), above and 90 m, 0·6 km and 2·4 km below the entry of a sewage effluent. After Hawkes (1964).

peregra and *Ancylus fluviatilis* together with certain limnephilid Trichoptera.

The numbers of *Gammarus* were similar at all four stations until 1957 when discharges of sewage effluent increased. In that year the population declined at station 2 and by the following year numbers had declined at station 3. When the organic load decreased in 1955 *Gammarus* populations recovered and this occurred earlier at station 3. At station 2 there was a considerable delay in recovery while only slight suppression of numbers could be observed at station 4 during this period. *Baetis rhodani* was observed to respond in a very similar manner to *Gammarus* at the four stations. Other species of mayfly, stoneflies and the freshwater limpet *Ancylus* were rarely taken at stations 2 and 3 but were present at station 4 where their numbers declined during the period of increased organic enrichment.

The variations in abundance of *Simulium* larvae responded beneficially to increased organic loading and differences in population numbers reflected the degree of self-purification. Its occurrence at station 1 was sporadic but at station 2 its maximum abundance occurred before and after the period of highest pollution load. At station 3 its numbers were greatest during the recovery period while at station 4 its highest abundance occurred throughout the time when organic input increased.

Another species which reflected changes in organic load and self-purification was *Asellus aquaticus*. It was almost always absent at station 1 yet always present at stations 2 and 3. Numbers increased as the loading increased until the period of maximal polluting load when numbers declined at station 2 indicating that conditions immediately below the discharge had exceeded the optimum for this species. At station 4, where *Asellus* was rare until organic loadings increased, numbers varied with the pollution load. When the organic load decreased, the *Asellus* population did not decline immediately but was maintained until *Gammarus* numbers increased. It was postulated that this relationship might indicate some interspecific competition. The fortunes of another organic pollution 'opportunist' species, *Chironomus riparius*, were very similar to those of *Asellus* but the changes in population numbers, both increases and decreases, occurred more rapidly. The response to increased loading was dramatic but it was observed mainly at station 2 with little increase at station 3, and this species was virtually absent from station 4. The incidence of *Chironomus riparius* was thus more restricted, both temporally and spatially, than *Asellus*.

The responses of different macroinvertebrates to the organic loading at a given time (1953) are indicated in Fig. 6.13, where the percentage distribution, expressed as a fraction of the total number found at all stations, is plotted against the distance from the outfall.

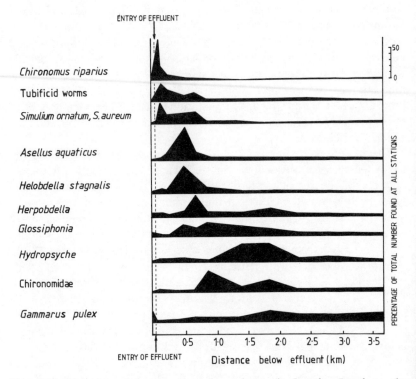

FIG. 6.13 Responses of benthic macroinvertebrates in Langley Brook to the organic loading from a sewage effluent in 1953. After Hawkes (1964).

Floral changes also tended to follow the classical patterns associated with organic pollution. The dominant alga above the effluent was *Cladophora* but it did not grow profusely and over-wintered as small tufts on the larger stones. 'Sewage fungus' and the alga *Stigeoclonium* rarely occurred above the effluent outfall but rich growths of the red alga *Batrachospermum* were sometimes present. During the period of highest organic loading, Cladophora disappeared from station 2 to be replaced by *Stigeoclonium* and 'sewage fungus' which were otherwise rare at this point. The impact of high loadings on *Cladophora* was less at station 3 and hardly discernible at station 4. *Stigeoclonium* was virtually as abundant at station 3 as it was at 2 and 'sewage fungus' was only slightly less abundant. At station 4 *Stigeoclonium* appeared only briefly and 'sewage fungus' was confined to the underside of stones.

In addition to the filamentous algae, diatoms were commonly abundant at all stations. Seasonal changes tended to be more evident than spatial changes. At all stations *Rhoicosphenia curvata*, *Cocconeis placentula* and *Gomphonema olivaceum* were most common during summer while *Navicula radiosa*, *Surirella ovata* and *Synedra ulna* were more common at other seasons. These changes were associated largely with the epiphytic habit of the first three species which were associated with the summer incidence of *Cladophora*. Some spatial changes could be discerned, however. For example, *Rhoicosphenia* and *Navicula* formed a lower proportion of the diatom flora at station 2 where *Surirella* and *Gomphonema* assumed greater importance. The reaction of *Synedra* to the maximal loading at station 2 varied seasonally: it tended to form a greater proportion of the diatom population at this station in spring and summer than it did in autumn or winter.

6.5.2 River Cole (Hawkes and Davies, 1971)

The River Cole is a small river which rises to the south-west of Birmingham, flowing first through a rural area and then a residential area after which it enters a highly industrialised zone of the city. Within the city it received sewage effluent before it joins the River Blythe almost immediately upstream of that river's confluence with the River Tame.

The section considered in detail below is the upper reach (before it enters the industrial region of Birmingham) where the river received an organic discharge from an overloaded sewage works. Six sampling stations were established at riffles, one above the discharge and five downstream, at approximately 1 km intervals.

Hawkes and Davies (1971) published a table of the mean population densities of the more important taxonomic components. Their data are reproduced in Fig. 6.14 where the changes in community composition following self-purification can be seen. Trends noted above are also seen here. For example, the crustacean *Gammarus pulex* was abundant upstream of the discharge but was virtually absent from all downstream stations, except the last. Here, numbers were greater but much less than upstream of the discharge. In contrast, *Asellus aquaticus* was unimportant until station 5 (about 4·5 km below the outfall) where it dominated

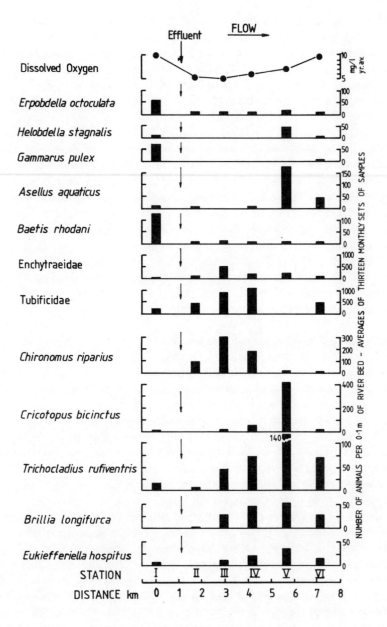

FIG. 6.14 Changes in the composition of the benthic invertebrate community in the River Cole, England, in relation to an organic discharge. After Hawkes and Davies (1971).

the fauna, along with several species of Chironomidae. The numbers of tubificid worms and the chironomid *Chironomus riparius* first increased and then declined as one moved downstream reaching their peaks at stations 4 and 3, respectively. Numbers of the mayfly *Baetis rhodani* declined rapidly below the outfall and remained low at all stations. Full recovery to a clean-water fauna was not observed in the section described here.

Several seasonal changes in the fauna were apparent: these are naturally not evident in the description above which is based on mean annual data. As an example of this one may take the numbers of larvae of the chironomid *Eukiefferiella hospitus*. At the lowest stations (4–6) it reached peaks of abundance three times in the year, possibly caused by the emergence of three generations. At station 3, where oxygen levels were more severely depleted in the summer months, the summer generation was virtually eliminated although the spring and autumn generations were present. It is thought that aerial dispersal of the adults enabled this species to re-establish itself after the summer generation had been eliminated. At station 2, immediately below the point of effluent discharge, the summer generation was also eliminated and the spring and autumn peaks were greatly depressed. The chironomid *Prodiamesa olivacea* was affected similarly but another species, *Brillia longifurca*, which is normally most abundant during winter and spring was largely unaffected. Nymphs of *Baetis rhodani* were evident at stations 2 to 5 only in spring but were also abundant in summer above the outfall. Another chironomid, *Chironomus riparius*, extended its range downstream and attained maximum densities further downstream in summer than in winter.

Perhaps partly as a result of poorer powers of dispersion when compared with insects having an aerial phase, the crustacean *Gammarus* did not increase numerically in winter even though measurements of daytime oxygen levels suggested that conditions were suitable for its survival. However, on further investigation it was discovered that, in the recovery zone (station 5), where algae had re-established themselves and daytime oxygen levels were adequate, during the hours of darkness oxygen levels were severely depleted and were too low to allow *Gammarus* survival.

Although the spatial distribution of many of the dominant components of the benthic invertebrate fauna could be related to the prevailing dissolved oxygen levels, for some species it was determined by their food: for example, the leech *Helobdella stagnalis* followed closely the distribution of its prey, *Asellus aquaticus*.

6.5.3 River Lee (Bryce *et al.*, 1978)

This example was chosen because it typifies a condition which is nearer to that now prevailing in many British rivers (and probably elsewhere) than that represented by the more 'classical' organic enrichment sequences of the two previous examples.

The River Lee rises near Luton and flows in a generally south-easterly direction to join the River Thames. In flowing through Luton the river receives only urban run-off but below Luton it receives a large volume of high quality sewage effluent from Luton (East Hyde) sewage works. In

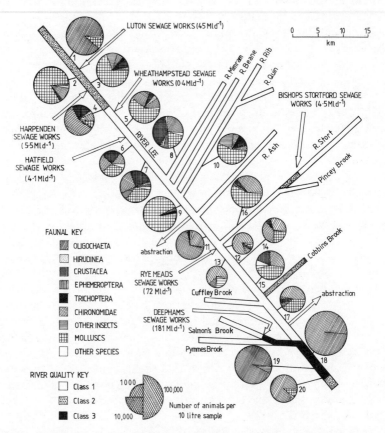

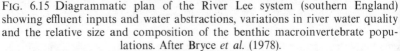

FIG. 6.15 Diagrammatic plan of the River Lee system (southern England) showing effluent inputs and water abstractions, variations in river water quality and the relative size and composition of the benthic macroinvertebrate populations. After Bryce *et al.* (1978).

dry summer weather the river consists largely of high quality sewage effluent (Hynes, 1960). By the time the river has reached Hertford, a distance of about 30 km, the river has recovered considerably from the effects of this effluent by 'self-purification'.

The enrichment of the upper river is probably responsible for the dense growths of alga *Cladophora* which occur on gravel substrates in shallows. The macrophyte *Ranunculus fluitans* is also abundant.

Chemical determinands show little deterioration in water quality downstream of the discharge from Luton sewage works and its impact is only evident from an increase in chloride, ammonia, nitrate and ortho-phosphate. A little further downstream a rise in BOD is evident, but this sharply declines. The water quality of the River Lee through Luton and below the sewage works remained within Class 2 (National Water Council, 1981) but attained unpolluted status (Class 1) about 8 km below the sewage outfall, presumably by self-purification.

Some unusual changes in the macroinvertebrates were evident (Fig. 6.15). Below the discharge a marked reduction in the density of olig-ochaetes was observed with an equally dramatic increase in the abun-dance of molluscs. The diversity of the invertebrate community decreased but two commonly used biotic indices, the Chandler Biotic Score and Trent Biotic Index (see Section 9.4.2 and Appendices 1 and 2) increased. When the communities at each station were compared by means of a similarity index the station above the sewage outfall had almost no similarity to any other and the station immediately below, although having some affinity with subsequent downstream stations, was also rather distinct. All other stations in the upper Lee had moderate similarity.

Bryce *et al.* (1978) concluded that in enriched, mildly polluted waters changes in the nature of the substratum may have at least as great an influence on the fauna as the mild organic pollution. Similarly, urban run-off may restrict the faunal diversity of a river more severely than well-treated sewage effluent.

CHAPTER 7

Effects of Toxic Materials

7.1 INTRODUCTION

Toxic materials considered here are largely the waste by-products of industrial activity or the residues of substances, such as pesticides, which have been manufactured for use in agriculture, horticulture or forestry. The nature and character of these poisons (toxicants) have changed over the years, as has the relative importance attached to them. Around the turn of the century and until quite recently, many poisons were associated with heavy industries and included ammonia, phenols, hydrocarbons, cyanide and tars from town-gas manufacture, coking plants, or heavy metals (for example, chromium, nickel, copper and cadmium in plating wastes). Reduction in the importance of coal as a source of energy and improvements in the treatment of effluents either by in-plant units or by discharge to sewer and combined treatment with domestic sewage has changed the significance of industrial effluents as a source of poisons. However, the recognition that diffuse sources of certain substances, notably pesticides and some heavy metals, which may be accumulated and concentrated in food chains, means that the emphasis on the short-term (acute) toxicity of poisons has given way to greater interest in their long-term (chronic) consequences.

The study of the effects of poisons is complicated by the fact that there are approximately half a million chemicals currently in use, although only about 10 000 or so are produced in amounts between 500 kg and 1000 tonnes (Goodman, 1974) and many of these eventually enter freshwaters, although perhaps for the majority the quantities are small. In addition to those which are very toxic, substances which are potentially significant in environmental terms are those which tend to be accumulated in organisms or which are persistent through inherent

TABLE 7.1

TOXIC SUBSTANCES PRESENT IN INDUSTRIAL EFFLUENTS

Substance	Source
Acids	Chemical industries, battery manufacture, minewaters, iron and copper pickling wastes, brewing, textiles, insecticide manufacture
Alkalis	Kiering of cotton and straw, cotton mercerising, wool scouring, laundries
Ammonia	Gas and coke production, chemical industries
Arsenic	Phosphate and fertiliser manufacture, sheep dipping
Cadmium	Metal plating, phosphate fertilisers
Chlorine (free)	Paper mills, textile bleaching, laundries
Chromium	Metal plating, chrome tanning, anodising, rubber manufacture
Copper	Plating, pickling, textile (rayon) manufacture
Cyanide	Iron and steel manufacture, gas production, plating, case hardening, non-ferrous metal production, metal cleaning
Fluoride	Phosphate fertiliser production, flue gas scrubbing, glass etching
Formaldehyde	Synthetic resin manufacture, antibiotic manufacture
Lead	Paint manufacture, battery manufacture
Nickel	Metal plating, iron and steel manufacture
Oils	Petroleum refining, organic chemical manufacture, rubber manufacture, engineering works, textiles
Phenols	Gas and coke production, synthetic resin manufacture, petroleum refining, tar distillation, chemical industries, textiles, tanning, iron and steel, glass manufacture, fossil fuel electricity generation, rubber processing
Sulphides	Leather tanning and finishing, rubber processing, gas production, rayon manufacture, dyeing
Sulphites	Pulp processing and paper mills, viscose film manufacture
Zinc	Galvanising, plating, rubber processing, rayon manufacture, iron and steel production

chemical stability or poor biodegradability. Substances which do not form complexes with organic matter or soils and which are readily soluble and, therefore environmentally mobile, are also likely to be important.

A selection of substances which are generally toxic to freshwater organisms together with an indication of common sources of these materials are given in Table 7.1. It should be noted that this list is by no means exhaustive and is intended merely to show the variety of chemicals which may be involved. This list may be compared with the EEC

TABLE 7.2
EEC 'BLACK' AND 'GREY' LIST SUBSTANCES

List No. 1 ('Black List')

The substances on this list were selected mainly for their toxicity, persistence or bioaccumulation

1. Organohalogen compounds and substances which may form such compounds in the aquatic environment
2. Organophosphorus compounds
3. Organotin compounds
4. Substances, the carcinogenic activity of which is exhibited in or by the aquatic environment (substances in List 2 which are carcinogenic are included here)
5. Mercury and its compounds
6. Cadmium and its compounds
7. Persistent mineral oils and hydrocarbons of petroleum
8. Persistent synthetic substances

List No. 2 ('Grey List')

These substances are regarded as less dangerous than those of List 1 and the impact of which may be local

1. The following metalloids/metals and their compounds:

1. Zinc	6. Selenium	11. Tin	16. Vanadium
2. Copper	7. Arsenic	12. Barium	17. Cobalt
3. Nickel	8. Antimony	13. Beryllium	18. Thalium
4. Chromium	9. Molybdenum	14. Boron	19. Tellurium
5. Lead	10. Titanium	15. Uranium	20. Silver

2. Biocides and their derivatives not appearing in List 1
3. Substances which have a deleterious effect on the taste and/or smell of products for human consumption derived from the aquatic environment and compounds liable to give rise to such substances in water
4. Toxic or persistent organic compounds of silicon and substances which may give rise to such compounds in water, excluding those which are biologically harmless or are rapidly converted in water into harmless substances
5. Inorganic compounds of phosphorus and elemental phosphorus
6. Non-persistent mineral oils and hydrocarbons of petroleum origin
7. Cyanides, fluorides
8. Certain substances which may have an adverse effect on the oxygen balance, particularly ammonia and nitrites

---

'Black List' and 'Grey List' substances of Table 7.2. The responses of organisms to exposure to these substances are complex, being determined, in part, by the nature of the substance, its concentration, the duration of exposure and the sensitivity of the organism to that substance. These aspects are dealt with more fully below, but the important consequence of these variable factors is that it is not possible to predict with desirable accuracy the responses of biota to given toxic pollutants. Indeed, the term 'toxic' itself is an omnibus term, since responses to a given concentration of a substance may range from rapid death to no observable effect, depending on the organism involved. There are also variations in the responses of individuals of the same species to a given level of toxicant.

In order to try to overcome these problems, standardised toxicity tests have been devised (see Section 8.2.2.1). In using these tests it is hoped that variability of responses caused by such factors as varying concentrations of toxicant, variability in the organisms used and the conditions under which the determinations are made (for example, pH, temperature, dissolved oxygen, etc.) are minimised. Under ideal conditions, similar toxicities should be manifested whenever the test is carried out. This facilitates comparisons of the effect of different concentrations of one substance or of the relative toxicity of different substances to be made under different conditions.

Problems often arise when one tries to relate the toxicities determined under controlled experimental conditions in the laboratory to the circumstances which prevail in the environment. Here the toxicity data serve as a general guide to the probable effects of the release of poisons, but natural waters may contain organic substances which form complexes and modify their effects or the presence of other materials may heighten or reduce the effects observed in the laboratory. Rarely are poisons present at fixed concentrations and as the effect varies with concentration it may be impossible to predict the likely outcome from mean levels or even maximum levels unless their duration is known, and even then not with any degree of certainty.

These warnings are necessary because many attempts have been made to apply toxicological data as though they were precise determinations of the kind which analysts can readily furnish and without regard to the range of variable responses exhibited by organisms and with which biologists are only too familiar.

In subsequent sections consideration is given to the processes which contribute to the effects of poisons, that is their physiological consequences, the kinds and degrees of response and a review of the toxicities to different organisms of common poisons. As might be expected, more data are available for the effects of poisons on fishes.

7.2 PHYSIOLOGICAL RESPONSES TO POISONS

Many aquatic organisms are relatively permeable to their surrounding aquatic medium and so, unlike terrestrial organisms, poisoning need not depend on swallowing or inhaling the toxicant but it may simply be absorbed through the body surface.

The effect of a poison depends upon its concentration and the duration of exposure. Over a wide range of concentrations of the poison these two factors are often (though not necessarily) reciprocal but, as will be seen shortly, at high concentrations the time of exposure required for the observed effect (for example, death) may be the same even when the concentration is increased and at very low concentrations natural death may occur before the toxicant has any effect.

With many poisons, at very low concentrations or short periods of exposure, the normal homeostatic mechanisms of the organism are able to cope and the substance is excreted or detoxified so that no damage is sustained. Increasing concentrations or longer exposure may initially cause reversible physiological changes and subsequently some impairment of function. As concentrations are increased or exposure prolonged a critical point is reached where significant changes are induced and even if normality is restored some permanent damage may have been suffered. Ultimately, increasing concentrations or period of exposure will cause

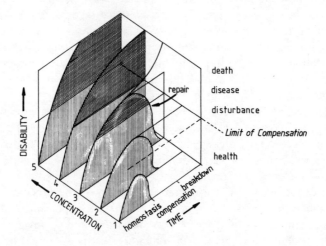

FIG. 7.1 Effect of increasing dose of a poison on the physiological processes of an organism. From Tooby, unpublished.

disease and death. These stages are shown diagrammatically in Fig. 7.1. Depending on the poison involved and the extent to which the organism is exposed to its effects, that is the dose, the response may vary from no perceptible change to death, with sub-lethal effects such as impaired behavioural responses, sterility, disturbed metabolism and formation of lesions as possible intermediate consequences.

Much of the early experimental work on the effects of poisons was devoted to investigating the rapid effects, usually death, of relatively high concentrations which were administered for short periods. Although such acute toxicity studies are still important, especially in establishing the significance of new substances, attention is now focused on the long-term (chronic) effects, particularly the sub-lethal responses, since it is now recognised that in ecological terms disturbed behaviour patterns, impaired physiology or induced sterility can have more or less the same ultimate effect on populations or organisms as the fairly rapid death of individuals.

7.3 TERMINOLOGY AND CRITERIA

Details of toxicity tests and the terms used in deriving and expressing the results of toxicity tests are covered more fully in Section 8.2.2.1. It is necessary, however, to understand something of the terminology employed in order to be able to interpret the data and observations on ecological effects. These terms are given and defined in Table 7.3. The most commonly encountered European term in studies of acute toxicity is the median lethal concentration (LC_{50}), which is equivalent to the term median tolerance limit (TL_m) which is more commonly used in North America. For the long-term survival of populations there is more difficulty in finding an appropriate measure of the concentration of poisons. What is required is a level which is 'safe', but this is difficult to obtain since all populations have variable natural mortality. The 'threshold concentration' concept is useful in that it is applied in long-term toxicity studies to the level at which there is no significant difference in survival between an experimental population exposed to this concentration and the control population. This is equivalent to the MATC or Maximum Acceptable Toxicant Concentration, that is the environmental level which will permit the indefinite survival of a population of a given species. A penetrating review of concepts and terminology used in

TABLE 7.3
TERMS USED IN EXPRESSING TOXICITY DATA

Term	Definition	Abbreviation/example
Acute (short-term) tests		
Median effective concentration	Concentration at which a specified effect is observed in half the population within a given time	48 hr EC_{50}
Median lethal concentration	Concentration at which half the population will die in a specified time	96 hr LC_{50}
Median tolerance limit	Concentration at which half the population will die in a specified time	96 hr TL_M (TL_{50})
Median effective time	Time taken for an observed effect to occur in half the test population at a given concentration	ET_{50}
Median survival time or median period of survival	Time taken for half of test organisms to die at a given concentration	LT_{50}
Chronic (long-term) tests		
Incipient lethal level	Concentration of toxic substance or other potentially lethal condition (e.g. low oxygen) which organisms could tolerate indefinitely	ILL
Maximum acceptable (allowable) toxicant concentration	(Maximum) concentration of the toxic substance which is acceptable within the environment in order to ensure no harm to organisms	MATC

toxicology has been provided by Brown (1973). Particular stress is laid upon the necessity for accurate terminology and recognition of the variability of organism responses and limitations of experimental methods.

More recently the concept of 'quality criteria' has come into favour.

TABLE 7.4

WATER QUALITY CRITERIA FOR FRESHWATER FISH AND AQUATIC LIFE BASED ON EPA (USA) AND EIFAC (EUROPE) GUIDELINES

Substance	Criteria for protection of fish		Criteria for protection of aquatic life
	EIFAC	EPA	EPA
Aldrin/Dieldrin		0·003	0·003
Ammonia (un-ionised)	25	(20)	20
Arsenic			100
Cadmium hardwater	1·5[a]	1·2	12
softwater	0·9[a]	0·4	4
Chlordane		0·01	0·01
Chlorine	4	2	10
Chromium		(100)	100
Copper hardwater	112[a]	0·1 96 hr LC_{50} (c. 10)	0·1 96 hr LC_{50}
softwater	22[a]	0·1 96 hr LC_{50} (c. 2)	0·1 96 hr LC_{50}
Cyanide		5	5
DDT		0·001	0·001
Endosulphan		0·003	0·003
Endrin		0·004	0·004
Heptachlor		0·001	0·001
Lead hardwater		0·01 96 hr LC_{50} (c. 500)	0·01 96 hr LC_{50}
softwater		0·01 96 hr LC_{50} (c. 5)	0·01 96 hr LC_{50}
Lindane (HCH)		0·01	0·01
Malathion		0·1	0·1
Mercury		0·05	0·05
Methoxychlor		0·03	0·03
Nickel		0·01 96 hr LC_{50} (c. 100)	0·01 96 hr LC_{50}
Oil		0·01 96 hr LC_{50} (c. 100)	0·01 96 hr LC_{50}
Parathion		0·04	0·04
PCBs		0·001	0·001
Phenols	1 000	1[b]	
Phthalate esters		3	3
Selenium		0·01 96 hr LC_{50} (c. 20)	0·01 96 hr LC_{50}
Silver		0·01 96 hr LC_{50} (c. 0·1)	0·01 96 hr LC_{50}
Sulphide (undissociated $H_2 S$)		2	2
Zinc hardwater	50[a]	0·01 96 hr LC_{50} (c. 25)	0·01 96 hr LC_{50}
softwater	20[a]	0·01 96 hr LC_{50} (c. 3)	0·01 96 hr LC_{50}

[a] 95 percentile values.
[b] Value selected to avoid tainting of flesh.
Values are in µg litre^{-1}. Where application factors of the 96 hr LC_{50} are specified approximate values are given in brackets.

These are concentrations of substances which, if not exceeded, provide long-term protection to aquatic organisms, fish or even man when applied to potable water. Quality criteria are based on toxicity tests, both chronic and acute, the latter being used to assess the relative toxicity from which a long-term value may be predicted by using an 'application factor', for example one-hundredth of the 96 hr LC_{50}. The magnitude of the apparently arbitrary application factor is derived from previous experience of similar substances and any other relevant data but usually erring on the side of safety. As more is known of the effect of a substance so the quality criteria may be modified, either by relaxing the stringency imposed initially for safety reasons or by further lowering of acceptable levels when evidence of bioaccumulation or persistence accrues. Examples of quality criteria for fisheries and protection of other aquatic life are given in Table 7.4. These are based on EIFAC recommendations (Alabaster and Lloyd, 1980) and USA criteria from the EPA (Train, 1979).

7.4 FACTORS WHICH AFFECT TOXICITY

In applying the results of toxicity tests and published toxicity data to field situations, it is necessary to take into account a number of factors which may influence the toxicity of a given poison. These are set out below.

7.4.1 Biological Variability

Organisms are not identical: individuals of the same species differ and these differences are evident in toxicity tests. This is one reason why median values are often used when reporting the results of such tests, and accounts, in part, for variations in the values quoted in the literature for measurements of the toxicity of a given substance to apparently similar test populations. In an attempt to overcome this problem it has been proposed that reference toxicants should be used to identify differences in the sensitivity of test populations (Davis and Hoos, 1975). A number of substances have been proposed as reference toxicants and although Alexander and Clarke (1978) found that phenol was better than four other substances tested for detecting differences in sensitivity among rainbow trout, they questioned the value of choosing a single reference toxicant. They suggested that a series of physiological and behavioural tests coupled with diagnostic health checks might be more useful than reference toxicants. A balance must be struck, however, between devoting attention to the fine details of toxicity tests to make them more reproducible and the relevance of this to the use of toxicity tests in regulating field conditions where such controls are impossible, as noted above.

An alternative approach, which is simple and most attractive, is to use a test species such as the crustacean *Daphnia* which can reproduce by pathenogenesis (Scourfield and Harding, 1966) and thereby can be cloned to obtain genetically identical individuals. Again, this technique improves the test procedure but does not necessarily assist one when applying the data to field conditions.

7.4.2 Age or Development Stage

Different stages in the life histories of organisms may differ in their susceptibility to given concentrations of poisons. It is often considered that eggs or larval stages are more sensitive than adults but often the reverse is true, especially where eggs have protective or impermeable membranes. For example, many fish eggs are less sensitive to heavy metals and phenol than adults of the same species (Jones, 1964).

7.4.3 The Chemical Quality of the Environment

Water quality, principally hardness and pH, is known to modify the toxicity of a number of poisons. In general, heavy metals are more toxic to fish in soft water than in hard water. For example, the 60 hour median lethal concentrations of zinc for rainbow trout increases eightfold when hardness is raised from 12 to 320 mg litre^{-1} as $CaCO_3$ (i.e. 0·5 and 5 mg Zn per litre, respectively) (Lloyd, 1960).

The variation in toxicity of ammonia with pH has been described above (6.4.7.2) and is attributable to the degree of ionisation. As an example, an increase of pH from 7·0 to 7·3 would double the concentration of the toxic un-ionised ammonia in an ammonia solution (Alabaster and Lloyd, 1980), although this effect is reduced above pH 8·5. Nitrite toxicity is also pH-dependent and varies with presence of other anions (Russo *et al.*, 1981). The interrelationships of factors which make up water quality in affecting the toxicity of a poison are well illustrated by the effects of hardness, alkalinity and pH on the toxicity of copper to rainbow trout (Fig. 7.2, after Miller and Mackay, 1980).

7.4.4 Other Stresses

Other factors which influence the toxicity of poisons include dissolved oxygen and carbon dioxide levels and temperature. Examples of these effects include the increase in toxicity under lowered oxygen levels of cyanide (Downing, 1954), ammonia (Merkens and Downing, 1957), monohydric phenols, lead and copper (Lloyd, 1961). The toxicity of un-ionised ammonia decreases with the concentration of free carbon dioxide

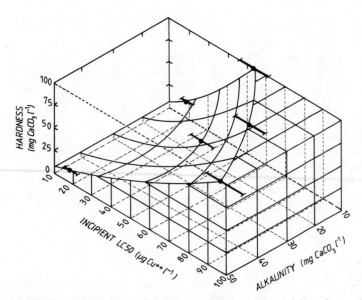

FIG. 7.2   Effects of hardness, alkalinity and pH on the toxicity of copper to rainbow trout (*Salmo gairdneri*). After Miller and Mackey (1980).

(Lloyd and Herbert, 1960). Temperature may also affect the degree of toxic effect of a poison. For example, the resistance of rainbow trout to poisoning by phenol increases with increasing temperature up to 18 °C: at this temperature the 48 hr LC_{50} is almost twice that at 6 °C (Brown *et al.*, 1967). Similarly, the toxicities of ammonia and cyanide are greater at lower temperatures (3 °C) than at higher temperatures (13 °C) (Brown, 1968).

Salinity is another possible source of stress, especially in brackish environments such as estuaries. For example, Jones *et al.* (1976) found that the toxicity of copper to the polychaete worm *Nereis diversicolor* was increased at high and lower salinities.

7.4.5 Chemical Species and Complexes

The importance of the identity of the chemical species in determining toxicity and the possibility of the formation of complexes or adsorption onto particulate matter is well illustrated by reference to copper. The more commonly used methods of analysis measure the total amount of copper in the sample. The metal is readily adsorbed onto solid particles and may not, therefore, be available as a poison. The amount so adsorbed may be estimated by difference on filtration of part of the sample. The formation of complexes may also influence toxicity: those formed in the presence of organic, and especially humic, substances may be particularly stable. Toxicity appears to be related to the soluble copper present and there is evidence that the copper ion (Cu^{2+}) and perhaps the hydroxide ($CuOH^+$) are the only toxic forms of copper (Stiff, 1971b; Alabaster and Lloyd, 1980; Waiwood and Beamish, 1978).

It seems quite likely that other heavy metals behave similarly. Certainly there is evidence that zinc toxicity may be reduced by chelating agents such as EDTA (ethylene diaminetetra-acetic acid) and NTA (nitrilotriacetic acid), and possibly also by the presence of organic matter (Grande, 1967; Sprague, 1968).

7.4.6 Presence of Other Poisons

Often the toxicity of a mixture of poisons is additive (see Section 7.5), that is, equivalent to the sum of the proportions of the acutely toxic concentrations of each poison present (Brown, 1968), but occasionally the presence of more than one poison may give a toxicity which differs

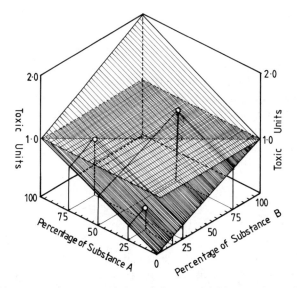

Fig. 7.3 Diagrammatic representation of the potential interactions of toxicants. Inclined plane indicates combined toxicity of different proportions of two substances of equal toxicity and with purely additive effects. The horizontal plane indicates the level of one toxic unit.

from that which might be expected if their individual toxicities were simply additive. Sometimes the effect of a mixture is greater than that expected from the sum (synergistic) while with other combinations the effect is less than expected (antagonism). These relationships are shown diagrammatically in Fig. 7.3 and the principles have been discussed by Jones (1964), Warren (1971) and Calamari and Alabaster (1980). Examples of these effects include synergism between nickel and chromium in which nickel toxicity increases about ten-fold in the presence of chromium, and antagonism between nickel and strontium where solutions of equal individual toxicity may have only a third of their expected toxicity when combined (Jones, 1939).

7.4.7 Using Published Toxicity Data

It will be abundantly evident from the examples provided in the preceding six sections that great care must be exercised when using published data since many factors influence toxicity. Such a warning implies no criticism of the data or the methods by which they are obtained. To do so would be unreasonable in view of the care with which toxicities are usually determined. The problem is largely one of matching exactly the water quality (hardness, pH, temperature, etc.) and biological factors (species, age, environment prior to exposure, etc.) which obtained under the experimental conditions. In the laboratory these variables are measured and controlled; in the field they are seldom under control and often inadequately measured. A further complication which must be recognised is that many pollutants do not occur in isolation: many wastes and effluents contain several toxic substances. These may interact, as explained in Section 7.4.6. Thus, in using published data to estimate the likely effects of many effluents, some means of predicting the toxicity of combinations of toxins is required. This is discussed in the next section.

7.5 PREDICTING THE TOXICITY OF COMBINATIONS OF POISONS

Poisons in effluents rarely occur singly so that for purposes of environmental protection it is necessary to know the probable toxicity of combinations. One may do this by testing appropriate mixtures as though they were single substances but the value of such work is limited by the large range of combinations which could be encountered. The literature on single substances is extensive enough: tests of all likely permutations would be a daunting prospect. Fortunately, it is possible to make reasonable predictions of the toxicity of combined poisons from their individual toxicities.

For example, the acute toxicity of mixtures of several pairs of poisons to the rainbow trout, *Salmo gairdneri*, has been shown to be predictable from the sum of the threshold concentration of median resistance (LC_{50}) of the individual poisons (Brown, 1968). That is:

$$\frac{A}{T_A} + \frac{B}{T_B} = 1$$

where A = concentration of poison A (mg litre^{-1}); B = concentration of poison B (mg litre^{-1}); T_A = median lethal concentration of poison A (e.g. 48 hr LC_{50}, mg litre^{-1}); and T_B = median lethal concentration of poison B (e.g. 48 hr LC_{50}, mg litre^{-1}). This relationship has been found to hold for mixtures of zinc and copper (Lloyd, 1961), ammonia and phenol (Herbert, 1962), ammonia and zinc (Herbert and Shurben, 1964), phenol and zinc, copper and ammonia (Herbert and Vandyke, 1964) and copper and phenol (Brown and Dalton, 1970). It also holds for mixtures of three poisons, for example ammonia, phenol and zinc, provided that the zinc does not predominate (more than 70%), and this was also applicable for fluctuating concentrations within $\pm 50\%$ with equal intervals of time (Brown *et al.*, 1969); copper, zinc and phenol and also copper, zinc and nickel (Brown and Dalton, 1970). It seems probable that, with a few exceptions, the sum of the fractional toxicities will be a reasonable estimate of the acute toxicity of combinations of most poisons.

Chronic toxicity prediction is less well developed, although it is now common to estimate long-term effects from the sum of fractional threshold toxicities, 'safe' concentrations or 'no-effect' levels. However, such predictions are liable to greater errors than acute toxicity estimates simply because chronic toxicity is less easily defined.

A promising empirical approach to predicting chronic toxicity has developed in the United Kingdom in which fractions of acute toxicity determinations (application factors) are used to estimate chronic toxicity, both for single poisons and mixtures, the appropriate fractions being derived from studies of field populations. In studies of fishless rivers in the British industrial Midlands, Herbert *et al.* (1965) placed rainbow trout in cages in rivers or in aerated river water in aquaria on the river bank and noted their survival. They concluded that although an insignificant proportion of fish as sensitive as rainbow trout would be killed if the concentrations never exceeded 0·7 of their expected thresholds, lower toxicity would be necessary in order to maintain fish populations indefinitely, probably below 0·2 of the expected threshold. Further, similar, work by Brown *et al.* (1970) showed that there was a tendency to underestimate toxicity using laboratory derived data: the observed 48 hr LC_{50} for river waters was only 0·6–0·7 times the predicted value. They reviewed the problem of chronic toxicity predictions based on acute tests and noted that 0·2 of the 48 hr LC_{50} of many poisons would be an inadequate criterion for long-term survival of fish populations. Their work did suggest, however, that it would often be possible to predict acute toxicity to non-salmonid fish: the 48 hr LC_{50} appears to lie between 1·1 and 1·4 times the predicted 48 hr LC_{50} for rainbow trout. It should be stressed, however, that some substances are more toxic to non-salmonid fishes. For example, perch, *Perca fluviatilis*, are more sensitive with high concentrations of carbon dioxide and oxygen than trout (Alabaster, 1971). Examination of water quality data for stations within the River Trent catchment led Edwards and Brown (1967) to postulate that fish populations were found in rivers in which the total concentration of poisons present did not exceed 0·4 of their combined acutely toxic levels. These data were re-examined by Alabaster *et al.* (1972) and it was found that where trout and other fish were present the sum of the 48 hr LC_{50} was always less than 0·3, with a median value of about 0·1. On reviewing other data Alabaster *et al.* (1972) concluded that the demarcation between rivers containing fish and those which are fishless corresponds with a median value of about 0·28 of the 48 hr LC_{50} to rainbow trout.

Recently, in a comprehensive review (EIFAC, 1980) of laboratory and field data on mixtures of toxic substances, including pesticides, it was concluded that for most sewage and industrial wastes the acute lethal toxicity is between 0·4 and 2·6 times that predicted from the proportions of constituents with a median value close to unity. Mixtures of pesticides were usually more toxic than predicted, the median value being about 1·3. Although few data exist on the relationships between the effects of short-term and chronic exposures to mixtures, there is some evidence that mixtures of mercury, cadmium, chromium and nickel are more toxic to fish after prolonged exposure than might be predicted from their additive toxicity under shorter exposures.

On the other hand, sub-lethal effects of poisons (changes in growth or production) are less than additive and it was postulated that at concentrations approaching 'no observable effect' levels their additive effects might be reduced.

Of particular interest was the conclusion that concentrations below the level of 'no effect' do not contribute to toxicity, even in mixtures. This was contrasted with the North American approach (EPA, 1973) in which an arbitrary 0·2 'no effect' concentration was chosen as the probable upper limit for positive contribution to the toxicity of mixtures.

7.6 INORGANIC POISONS

The inorganic poisons to be considered here are those other than the heavy metals which are dealt with in Section 7.8.1, and include ammonia, arsenic, chloride, chlorine, nitrite, nitrogen supersaturation, ozone and sulphide. Their chemical and toxicological properties differ: the decision to consider them together in one section was made purely on grounds of convenience since they do not fit easily elsewhere.

7.6.1 Ammonia

Ammonia is readily soluble in water and is formed as a product of degradation of nitrogenous organic matter, for example in sewage treatment, or as an industrial by-product. In aqueous solution ammonia forms ammonium ions and an equilibrium is reached between the un-ionised ammonia (NH_3), ionised ammonia (NH_4^+) and the hydroxide ions (OH^-):

$$NH_3 + H_2O \rightleftharpoons NH_4^+ + OH^-$$

The toxicity of the solution depends on the amount of un-ionised ammonia present and this in turn is related to pH and temperature.

The un-ionised ammonia increases with rising pH and temperature (see Fig. 6.7, Section 6.4.7.2). Two criteria for the maintenance of aquatic life have been proposed. The North American EPA standard is 0·020 mg NH_3 per litre (Train, 1979) while the European EIFAC standard is slightly higher at 0·025 mg NH_3 per litre. Table 7.5 gives the concentrations of total ammonia (NH_3 and NH_4^+) which contain the appropriate standard: if the total observed exceeds the value in the table at the specified temperature and pH then the water will contain more un-ionised ammonia than the standard, although at pH values above 8·0 these ammonia concentrations may be too stringent if free carbon dioxide concentrations are very low. Low dissolved oxygen concentrations can have a marked effect on ammonia toxicity (Lloyd, 1961) but in field situations, where low oxygen is likely to be accompanied by high

TABLE 7.5

CONCENTRATIONS OF TOTAL AMMONIA (NH_3^+, NH_4^+) WHICH WILL CONTAIN AN UN-IONISED AMMONIA CONCENTRATION OF 0·020 (EPA) OR 0·025 (EIFAC) mg NH_3 PER LITRE UNDER THE STATED CONDITIONS OF TEMPERATURE AND pH

(After Train, 1979 and Alabaster and Lloyd, 1980)

Temperature (°C)		pH value								
		6·0	6·5	7·0	7·5	8·0	8·5	9·0	9·5	10
5	EPA	160	51	16	5·1	1·6	0·53	0·18	0·071	0·036
	EIFAC	—	63·3	20·0	6·3	2·0	0·66	0·23	—	—
10	EPA	110	34	11	3·4	1·1	0·36	0·13	0·054	0·031
	EIFAC	—	42·4	13·4	4·3	1·4	0·45	0·16	—	—
15	EPA	73	23	7·3	2·3	0·75	0·25	0·09	0·043	0·027
	EIFAC	—	28·9	9·2	2·9	0·94	0·31	0·12	—	—
20	EPA	50	16	5·1	1·6	0·52	0·18	0·07	0·036	0·025
	EIFAC	—	20·0	6·3	2·0	0·66	0·22	0·09	—	—
25	EPA	35	11	3·5	1·1	0·37	0·13	0·06	0·031	0·024
	EIFAC	—	13·9	4·4	1·4	0·46	0·16	0·07	—	—
30	EPA	25	7·9	2·5	0·81	0·27	0·10	0·05	0·028	0·022
	EIFAC	—	9·8	3·1	1·0	0·34	0·12	0·06	—	—

At pH values above 8·0 the stated values may be too stringent.

levels of free carbon dioxide, the resulting lowering of pH can more than compensate for increased toxicity caused by low oxygen (Alabaster and Lloyd, 1980).

Laboratory determinations have revealed that over a short period of exposure the lethal concentrations vary widely between species and, in general, coarse fish, such as carp, survive longer in toxic solutions than salmonids, such as trout. However, the sensitivity to prolonged exposure to less toxic solutions is remarkably similar. For example, when Ball (1967a) studied the toxicity of ammonia to four coarse fish species (see Table 7.6) he found the threshold LC_{50} values over periods of 2–4 days to be between 0·35 and 0·50 mg NH_3 per litre, which are similar to the 24 hr values for rainbow trout under the same conditions. Clearly, these results have significance for short periods which would prove lethal to salmonids. Long-term exposure to lower levels may have much the same result whatever the species.

Exposure of fish to sub-lethal levels of ammonia increases their subsequent tolerance of the lethal concentration. For example, rainbow trout showed prolonged resistance to lethal ammonia concentrations after one day's exposure to half the lethal concentration but this resistance was lost after three days (Lloyd and Orr, 1969).

TABLE 7.6
TOXICITY OF AMMONIA TO SEVERAL SPECIES OF FISH

Species	Test	Value (mg NH_3 per litre)	Reference
Rainbow trout (*Salmo gairdneri*)	24 hr LC_{50}	0·5	Herbert and Shurben (1963) Ball (1967)
	24 hr LC_{50}	0·7	Herbert and Shurben (1965)
Salmon (*Salmo salar*)	24 hr LC_{50}	0·28	Herbert and Shurben (1965)
Perch (*Perca fluviatilis*)	Threshold LC_{50}	0·29	Ball (1967a)
Roach (*Rutilus rutilus*)	Threshold LC_{50}	0·35	Ball (1967a)
Rudd (*Scardinius erythrophthalmus*)	Threshold LC_{50}	0·36	Ball (1967a)
Bream (*Abramis brama*)	Threshold LC_{50}	0·41	Ball (1967a)

Interesting behavioural responses to ammonia have been observed. Sticklebacks (*Gasterosteus*) and green sunfish (*Lepomis*) discriminate, and are repelled by, lethal concentrations of ammonia, but the sunfish did not avoid concentrations in which they were distressed and sticklebacks were actually attracted to sub-lethal concentrations (Jones, 1948; Summerfelt and Lewis, 1967).

It is thought that ammonia increases the permeability of fish to water and may damage the vascular system (Lloyd and Orr, 1969; Alabaster and Lloyd, 1980).

Little work has been carried out to study the effects of ammonia on invertebrates but the few data which are available show that most invertebrates are more resistant than trout and even sensitive genera such as *Daphnia* are only as sensitive as trout. The proposed environmental quality criteria (0·020 or 0·025 mg NH_3 per litre) will, therefore, safeguard most invertebrates and may even enhance productivity (Alabaster and Lloyd, 1980).

7.6.2 Arsenic

Arsenic is important in a wide range of industrial processes including glass, paint and textile manufacture, hardening of metals, the formulation of wood preservatives and herbicides and the production of electrical semiconductors. The natural cycling of arsenic has been reviewed by Ferguson and Gavis (1972) and the toxicity and accumulation of arsenic compounds in fish and invertebrates has been considered by Spehar *et al.* (1980).

In a study of the effects of sodium arsenite on bluegills (*Lepomis macrochirus*), Gilderhus (1966) noted that at concentrations of 4 mg per litre^{-1} of $NaAsO_2$ (2·31 mg As per litre) survival and growth were reduced. This author noted that 96 hr LC_{50} values of $NaAsO_2$ at 12 °C were 25·6 mg litre^{-1} for rainbow trout, 34·0 mg litre^{-1} for goldfish and 35·0 mg litre^{-1} for bluegills, well above the concentrations used for control of aquatic vegetation (2·5–12·5 mg litre^{-1} equivalent of sodium arsenite).

The 6 day LC_{50} to rainbow trout of arsenic (as arsenite) in hard water (385 mg litre^{-1} as $CaCO_3$) was found to be 13·3 mg As per litre but this could be increased by 47% to 19.7 mg As per litre after 21 days of pre-exposure to 0·22 of the incipient lethal level (Dixon and Sprague, 1981).

Little effect on the growth of the alga (diatom) *Asterionella formosa* in culture was observed at concentrations of up to 160 mg As per litre (Conway, 1978).

7.6.3 Chloride

Chloride is present in sewage effluents but high levels are associated with brine discharges either from working geological deposits of rock-salt (halite) or from mining marine coals. Minewaters may have a chloride content up to 36 000 mg litre^{-1} and streams which receive these discharges may have chloride levels of 1200 mg litre^{-1} (Lester, 1975).

Chloride is not very toxic, but can influence osmotic balance. As an example, the 96 hr LC_{50} for fathead minnow (*Pimephales promelas*) was 7650 mg litre^{-1} and for goldfish (*Carassius auratus*) it was 7341 mg litre^{-1} (Adelman *et al.*, 1976). The channel catfish, *Ictalurus punctatus*, tolerated 10 mg litre^{-1} for 13 days (Davis and Simco, 1976). The 96 hr LC_{50} for *Gambusia affinis* was 15 000 mg litre^{-1} in sodium chloride and the 48 hr LC_{50} for calcium chloride was 10 000 mg litre^{-1}. Survival in potassium chloride was shorter: 50% mortality occurred in a solution of 10 000 mg litre^{-1} in 8 hours (Al-Daham and Bhatti, 1977).

Shaw *et al.* (1975) found the growth of Atlantic salmon parr to be similar at chloride levels of 0·1 (freshwater) to 20%, although 20% was isosmotic with parr plasma. Parr suffer high mortality in water above 22% (Saunders and Henderson, 1969).

In a study of the effects of road deicing salt on streams, Crowther and Hynes (1977) noted chloride levels up to 1770 mg Cl^{-} per litre in urban streams. Laboratory tests of the response of three invertebrates (*Hydropsyche betteni*, *Cheumatopsyche analis* and *Gammarus pseudo limnaeus*) showed that pulses of salt at concentrations up to 800 mg Cl^{-} per litre had no effect on drift patterns. In field trials, levels as high as 750 mg Cl^{-} per litre produced no differences in drift but a pulse of 2165 mg Cl^{-} per litre increased the drift of all organisms. The effect became apparent only when concentrations exceeded about 1000 mg Cl^{-} per litre.

7.6.4 Chlorine

Chlorine and its compounds have been important bactericides for many years. Their principal use has been in disinfection of drinking water but they have also been used to reduce bacterial slimes in industry and in bleaching textiles or in paper manufacture. Chlorination has been used to prevent 'fouling' of pipes and reduce algal or bacterial growths in cooling towers or similar structures. Brungs (1973) has provided a useful review of literature on the effects of chlorine on aquatic organisms.

In water, chlorine forms hydrochloric and hypochlorous acids, the latter dissociating to yield hydrogen and hypochlorite ions:

$$Cl_2 + H_2O \rightarrow H^+ + Cl^- + HOCl$$
$$HOCl \rightarrow H^+ + OCl^-$$

The proportions of hypochlorous acid and hypochlorite ions in solution are a function of temperature and pH: the amount of hypochlorous acid decreases as pH and temperature increase. It is thought that the most toxic species in chlorine poisoning is hypochlorous acid and at normal temperatures most of the chlorine present will be in this form and the toxicity of chlorine solutions will be markedly influenced by pH.

When hypochlorites are used instead of chlorine gas, the hypochlorite ion establishes an equilibrium as indicated above, e.g.

$$Ca(OCl)_2 \rightarrow Ca^{2+} + 2OCl^-$$
$$OCl^- + H^+ \rightarrow HOCl$$

In wastewaters, chlorine and hypochlorites react with several substances but principally with ammonia to form chloramines. Although chloramines may be slightly less toxic they are more persistent. A study of the effects of monochloramine on several species of fish (mostly North American) has been made by Seegeert *et al.* (1979), which supports this view. The mechanism of chlorine poisoning is not well understood but it is thought that sulphydryl (–SH) groups of enzymes are irreversibly oxidised and enzymatic activity is thereby destroyed: fish which lose equilibrium in chlorine do not recover when transferred to clean water. Some doubt exists as to the site of action although gills are favoured by several authors (Alabaster and Lloyd, 1980).

Since chlorine is highly reactive and most surface waters are likely to contain organic matter with which the chlorine will combine, the proposed EIFAC criterion of $4 \mu g$ HOCl per litre has been set quite close to lethal levels (Alabaster and Lloyd, 1980). This may be compared with the EPA criterion (Train, 1979) of $2 \mu g$ litre^{-1} total residual chlorine for salmonid fish and $10 \mu g$ litre^{-1} for other freshwater organisms. In a recent study (Arthur *et al.*, 1975) it was concluded that a total residual chlorine level as low as $10 \mu g$ litre^{-1} had a detrimental effect on the survival of aquatic organisms. For example, the highest concentration which had no long-term adverse effect on *Daphnia* was $2-4 \mu g$ litre^{-1}, although the equivalent for the amphipod *Gammarus pseudolimnaeus* and the fathead minnow were 12 and $14 \mu g$ litre^{-1}, respectively. The lowest mean total residual chlorine concentration having a measurable long-

TABLE 7.7
ACUTE TOXICITY OF RESIDUAL CHLORINE TO FRESHWATER ORGANISMS

Species	Temperature (°C)	Test	Value (mg litre^{-1})	Reference
Fish				
Coho salmon, *Oncorhynchus kisutch*	12	96 hr LC$_{50}$	289	Marking and Bills (1977)
	14	96 hr LC$_{50}$	102	Arthur *et al.* (1975)
Rainbow trout, *Salmo gairdneri*	12	96 hr LC$_{50}$	172	Marking and Bills (1977)
	15	96 hr LC$_{50}$	40	Cairns and Conn (1979)
Brook trout, *Salvelinus fontinalis*	14	96 hr LC$_{50}$	135	Arthur *et al.* (1975)
	14	7 day LC$_{50}$	83	Arthur *et al.* (1975)
Fathead minnow, *Pimephales promelas*	12	96 hr LC$_{50}$	998	Marking and Bills (1977)
	17–18	96 hr LC$_{50}$	86–130	Arthur *et al.* (1975)
	17–18	7 day LC$_{50}$	82–115	Arthur *et al.* (1975)
Carp, *Cyprinus carpio*	12	96 hr LC$_{50}$	800	Marking and Bills (1977)
Goldfish, *Carassius auratus*	12	96 hr LC$_{50}$	1 180	Marking and Bills (1977)
Black bullhead, *Ictalurus melas*	12	96 hr LC$_{50}$	1 410	Marking and Bills (1977)
White sucker,	12	96 hr LC$_{50}$	379	Marking and Bills (1977)
Catostomus commersoni	16	96 hr LC$_{50}$	138	Arthur *et al.* (1975)
Yellow perch, *Perca flavescens*	17	96 hr LC$_{50}$	205	Arthur *et al.* (1975)
	12	96 hr LC$_{50}$	558	Marking and Bills (1977)
Invertebrates				
Amphipod, *Gammarus pseudolimnaeus*	17–18	96 hr LC$_{50}$	215–330	Arthur *et al.* (1975)
Stonefly, *Pteronarcys* sp	18	96 hr LC$_{50}$	400	Arthur *et al.* (1975)
Caddis, *Hydropsyche* sp	18	96 hr LC$_{50}$	> 740	Arthur *et al.* (1975)
Crayfish, *Orconectes virilis*	17	96 hr LC$_{50}$	> 780	Arthur *et al.* (1975)
Snail, *Physa integra*	18	96 hr LC$_{50}$	> 810	Arthur *et al.* (1975)
Snail, *Campeloma decisum*	18	96 hr LC$_{50}$	> 810	Arthur *et al.* (1975)

term effect on fathead minnows was $42 \mu g$ litre^{-1}, which compares with the sub-lethal stress and threshold concentrations reported by Zillich (1972) of 0·04–0·09 and 0·04–0·05 mg litre^{-1}, respectively.

Studies of the toxicity of chlorine to twelve species of fish indicated that susceptibility varied widely between closely related species (Marking and Bills, 1977). For example, the 96 hr LC_{50} for channel catfish (*Ictalurus punctatus*) was 0·156 mg litre^{-1} while for black bullheads (*I. melas*) the equivalent value was 1·41 mg litre^{-1}. In general, however, salmonid species seem to be more sensitive to chlorine than other fishes (Alabaster and Lloyd, 1980). A selection of published data on the toxicity of chlorine to fish and invertebrate species is given in Table 7.7. A review of the toxicity of chlorine to invertebrates, including early work and especially species associated with water distribution systems, has been made by Evins (1975).

7.6.5 Nitrite

In aqueous solution nitrite is present as the ion or the free acid:

$$NO_2^- + H^+ \rightarrow HNO_2$$

Nitrite oxidises methaemoglobin and when this reaches elevated levels cyanosis and tissue hypoxia may result. Both nitrite species (NO_2^- and HNO_2) are toxic (Russo *et al.*, 1981).

Westin (1974) reported a 96 hr LC_{50} of 0·88 mg NO_2^-N per litre for chinook fingerlings (*Onchorhyncus tshawytscha*) while Russo *et al.* (1974) give a 96 hr LC_{50} for rainbow trout of 0·19 mg NO_2^-N per litre which compares with the equivalent value of 0·23 NO_2^-N mg per litre given by Brown and McLeay (1975).

Perrone and Meade (1977) found that tolerance of nitrate by yearling Coho salmon (*O. kisutch*) was increased by the presence of high chloride levels and postulated that chloride competes with nitrite, inhibiting the onset of methaemoglobinemia. The toxicity of nitrite to rainbow (steelhead) trout (*Salmo gairdneri*) was reduced by increased hardness or by increased chloride. For example, the 96 hr LC_{50} in soft water (25 mg litre^{-1} as $CaCO_3$) was 0·9 mg NO_2^-N per litre while in hardwater (300 mg litre^{-1}, $CaCO_3$) the 96 hr LC_{50} was 12·1 mg NO_2^-N per litre (Wedemeyer and Yasutake, 1978). A similar inhibition of nitrite-induced methaemoglobinemia by chloride has been observed in channel catfish *Ictalurus punctatus* by Tomasso *et al.* (1979).

Toxicity of nitrite to rainbow trout (*Salmo gairdneri*) is pH dependent within the normally encountered range (pH 6·5–9·0).

7.6.6 Nitrogen

Nitrogen is not actually toxic itself, but when present to excess in the form of a supersaturated solution, it may give rise to 'gas bubble disease'. This problem arises when water under pressure, for example in the depths of a reservoir or in groundwater, is released at lower pressure (usually atmoshperic) or is subjected to a sudden increase in temperature. Examples of the sources of supersaturated water are hydroelectric projects, river augmentation by groundwater or the discharge of cooling water in power generation.

As the pressure is reduced or temperature increased, the dissolved nitrogen comes out of solution and gas bubbles form on the surface of organisms or in tissues. In fish, serious consequences may follow if they cause emboli in gill blood vessels.

Oxygen in solution behaves in exactly the same way as nitrogen but since it may be metabolised it causes few internal problems, although it does cause severe external signs of gas bubble disease. However, high oxygen : nitrogen ratios appear to mitigate against the most acute effects of supersaturated nitrogen (Nebeker *et al.*, 1979). A most comprehensive review of dissolved gas supersaturation and its effects on fishes and invertebrates has been provided by Weitkamp and Katz (1980).

7.6.7 Ozone

Sterilisation by means of ozone is employed in potable water (as an alternative to chlorine and its alleged hazards), sewage effluent treatment and as a biocide (again as a replacement for chlorine) in pipes, for example in power stations.

Wedemeyer *et al.* (1979) observed a 96 hr LC_{50} of 9·3 µg litre^{-1} to rainbow trout in soft water at 10 °C. The lethal threshold level was 8 µg litre^{-1} and in a three month study of chronic toxicity 5 µg litre^{-1} caused pathological changes to gill tissue and reduced feeding. They suggested, therefore, that a provisional maximum safe exposure level for salmonids would be 2 µg litre^{-1}, a level which caused no significant damage in the chronic tests.

In a study of the effects of exposure of bluegills (*Lepomis macrochirus*) to ozone, Paller and Heidinger (1980) examined fish which had survived acute toxicity tests and noted gross damage to the gills, in particular the destruction, and loss, of epithelial cells, and often the entire tip of the lamella was missing.

Asbury and Coler (1980) have provided data on the toxicity of ozone to the eggs and fry of several North American species (Table 7.8). They

TABLE 7.8

TOXICITY OF OZONE TO EGGS AND FRY OF FISHES

(From Asbury and Coler, 1980)

Species	Stage	Temperature (°C)	Exposure time (mins)	LC_{50} (mg litre^{-1})
Yellow perch,				
Perca flavescens	eggs	14	80	1·14–1·98
		14	40	1·28–1·91
	larvae	17	1	0·06
			0·5	0·11
Fathead minnow,				
Pimephales promelas	eggs	24	160	0·6–1·1
			20	2·5–4·4
	larvae	24	5	<0·1
Common sucker	eggs	14	80	1·43
Catostomus commersoni			5	>5·9
Bluegill sunfish	larvae	26	0·25	0·15
Lepomis macrochirus				

concluded that the eggs of most species would be protected by short exposures to moderate residual levels of ozone (< 1 mg litre^{-1}) but since larvae were killed by virtually any measurable residual at even very short exposures a safe concentration was thought to be less than 50 µg litre^{-1}.

7.6.8 Sulphide

Sulphides arise from anaerobic degradation of organic matter in sediments, which may be augmented by sewage effluents, or from industrial wastes from paper mills, tanneries or chemical processing.

Concentrations of sulphide are often high within the sediment and this has serious implications for the survival of fish eggs, their fry and benthic invertebrates.

The 96 hr LC_{50} for brown trout (*Salmo trutta*) fry was 7 µgH$_2$S per litre, and fish which survived chronic exposure to sub-lethal concentrations grew better than the controls (possibly an antibiotic effect), but there was evidence of respiratory stress (Reynolds and Hains, 1980). Broderius *et al.* (1977) studied the toxicity of sulphide to the fathead minnow, and in particular the effect of pH on toxicity. The toxicity of H$_2$S increased with increasing pH (for example, the 96 hr LC_{50} was 57·2 µg litre^{-1} as H$_2$S at pH 7·1 and 14·9 µg litre^{-1} as H$_2$S at pH 8·7) while dissolved sulphide became less toxic (96 hr LC_{50} at pH 7·1 was 133 µg litre^{-1} as H$_2$S and at pH 8·7 was 806 µg litre^{-1}). Smith *et al.* (1976) measured the toxicity of hydrogen sulphide to each stage of the life history of bluegills (*Lepomis macrochirus*) and found the 72 hr LC_{50} for eggs to be $19 \cdot 0$ µg litre^{-1}. The 96 hr LC_{50} for subsequent stages was as follows: 35 day old fry, $13 \cdot 1$ µg litre^{-1}; juveniles, $47 \cdot 8$ µg litre^{-1}, and adults $44 \cdot 8$ µg litre^{-1}. Feeding fry were most sensitive in acute toxicity tests but on chronic exposure spawning adults were most susceptible. At 1 µg litre^{-1} spawning was limited and at $2 \cdot 2$ µg litre^{-1} no spawning occurred.

7.7 ORGANIC POISONS

Although there is an almost endless list of organic substances which may prove poisonous to aquatic life, attention must be limited here to those substances which are commonly encountered in effluents and which are released in significant quantities such that they may have an appreciable, though local, effect in environmental terms. Organic substances which are used as pesticides are dealt with elsewhere (herbicides in Section 7.9.2 and the other pesticides in Sections 7.9.3 to 7.9.5). Two important organic poisons of early significance are cyanide and phenols, which were followed in historical sequence by detergents. More recently polychlorinated biphenyls (PCBs) have assumed considerable importance as environmental contaminants (see Section 7.10).

7.7.1 Cyanide

Cyanide is a common constituent of many industrial effluents, being important for case-hardening of metals, electroplating and metal cleaning.

In addition to hydrocyanic acid (HCN), the metallic salts, potassium (KCN) and sodium (NaCN) cyanide, are also commonly encountered. Cyanide rapidly dissociates at low pH and when the salts dissociate in aqueous solution the toxic hydrocyanic acid is formed. This is sometimes referred to as 'free cyanide' and is regarded as the primary toxic component (Leduc et al., 1982). Cyanide ions also form complexes with heavy metal ions, their stability depending on the metal involved.

Poisoning by cyanide is essentially by inhibition of respiration, or more specifically of oxygen metabolism, by binding porphyrins containing Fe^{3+} such as cytochrome oxidase, hydroperoxidases and methaemoglobin. Its acute toxicity is modified by several factors, although temperature seems to be the most important.

Concentrations from 30 to 150 µg HCN per litre are usually acutely

TABLE 7.9
TOXICITY OF CYANIDE TO FISH

Species	Temperature (°C)	Hardness as $CaCO_3$ (mg litre^{-1})	Test	Value mg CN per litre	Reference
Fathead minnow (*Pimephales promelas*)	20	230	96 hr LC_{50}	0·23	Doudoroff (1956)
Bluegill (*Lepomis macrochirus*)	25	91	24 hr LC_{50}	0·091	Renn (1955)
Largemouth bass (*Micropterus salmoides*)	25	110	24 hr LC_{50}	0·11	Renn (1955)
Rainbow trout (*Salmo gairdneri*)	17·5	70	24 hr LC_{50}	0·07	Herbert and Merkens (1952)
Rainbow trout (*Salmo gairdneri*)	6	127	96 hr LC_{50}	0·028	Kovacs and Leduc (1982)
Rainbow trout (*Salmo gairdneri*)	12	127	96 hr LC_{50}	0·042	Kovacs and Leduc (1982)
Rainbow trout (*Salmo gairdneri*)	18	127	96 hr LC_{50}	0·068	Kovacs and Leduc (1982)
Rainbow trout (*Salmo gairdneri*)	15	236	144 hr LC_{50}	0·096 to 0·098	Dixon and Sprague (1981)
Tench (*Tinca tinca*)	15	200	48 hr LC_{50}	0·2	Wuhrmann (1952)
Chub (*Leuciscus cephalus*)	15	220	48 hr LC_{50}	0·22	Wuhrmann (1952)

toxic to most fishes (Table 7.9). At high rapidly lethal concentrations (i.e. above 0·1 mg HCN per litre), elevated temperature enhances toxicity to rainbow trout while at lower and more slowly lethal concentrations the reverse is observed (Kovacs and Leduc, 1982). Toxicity is also influenced by pH. In studies of the toxicity of cyanide to fathead minnows, as free cyanide and molecular HCN, Broderius *et al.* (1977) noted that the 96 hr LC_{50} for both forms were similar (115–133 µg litre^{-1} as HCN) at pH values from 6·8 to 8·3 but above this, to pH 9·3, the free cyanide became less toxic while molecular HCN became more toxic. In chronic exposures physiological systems other than respiratory may be impaired. Exposure of fishes to concentrations as low as 4 or 5 µg HCN per litre has caused impairment of reproduction and significant detrimental effects on reproductive processes have been observed following exposure to 10 µg HCN per litre. Newly fertilised fish eggs are relatively tolerant of free cyanide but developmental anomalies and increased mortality occur at concentrations of 60–100 µg HCN per litre (Leduc *et al.*, 1982).

In a study of the effects of cyanide on early life stages of the Atlantic salmon, Leduc (1978) noted a high incidence of abnormalities, ranging from about 6% at 0·01 mg HCN per litre to 19% at 0·1 mg HCN per litre. At the higher levels tested (0·8 and 0·1 mg litre^{-1}) hatching was delayed but all concentrations reduced hatching success by 15–40%. Leduc concluded that the maximum acceptable toxicant concentration (MATC) was 0·005 mg HCN per litre.

Pre-exposure of rainbow trout to 0·35 of the incipient lethal level of cyanide decreased their tolerance by 32% after one week and 15% after two weeks but this initial sensitisation disappeared after three weeks (Dixon and Sprague, 1981).

7.7.2 Phenols

Phenolic compounds are found in a wide range of effluents but are commonly associated with the carbonisation of coal (Herbert, 1962), oil refining and chemical industries. There are a number of compounds included within the omnibus term 'phenols': monohydric phenols include phenol, cresols and xylenols, and dihydric phenols include catechols and resorcinols. The wastes which contain phenols, 'phenolic wastes', often contain other compounds such as ammonia, cyanide, organic acids alcohols, aldehydes and ketones, Most investigations, and especially laboratory studies, have concentrated on monohydric phenols. The toxicity of phenols, cresols and xylenols is similar (Alabaster and Lloyd, 1980) and in many studies phenol was used as a model substance.

TABLE 7.10

TOXICITY OF PHENOL TO FISH

Species	Temperature (°C)	Test	Value (mg litre^{-1})	Reference
Rainbow trout, *Salmo gairdneri*	15–18	48 hr LC_{50}	9·4	Brown and Dalton (1970)
		48 hr LC_{50}	9·0–10·4	Alexander and Clarke (1978)
	12	48 hr LC_{50}	8·0	Brown et al. (1969)
	12·6	48 hr LC_{50}	7·5	Mitrovic et al. (1968)
	6·3	48 hr LC_{50}	5·4	Brown et al. (1967)
	11·8	48 hr LC_{50}	8·0	Brown et al. (1967)
	18·1	48 hr LC_{50}	9·8	Brown et al. (1967)
	15	48 hr LC_{50}	9	Swift (1978)
Bluegill, *Lepomis macrochirus*	18	96 hr LC_{50}	13·5	Patrick et al. (1968)
Goldfish, *Carassius auratus*	25	48 hr LC_{50}	44·5	Pickering and Henderson (1966)
Gudgeon, *Gobio gobio*	10	48 hr LC_{50}	25	Alabaster and Lloyd (1980)

Examples of the toxicity of phenol to a range of fish are given in Table 7.10.

The toxicity of phenol is temperature-dependent: at high concentrations reaction and survival periods decrease with temperature but at lower concentrations resistance of rainbow trout increases with increasing temperatures up to 18 °C, at which temperature the 48 hr LC_{50} (9·8 mg litre^{-1}) is almost twice that at 6 °C (5·4 mg litre^{-1}). It has been postulated that the rate of excretion at lower temperatures decreases more rapidly than the rate of absorption (Brown et al., 1967). The pathological effects of phenol on fish appear to be extensive. Mitrovic et al. (1968) found that rainbow trout exposed to a concentration about 25% greater than the 48 hr LC_{50} (7·5 mg litre^{-1}) and which died within a few hours had internal haemorrhages, enlarged spleens, inflammation and necrosis of the pharynx and gills. Less severe damage to liver and kidney, which was also evident in fish surviving for seven days, was also noted. Damage to the gills occurred within a few hours; first there was inflammation followed by loss of the epithelium from the secondary lamella, and ultimately complete destruction of the distal parts of the filaments.

Swift (1978) studied the physiological responses of rainbow trout exposed for 24 hours to sub-lethal concentrations of phenol in hard water and found no effect on urine flow rate or haemocrit. Tissue uptake of phenol was rapid and reached equilibrium within three hours and loss after exposure was also rapid. Close to or above the lethal threshold concentration, phenol concentration in the tissues, and notably the brain, was higher than ambient which gives support to the belief that phenol poisoning is principally an effect on the central nervous system. Evidence of behavioural changes induced by phenol has been reported by Schneider *et al.* (1980), who noted increased susceptibility to predation in juvenile rainbow trout exposed to phenol at concentrations greater than 7 mg litre^{-1} but less than 10 mg litre^{-1} (the level at which significant mortalities occurred), although the effect was not large.

7.8 HEAVY METALS

The heavy metals which are of principal interest in pollution biology are zinc, copper, lead, mercury, cadmium, nickel and chromium. Historically, lead, zinc and copper were the most significant but more recently mercury and cadmium have become prominent.

7.8.1 Early Case Histories

Early 'classical' work on the effects of heavy metal pollution centred on the wastes produced by lead mining in Cardiganshire (now Dyfed) in central coastal Wales. Mining virtually ceased by the end of the nineteenth century except for a temporary reprieve during the First World War. The rivers Rheidol and Ystwyth, together with sections of neighbouring rivers, were fishless and lacking a normal flora and invertebrate fauna. During active mining it was accepted that mining operations and especially the wastes from ore washing and flotation were responsible. Although some recovery of the biota had followed cessation of operations, fish remained absent. It was postulated that the cause of this fishlessness and the paucity of flora and fauna was lead leaching from the weathering spoil heaps and being carried into the streams by rainwater (Carpenter, 1924, 1925, 1926), a conclusion supported by experimental work using caged trout in the river and minnows in the laboratory

(Carpenter, 1927). The presence of zinc was also suspected but analytical methods then available were inadequate. Jones (1964) has suggested that in the period of Carpenter's investigations it is possible that the rivers were affected more by zinc than lead.

Progressive recovery of these rivers has given insights into the effects of heavy metal pollution. Prior to 1922 only fourteen species of invertebrate were found, all arthropods and predominantly insects, including the chironomid *Tanypus nebulosus*, the blackfly *Simulium latipes* and the mayfly *Cloeon simile*. Groups such as Platyhelminthes, Mollusca, Malacostraca, Trichoptera, Hirudinea and Oligochaeta were absent (Carpenter, 1924). The absence of molluscs and malacostracan crustacea was later thought to be a result of high metal concentrations of heavy metal but the absence of the rest was attributed to harsh physical conditions (Jones, 1940, 1958). Later in 1922 and 1923 the fauna had risen to 29 species: these included more species of previously represented groups together with species of new groups such as Trichoptera and the platyhelminth *Polycelis nigra*. Where previously the flora had been restricted to the algae *Batrachospermum* and *Lemanea*, the macrophytes *Ranunculus* and *Callitriche* appeared.

By the 1930s (Laurie and Jones, 1938) the total number of macroinvertebrates had risen to 103 which included both Mollusca and fish (four species), both groups having previously been absent. Later still (1947 and 1948) almost 200 species were recorded but only 130 were recorded in the main stream (Jones, 1949). The latest study (Brooker and Morris, 1980) revealed 134 species in the Rheidol but this was restricted to the fauna of stony riffles while earlier studies had included collections from pools and back waters. No evidence was obtained that the distribution and density of the macroinvertebrates was related to heavy metal concentrations. One may conclude that invertebrate recovery has probably been completed since the present fauna is generally similar to that of other unpolluted base-poor upland rivers.

The recovery of the fisheries has been documented (Jones and Howells, 1969) to the point where a healthy salmon stock is present although the water quality is not yet regarded as safe for human consumption (Jones and Howells, 1975).

Another, early, example of heavy metal toxicity (Butcher, 1946, 1955) concerns discharges of copper works' effluent into the River Churnet. Organic wastes from Leek sewage works and a dyeworks caused typical changes in the benthic community. About 11 km downstream the river had recovered considerably when it received an effluent from a copper

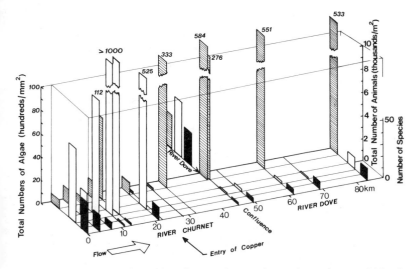

FIG. 7.4 Distribution of the algal flora and macroinvertebrate fauna within the River Churnet, England, in relation to the outfall from a copper works. Redrawn from Bradfield and Rees (1978) after Butcher (1946).

works which raised the copper level of the river to 1 or 2 mg litre^{-1}. Below this discharge the algae associated with recovery from organic enrichment (*Stigeoclonium, Nitschia, Gomphonema, Cocconeis* and *Chamaesiphon*) disappeared and for about 5 km cell numbers were greatly reduced. After this algal numbers recovered dramatically with *Achnanthes affinis, Chlorococcum* and *Stigeoclonium tenue* as the principal components of the algal flora. Pronounced changes were also observed in the macroinvertebrate fauna: the tubificid worms, chironomid larvae, molluscs, leeches and *Asellus*, present upstream of the copper works, disappeared and no animals were found until the Churnet joined the River Dove where the copper became diluted. Even here animals were scarce and the impact of the copper was evident for a further 30 km. These changes are shown in Fig. 7.4.

7.8.2 Trends in Metal Toxicity

7.8.2.1 Relative toxicities
Some metal ions have low toxicity; others are toxic to many organisms at very low concentrations (μg to mg litre^{-1}). Several authors have arranged

metals in descending order of toxicity and attempts to explain the fairly consistent 'league table' have tended to be based on the, now obsolete, concept of 'solution pressure' (Jones, 1964). This is based on the ease with which they pass into solution as positively charged ions or the ease with which metal ions in solution give up their charge to combine with other ions or compounds. An approximate series of decreasing toxicity is tentatively given in Table 7.11.

TABLE 7.11

TENTATIVE TABLE OF THE APPROXIMATE ORDER OF TOXICITY OF METALS, BASED ON PUBLISHED DATA

Highly toxic					Decreasing toxicity→					
Hg										
	Cu	Cd	Au?	Ag?	Pt?					
		Zn								
			Sn	Al						
				Ni	Fe^{3+}					
					Fe^{2+}					
						Ba				
							Mn	Li		
							Co	K	Ca	Sr
									Mg	Na

Using published data on the relative toxicity of metal ions to aquatic organisms, and *Daphnia magna* in particular, Kaiser (1980) investigated their correlation with ion specific physico-chemical parameters. Significant correlations between toxicity and the electron configurations of their outer orbitals were seen in three groups of ions. The first group included Na, Be^{2+}, Ba^{2+}, Al^{3+} and Cr^{6+} which have electron configurations like inert gases. The second groups includes Cr^{3+}, Mn^{2+}, Fe^{3+}, Co^{2+}, Ni^{2+}, Cu^{2+}, Zn^{2+} As^{5+}, Cd^{2+}, Pt^{4+}, Au^{3+} and Hg^{2+} with partially or completely filled d electron orbitals. The third group includes Sn^{2+}, As^{3+}, Se^{4+} and Pb^{2+} with filled d and s, but unfilled p electron orbitals. An equation was derived for predicting the toxicity of other ions under similar conditions when constants (a_0, a_1 and a_2) which depend on the ionic group, the biota and the particular effect have been determined. Within the group the equation is

$$pT = a_0 + a_1 \log \frac{AN}{IP} + a_2 E_0$$

where pT is the negative logarithm of a metal ion concentration with a given toxicity in mol litre^{-1}, AN is the metal's atomic number, IP is the difference between ion's ionisation potential with the oxidation number (OX) and the ionisation potential of the next lower oxidation number (OX -1) in electron volts (eV), and E_0 is the absolute value of the electrochemical potential between the ion and the first stable reduced state.

As the responses to different heavy metals vary according to the groups of organisms involved, it has not proved possible to provide an adequate general qualitative model similar to that which is available for organic pollutants. Indeed, when surveys generate results which differ from those associated with various degrees of organic enrichment (see Chapter 6) it is usual to suspect the presence of some toxicant. However, recently Winner *et al.* (1980) have suggested that there is a predictable, graded response to heavy metal pollution. They studied the effects of continuously introducing low concentrations of copper (120 µg Cu per litre) into an experimental stream and also observed changes in the fauna of a similar stream which received heavy metals (Cu, Cr and Zn) from the metal-plating industry. The effects of a low continuous dose of one metal and larger but fluctuating inputs of several metals were remarkably similar. It was noted that there was a consistent relationship between the intensity of the pollution stress and the ratio of the number of chironomids to total insects in the community. Three stages of response were recognised. In the most heavily polluted reaches the normal macroinvertebrate fauna (including the bivalve *Pisidium*, gastropod *Physa*, isopod *Lirceus*, flatworm *Dugesia*, crayfish *Orconectes*), other than tubificid worms and chironomid larvae, were virtually eliminated. Moderate metal pollution reduced the invertebrate fauna but Trichoptera (caddis flies) were present along with chironomids. Mild pollution was characterised by a fauna which also included Ephemeroptera (mayflies). The number of chironomid species was also influenced by the metal pollution, the greater number of species being associated with mildest pollution. Some insects, such as the neuropteran *Sialis* and naiad damselflies, appeared to be largely unaffected by metal pollution.

7.8.3 Effects of Metals on Different Taxa

Although the characteristics of individual heavy metals are considered below, it is convenient to review briefly the literature on different taxa. General reviews include Whitton and Say (1975). Effects on plants have been documented extensively. Whitton (1970a, b) reported toxicity of heavy metals to freshwater algae, in field studies and under experimental conditions. The flora of a stream polluted by zinc from mining wastes (Say and Whitton, 1981) showed a response to the gradient of zinc concentrations (1·2 to 25·6 mg Zn per litre) in which 25 species were present at the most heavily polluted site compared with a maximum of 41 species at other sites with elevated zinc levels. A nearby unpolluted stream had 61 species. Changes in the species composition of the community were a more evident response to zinc pollution than the total number of species present. Species which were associated with high levels of zinc included *Schizothrix delicatissima*, *Synechococcus* sp., *Eunotia tenella*, *Pinnularia borealis* and *Euglena mutabilis*. Tolerances of diatoms to copper and zinc pollution in Canada have been described by Besch *et al.* (1972). The red macro-alga *Lemanea* is also able to survive high levels of heavy metals, including zinc levels up to 1·16 mg Zn per litre (Harding and Whitton, 1981).

Aquatic bryophytes, in particular *Rhynchostegium* (*Eurhynchium*) *riparioides*, *Fontinalis antipyretica*, *F. squamosa*, *Scapania undulata* and *Amblystegium riparium*, are tolerant of enhanced levels of heavy metals (Bensen-Evans and Williams, 1976; Say *et al.*, 1981).

Aquatic bryophytes also accumulate high levels (Burton and Peterson, 1979) and may be used as indicators of heavy metal environmental contamination (Say *et al.*, 1981). Examples of the levels of accumulation of metals are given in Table 7.12. The enrichment factor (ratio of metal in the plant to that in the water) may be of the order of 30 000.

Higher, flowering plants seem less able to survive heavy metal pollution, although few studies of their responses are reported in the literature (Whitton and Say, 1975). Two genera which returned to the metal polluted rivers studied by Carpenter (1926) were *Ranunculus* and *Callitriche*. In studies of the relative toxicity of metals to *Elodea canadensis* and *Myriophyllum spicatum* the descending order of importance was

TABLE 7.12

MAXIMUM CONCENTRATIONS OF HEAVY METALS IN AQUATIC BRYOPHYTES, EXPRESSED AS ppm (μg litre^{-1} OR mg kg^{-1})

Species	Lead	Zinc	References
Fontinalis squamosa	—	5 430	Say *et al.* (1981)
Philonotis fontana	5 965	7 023	Burton and Peterson (1979)
Rhynchostegium riparioides	—	6 705	Say *et al.* (1981)
Scapania undulata	14 825	1 950	Maclean and Jones (1975)
Scapania undulata	8 902	3 558	Burton and Peterson (1979)

found to be copper, mercury, arsenic, cadmium, zinc and lead (Brown and Rattigan, 1979; Stanley, 1974).

Different macroinvertebrate groups, as might be expected, respond to differing degrees when exposed to metal pollution, so that generalisations are difficult to make. However, resistant groups include many insect orders such as Diptera and Trichoptera (Warnick and Bell, 1969) while oligochaetes are intermediate; molluscs and malacostracan Crustacea seem to be most sensitive (Whitton and Say, 1975).

Fish also vary in their responses; perhaps the only generalisation one can make is that salmonids tend to be more sensitive than cyprinids (Alabaster, 1971). An early hypothesis that heavy metal toxicity is caused by asphyxia, arising from coagulation of the mucus on fish gills ('coagulation film anoxia'—Jones, 1964) may be justified at high concentrations but for lower and environmentally encountered concentrations this hypothesis is almost certainly an over-simplification (Alabaster and Lloyd, 1980).

7.8.4 Aluminium

The chemistry, toxicology and aquatic distribution of aluminium has been extensively reviewed by Burrows (1977). The chemistry of aluminium is complex and this has considerable bearing on the interpretation of field studies (Hunter *et al.*, 1980). Even experimentally determined toxicities reported in the literature are contradictory, probably because of wide variations in pH caused by the presence of the aluminium and not always measured. One important feature of the chemistry of aluminium is the tendency to form soluble salts (Al^{3+}) at acidic pH values and soluble aluminate ($Al(OH)_4^-$) compounds at alkaline pH. Between these, at a pH range of about 6·5–7·5, the insoluble hydroxide $Al(OH)_3$ is formed. Since the soluble aluminium is the principal factor in acute toxicity, actual toxicity of aluminium compounds will depend on pH, among other factors (Freeman and Everhart, 1971). This means that aluminium may be relatively unimportant in neutral solutions but quite toxic in acid or alkaline waters. It has been suggested that much of the mortality of fish in waters subjected to acid precipitation may be attributed to the increased toxicity of naturally occurring aluminium compounds (Parmann, 1981).

7.8.5 Cadmium

A recent review, containing references to much unpublished work, by Alabaster and Lloyd (1980) has indicated that data are sufficient to determine only very tentative values for control of cadmium discharges, but it is clear that this metal is toxic to many organisms after exposure to low concentrations ($<1\,\mu g\,Cd$ per litre), although this may be for extended periods (months). Cadmium accumulates in the tissues of organisms and is thought to damage ion-regulating mechanisms rather than respiratory or nervous functions. Recent evidence from studies of cadmium-induced hypocalcemia suggests that the decline in plasma calcium concentrations may be the direct cause of mortality (Roch and Maly, 1979). Although cadmium may be lost from the tissues after exposure to single or intermittent doses, large amounts absorbed may be acquired rapidly but lost slowly. Thus organisms may survive exposure for a short period to large doses, but die subsequently. The phenomenon may account, in part at least, for the apparent marked discrepancy between acutely lethal doses and chronic lethal toxicity. For example, Ball (1967c) reported that the 96 hr LC_{50} for rainbow trout in hard water ($290\,mg\,litre^{-1}$ as $CaCO_3$) was about 2 or 3 mg Cd per litre, while the 7 day LC_{50} was between 8 and $10\,\mu g\,Cd$ per litre. He noted that fish exposed to concentrations in the range 1 to 64 mg Cd per litre showed a linear relationship between log concentration and log survival but no increase in median survival period was observed at concentrations between 0·01 and 1 mg Cd per litre (Fig. 7.5). A similar curve, having a plateau around 1 mg Cd per litre, has been observed in a study of cadmium toxicity in the stickleback, *Gasterosteus aculeatus*, by Pascoe and Cram (1977), which they interpreted as evidence of two toxic mechanisms. In studies of the uptake of cadmium in perfused gills, Part and Svanberg (1981) showed that at low concentrations ($<56\,\mu g\,Cd$ per

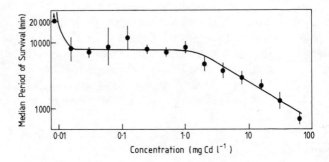

FIG. 7.5 Toxicity of cadmium to rainbow trout (*Salmo gairdneri*) in hard water. Points indicate median period of survival and vertical lines the 95% confidence intervals. After Ball (1980).

litre) a tenfold increase in cadmium resulted in a hundredfold increase in transfer of cadmium, but at higher concentrations the uptake rate diminished after about 30 minutes. Cadmium does not appear to damage gills of rainbow trout (Hughes *et al.*, 1979). Evidence of a biological mechanism dependent on calcium (water 'hardness') and probably gill permeability was presented by Calamari *et al.* (1980) who showed that fish acclimated in hard water were less susceptible to cadmium when tested in soft water than those acclimated and tested in soft water.

Later work (Department of the Environment, 1972) confirmed that in slightly softer water the 50 day LC_{50} for rainbow trout was $10 \,\mu g$ Cd per litre. The recommended 95 percentile concentration for survival of rainbow trout in hard water ($300 \,mg \,litre^{-1}$ as $CaCO_3$) is $1.5 \,\mu g$ Cd per litre (Alabaster and Lloyd, 1980). Acute toxicity of cadmium to brook trout (*Salvelinus fontinalis*) could not be determined by Benoit *et al.* (1976) since a flat mortality–concentration curve was obtained.

The maximum acceptable toxicant concentration (MATC) for brook trout (*Salvelinus fontinalis*), determined over three generations in soft water ($44 \,mg \,litre^{-1}$ as $CaCO_3$) has been determined at between 1.7 and $3.4 \,\mu g$ Cd per litre (Benoit *et al.*, 1976).

Exposure to $0.1 \,mg$ Cd per litre promoted premature hatching, retarded growth and increased the rate of mortality and the occurrence of developmental abnormalites such as spinal curvature and blood clots in eggs and alevins of rainbow trout (Woodworth and Pascoe, 1982). Examples of toxicity data in the literature are given in Table 7.13.

McCarty *et al.* (1978) investigated the effects of cadmium on the goldfish (*Carassius auratus*) in hard ($140 \,mg \,litre^{-1}$ as $CaCO_3$) and soft ($20 \,mg \,litre^{-1}$ as $CaCO_3$) water. In soft water LC_{50} values were for 48 hr $2.76 \,mg$ Cd per litre and 96 hr $2.13 \,mg$ Cd per litre, while in hard water the equivalent values were 46.9 and $46.8 \,mg$ Cd per litre, respectively.

Increased vulnerability of fathead minnows to predation by large mouth bass was observed after exposure to cadmium at acute (24 hr of $0.375 \,mg$ Cd per litre) and sub-acute (21d of $0.025 \,mg$ Cd per litre) levels (Sullivan *et al.*, 1978). The latter level is well below the maximum acceptable toxicant concentration (MATC) to fatheads of $57–37 \,\mu g$ Cd per litre of Pickering and Gast (1972).

Some macroinvertebrates are particularly sensitive to cadmium. For example, the acute toxicity (48 hr LC_{50}) in soft water ($41–50 \,mg \,litre^{-1}$) to *Daphnia magna* was found to be $65 \,\mu g \,litre^{-1}$ while in a three week chronic test, in which the criterion was 16% reproductive impairment, the effective concentration was $0.17 \,\mu g \,litre^{-1}$. The three week LC_{50} was $5 \,\mu g \,litre^{-1}$ (Biesinger and Christensen, 1972). Under similar conditions

TABLE 7.13
TOXICITY OF CADMIUM: SELECTION OF VALUES FROM LITERATURE

Species	Hardness (mg litre^{-1} as $CaCO_3$)	Test	Value (μg litre^{-1})	Reference
Coho salmon (*Oncorhynchus kisutch*)	22	215 hr LC_{50}	3·7	Chapman and Stevens (1978)
Chinook salmon (*O. tshawytscha*) juveniles	24	96 hr LC_{50}	3·5	Chapman (1978)
Rainbow trout (*Salmo gairdneri*)	54	408 hr LC_{50}	5·2	Chapman and Stevens (1978)
(juveniles)	24	96 hr LC_{50}	1·0	Chapman (1978)

the 3 and 4 day LC_{50} to *Daphnia pulex* were 62 and 47 μg Cd per litre, respectively (Bertram and Hart, 1979). The average longevity was not affected by exposure to 1 μg Cd per litre but was reduced, in a dose-dependent manner, at concentrations between 5 and 30 μg Cd per litre. On the other hand, studies of the acute toxicity of cadmium to nine species of North American aquatic insect showed that most were relatively insensitive to cadmium. For two of the more sensitive species, the stonefly *Pteronarcella badia* and the mayfly *Ephemerella grandis*, the 96 hr LC_{50} values were 18 and 28 mg Cd per litre, respectively, in hard water (Clubb *et al.*, 1975). Insects transferred to clean water after exposure to 5 mg Cd per litre for four days lost cadmium at a steady rate, a feature which may contribute to the survival of populations exposed to intermittent doses. Similar resistance to cadmium has been observed in *Chironomus* larvae: the 48 hr LC_{50} was found to be 25 mg litre^{-1} in moderately hard water (Rao and Saxena, 1981). On exposing larvae of *Chironomus tentans*, Rathore *et al.* (1979) found the LD_{100} for cadmium chloride was 20 ppm, and the mean survival time in 100 ppm was about 24 hr. Studies by Brkovic-Popovic and Popovic (1977a) on the toxicity of cadmium to the oligochaete worm *Tubifex tubifex* revealed that in soft water (34·2 mg litre^{-1} $CaCO_3$) the 24 hr LC_{50} was 63–77 μg Cd per litre and the 48 hr LC_{50} was 31–45 μg Cd per litre. In hard water (261 mg litre^{-1} $CaCO_3$) the 24 hr LC_{50} was 1·2 mg Cd per litre and the 48 hr LC_{50} was 0·72 mg Cd per litre.

Spehar *et al.* (1978) observed that no significant effect was detectable after 28 days exposure of the stonefly *Pteronarcys dorsata* to 238 μg Cd per litre but at this level (and also at 85·5 μg Cd per litre) the caddis *Hydropsyche betteni* showed abnormal behaviour. The 28 day LC_{50} for the mayfly *Ephemerella* sp. was less than 3·0 μg Cd per litre while for the snail *Physa integra* the 7 and 28 day LC_{50} values were 114 and 10·4 μg Cd per litre, respectively. Cadmium residues were up to 30 000 times higher than in the water.

In experimental studies of the effect of cadmium on cultures of the alga (diatom) *Asterionella formosa*, it was found that 2 μg Cd per litre reduced the growth rate by an order of magnitude while populations exposed to more than 10 μg Cd per litre ceased growth in about one day (Conway, 1978). In field studies of cadmium-stressed plankton communities, phytoplankton photosynthesis and primary production were reduced by very low concentrations (0·2 μg Cd per litre) and zooplankton community structure was affected by concentrations of 5 μg Cd per litre or less (Marshall and Mellinger, 1980).

7.8.6 Chromium

Chromium has a wide range of oxidation states and although the trivalent form (Cr^{3+}) is most commonly found in nature, hexavalent chromium is also important. Fish are generally not very sensitive to chromium but some invertebrates are relatively intolerant. Pickering and Henderson (1966) studied the effects of trivalent and hexavalent chromium on several species of North American warm water fishes while Benoit (1976) derived data for hexavalent chromium toxicity to rainbow and brook trout (Table 7.14). Hughes *et al.* (1979) observed no significant changes in the gills of rainbow trout exposed to 0·2 of the 48 hr LC_{50} for several days.

Toxicity of trivalent chromium to two insects in soft water (44 mg litre^{-1} $CaCO_3$) differed quite markedly: the 96 hr LC_{50} for the mayfly *Ephemerella subvaria* was 2 mg Cr per litre while that for the caddisfly *Hydropsyche betteni* was 64 mg Cr per litre (Warnick and Bell, 1969).

Chromium toxicity to the tubificid worm *Tubifex tubifex* was measured in hard and soft water by Brkovic-Popovic and Popovic (1977a). In hard water (260 mg litre^{-1} $CaCO_3$) the 24 hr and 48 hr LC_{50} values were 86 and 4·6 mg Cr per litre, respectively. The values in soft water

TABLE 7.14

TOXICITY OF CHROMIUM TO FISH: 96 hr LC_{50} VALUES FOR TRI- AND HEXAVALENT CHROMIUM

(After Pickering and Henderson, 1966, and Benoit, 1976)

Species	Hardness (mg litre^{-1} as $CaCO_3$)	96 hr LC_{50} value (mg Cr per litre)	
		Trivalent	Hexavalent
Fathead minnow	20	5·1	17·6
(*Pimephales promelas*)	360	67·4	27·3
Bluegill	20	7·5	118
(*Lepomis macrochirus*)	360	71·9	133
Goldfish	20	4·1	37·5
(*Carassius auratus*)			
Guppy	20	3·3	30
(*Poecilia reticulata*)			
Rainbow trout	45	—	69
(*Salmo gairdneri*)			
Brook trout	45	—	59
(*Salvelinus fontinalis*)			

(34 mg litre^{-1} as $CaCO_3$) were much lower: the 24 hr LC_{50} was between 10 and 15 mg Cr per litre and the 48 hr LC_{50} was 1·4–1·5 mg Cr per litre. Chromium has been found to be moderately toxic to several algae (*Palmella, Oedogonium, Hydrodictyon* and *Palmellococcus*) but less so to the floating macrophytes *Lemna* and *Spirodela* (Mangi *et al.*, 1978). Growth of all these plants was inhibited by 10 ppm.

7.8.7 Cobalt
The toxicity of cobalt is not well documented. Its toxicity to *Daphnia magna* is moderate: the 48 hr LC_{50} in soft water (about 50 mg litre^{-1} as $CaCO_3$) was 1·1 mg Co per litre (Biesinger and Christensen, 1972). Warnick and Bell (1969) found the 96 hr LC_{50} of cobalt to the mayfly *Ephemerella* to be 16 mg Co per litre in soft water (44 mg litre^{-1} as $CaCO_3$). Data given by Jones (1964) indicate that cobalt is not very toxic to fish. In exposures of about 7 days the lethal concentrations, given in ppm, were for goldfish 10, stickleback 15 (as Co) and rainbow trout 30 (as Co).

7.8.8 Copper
Copper is one heavy metal, along with zinc, for which the aquatic toxicological effects are fairly well documented. Its toxicity is thought to be largely attributable to the Cu^{2+} ion. Copper readily forms complexes with a wide range of other substances commonly found in clean and polluted waters and is also readily adsorbed onto suspended solids. It is thought that these properties, together with the attendant difficulties in separating the chemical species of copper, may account for some of the variability of results in the literature. Some aspects of the chemistry of copper, its speciation and methods for differential analysis will be found in Shaw and Brown (1974) and Stiff (1971a, b). One further difficulty encountered in assessing the toxicity of copper is that the values obtained in the laboratory are not usually applicable to field situations, except, perhaps, in very soft water which is free from organic matter or organic solids (Alabaster and Lloyd, 1980). Factors which influence copper toxicity, in addition to hardness and the presence of organic matter, include temperature, oxygen concentration and pH value. An equation for predicting the 96 hr LC_{50} for rainbow trout from the total dissolved copper, hardness and pH has been provided by Howarth and Sprague (1978). Copper is relatively quite acutely toxic to fish as indicated by the selection of values shown in Table 7.15 (after Alabaster and Lloyd, 1980). The recommended EIFAC values for annual maximum 95 percentile

TABLE 7.15

ACUTE TOXICITY OF COPPER TO FISH: SELECTED EXAMPLES FROM THE LITERATURE
(Mainly after Alabaster and Lloyd, 1980 and Chapman and Ṣtevens, 1978)

Species	Hardness (mg litre^{-1} as CaCO)	Test	Value (mg Cu per litre)
Brook trout (*Salvelinus fontinalis*)	45	96 hr LC$_{50}$	0·1
Atlantic salmon (*Salmo salar*) parr	10	96 hr LC$_{50}$	0·03
Rainbow trout (*Salmo gairdneri*)	250	96 hr LC$_{50}$	0·9
	42	96 hr LC$_{50}$	0·06
	250	8 day LC$_{50}$	0·5
	12	7 day LC$_{50}$	0·03
	42	7 day LC$_{50}$	0·08
Coho salmon (*Oncorhynchus kisutch*)	20	96 hr LC$_{50}$	0·046
Goldfish (*Carassius auratus*)	220	96 hr LC$_{50}$	0·46
Carp (*Cyprinus carpio*)	53	96 hr LC$_{50}$	6·4
	250	96 hr LC$_{50}$	0·6
Fathead minnow (*Pimephales promelas*)	31	96 hr LC$_{50}$	0·075
	360	96 hr LC$_{50}$	1·5
Rudd (*Scardinius erythrophthalmus*)	250	96 hr LC$_{50}$	0·6
Stone loach (*Nemacheilus barbatulus*)	250	96 hr LC$_{50}$	0·76
Pike (*Esox lucius*)	250	96 hr LC$_{50}$	3·0
Perch (*Perca fluviatilis*)	250	96 hr LC$_{50}$	0·3
Bluegill (*Lepomis macrochirus*)	360	96 hr LC$_{50}$	10·2
Eel (*Anguilla anguilla*)	53	96 hr LC$_{50}$	0·81
	260	96 hr LC$_{50}$	4·0

concentrations for rainbow trout in water low in organic matter are, for soft water (50 mg litre^{-1} as CaCO$_3$) 22·0 µg Cu per litre and, for hard water (300 mg litre^{-1} CaCO$_3$), 112 µg Cu per litre.

The toxicity of copper to North American species is of the same order. For example, the 96 hr LC$_{50}$ for rainbow trout in soft acid water was 20 µg Cu per litre and in hard alkaline water 520 µg Cu per litre (Howarth and Sprague, 1978). The 96 hr LC$_{50}$ for brook trout (*Salvelinus fontinalis*) was 100 µg Cu per litre in soft water (45 mg litre^{-1} as CaCO$_3$) while the maximum acceptable concentration was only 17·4 µg Cu per litre (McKim and Benoit, 1971).

The 96 hr LC_{50} for fathead minnows (*Pimephales promelas*) was about 23 µg Cu per litre and about 150 µg Cu per litre in soft (20 mg litre^{-1} $CaCO_3$) and hard (360 mg litre^{-1} $CaCO_3$) water, respectively (Pickering and Henderson, 1966), while in moderately hard water (200 mg litre^{-1} $CaCO_3$) the chronic maximum acceptable toxicant concentration for this species was estimated to be 32 µg Cu per litre (Pickering *et al.*, 1977). Bluegills seem to be less sensitive to copper: the 96 hr LC_{50} in hard (360) and soft water (20) was 10 and 0·7 mg Cu per litre, respectively (Pickering and Henderson, 1966). The flagfish (*Jordanella floridae*) exhibits intermediate sensitivity, the 96 hr LC_{50} was found to be 1·27 mg Cu per litre in hard water (350–375 mg litre^{-1} as $CaCO_3$) although in their discussion Fogels and Sprague (1977) stress that care must be exercised in comparisons of acute results and that only differences of orders of magnitude are likely to be significant.

Rainbow trout exposed for a few weeks to concentrations of copper from about 0·3 to 0·6 of the incipient lethal level (ILL) showed significant increases in lethal tolerance, both in concentrations and exposure (Dixon and Sprague, 1981). Fish exposed to less than 0·1 ILL became sensitised with about 20% reduction in tolerance. Exposure to copper resulted in sensitisation to lethal levels of zinc. Acclimation was not retained: on being returned to clean water most of the increased tolerance was lost in 7 days.

Arthur and Leonard (1970) investigated the effects of copper on the amphipod *Gammarus* and the molluscs *Campeloma* and *Physa*, for which the 96 hr LC_{50} concentrations were 0·02, 1·7 and 0·039 mg total Cu per litre. The soft water (45 mg litre as $CaCO_3$) used had low levels of dissolved substances present and it is thought that the amounts of total copper are close to the concentrations which are biologically effective. As copper is commonly used as a molluscicide it is interesting to note the low toxicity to *Campeloma* compared with *Physa*. This may be explained by the closure of the operculum in *Campeloma*. Death was taken to have occurred when the molluscs did not respond to probing. The six-week 'no-effect' concentrations were, however, the same for all three species (about 8–15 µg Cu per litre). In harder water (100 mg per litre as $CaCO_3$) the gastropod *Biomphalaria glabrata* showed an almost identical response to *Physa*, the 96 hr LC_{50} being 0·04 mg Cu per litre (Bellavere and Gorbi, 1981). These authors also found the 24 hr LC_{50} for *Daphnia magna* was 0·05 and 0·07 mg Cu per litre for waters of 200 and 100 mg litre^{-1} as $CaCO_3$ hardness, respectively, which may be compared with 72 hr LC_{50} of 54–87 µg Cu per litre to four species of *Daphnia* in water of similar hardness (130–160 mg litre^{-1} as $CaCO_3$) reported by Winner and Farrell (1976).

For the oligochaete worm *Tubifex tubifex* in soft water (34 mg litre^{-1} $CaCO_3$) the 24 hr LC_{50} was found to be 0·36 mg Cu per litre and the 48 hr LC_{50} 0·21 mg Cu per litre. In hard water (261 mg litre^{-1} $CaCO_3$) the respective values were 1·38 and 0·89 mg Cu per litre (Brkovic-Popovic and Popovic, 1977a).

Copper has been used extensively as an algicide, especially in the United States where it is commonly applied to reservoirs to control algal blooms (Effler *et al.*, 1980; McKnight, 1981) and also to control growth of macrophytes (Newbold, 1975). Experimental studies of the effects of copper on the macrophyte *Elodea canadensis* showed that 0·5 mg Cu per litre completely inhibited photosynthesis but only 50% damage was observed in plants growing in soil after 28 days exposure to 3·25 mg Cu per litre (Brown and Rattigan, 1979). This supports the commonly observed erratic effectiveness of copper as a herbicide.

7.8.9 Lead

As explained above, early interest in heavy metal toxicity was concerned with lead (Carpenter, 1924–1927). Most lead salts have low solubility and it is unlikely that acute toxicity will be observed under natural conditions. Toxicity is affected by pH, hardness and the presence of organic materials. Data on acute toxicity to fish is given in Table 7.16. Lead is most likely to cause sub-lethal effects: principally lead uptake and tissue contamination, changes in heamatological parameters (Hodson, 1976) and spinal curvatures (Holcombe *et al.*, 1976; Hodson *et al.*, 1978). In hard water dissolved lead levels which do not appear to have effects on rainbow trout are of the order of 20–30 µg Pb per litre while in soft water the appropriate values are about 5–10 µg Pb per litre. These values would almost certainly protect coarse fish populations.

For brook trout (*Salvelinus fontinalis*) the maximum acceptable toxicant concentration (MATC) in water with a hardness of 44 mg litre^{-1}, as $CaCO_3$, was between 58 and 119 µg litre^{-1} total lead and 39–84 µg litre^{-1} dissolved lead. The 96 hr LC_{50} for brook trout was 4·1 mg Pb per litre (total) and 3·7 mg Pb per litre (dissolved lead) (Holcombe *et al.*, 1976). Merlini and Pozzi (1977) found that up to three times as much lead was accumulated by sunfish (*Lepomis gibbosus* L.) at pH 6 than at pH 7·5.

In experimental studies of the effects of lead on insects, Warnick and Bell (1969) were unable to determine a 96 hr LC_{50} since at the maximum

TABLE 7.16
TOXICITY OF LEAD TO FISH

Species	Hardness (mg litre^{-1} CaCO$_3$)	Test	Value (mg Pb per litre)	Reference
Fathead minnow (*Pimephales promelas*)	20	96 hr LC$_{50}$	6·5	Pickering and Henderson (1966)
	360	96 hr LC$_{50}$	482	Pickering and Henderson (1966)
Bluegill (*Lepomis macrochirus*)	20	96 hr LC$_{50}$	23·8	Pickering and Henderson (1966)
	360	96 hr LC$_{50}$	442	Pickering and Henderson (1966)
Goldfish (*Carassius auratus*)	20	96 hr LC$_{50}$	31·5	Pickering and Henderson (1966)
Guppy (*Poecilia reticulatus*)	20	96 hr LC$_{50}$	20·6	Pickering and Henderson (1966)
Rainbow trout (*Salmo gairdneri*)	28	96 hr LC$_{50}$	1·32	Davies *et al.* (1976)
	50	96 hr LC$_{50}$	1	Brown, (1968)
	353	96 hr LC$_{50}$	471 (1·38)	Davies and Everhart (1973)

Values given are total lead but, where known, free lead is indicated in brackets. In soft water total and free lead may be considered to be the same.

concentration used $(64 \, \text{mg litre}^{-1})$ no deaths occurred within 4 days. Their data show that medial survival of the stonefly *Acroneuria* at $64 \, \text{mg litre}^{-1}$ nominal $(2·2 \, \text{mg litre}^{-1}$ actual) exceeded 14 days. For the mayfly *Ephemerella* the 7 day LC_{50} was $32 \, \text{mg litre}^{-1}$ nominal, $3·2 \, \text{mg litre}^{-1}$ actual. The differences no doubt arose from problems relating to the solubility of the lead sulphate although water hardness was low at 40–50 mg litre^{-1} as $CaCO_3$. The 'actual' figures are based on analysis of the test solution after two weeks.

In fairly soft water (hardness $45 \, \text{mg litre}^{-1}$ $CaCO_3$) the crustacean *Daphnia magna* exhibited a 3 week LC_{50} of $0·3 \, \text{mg Pb}$ per litre although 16% reproductive impairment was shown at $30 \, \mu\text{g Pb}$ per litre (Biesinger and Christensen, 1972). Borgmann *et al.* (1978) studied the effects of chronic exposure to lead on the freshwater snail *Lymnaea palustris* and found increased mortality at $19 \, \mu\text{g Pb}$ per litre in moderately hard $(139 \, \text{litre}^{-1}$ as $CaCO_3)$ water and a 20 week LC_{50} of about $22 \, \mu\text{g Pb}$ per litre.

Larvae of *Chironomus tentans* exposed to lead nitrate exhibited a LD_{100} of 70 ppm and the mean survival time in 100 ppm was a little over one day (Rathore *et al.*, 1979).

Spehar *et al.* (1978) studied toxicity and accumulation of lead in soft water $(44–48 \, \text{mg litre}^{-1}$ $CaCO_3)$ using several species of macroinvertebrate. The stonefly *Pteronarcys dorsata*, the caddis *Brachycentrus* sp. and the snail *Physa integra* were unaffected by $565 \, \mu\text{g litre}^{-1}$ after 28 days but for the amphipod *Gammarus pseudolimnaeus* a 4 day LC_{50} of $124 \, \mu\text{g litre}^{-1}$ and a 28 day LC_{50} of $28·4 \, \mu\text{g litre}^{-1}$ were observed. Lead residue concentrations in the tissues were up to 9000 times greater than in the water.

Evidence is accumulating that in some crustaceans (*Asellus meridianus* and *A. aquaticus*) populations exposed to lead pollution have acquired increased resistance through genetic adaptation (Brown, 1976; Fraser *et al.*, 1978).

The formation of highly toxic organo-lead compounds in the environment by microorganisms, in addition to their use as 'anti-knock' additives in petroleum, and the accumulation of these in fish (Wong *et al.*, 1981), is an interesting recent aspect of lead toxicology. It is to be hoped that it does not parallel the developments observed for organo-mercury compounds.

7.8.10 Manganese
Manganese does not appear to have much significance as a pollutant. It is

rarely found at concentrations exceeding 1 mg litre^{-1} and since tolerances of aquatic life range from 1·5 to 1000 mg litre^{-1} (Train, 1979) it is
not considered to be a problem in fresh waters. Although permanganates
have been observed to be acutely lethal to fish at concentrations of
several mg litre^{-1}, they readily oxidise organic matter and are thus
reduced and made non-toxic. In studies of the toxicity of five heavy
metals to aquatic insects Braginskyi and Shcherban (1978) found manganese to be the least toxic.

7.8.11 Mercury

Mercury has received considerable attention as a toxicant recently,
partly as a result of serious poisoning of man ('Minamata' disease) and
partly through the discovery of microbial conversion of inorganic mercury to highly toxic methyl mercury which has increased the potential
toxicity of mercury pollution (Jensen and Jernelov, 1969). The problem is
compounded by the ability of algae and other aquatic plants to accumulate and, therefore, concentrate environmentally low levels of mercury
which are further bioaccumulated in food chains. Mercury is accumulated rapidly and eliminated slowly. Concentration factors for fish and
eggs (Heisinger and Green, 1975) may exceed 10^4, in comparison with
ambient water levels.

Inorganic mercury is acutely toxic to fish at about 1 mg Hg per litre
and exposures for more than ten days of 10–20 μg Hg per litre are known
to be fatal. However, published data vary widely (Table 7.17) and are
often conflicting (Jones, 1964). For example, the acute toxicity to rainbow trout of mercuric chloride ranges from 0·01 mg Hg per litre after 8
days (Jones, 1939) to 1·0 mg Hg per litre after 25 days (Boetius, 1960).

In studies of the toxicity of mercuric chloride to juvenile fathead
minnows (*Pimephales promelas*), the LC$_{50}$ for 96 hours and seven days
were found to be 168 and 74 μg Hg per litre, respectively (Snarski and
Olson, 1982). Growth was retarded and some fish were severely stunted
and scoliotic on exposure to less than 4 μg Hg per litre for 41 weeks.
Spawning was inhibited on exposure to 1 μg Hg per litre and much
reduced at 0·3 μg Hg per litre.

Mercury ions are taken up by rainbow trout more rapidly from methyl
mercury chloride than from mercuric chloride, and severe epithelial
necrosis was produced in the gills by mercuric chloride, while methyl
mercury chloride caused epithelial hyperplasia, degeneration and de-
squamation of gill lamellae (Wobeser, 1975). In a study of the mosquitofish, *Gambusia affinis*, subjected to sub-lethal concentrations of mercury

TABLE 7.17

TOXICITY OF MERCURY AND METHYLMERCURY[a] TO FISH

Species	Hardness (mg litre^{-1} CaCO$_3$)	Test	Value (µg Hg per litre)	Reference
Fathead minnow (*Pimephales promelas*)	40–48	48 hr LC$_{50}$	400	Curtis et al. (1979)
Fathead minnow (*Pimephales promelas*)	40–48	96 hr LC$_{50}$	170	Curtis et al. (1979)
Fathead minnow (*Pimephales promelas*)	45–47	96 hr LC$_{50}$	168	Snarski and Olson (1982)
Rainbow trout (*Salmo gairdneri*)	—	24 hr LC$_{50}$	903	Wobeser (1975)
	—	48 hr LC$_{50}$	66[a]	Wobeser (1975)
	—	96 hr LC$_{50}$	42[a]	Wobeser (1975)
	90	96 hr LC$_{50}$	280	Macleod and Pessah (1973)

TABLE 7.18

TOXICITY OF NICKEL TO FISH: SELECTION OF EXAMPLES FROM THE LITERATURE

Species	Hardness (mg litre^{-1} CaCO$_3$)	Test	Value (mg Ni per litre)	Reference
Rainbow trout (*Salmo gairdneri*)	240	48 hr LC$_{50}$	32	Hughes et al. (1979)
Fathead minnow (*Pimephales promelas*)	20	96 hr LC$_{50}$	4·6–5·2	Pickering and Henderson (1966)
	360	96 hr LC$_{50}$	42–45	Pickering and Henderson (1966)
Bluegills	20	96 hr LC$_{50}$	5·2–5·4	Pickering and Henderson (1966)
(*Lepomis macrochirus*)	360	96 hr LC$_{50}$	39·6	Pickering and Henderson (1966)
Goldfish (*Carassius auratus*)	20	96 hr LC$_{50}$	9·8	Pickering and Henderson (1966)
Guppy (*Poecilia reticulata*)	20	96 hr LC$_{50}$	4·5	Pickering and Henderson (1966)

(0·01–0·1 ppm), it was found that avoidance of a predator was impaired (Kania and O'Hara, 1974).

Identical 96 hr LC_{50} values of 2·0 mg Hg per litre were reported by Warnick and Bell (1969) for three insect species (the stonefly *Acroneuria*, the mayfly *Ephemerella* and the caddisfly, *Hydropsyche*). Mercury was also found to be very toxic to the tubificid worm *Tubifex tubifex*: the 48 hr LC_{50} value for soft water (34 mg litre^{-1} $CaCO_3$) and for hard water (260 mg litre^{-1} $CaCO_3$) were 0·08 and 0·1 mg Hg per litre, respectively (Brkovic-Popovic and Popovic, 1977a). The crustacean *Daphnia magna* was even more sensitive to mercury. In soft water (45 mg litre^{-1} $CaCO_3$) the 48 hr LC_{50} was 5 µg Hg per litre while in a three week test 16% reproductive impairment was observed at a level of 3·4 µg Hg per litre (Biesinger and Christensen, 1972).

In studies using an artificial stream, Sigmon *et al.* (1977) observed that 0·1 and 1·0 µg Hg per litre caused a reduction in algal numbers, biomass and community diversity, but no effect could be detected on the invertebrates present. The toxicity of mercury to the alga *Scenedasmus acutus* in culture increased with temperature (Huisman *et al.*, 1980).

7.8.12 Nickel

Nickel is not very toxic to rainbow trout (the 48 hr LC_{50} is about 32 mg Ni per litre in hard water) but in a study of the effects of heavy metals on gill morphology (Hughes *et al.*, 1979) only nickel appeared to have a harmful effect. The toxicity of nickel to several species of fish, including North American species, is given in Table 7.18.

The 96 hr LC_{50} for the stonefly *Acroneuria lycorias* in moderately soft water (44 mg litre^{-1} $CaCO_3$) was high at 33·5 mg Ni per litre, while the mayfly *Ephemerella subvaria* was more sensitive, the 95 hr LC_{50} for this species being 4 mg Ni per litre (Warnick and Bell, 1969). The tubificid worm *Tubifex tubifex* exhibited a nickel tolerance closer to that of the mayfly with a 48 hr LC_{50} of 7·0–8·7 mg Ni per litre in soft water (34 mg litre^{-1} as $CaCO_3$). In harder water (260 mg litre^{-1} as $CaCO_3$) the 48 hr LC_{50} was much higher at 61·4 mg Ni per litre (Brkovic-Popovic and Popovic, 1977a).

In studies of the effects of heavy metals on the inhibition of photosynthesis and toxicity in macrophytes, Brown and Rattigan (1979) found that, unlike other metals, nickel had little short-term effect on supression of photosynthesis but was highly phytotoxic to *Elodea* and *Lemna* after four weeks.

In a study of the effects of nickel on algae, Spencer and Green (1981) found that growth of several species of green alga was inhibited by concentrations of 100 µg Ni per litre while the blue-green alga *Anabaena flos-aquae* was unaffected by concentrations up to 600 µg Ni per litre. Another blue-green alga *A.cylindrica* was not affected by this higher concentration. It was suggested that the extracellular products of blue-green algae may form complexes with the nickel, as has been demonstrated for copper in *A. cylindrica* (Fogg and Westlake, 1955) and *A. flos-aquae* (McKnight and Morel, 1979).

7.8.13 Selenium

Although selenium is an essential element in animal nutrition its increasing presence in the environment as a result of burning fossil fuels and its use in electronics, glass and photographic industries makes it a potential pollutant.

Selenium interacts with mercury: it is an antidote to methyl mercury poisoning and retards the rate of mercury bioaccumulation in fish and other biota (Rudd *et al.*, 1980). The toxicity of selenium to fish is indicated in Table 7.19. Carp (*Cyprinus carpio*) appear to be more tolerant than rainbow trout (*Salmo gairdneri*). In the experimental enclosure used by Rudd *et al.* (1980) young perch (*Perca flavescens*) began to die after four days exposure to 1 mg Se per litre but pearl dace (*Semotilus margarita*) were unaffected. At 0·1 mg Se per litre neither species showed increased mortality. Hodson *et al.* (1980) have suggested

TABLE 7.19
TOXICITY OF SELENIUM TO FISH

Species	Hardness (mg litre^{-1} CaCO$_3$)	Test	Value (mg Se per litre)	Reference
Rainbow trout (*Salmo gairdneri*)	135	96 hr LC$_{50}$	8·1	Hodson *et al.* (1980)
		9 day LC$_{50}$	6·5	Hodson *et al.* (1980)
Carp (*Cyprinus carpio*)	(not given)	24 hr LC$_{50}$	72	Sato *et al.* (1980)
		48 hr LC$_{50}$	50	Sato *et al.* (1980)
		96 hr LC$_{50}$	35	Sato *et al.* (1980)
Goldfish (*Carassius auratus*)	—	7 day LC$_{50}$	12	Weir and Hine (1970)

that 50 µg Se per litre might be a useful working environmental limit to safeguard fish stocks. In moderately hard water (135 mg per litre as $CaCO_3$) the 96 hr and 9 day LC_{50} values for rainbow trout were 8·1 and 6·5 mg Se per litre, respectively. Sub-lethal responses of eggs and fry were observed after 44 weeks post-hatch with exposure to levels as low as 4–16 µg Se per litre.

In studies of the effects of prolonged exposure (120 days) of juvenile bluegill (*Lepomis macrochirus*) and large mouth bass (*Micropterus salmoides*) to sub-lethal levels of selenium (10 µg litre^{-1}), Lemly (1982) observed that accumulation was greatest in the spleen, heart, liver and kidney, and least in gonads, gut and brain. After 30 days of elimination, levels remained unchanged in spleen, liver, kidney and white muscle but some loss was observed in other tissues. It was concluded that, under natural conditions, levels of selenium approaching 10 µg litre^{-1} were potentially toxic.

7.8.14 Silver
Silver is likely to be present in freshwaters as a result of heavy metal mining or extraction but is also used in electroplating photographic and ink manufacturing industries. It is highly toxic to fish: the 96 hr LC_{50} for rainbow trout in soft (26 mg litre^{-1} as $CaCO_3$) and hard (350 mg litre^{-1} as $CaCO_3$) water was 6·5 and 13·0 µg Ag per litre. The apparent 'no effect' concentration in soft water was 0·09 µg Ag per litre. At concentrations of 0·17 µg Ag per litre or greater, premature hatching of eggs and retarded growth of fry was observed (Davies *et al.*, 1978). Large mouth bass (*Micropterus salmoides*) survived 7 µg Ag per litre for six months but 70 µg Ag per litre proved lethal in one day, yet bluegills (*Lepomis macrochirus*) survived this concentration for six months (Coleman and Cearley, 1974). Jones (1939, 1964) reported a threshold of 3 µg Ag per litre for sticklebacks (*Gasterosteus*) in soft water.

7.8.15 Tin
Few published data exist for tin, although its toxicity appears to be low: the 48 hr LC_{50} (in soft water, 50 mg litre^{-1} as $CaCO_3$) for *Daphnia magna* was 55 mg Sn per litre (Biesinger and Christiansen, 1972).

7.8.16 Vanadium
Although vanadium is a fairly abundant element it has rarely attracted much attention as a pollutant. Fossil fuels are a major source but little seems to be known about the environmental significance of vanadium.

Vanadium forms a wide range of ionic forms which do not complex with carbonates or bicarbonates as do most other heavy metals. This means that vanadium may be expected to differ in its toxic properties from the other metals and show considerable variation depending on the ionic species present. Toxicity to fish has been measured: the 96 hr LC_{50} for flagfish (*Jordanella floridae*) was found to be 11·2 mg V per litre in hard water at pH 8·2 (Holdway and Sprague, 1979). This value is similar to that observed for rainbow trout by Stendahl and Sprague (1982). These authors investigated the effect of water hardness and pH on vanadium toxicity and found that the 7 day LC_{50} showed only a small variation over a very wide range of hardness and pH. Vanadium was most toxic at pH 7·7 and decreased with increasing hardness. Small fish were more resistant to vanadium than larger ones, at least up to 12 g wet weight. Cumulative mortality continued to increase until 2 weeks of exposure, suggesting a slow uptake. Studies of the uptake and loss of radioactive vanadium by the eel (*Anguilla anguilla*) showed that the rate of total body burden loss was about one tenth as slow as the rate of uptake (Bell *et al.*, 1981). Liver and kidney accumulated most vanadium.

Giles and Klaverkamp (1982) investigated the acute toxicity of vanadium in soft water to eyed eggs of rainbow trout (*Salmo gairdneri*). The LC_{50} was 118 mg V per litre but premature hatching was observed at concentrations of 44 mg V per litre.

7.8.17 Zinc

Zinc is a common constituent of industrial and mining effluents. It is mainly used to galvanise iron or steel and in the formation of alloys (brass and bronze). Many compounds are important industrial raw materials. An extensive literature exists of the toxicity of zinc, especially to fishes, and reviews have been prepared by Doudoroff and Katz (1953), Skidmore (1964) and, more recently, by Alabaster and Lloyd (1980) and Spear (1981).

Zinc is unusual in that it has low toxicity to man but high toxicity to fish, such that adequate standards for domestic water supply (maximum around 5 mg Zn per litre) would be highly toxic to many fish species.

As with other heavy metals, toxicity of zinc is modified by environmental factors including hardness, temperature, dissolved oxygen levels and the presence of suspended solids or organic matter. Some of these factors have complex effects. For example, above 15 °C survival of salmonids is curtailed but at lower temperature (about 5 °C) the lethal threshold value is reduced and, therefore, toxicity increases. Hodson and

Sprague (1975) concluded that, in polluted rivers, zinc concentrations would be more damaging to fish in the winter months. In unacclimated fish, lower dissolved oxygen concentrations reduce survival while the presence of organic matter or other suspended material, on which zinc may become adsorbed, tends to enhance survival.

Possibly the most important factor which influences zinc toxicity is water hardness. As a general indication of the relationship there appears to be an inverse linear correlation between the logarithms of the hardness and the toxicity, i.e. a tenfold increase in hardness is accompanied by a tenfold increase in the LC_{50}, for salmonid species at least. It has been postulated that the presence of calcium ions provides some protection to fish exposed to zinc, perhaps by competitively occupying sites on enzyme molecules which would otherwise be taken by zinc with adverse consequences. Some support for this hypothesis was provided by Lloyd (1965) who observed that fish reared in soft water were less tolerant of zinc than fish reared in hard water but tested immediately on transfer to soft water, but this apparent protection was lost if the fish were maintained in softer water for about five days. Matthiessen and Brafield (1977) investigated zinc uptake and loss in the stickleback (*Gasterosteus*) and showed that, in hard water, after an initial influx of zinc the fish were able to reduce their internal zinc levels to near-normal in spite of external concentrations of 1–4 mg Zn per litre. Fish exposed to zinc in calcium-free water were unable to reduce internal zinc and they postulated that calcium ions might facilitate zinc excretion thereby enhancing survival in hard water. Initially, however, zinc uptake rates were higher in hard water and it was further postulated that, in soft waters containing high levels of zinc, precipitation and loss of colloidal mucus might afford some protection at first. This is an interesting development, especially when one considers that early investigators attributed death in zinc solutions to suffocation caused by coagulation of mucus!

Data on the acute toxicity of zinc to a selection of freshwater fish species are given in Table 7.20. On reviewing the available experimental and field data the EIFAC tentative criteria, as 95 percentile concentrations, have been selected as shown in Table 7.21 (Alabaster and Lloyd, 1980). In general, non-salmonid species are more resistant to zinc poisoning than salmonids. Variations in susceptibility occur through the life history of some species. For example, Zitko and Carson (1977) noted that in soft water (14 mg litre^{-1} as $CaCO_3$) the incipient lethal level (ILL) varies from 150 to 1000 μg Zn per litre as a function of the seasonal

TABLE 7.20

ACUTE TOXICITY TO FISH OF ZINC: SELECTED EXAMPLES FROM THE LITERATURE

Species	Hardness (mg litre^{-1} CaCO$_3$)	Test	Value (mg Zn per litre)	Reference
Brook trout (*Salvelinus fontinalis*)	45	96 hr LC$_{50}$	2·0	Holcombe et al. (1979)
Rainbow trout (*Salmo gairdneri*)	26	96 hr LC$_{50}$	0·4	Sinley et al. (1974)
	44	48 hr LC$_{50}$	0·9	Herbert and Shurben (1964)
	83	96 hr LC$_{50}$	1·76	Chapman and Stevens (1978)
	240	48 hr LC$_{50}$	2·7	Herbert and Shurben (1963)
	300	48 hr LC$_{50}$	3·2	Brown (1968)
	320	5 day LC$_{50}$	46	Ball (1967b)
	333	96 hr LC$_{50}$	7·2	Sinley et al. (1974)
	500	24 hr LC$_{50}$	5·3	Solbé (1974)
	500	48 hr LC$_{50}$	4·8	Solbé (1974)
Atlantic salmon (*Salmo salar*)	20	24 hr LC$_{50}$	0·7	Sprague (1964)
	18	21 day LC$_{50}$	0·34–1·60	Farmer et al. (1979)
Coho salmon (*Oncorhynchus kisutch*)	25	96 hr LC$_{50}$	0·91	Chapman and Stevens (1978)
Chinook salmon (*Oncorhynchus tshawytscha*)	24	96 hr LC$_{50}$	0·46	Chapman (1978)
Fathead minnow (*Pimephales promelas*)	20	96 hr LC$_{50}$	0·8–1·0	Pickering and Henderson (1966)
	360	96 hr LC$_{50}$	33·4	Pickering and Henderson (1966)
Bluegills (*Lepomis macrochirus*)	20	96 hr LC$_{50}$	4·9–5·8	Pickering and Henderson (1966)
	360	96 hr LC$_{50}$	41	Pickering and Henderson (1966)
Perch (*Perca fluviatilis*)	320	5 day LC$_{50}$	16	Ball (1967b)
Stoneloach (*Nemacheilus barbatulus*)	290	5 day LC$_{50}$	2·5	Solbé and Flook (1975)
Gudgeon (*Gobio gobio*)	320	7 day LC$_{50}$	8·4	Ball (1967b)
Roach (*Rutilus rutilus*)	320	5 day LC$_{50}$	17·3	Ball (1967b)
Bream (*Abramis brama*)	320	5 day LC$_{50}$	14·3	Ball (1967b)

TABLE 7.21

EIFAC TENTATIVE CRITERIA FOR THE MAINTENANCE OF FISH POPULATIONS BASED ON 95 PERCENTILE CONCENTRATIONS OF METALS (μg litre^{-1}) AND VARYING DEGREES OF WATER HARDNESS (mg litre^{-1} as $CaCO_3$)

	Hardness			
	10	50	100	500
Salmonids				
Cadmium	0·6	0·9	1·0	1·5
Copper	5	22	40	112
Zinc	30	200	300	500
Coarse fish				
Cadmium	20	30	38	50
Copper[a]	—	—	—	—
Zinc[b]	300	700	1 000	2 000

[a]Data on coarse fish are considered inadequate even for tentative criteria.
[b]More stringent criteria are required to protect the minnow (*Phoxinus*).

and developmental stage of the Atlantic salmon (*Salmo salar*). The most sensitive stage was thought to be the parr-smolt transformation.

Zinc was not as toxic to aquatic insects as some other heavy metals tested by Warnick and Bell (1969). They were unable to determine a 96 hr LC_{50} since all specimens survived the highest concentrations (64 mg litre^{-1}, nominal) tested. However, longer tests in soft water (44 mg litre^{-1} as $CaCO_3$) showed that the 14 day LC_{50} for the stonefly *Acroneuria* was 32 mg Zn per litre nominal (5·5 mg Zn per litre residual after 14 days), the 10 day LC_{50} for the mayfly *Ephemerella* was 16 mg Zn per litre nominal (7·9 mg Zn per litre residual) and for the caddis, *Hydropsyche*, 11 days LC_{50} was 32·0 mg Zn per litre nominal (4·7 mg Zn per litre residual). The oligochaete *Tubifex tubifex* seems to be more sensitive to zinc; the 48 hr LC_{50} in soft (34 mg litre^{-1} as $CaCO_3$) and hard (260 mg litre^{-1} as $CaCO_3$) water was 3·0 and 60 mg Zn per litre, respectively (Brkovic-Popovic and Popovic, 1977a).

7.9 PESTICIDES

Pesticides are poisons which are used to destroy or control unwanted organisms, usually species which otherwise might cause severe economic

TABLE 7.22

LIST OF PRINCIPAL INSECTICIDES, GROUPED ACCORDING TO THE THREE MAIN TYPES (ORGANOCHLORINE, ORGANOPHOSPHORUS, CARBAMATE) AND HERBICIDES, WITH EXAMPLES OF PROPRIETARY NAMES, TOGETHER WITH THEIR FULL CHEMICAL NAMES

Common name	Proprietary name(s)	Chemical name
Insecticides		
Organochlorine		
Aldrin[*] (HHDN)		(1R,4S,5S,8R)-1,2,3,4,10,10-hexachloro-1,4,4a,5,8,8a-hexahydro-1,4:5,8-dimethanonaphthalene
BHC (see HCH)		
Camphechlor	Toxaphene	mixture of chlorinated camphenes
Chlordane		1,2,3,4,5,6,7,8-octachloro-2,3,3a,4,7,7a-hexahydro-4,7-methanoindene
Chlordecone	Kepone	dodecachloro octahydro-1,3,4-methano-2H-cyclobuta[cd]pentalen-2-one
DDD (see TDE)		
DDE		1,1-dichloro-2,2-bis (4-chlorophenyl) ethylene, formerly dichlorodiphenyldichloroethylene
DDT		1,1,1-trichloro-2-2-bis (4-chlorophenyl) ethane, formerly (dichlorodiphenyltrichloroethane)
Dieldrin		(1R,4S,5S,8R)-1,2,3,4,10,10-hexachloro-1,4,4a,5,6,7,8,8a-octahydro-6,7-epoxy-1,4:5,8-dimethanonaphthalene
Endosulphan	Thiodan	1,4,5,6,7,7-hexachloro-8,9,10-trinorborn-5-en-2,3-ylene dimethyl sulphite
Endrin		(1R,4S,5R,8S)-1,2,3,4,10,10-hexachloro-1,4,4a,5,6,7,8,8a-octahydro-6,7-epoxy-1,4:5,8-dimethanonaphthalene
HCH (gamma-HCH)		1,2,3,4,5,6-hexachlorocyclohexane
Heptachlor	Lindane	γ-1,2,3,4,5,6-hexachlorocyclohexane 1,4,5,6,7,8,8-heptachloro-3a,4,7,7a-tetrahydro-4,7-methanoindene
Lindane (see HCH)		
Methoxychlor		1,1,1-trichloro-2,2-bis (4-methoxyphenyl) ethane
Mirex		dodecachloro octahydro-1,3,4-methano-2H-cyclobuta [cd]pentalene
TDE (DDD)		1,1-dichloro-2,2-bis (4-chlorophenyl) ethane
Thiodan (see Endosulphan)		
Toxaphene		

Organophosphorus

Azinphos methyl	Guthion	S-(3,4-dihydro-4-oxobenzo[d]-[1,2,3]-triazin-3-yl methyl O,O-dimethyl phosphorodithioate
Chlorthion	Bayer 22/190	O-(3-chloro-4-nitrophenyl)O,O-dimethyl phosphorothioate
Chlorpyrifos	Dursban	O,O-diethyl O-(3,5,6-trichloro-2-pyridyl) phosphorothioate
Coumaphos	Asuntol, Co-Ral, Muscatox, Resitox	O-3-chloro-4-methylcoumarin-7-yl O,O-diethyl phosphorothioate
Demeton	Systox	O,O-diethyl O-2-ethylthioethyl phosphorothioate and O,O-diethyl S-2-ethylthioethyl phosphorothioate
Diazinon	Basudin	O,O-diethyl-O-2-isopropyl-6-methylpyrimidin-4-yl phosphorothioate
Dichlorvos (DDVP)	Vapona, Dedevap	2,2-dichlorovinyl dimethyl phosphate
Dimethoate	Rogor	O,O-dimethyl S-methylcarbamoylmethyl phosphorodithioate
Dioxathion	Delnav	S,S'-(1,4-dioxane-2,3-diyl) O,O,O',O'-tetraethyl di (phosphorodithioate)
Disulfoton	Bayer 19639 Di-syston Di-thiosystox	O,O-diethyl S-2-ethylthioethyl phosphorodithioate
Ethion	Embathion	O,O,O'-tetraethyl S,S'-methylene di(phosphorodithioate)
Fenitrothion	Folithion, Sumithion	O,O-dimethyl O-4-nitro-m-tolyl phosphorothioate
Fenthion	Bayer 29493, Baytex	O,O-dimethyl O-4-methylthio-m-tolyl phosphorothioate
Malathion		S-1,2-bis (ethoxycarbonyl) ethyl O,O-dimethyl phosphorodithioate
Methyl Parathion		O,O-dimethyl O-4 nitrophenyl phosphorothioate
Mevinphos	Phosdrin	2-methoxycarbonyl-1-methylvinyl dimethyl phosphate
Naled	Dibrom	1,2-dibromo-2,2-dichloroethyl dimethyl phosphate
Parathion		O,O-diethyl-O-4-nitrophenyl phosphorothioate
Phorate	Thimet	O,O-diethyl S-ethyl thiomethyl phosphorodithioate
Phosmet	Imidan	O,O-dimethyl S-phthalimidomethyl phosphorodithioate
Phosphamidon	Dimecron	2-chloro-2-diethylcarbomoyl-1-methylvinyl dimethyl phosphate
Temephos	Abate	O,O,O'-tetramethyl O,O'-thiodi-p-phenylene bis (phosphorothioate)
Trichlorophon	Dipterex Dylox	dimethyl 2,2,2-trichloro-1-hydroxyethyl phosphonate

TABLE 7.22—contd.

Common name	Proprietary name(s)	Chemical name
Carbamates		
Aminocarb	Metacil	4-dimethylamino-m-tolyl methyl carbamate
Carbaryl	Sevin	1-napthyl methylcarbamate
Carbofuran		2,3-dihydro-2,2-dimethyl-benzofuran-7-yl methylcarbamate
Methiocarb (Mercaptodimethur)	Mesurol	4-methylthio-3,5-xylyl methylcarbamate
Mexacarbate	Zectran	4-dimethylamino-3,5-xylyl methylcarbamate
Propoxur	Baygon	2-isopropor oxyphenyl methyl carbamate
Herbicides		
Asulam	Asulox	methyl 4-amino-phenylsulphonylcarbamate
Atrazine	Weedex, Residox	2-chloro-4-ethylamino-6-isopropyl-amino-1,3,5-triazine
Chlorthiamid	Prefix	2,6-dichlorothio-benzamide
2,4-D	Dormone, Fernimine	2,4-dichlorophenoxyacetic acid, esters or salts especially amine salts
2,4,-DB	Embutox	4-(2,4-dichlorophenoxy) butyric acid, esters or salts
Dalapon	Dowpon, Radapon	2,2-dichloropropionic acid
Dichlobenil	Casoron	2,6-dichlorobenzonitrile
Dichlone	Phygon	2,3-dichloro-1,4-naphthoquinone
Diquat	Reglone, Midstream,	1,1'-ethylene-2,2'-bipyridyldiylium dibromide
Diuron	Karmex	3-(3,4-dichlorophenyl)-1,1-dimethylurea
Glyphosate	Round-up, Spasor	N-(phosphoromethyl) glycine (iso-propylamine salt)
IPC (see Propham)		
Molinate	Ordram	S-ethyl N,N-hexamethylenethiocarbamate
Paraquat	Gramoxone, [Weedol]	1,1'-dimethyl-4,4'-bipyridyldiylium dichloride
Picloram		4-amino,-3 5,6-trichloropyridine-2-carboxylic acid
Propanil (DCPA)		3',4'-dichloroproprionanilide
Propham (IPC)		isopropyl phenylcarbamate
Simazine		2,chloro-4,6-bis (ethylamino)-1,3,5-triazine
2,4,5-T		(2,4,5-trichlorophenoxy) acetic acid
Terbutryne	Clarosan	2-tert-butylamino-4-ethylamino-6-methylthio-1,3,5-triazine
Trifluralin	Treflan	α,α,α-trifluoro-2,6-dinitro-N,N-dipropyl-p-toluidine
Vernolate		S-propyl dipropylthiolcabamate

loss to crops or property or be responsible for the transmission of disease. Since they are biocides they are intrinsically potentially dangerous to other, non-target species and may have unwelcome and perhaps unanticipated effects if used carelessly or improperly. In this section a brief review is provided of the principal pesticides and their toxicity. A vast literature exists on pesticides and pollution caused by them (Corbett, 1974; Edwards, 1973a, b; Holden, 1973; Johnson, 1973; Kerr and Vass, 1973; Khan, 1977; Muirhead-Thompson, 1971; Perring and Mellanby, 1977), and it is only possible to give attention to those which form a significant part of the pollution burden in freshwater.

For convenience, the pesticides have been considered under two major sections, namely the insecticides (chlorinated hydrocarbons, organophosphates and carbamates) and herbicides (including algicides) with smaller sections dealing with molluscicides and piscicides. Fungicides and nematocides are potentially dangerous substances in water but, apart from wood preservatives, are unlikely to be released in sufficient quantities as to pose a serious problem. Some compounds are effective against a number of kinds of pest. A list of common pesticides, classified under general headings, is given in Table 7.22. Many pesticides are available in several formulations and may be referred to in the literature under their trade or registered name, the common name of the active ingredient or even their chemical name. The table provides a useful checklist and should facilitate identification of the compound and the group to which it belongs.

7.9.1 Environmental Significance

The significance of pesticides in the environment is twofold. First, unlike substances in industrial or domestic effluents, which may or may not have a significant toxic effect on organisms, pesticides are compounds which have been selected for their biocidal properties and are applied in order to kill or control certain organisms. Few are absolutely specific to their target organisms and so other related species are at risk and many unrelated groups may also be affected, to a varying extent. Another significant factor stems from this, namely that sub-lethal concentrations of pesticides may impart a significant change in the physiology, reproduction or behaviour of non-target groups which ultimately impairs their ability to survive in the affected habitat. Second, many pesticides or their immediate degradation products persist in the environment. At one time this would be regarded as an advantage since their pesticidal action would be enhanced by their stability and, therefore, continued toxicity to target organisms. Such persistence is now regarded as undesirable: the ideal pesticide should persist only long enough to destroy the target organism, after which it should degrade rapidly to harmless metabolites. Environmental persistence is facilitated by the low water solubility of many pesticides and high solubility in lipid or fat tissues or organisms. This affinity for lipids in biota enhances their stability and ensures their persistence in the aquatic environment since the residues are shielded from degradation by microorganisms and may be passed from prey to predator along food chains. It has been suggested, however, that the 'bioamplification' of organochlorine insecticides in food chains is not essentially caused by increasing concentration of material through predation but is a product of the partitioning of the pesticides between water and lipid, resulting in increased organochlorine residues in organisms at higher trophical levels simply because their fat content is greater (Hamelink *et al.*, 1971; Hamelink and Waybrant, 1976). Many studies have shown that the organochlorine residues found in different species and in different tissues of the same species correlate well with their lipid contents. Using data from Reinert and Bergman (1974) on the total DDT content and lipid content of coho salmon (*Oncorhynchus kisutch*) taken at different seasons, Phillips (1978) demonstrated that the high variability observed in DDT residues, on a wet-weight basis, was almost entirely due to variations in the lipid content of the tissue examined and when concentrations are based on lipid weight the results were consistent and seasonal fluctuations were smoothed. Similarly, Anderson and Everhart (1966) postulated that total DDT residues in a land-locked population of salmon (*Salmo salar*) depended on the 'condition' of the fish which is related to the accumulated lipid. The kind of lipid may also be important: salmon brain tissues contained extremely low levels of pesticide compared with other tissues (Anderson and Fenderson, 1970).

The persistence of organochlorine (and other) pesticide residues in organisms enables them to be sampled and analysed in order to detect and assess the degree of environmental contamination by these substances, although a large number of factors affect the residue content of organisms. Increase in body lipids as a response to exposure to sub-lethal pesticide levels (Macek *et al.*, 1970; Grant, 1976) may complicate interpretation, as do variations in the age or size of the organism, season and reproductive state, temperature, toxicity of the substance and interactions between pesticides (Phillips, 1978).

7.9.2 Herbicides

Herbicides are selected primarily for their phytotoxicity and, therefore, may be expected to have lower toxicity to animals than many other pesticides. In assessing their impact on freshwaters it is necessary to distinguish between those which are applied to water for the control of aquatic weeds and those which enter water indirectly in surface run-off from agricultural land or accidentally from spray drift or spillages. Aquatic herbicides have the greatest potential for causing ecological changes in surface waters since they are applied at concentrations which are intended to kill or control the target plants. Herbicides entering indirectly from agricultural applications will almost always be at lower concentrations, except, of course, in serious accidental spillages. Indeed, Mullison (1970) believes there is little evidence that herbicides from agriculture or industry constitute a pollution problem. In the United Kingdom, herbicides cleared for aquatic use are restricted to those which are harmless to fishes and their food organisms and have little toxic impact on wildlife in general at doses which afford effective control of

TABLE 7.23

LIST OF CURRENTLY APPROVED HERBICIDES FOR USE IN OR NEAR WATER IN THE UNITED KINGDOM (HERBICIDES FORMERLY APPROVED ARE INDICATED IN BRACKETS)

Herbicide	Formulation	Proprietary names
Asulam	Liquid	Asulox
[Chlorpropham	—	—]
Chlorthiamid	Granules	Prefix
[Cyanatryn	Slow-release pellets	Aqualin]
2,4-D amine	Liquid	Chipman 2,4-D, Dormone Fernimine M
Dalapon	Wettable powder	BH Dalapon, Dow Dalapon, PP Dalapon, Dowpon
Dichlobenil	Granules	Casoron G and Casoron GSR
Diquat	Liquid	Reglone 40
Fosamine ammonium	Liquid	Krenite
Glyphosate	Liquid	Round-up, Spasor
Maleic hydrazide	Liquid	Regulox, Vondalhyd
Maleic hydrazide with 2,4-D	Liquid	BH43
[Paraquat	Liquid	Gramoxone S]
Terbutryne	Granules	Clarosan

TABLE 7.24

PRINCIPAL PROPERTIES OF THE MAJOR AQUATIC HERBICIDES AND THE PLANTS FOR WHICH THEY ARE EFFECTIVE
(After Hellawell and Bryan, 1982)

Herbicide	Mode of phytotoxic action	Susceptible plants	Non-susceptible plants	Comments
Chlorthiamid	Affects growing points of shoots and roots	Rooted, submerged or emergent broad-leaved plants	Ineffective for narrow-leaved plants, free-floating plants or algae	Similar action to dichlobenil to which it breaks down
2,4-D amine	Distorts new growth on translocation from leaves or roots	Some emergent, especially broad-leaved, plants	Selective action—ineffective for grasses	
Dalapon		Emergent Graminae (grasses)	Selective herbicide—ineffective for other groups, i.e. floating and sub-merged plants and algae	
Dichlobenil	Affects growing points of roots and shoots by inhibiting cell division	Rooted submerged or rooted emergent broad-leaved plants	Narrow-leaved plants and free-floating, rootless plants and algae	

Diquat	Contact herbicide, rapid defoliant	Wide range of aquatic plants but especially submerged and floating plants including filamentous alga *Cladophora*	Emergent vegetation and most algae, except *Cladophora*	Rapidly deactivated in soil and water
Glyphosate	Foliar acting herbicide which is translocated from treated vegetation (above water) to underground parts	Emergent and floating species	Algae and submerged plants	
Terbutryne	Translocated herbicide, absorbed by foliage and roots; photosynthesis inhibitor	Wide range of submerged and floating plants and algae	Emergent plants (including reeds) and water lilies	Not persistent but relatively slow formulation controls regrowth; appears to be ineffective below 5–8°C

macrophytes. Screening procedures are not simply confined to the active ingredients but include any other substances (for example, wetting agents, emulsifiers or other additives) in the proprietary formulations.

Herbicides approved for use in or near water in the United Kingdom include asulam, chlorthiamid, 2,4-D amine, dalapon, dichlobenil, diquat, fosamine ammonium, glyphosate, maleic hydrazide and terbutryne (Table 7.23). Approval for some of these herbicides is confined to certain proprietary formulations and for specific uses such as bracken control (asulam) or control of woody plants near water (fosamine ammonium). Some herbicides would be acceptable for aquatic use, were they available in a suitable formulation. For example, paraquat has been approved as 'Gramoxone S' but this is no longer available, while the readily available terrestrial formulation 'Gramoxone' is not acceptable.

7.9.2.1 Properties of the principal aquatic herbicides

Reviews of herbicides used in aquatic environments will be found in Brooker and Edwards (1975), Newbold (1975), Robson and Fearon (1976) and Hellawell and Bryan (1982).

There are five main groups of macrophytes, determined largely on the basis of their growth habits, each of which requires different herbicides for its control since no herbicide is effective in the treatment of all groups, although some herbicides have a broader spectrum of efficacy than others. These groups are bankside broad-leaved plants, bankside grasses, submerged plants, floating or emergent plants and emergent reeds or grasses. The principal properties of the more important herbicides and the groups of plants which they treat are set out in Table 7.24. Filamentous algae and planktonic algae are two other categories of aquatic plants. Some members of the former group are susceptible to control but planktonic algae are difficult to control with UK approved herbicides. Copper salts are known to be effective for algal control (Effler et al., 1980; McKnight, 1981) but their margin of safety for fish species is considered to be inadequate (see below).

7.9.2.2 Herbicides used in water

True aquatic herbicides are approved for plants growing in water and include those herbicides which are applied to emergent vegetation above the water but which are considered safe if spray enters the water.

Chlorthiomid and Dichlobenil

These herbicides are considered together since chlorthiamid breaks down into dichlobenil. They are total herbicides which are absorbed through the roots and translocated to other parts of the plant. Death is not rapid, taking several weeks, and the exact mechanism of phytotoxicity is not known. At the recommended dose (1 mg litre^{-1}) there is a tendency to kill all the submerged macrophytes present and only a few less-susceptible floating plants (*Utricularia, Lemna minor*), together with filamentous algae (*Cladophora, Vaucheria, Spirogyra*), remain. The yellow and white water lilies (*Nuphar lutea* and *Nymphaea alba*) are resistant, as are a number of emergent plants (for example, *Polygonum amphibium* and *Sparganium erectum*). Algal blooms often follow dichlobenil treatment (Newbold, 1974). Dichlobenil persists in mud for up to six months and at levels used for weed treatment it has less margin of safety than other aquatic herbicides (see below). Wilson and Bond (1969) have drawn attention to the possibility that since dichlobenil may be added as a granular formulation and may take three days to reach the effective concentration, the consequences for invertebrates living on or in substrate may be greater than for free-swimming organisms.

Newbold (1974) has suggested that it might be more effective to use a lower concentration in order to inhibit macrophyte growth instead of achieving virtually complete destruction of plants with subsequent algal blooms and marked habitat changes.

Dalapon

Dalapon is used to control 'reeds' or, more exactly, emergent monocotyledons, and especially the common reed *Phragmites communis* which grows at the margins of drainage channels. Other species controlled include *Phalaris, Sparganium* and *Typha* species. It is often combined with 2,4-D and also, previously, paraquat was added to obtain better control of the majority of plants which commonly tend to choke drainage channels. Its persistence is limited to a few days since it does not bind to soil and is rapidly degraded in water by bacteria (Newbold, 1975; Brooker, 1976a). Toxicity to fish and other organisms is low and there appears to be little evidence that it has a harmful effect on the aquatic environment, although the destruction of the emergent reeds has serious consequences for animals associated with these and reed-nesting birds in particular (Brooker, 1976b).

Diquat

Diquat is a suitable herbicide for the control of many submerged macrophytes although some, for example *Ceratophyllum, Chara* and

Ranunculus, are resistant. The toxic action of this herbicide is through the formation of hydrogen peroxide within the plant which then kills it. Light is essential for this process. Diquat persists in water for about 10 days but in mud it may remain for up to six months. It has an affinity for organic matter and appears to be less effective in water with a perceptible suspended solids load. Although most filamentous algae are fairly resistant (but not *Cladophora*—fortunately this is controlled by diquat) there is conflicting evidence with respect to its effects on planktonic algae (Newbold, 1975). After treatment, the decomposing weeds release the herbicide sorbed on them and microbial attack releases water-soluble degradation products within about 22 days. Accumulation of the herbicide in the sediment is minimal but the degradation products may persist for up to 180 days (Simisan and Chesters, 1976).

Glyphosate

Glyphosate is a relatively new aquatic herbicide, the toxic action of which is not understood, but it may interfere with protein synthesis. It has a very broad spectrum of susceptible species. Glyphosate is approved for late summer or autumn treatment of emergent and floating macrophytes to provide control in the following year. The herbicide is readily translocated from leaves to the roots, especially when these are dying back and material is being transported to the roots for storage over winter. It is, therefore, particularly useful for controlling plants with underground stems such an *Nuphar lutea* and *Nymphaea* sp. and the common reed-like, rush and emergent grass species (*Phragmites, Glyceria, Phalaris, Typha, Juncus, Carex*), several of which have well-developed rootstocks.

Glyphosate is readily biodegraded, with a half-life of less than two months, residues being bound to mud particles (Tooby, 1976).

Paraquat

Paraquat is similar in action to diquat, forming hydrogen peroxide within the plant under the influence of sunlight. Paraquat is thought to be more persistent than diquat. Several studies have been made of the ecological effects of paraquat (Newman and Way, 1966; Newman, 1967; Yeo, 1967; Way *et al.*, 1971; Brooker and Edwards, 1973a,b, 1974). The application of paraquat to a reservoir with growths of *Potamogeton pectinatus* and *Myriophyllum spicatum* resulted in the disappearance of these two species and their replacement by *Chara globularis*. Paraquat was quickly lost from the water but subsequent analysis showed that while up to 6% of the initial dose was absorbed by the macrophytes about 36% of the total amount of paraquat was not recovered (Brooker and Edwards, 1973a). Although gross oxygen production declined following herbicide application, the respiratory demand did not increase and so the ratio of the gross production:community respiration (P:R ratio) remained fairly constant (Brooker and Edwards, 1973b). The main impact on the fauna was loss, or a reduction in density, of species intimately associated with the macrophytes (Brooker and Edwards, 1974).

Terbutryne

Terbutryne is a broad-spectrum herbicide which controls submerged and floating macrophytes and also filamentous algae. This substance inhibits photosynthesis but does not suppress respiration. Emergent macrophytes are not affected by this herbicide since photosynthesis can continue in the aerial parts of the plant. It persists for a considerable time; the half-life in water is about 25 days and it may be detected in mud for five months.

The effects of terbutryne treatment on the aquatic flora and chemistry in navigable canals have been described by Murphy *et al.* (1981) while the effects of such treatment on the flora have been described by Hanbury *et al.* (1981). Some difficulty was experienced in obtaining optimum dosage rates when using the granular formulation. Sub-optimal concentrations appeared to have little long-term effect on the aquatic flora and even repeated treatment at adequate levels allowed subsequent recovery, although the immediate effects were more pronounced. Two immediate effects of terbutryne treatment were a tendency for filamentous algae to dominate, at the expense of submerged vascular plants, and for the duckweed *Lemna* to exploit rapidly the areas cleared by the herbicide. Although *Lemna* is susceptible to the herbicide, a small surviving inoculum was able to recover quickly. Marked changes in water chemistry, but of short duration, followed severe terbutryne treatments. Dissolved oxygen was depressed as community respiration increased (Murphy *et al.*, 1981).

Evidence from repeated treatments suggests that terbutryne use may lead to a reduction in the species of macroinvertebrates. When treatments were effective, invertebrates associated with submerged macrophytes were invariably lost, except when replacement growths of filamentous algae afforded shelter for otherwise disposed species. Some members of the benthic fauna increased in abundance, notably

Lumbriculidae and Nematoda. These groups are benthic detritivores and clearly their increase is a reflection of the large amounts of decaying vegetation which becomes available after herbicide treatments (Hanbury *et al.*, 1981).

7.9.2.3 Herbicides used near water
These herbicides are for use on weeds near water but are not usually applied directly to the water. Some are more strictly used as growth regulators although they can be used to kill plants if necessary.

2,4-D amine
The amine salt of 2,4-D (2,4-dichlorophenoxyacetic acid) is used principally to control broad-leaved weeds on the banks and margins of watercourses and drainage channels. It is also effective in controlling the yellow lily, *Nuphar lutea*. This herbicide acts as a growth regulator with a hormone-like effect causing malformed growth, swelling and splitting of stems followed by slow necrosis.

In a pond treated with 2,4-D at 0.8 mg litre^{-1}, water residues were very small (0.005 mg litre^{-1}) after two weeks and no residues were found in mud after eight weeks (Schultz and Harman, 1974), while Robson (1968) has estimated that several months might be required for a suitable microflora to develop in soil. Biodegradation of 2,4-D in river water was dependent on bacterial activity which in turn was related to nutrient concentration, dissolved organic carbon and sediment load in the water (Nesbitt and Watson, 1980a,b). Light probably also plays a part in its decomposition (Aly and Faust, 1964).

Asulam
Although asulam is known to be effective for the control of the yellow lily, *Nuphar lutea* (Newbold, 1975), it is cleared for use in the United Kingdom only for bracken (*Pteridium aquilinum*) control near water. Its acceptability near water is based on its very low toxicity to fish and aquatic invertebrates.

Maleic hydrazide
Maleic hydrazide is approved for use as a grass-retardant on banks of watercourses where the likelihood of entry into the water is minimal.

7.9.2.4 Environmental significance of herbicides
For reasons outlined above, the indirect effects of aquatic herbicides are

likely to be of greater ecological significance than direct toxicity. For example, if a herbicide can be shown to have very low toxicity to fish fry but it destroys the vegetation in which they find shelter from predators or which supports their food organisms, then, although the direct toxic effects can be ignored, the ultimate effect of the use of that herbicide will be to decrease the survival of that fish species.

The general major effects of herbicide treatment tend to follow in succession as first the physiology of the plants is impaired, they then subsequently die and decay and their absence allows other factors to come into operation causing changes in the habitat and leading to the establishment of an otherwise foreign community (Fig. 7.6). The typical sequence of events may be illustrated by following the consequences of treating submerged weeds (for example, *Elodea*) with herbicide which acts by inhibiting photosynthesis (e.g. terbutryne). The first effect would be a loss of photosynthetic activity during daytime but with respiration maintained day and night, so that mean dissolved oxygen levels would be depressed and the normal diurnal variation suppressed or, at least, attenuated. Lowered oxygen levels may affect the survival of sensitive animal species. The effect is more pronounced when the plants' biomass is large. Further lowering of oxygen levels follows the death and decay of

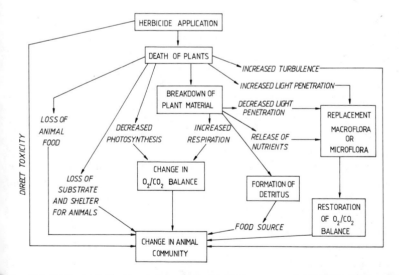

FIG. 7.6 Ecological consequences of herbicide treatment. After Brooker and Edwards (1975).

the plants as bacterial respiration increases. Deoxygenation following herbicide treatment is one of the most serious indirect effects and may result in large fish kills. In order to avoid this complication, it is usually recommended that herbicides should be applied early in the season when weed growths are small or to provide partial treatment (that is, to a quarter or a third of the area to be controlled) at intervals of two or three weeks.

As oxygen levels fall, carbon dioxide tends to increase, through plant and bacterial respiration, and there may also be a lowering of pH (Wojtalik *et al.*, 1971; Brooker and Edwards, 1973a). The breakdown of plant material may release some nutrients (Boyd, 1970; Jewell, 1970; Nichols and Keenay, 1973) and will certainly increase the amount of detritus present.

Any non-susceptible member of the plant community may be stimulated into more active growth by the ensuing lack of competition for space, light and nutrients, even becoming as abundant as the previously dominant plants, an effect which, no doubt, is further promoted by the release of inorganic nutrients and higher carbon dioxide levels. Often, planktonic algae are able to exploit the changed conditions quite rapidly and an algal bloom may follow the demise of susceptible macrophytes (Walsh *et al.*, 1971; Way *et al.*, 1971). This phenomenon may also be attributable, in part at least, to the fact that planktonic algae are largely resistant to aquatic herbicides.

The loss of macrophytes leads to a loss of shelter for many invertebrates and of a substratum for other organisms, especially the 'aufwuchs' or microflora and microfauna which inevitably covers the surface of submerged macrophytes. Most herbivorous macroinvertebrates feed on the minute epiphytes rather than on the tissues of higher plants. Many fish species feed directly on macrophytes, including common European cyprinids (Hellawell, 1971, 1972, 1974), although most are sufficiently catholic in their tastes to be able to take advantage of the many macroinvertebrates which are rendered homeless after the destruction of the plants. These include many molluscs, chironomid larvae and Trichoptera and their reduced density may persist a year later when the plants have made a good recovery (Brooker and Edwards, 1974). The density of detritivorous invertebrates may increase as their food supply becomes more abundant on the decay of dead plants. The carnivorous invertebrates often increase in density although this and other changes must largely result from redistribution since few species can reproduce quickly enough to exploit the new situation. Some chironomids are a notable exception; this family seems capable of taking advantage of almost any situation with almost incredible speed! When an algal bloom follows the collapse of the macrophytes there may be an increased zooplankton also (Walsh *et al.*, 1971).

The application of herbicides to emergent vegetation has less immediate effect on the aquatic environment since •the photosynthetic oxygen contribution will hardly affect the water and the depletion of oxygen will depend on how much decay occurs within the water. The indirect effects of killing emergent or bankside vegetation on the aquatic community may, however, be considerable. As the bankside or emergent vegetation dies, the shading which it provided is removed, thus allowing submerged algae and macrophytes to grow, the degree of change depending to a large extent on the density of the original shade. The composition of the bankside and emergent floral community will change also and this will have implications for the associated aerial and bankside fauna.

7.9.2.5 Toxic effects of herbicides

Fortunately, the toxic effects on animals of the majority of herbicides are relatively small. Some substances, including copper salts and dichlobenil, have a smaller margin of safety than most other herbicides and it may be inadvisable to use these where a valuable fishery is at risk (Tooby, 1976, 1981b).

Some herbicides appear to have very low toxicity to invertebrates. For example, even when it was applied at one hundred times the field dose, glyphosate ('Roundup') had no effect on captive populations of *Daphnia magna* (Hildebrand *et al.*, 1980). The safety of this herbicide is enhanced in actual use by the way in which it is applied to emergent foliage so that little should enter the water. Similarly, asulam has a very high safety factor for fish and aquatic invertebrates and this is enhanced by confining its use to bracken control near water.

Many insects survived diquat concentrations at forty times the maximum field application rate but the amphipod *Hyalella* was found to be quite sensitive, having a 96 hr LC_{50} under experimental conditions of 0·048 mg litre^{-1}. However, on adding mud to the test vessel the 96 hr LC_{50} was 6·8 mg litre^{-1} (Wilson and Bond, 1969), which suggests that in the field the herbicide might have less serious consequences. This effect is similar to that observed for trifluralin: the compound is highly toxic to fish but when applied to soil it is strongly absorbed and so, apart from

TABLE 7.25
TOXICITY OF AQUATIC HERBICIDES TO FISH
(Mostly after Newbold, 1975)

Herbicide	Species	Test	Value (mg litre^{-1})	Reference
Asulam	Harlequin (*Rasbora heteromorpha*)	24 hr LC$_{50}$	5 200	Alabaster (1969b)
Chlorthiamid	Harlequin (*Rasbora heteromorpha*)	24 hr LC$_{50}$	41	Alabaster (1969b)
Copper	Rainbow trout (*Salmo gairdneri*)	48 hr LC$_{50}$	0·14	Shaw and Brown (1974)
Cyanatryn	Harlequin (*Rasbora heteromorpha*)	96 hr LC$_{50}$	7·5	Tooby *et al.* (1975)
Dalapon	Rainbow trout (*Salmo gairdneri*)	24 hr LC$_{50}$	350	Alabaster (1969b)
Dichlobenil	Rainbow trout (*Salmo gairdneri*)	96 hr LC$_{50}$	6·4	Tooby (1981b)
	Rainbow trout (*Salmo gairdneri*)	48 hr LC$_{50}$	22	Jones (1964)
	Roach (*Rutilus rutilus*)	10 day LC$_{50}$	1·6	Tooby (1972)
	Bluegill (*Lepomis macrochirus*)	24 hr LC$_{50}$	20	Jones (1964)
	Harlequin (*Rasbora heteromorpha*)	96 hr LC$_{50}$	4·2	Tooby *et al.* (1975)
	Harlequin (*Rasbora heteromorpha*)	96 hr LC$_{50}$	6·5	Tooby (1981b)
	Roach (*Rutilus rutilus*)	96 hr LC$_{50}$	8·0	Tooby (1981b)
	Bream (*Abramis brama*)	96 hr LC$_{50}$	7·5	Tooby (1981b)
Diquat	Rainbow trout (*Salmo gairdneri*)	24 hr LC$_{50}$	90	Alabaster (1969b)

Diuron	Mosquito fish (*Gambusia affinis*)	24 hr LC$_{50}$	723	Lueng et al. (1983)
	Mosquito fish (*Gambusia affinis*)	96 hr LC$_{50}$	289	Lueng et al. (1983)
	Rainbow trout (*Salmo gairdneri*)	48 hr LC$_{50}$	4·03	FWPCA (1968)
	Harlequin (*Rasbora heteromorpha*)	48 hr LC$_{50}$	152	Tooby et al. (1975)
2,4-D	Carp (*Cyprinus carpio*)	96 hr LC$_{50}$	96·5	Rehwoldt et al. (1977)
	Rainbow trout (*Salmo gairdneri*)	24 hr LC$_{50}$	250	Alabaster (1969b)
	Fathead minnow (*Pimephales promelas*)	96 hr LC$_{50}$	8·23	Holcombe et al. (1980)
	Harlequin (*Rasbora heteromorpha*)	96 hr LC$_{50}$	12	Tooby (1981b)
Glyphosate	Rainbow trout (*Salmo gairdneri*)	96 hr LC$_{50}$	54·8	Hildebrand et al. (1982)
	Harlequin (*Rasbora heteromorpha*)	96 hr LC$_{50}$	25–30	Tooby et al. (1975)
Maleic hydrazide	Harlequin (*Rasbora heteromorpha*)	24 hr LC$_{50}$	840	Alabaster (1969b)
Paraquat	Rainbow trout (*Salmo gairdneri*)	96 hr LC$_{50}$	3·5	Tyson (1974)
Terbutryne	Harlequin (*Rasbora heteromorpha*)	96 hr LC$_{50}$	1·8	Tooby et al. (1975)
	Harlequin (*Rasbora heteromorpha*)	96 hr LC$_{50}$	11	Tooby (1981b)
	Roach (*Rutilus rutilus*)	96 hr LC$_{50}$	5·5	Tooby (1981b)
	Rudd (*Scardinius erythrophthalmus*)	96 hr LC$_{50}$	6·7	Tooby (1981b)

TABLE 7.26
TOXICITY OF AQUATIC HERBICIDES TO INVERTEBRATES
(Mostly after Newbold, 1975)

Herbicide	Species	Test	Value (mg litre^{-1})	Reference
Asulam	*Gammarus* sp.	96 hr LC$_{50}$	1 500	Arthur and Leonard (1970)
Copper	*Gammarus pseudolimnaeus*	96 hr LC$_{50}$	0·02	Arthur and Leonard (1970)
	Campeloma decisum	96 hr LC$_{50}$	1·7	Arthur and Leonard (1970)
	Physa integra	96 hr LC$_{50}$	0·04	Arthur and Leonard (1970)
	Daphnia magna	24 hr LC$_{50}$	0·06	Bellavere and Gorbi (1981)
Dalapon	*Pteronarcys californica*	96 hr LC$_{50}$	>100	Sanders and Cope (1968)
	Daphnia magna	26 hr LC$_{50}$	6·0	Frear and Boyd (1967)
Dichlobenil	*Daphnia pulex*	48 hr LC$_{50}$	3·7	FWPCA (1968)
	Daphnia magna	48 hr LC$_{50}$	7·8	Tooby (1981)
	Gammarus lacustris	24 hr LC$_{50}$	16	Sanders (1969)
		48 hr LC$_{50}$	1·5	FWPCA (1968)
	Pteronarcys californica	24 hr LC$_{50}$	42	Sanders and Cope (1968)
	Pteronarcys californica	48 hr LC$_{50}$	8·4	FWPCA (1968)
	Hyalella azeteca	96 hr LC$_{50}$	8·5	Wilson and Bond (1969)
	Callibaetis sp.	96 hr LC$_{50}$	10·3	Wilson and Bond (1969)
	Limnephilus sp.	96 hr LC$_{50}$	13·0	Wilson and Bond (1969)

Compound	Species	Test	Value	Reference
Diquat	*Diaptomus* sp.	48 hr LC_{50}	19	Naqvi *et al.* (1980)
	Hyalella azeteca	96 hr LC_{50}	0·05	Wilson and Bond (1969)
	Callibaetis sp.	96 hr LC_{50}	16·4	Wilson and Bond (1969)
	Limnephilus sp.	96 hr LC_{50}	33·0	Wilson and Bond (1969)
Diuron	*Daphnia pulex*	48 hr LC_{50}	1·4	FWPCA (1968)
	Gammarus lacustris	24 hr LC_{50}	0·7	Sanders (1969)
	Gammarus lacustris	48 hr LC_{50}	0·38	FWPCA (1968)
	Pteronarcys californica	24 hr LC_{50}	3·6	Sanders and Cope (1968)
	Pteronarcys californica	48 hr LC_{50}	2·8	FWPCA (1968)
2,4-D	*Daphnia pulex*	48 hr LC_{50}	3·2	FWPCA (1968)
	Gammarus lacustris	24 hr LC_{50}	1·4–6·8	Sanders (1969)
	Pteronarcys californica	24 hr LC_{50}	8·5	Sanders and Cope (1968)
	Pteronarcys californica	48 hr LC_{50}	1·8	Sanders and Cope (1968)
Glyphosate	*Daphnia pulex*	48 hr LC_{50}	36	Tooby (1981)
Paraquat	*Daphnia pulex*	48 hr EC_{50}	4·0	Sanders and Cope (1966)
	Simocephalus serrulatus	48 hr EC_{50}	3·7	Sanders and Cope (1966)
	Gammarus lacustris	24 hr LC_{50}	38	FWPCA (1968)
	Gammarus lacustris	48 hr LC_{50}	18	FWPCA (1968)
	Pteronarcys californica	96 hr LC_{50}	$\gg 100$	Sanders (1969)
	Diaptomus mississippiensis	48 hr LC_{50}	5·3	Naqvi *et al.* (1981)
Terbutryne	*Daphnia magna*	48 hr LC_{50}	1·4	Tyson (1974)
	Daphnia magna	48 hr LC_{50}	1·4	Tooby (1981)

accidental spillages, the actual environmental risk is very small (Mullison, 1970).

The toxicity of a selection of aquatic herbicides to fish and invertebrates is given in Tables 7.25 and 7.26, respectively.

In contrast to the literature on accumulation of insecticide residues, little seems to be known of the mechanisms of bioaccumulation of herbicides. Gunkel and Streit (1980) have used ^{14}C ring-labelled atrazine, the unlabelled compound and tritiated water in order to study the uptake of a terrestrial non-selective soil-acting herbicide by the mollusc *Ancylus fluviatilis* and the white fish *Coregonus fera*. Uptake from water was rapid and equilibrium was reached within hours or days. There was no evidence of high concentration factors. Accumulation kinetics indicated that atrazine enters via the gills and other organs are contaminated by the blood. Investigation of the water balance revealed that water exchange is the main mechanism of atrazine accumulation and exchange.

7.9.3 Insecticides

7.9.3.1 Classification of insecticides
A large number of compounds having insecticidal properties are known. Some, such as derris, nicotine or pyrethrin, are of natural origin but the majority are synthetic organic chemicals. Perhaps the best-known synthetic insecticide is DDT, a chlorinated hydrocarbon compound. The organochlorine insecticides form the major group of insecticides which also includes aldrin, dieldrin, HCH (BHC), heptachlor and toxaphene. These substances are less widely used now because of a number of disadvantages, principally their environmental persistence, their bioaccumulation within the tissues of organisms, including man, and their toxic action upon the nervous system.

Organochlorine insecticides have been largely superseded by the organophosphorus and, more recently, by the carbamate insecticides. Examples of the former group are the parathions, malathion, fenitrothion and azinophos methyl, while carbaryl, aldicarb and carbofuran are representatives of the latter group. Both these classes of insecticide have shorter environmental persistence and lower propensity for bioaccumulation.

More recently still, analogues of the naturally occurring insecticide pyrethrin have been developed: these synthetic pyrethroids include allethrin, bioresmethrin, cypermethrin and permethrin. Their insecticidal

TABLE 7.27
COMPARISON OF MAIN CHARACTERISTICS OF SYNTHETIC INSECTICIDES

Characteristic	Insecticide		
	Organochlorine	Organophosphorus	Carbamate
1. Potential for entry into freshwater	strong	strong	moderate
2. Solubility in water	very low	low	low
3. Aquatic toxicity	high	moderate	moderate
4. Aquatic persistence	prolonged	short	short
5. Bioaccumulation potential	strong	weak	weak

activity tends to be greater than that of the natural compound but they still retain their very low mammalian toxicity.

7.9.3.2 Environmental hazards from insecticide use
Insecticides are intrinsically poisonous substances with particularly acute toxicity to insects but also having chronic or sub-lethal effects on other organisms. Early, widespread and sometimes indiscriminate use (or abuse) of these compounds, largely because of their low acute toxicity to man coupled with very powerful effects on insect pests, led to an environmental scandal of the first order, although perhaps the real tragedy was one of indifference or ignorance. As a consequence organochlorine residues appeared in organisms living far from the original sites of application and at concentrations greater than those to be expected from the normal rates of application, giving due allowance for dilution and so on. Furthermore, these residues were still detectable many years after the substances had fallen out of general use. Perhaps the most important benefit to be gained from these experiences, that is beyond the immediate control of vectors of disease or amelioration of crop damage, has been to heighten our awareness of the potential dangers of unrestricted usage of such powerful agents and to instigate a watching brief on all new chemicals which are used in, or may inadvertently enter, the environment at large.

In addition to the danger of killing non-target species, there is always some risk to users of organophosphorus compounds and certain carbamates since they have anticholinesterase activity and therefore interfere with the mechanism of nerve impulse transmission, although this danger is minimal if proper precautions are taken (see Sections 7.9.3.4 and 7.9.3.5).

7.9.3.3 Organochlorine insecticides

Organochlorine insecticides were the first synthetic organic pesticides to be developed and of these DDT and BHC (now HCH) are probably the best known. They were introduced in the 1940s and during their first decade made a considerable impact on the improvement of human health both by destroying insect vectors of disease and in protecting crops against the ravages of insect pests. Agricultural benefits also followed from the use of these insecticides in the control of vectors of animal disease and insect parasites.

Organochlorine insecticides were known to be persistent, and indeed this was one of their advantages in that their residual insecticidal activity allowed for a degree of prophylactic control of insects in dwellings and in cultivated soils, but little anxiety was felt for subsequent environmental problems. However, in the 1950s and 1960s it became evident that organochlorine pesticide residues were to be found in soils and water, sometimes in areas which were not thought to have received any direct applications. More sinister was the discovery that organochlorine residues were accumulating in tissues, especially adipose tissues, of animals occupying the upper links of trophic chains. Concern for carnivorous mammals and birds of prey was almost matched by that for residue levels in human fat. The low solubility of organochlorine insecticides in water but high solubility in lipids is an important factor in their 'bioconcentration', as the pesticide will partition preferentially in tissue lipids and so be retained.

Persistence of organochlorine residues, an advantage for insect control, is another factor which has important environmental implications. The 'half-life' of many exceeds one year and 5% of the original material

TABLE 7.28

WATER SOLUBILITY AND PERSISTENCE OF ORGANOCHLORINE INSECTICIDES IN SOIL
(After Edwards, 1973c)

Insecticide	Approximate water solubility (mg litre^{-1})	Approximate half-life (years)	Approximate mean time for 95% disappearance of residues (years)
DDT	0·001–0·04	2·8	10
Dieldrin	0·1–0·25	2·5	8
Endrin	0·1–0·25	2·2	7
Lindane	7·0–10·0	1·2	6·5
Chlordane	very low	1·0	4
Heptachlor	very low	0·8	3·5
Aldrin	0·01–0·2	0·3	3

may still be present in soil for up to 10 years (Table 7.28). The prolonged availability of these insecticides enhances their likelihood of accumulation within tissues of organisms.

The uptake of organochlorine insecticides by aquatic organisms may be indirect, from food or direct from the water, especially through respiratory surfaces. On transfer of organisms containing organochlorine residues to uncontaminated water, the pesticides are gradually lost (Edwards, 1973c).

The mode of action of organochlorine insecticides is not fully understood but appears to differ between species. A common factor appears to be interference with the normal functioning of the central or peripheral nervous system resulting in greater excitability or abnormal behaviour in mild cases to paralysis and eventually death in severe poisoning. The specific effect is thought to be interference with enzyme action or related functions, particularly at nerve synapses (Livingstone, 1977).

Organochlorine insecticides are less widely used now and their decline can be ascribed to three principal factors.

First, many target insects have developed resistance to the insecticides and control is therefore less effective. It is thought that the resistance is not acquired by repeated exposure to low doses at the periphery of the area of application but by the survival of a very few individuals having intrinsically higher tolerance to the insecticide. In time these genetically adapted strains form an increasing proportion of the population and eventually the insecticide ceases to have its original efficacy. Thus an alternative means of control of serious pests is needed and the redundant substances remain for limited and specialised uses.

Second, the persistence of organochlorine insecticide residues for long periods, often many years, facilitates their incorporation into organisms, usually by simple preferential partitioning between the aqueous environ-

TABLE 7.29

COMPARISON OF LINDANE (HCH) RESIDUES IN TISSUES OF DEAD ROACH (*RUTILUS RUTILUS*), FROM AN ACUTE TOXICITY TEST AND A RIVER MORTALITY
(From Tooby and Durbin, 1975)

Origin of material	Number of individuals	Range of Lindane residues in tissue (mg kg^{-1})		
		Muscle	Liver	Brain
Acute toxicity test	30	1·6–4·7	2·9–9·2	2·8–7·5
River mortality	3	1·6–2·0	3·0–3·7	2·8–3·3

TABLE 7.30

TOXICITY OF ORGANOCHLORINE INSECTICIDES CHLORDANE, DDT, HCH(BHC), LINDANE, HEPTACHLOR, METH-OXYCHLOR, AND TOXAPHENE TO FISH

Insecticide	Species	Test	Value (μg litre^{-1})	Reference
Chlordane	Rainbow trout (*Salmo gairdneri*)	24 hr LC$_{50}$	50	Mayhew (1955)
	Bluegill (*Lepomis macrochirus*)	96 hr LC$_{50}$	22	Henderson *et al.* (1959)
	Fathead minnow (*Pimephales promelas*)	96 hr LC$_{50}$	69	Henderson *et al.* (1959)
	Goldfish (*Carassius auratus*)	96 hr LC$_{50}$ 96 hr LC$_{50}$	50 82	Rudd and Gennelly (1956) Henderson *et al.* (1959)
DDT	Cut throat trout (*Salmo clarki*)	96 hr LC$_{50}$	0·85-1·37	Post and Schroeder (1971)
	Rainbow trout (*Salmo gairdneri*)	96 hr LC$_{50}$	1·7	Post and Schroeder (1971)
	Brook trout (*Salvelinus fontinalis*)	96 hr LC$_{50}$	7·4-11·9	Post and Schroeder (1971)
	Coho salmon (*Oncorhynchus kisutch*)	96 hr LC$_{50}$	11·3-18·5	Post and Schroeder (1971)
	Bluegill (*Lepomis macrochirus*)	96 hr LC$_{50}$	8	Macek and McAllister (1970)
	Goldfish (*Carassius auratus*)	96 hr LC$_{50}$	27	Henderson *et al.* (1959)
HCH or BHC	Bluegill (*Lepomis macrochirus*)	96 hr LC$_{50}$	790	Henderson *et al.* (1959)
	Fathead minnow (*Pimephales promelas*)	96 hr LC$_{50}$	2 300	Henderson *et al.* (1959)

	Species	Test	Value	Reference
Lindane (HCH)	Goldfish (*Carassius auratus*)	96 hr LC_{50}	230	Henderson *et al.* (1959)
	Fathead minnow (*Pimephales promelas*)	96 hr LC_{50}	62	Henderson *et al.* (1959)
	Bluegill (*Lepomis macrochirus*)	96 hr LC_{50}	77	Henderson *et al.* (1959)
	Brown trout (*Salmo trutta*)	96 hr LC_{50}	2	Macek and McAllister, 1970
Heptachlor	Rainbow trout (*Salmo gairdneri*)	24 hr LC_{50}	250	Mayhew (1955)
	Goldfish (*Carassius auratus*)	96 hr LC_{50}	230	Henderson *et al.* (1959)
	Fathead minnow (*Pimephales promelas*)	96 hr LC_{50}	94	Henderson *et al.* (1959)
	Bluegill (*Lepomis macrochirus*)	96 hr LC_{50}	19	Henderson *et al.* (1959)
Methoxychlor	Rainbow trout (*Salmo gairdneri*)	24 hr LC_{50}	52	Mayhew (1955)
	Goldfish (*Carassius auratus*)	96 hr LC_{50}	56	Henderson *et al.* (1959)
	Bluegill (*Lepomis macrochirus*)	96 hr LC_{50}	62	Henderson *et al.* (1959)
Toxaphene	Rainbow trout (*Salmo gairdneri*)	24 hr LC_{50}	50	Mayhew (1955)
	Goldfish (*Carassius auratus*)	96 hr LC_{50}	5·6	Henderson *et al.* (1959)
	Bluegill (*Lepomis macrochirus*)	96 hr LC_{50}	3·5	Henderson *et al.* (1959)

TABLE 7.31

TOXICITY OF ORGANOCHLORINE INSECTICIDES CHLORDANE, DDT, ENDOSULPHAN (THIODAN®), HCH (BHC), LINDANE, HEPTACHLOR, METHOXYCHLOR, TDE (DDD) AND TOXAPHENE TO INVERTEBRATES

Insecticide	Species	Test	Value (μg litre^{-1})	Reference
Chlordane	*Pteronarcys californica*	96 hr LC$_{50}$	15	Sanders and Cope (1968)
	Gammarus fasciatus	96 hr LC$_{50}$	40	Sanders (1972)
	Gammarus lacustris	96 hr LC$_{50}$	26	Sanders (1969)
DDT	*Pteronarcys californica*	96 hr LC$_{50}$	1 800	Jensen and Gaufin (1964)
	Pteronarcys californica	96 hr LC$_{50}$	7·0	Sanders and Cope (1968)
	Gammarus lacustris	96 hr LC$_{50}$	1·0	Sanders (1969)
	Gammarus lacustris	96 hr LC$_{50}$	9·0	Nebeker and Gaufin (1964)
	Gammarus lacustris	96 hr LC$_{50}$	0·8–3·2	Sanders (1972)
	Daphnia magna	50 hr LC$_{50}$	1·4	Livingstone (1977)
Endosulphan (Thiodan®)	*Pteronarcys californica*	96 hr LC$_{50}$	2·3	Sanders and Cope (1968)
	Gammarus fasciatus	96 hr LC$_{50}$	6·0	Sanders (1972)
HCH/BHC	*Chaoborus* spp.	48 hr LC$_{50}$	8	Bluzat and Seuge (1979)

Compound	Species	Test	Value	Reference
Lindane	Pteronarcys californica	96 hr LC$_{50}$	4·5	Sanders and Cope (1968)
	Cloeon sp.	48 hr LC$_{50}$	92	Bluzat and Seuge (1979)
	Gammarus lacustris	96 hr LC$_{50}$	48	Sanders (1969)
	Gammarus fasciatus	96 hr LC$_{50}$	10–11	Sanders (1972)
	Gammarus pulex	48 hr LC$_{50}$	30	
	Daphnia magna	48 hr LC$_{50}$	485	Macek et al. (1976)
	Lymnaea stagnalis	48 hr LC$_{50}$	7 300	Bluzat and Seuge (1979)
Heptachlor	Pteronarcys californica	96 hr LC$_{50}$	1·1	Sanders and Cope (1968)
	Gammarus lacustris	96 hr LC$_{50}$	29	Sanders (1969)
	Gammarus fasciatus	96 hr LC$_{50}$	40–56	Sanders (1972)
Methoxychlor	Pteronarcys californica	96 hr LC$_{50}$	1·4	Sanders and Cope (1968)
	Gammarus lacustris	96 hr LC$_{50}$	0·8	Sanders (1969)
	Gammarus fasciatus	96 hr LC$_{50}$	1·8–1·9	Sanders (1972)
	Daphnia magna	50 hr LC$_{50}$	3·6	Livingstone (1977)
TDE (DDD)	Pteronarcys californica	96 hr LC$_{50}$	380	Sanders and Cope (1968)
	Gammarus lacustris	96 hr LC$_{50}$	0·64	Sanders (1969)
	Gammarus fasciatus	96 hr LC$_{50}$	0·6–0·86	Sanders (1972)
Toxaphene	Pteronarcys californica	96 hr LC$_{50}$	2·3	Sanders and Cope (1968)
	Gammarus lacustris	96 hr LC$_{50}$	26	Sanders (1969)
	Gammarus fasciatus	96 hr LC$_{50}$	6–35	Sanders (1972)

TABLE 7.32

ACUTE TOXICITY OF KEPONE TO FISH AND INVERTEBRATES

Species	Test	Value (μg litre^{-1})	Reference
Fish			
Pimephales promelas	96 hr LC_{50}	340	Buckler *et al.* (1981)
Anguilla rostrata (juv)	96 hr LC_{50}	35	Roberts and Bendl (1982)
Lepomis macrochirus	96 hr LC_{50}	50	Roberts and Bendl (1982)
Ictalurus punctatus	96 hr LC_{50}	514	Roberts and Bendl (1982)
Invertebrates			
Gammarus pseudolimnaeus	96 hr LC_{50}	180	Sanders *et al.* (1981)
Daphnia magna	48 hr EC_{50}	260	Sanders *et al.* (1981)
Chironomus plumosus	48 hr EC_{50}	350	Sanders *et al.* (1981)

The EC_{50} for invertebrates was based on immobilisation.

ment and the lipid in cells or tissues. Animals which occupy positions at or near the apex of food chains are at serious risk when these insecticide residues are bioaccumulated. In short, these insecticides are environmentally unacceptable. However, it should be remembered that this is largely because less effective and more environmentally acceptable alternatives now exist.

Third, human health has been threatened because organochlorine residues can be bioaccumulated in food chains which lead to man (for example, in fish) or through crops or livestock raised for food. Of particular concern was the occurrence of residues in milk from cows fed on insecticide-treated foodstuffs. This aspect has assumed greater importance as it is now considered that DDT is a potential human carcinogen (Train, 1979).

Data on individual organochlorine insecticides are given below but acute toxicity values are presented in Tables 7.30–7.34.

DDT (dichloro-diphenyl-trichlorethane)

DDT, now 1,1,1-trichloro-2, 2-bis (*p*-chlorophenyl)-ethane, was first synthesised in the nineteenth century but its insecticidal properties were

TABLE 7.33

TOXICITY OF ORGANOCHLORINE INSECTICIDES ALDRIN, DIELDRIN AND ENDRIN TO FISH

Species	Test	Value (μg litre^{-1})	Reference
Aldrin			
Rainbow trout (*Salmo gairdneri*)	24 hr LC$_{50}$	50	Mayhew (1955)
Bluegill (*Lepomis macrochirus*)	96 hr LC$_{50}$	13	Henderson *et al.* (1959)
Goldfish (*Carassius auratus*)	24 hr LC$_{50}$ 96 hr LC$_{50}$	50 28	Rudd and Gennelly (1956) Henderson *et al.* (1959)
Dieldrin			
Rainbow trout (*Salmo gairdneri*)	24 hr LC$_{50}$	50	Mayhew (1955)
Bluegill (*Lepomis macrochirus*)	96 hr LC$_{50}$	8	Henderson *et al.* (1959)
Goldfish (*Carassius auratus*)	96 hr LC$_{50}$	37	Henderson *et al.* (1959)
Endrin			
Rainbow trout (*Salmo gairdneri*)	96 hr LC$_{50}$	0·41	Post and Schroeder (1971)
Coho salmon (*Oncorhyncus kisutch*)	96 hr LC$_{50}$	0·27	Katz and Chadwick (1961)
Coho Salmon (*Oncorhyncus kisutch*)	96 hr LC$_{50}$	0·77	Post and Schroeder (1971)
Bluntnose minnow (*Pimephales notatus*)	96 hr LC$_{50}$	0·27	Mount (1962)
Fathead minnow (*Pimephales promelas*)	96 hr LC$_{50}$	1·8	Henderson *et al.* (1959)
Goldfish (*Carassius auratus*)	96 hr LC$_{50}$	1·96	Henderson *et al.* (1959)
Carp (*Cyprinus carpio*)	48 hr LC$_{50}$	140	Iatomi *et al.* (1958)
Bluegill (*Lepomis macrochirus*)	96 hr LC$_{50}$	0·6	Henderson *et al.* (1959)

not recognised until the outbreak of the 1939–45 war when it was used to combat lice which infested the troops. Lice can carry typhus, relapsing fever and trench fever and it is claimed that these diseases, along with malaria (also spread by an insect vector) were responsible for more deaths during the 1914–18 war than casualties in battle.

When introduced, DDT was acclaimed for its high persistent insecticidal activity and low toxicity to man. When applied to water for mosquito control in an attempt to eradicate malaria the water could be drunk without apparent harm. As production of DDT increased after the

TABLE 7.34

TOXICITY OF THE ORGANOCHLORINE INSECTICIDES ALDRIN, DIELDRIN AND
ENDRIN TO INVERTEBRATES

Species	Test	Value (μg litre^{-1})	Reference
Aldrin			
Gammarus lacustris	96 hr LC$_{50}$	9 800	Sanders (1969)
Acroneuria pacifica	96 hr LC$_{50}$	143	Jensen and Gaufin (1964)
	30 day LC$_{50}$	22	Jensen and Gaufin (1966)
Pteronarcys californica	96 hr LC$_{50}$	180	Jensen and Gaufin (1964)
	96 hr LC$_{50}$	1·3	Sanders and Cope (1968)
	30 day LC$_{50}$	3	Jensen and Gaufin (1966)
Ephemerella grandis	96 hr LC$_{50}$	9	Gaufin *et al.* (1965)
Dieldrin			
Limnodrilus hoffmeisteri	96 hr LC$_{50}$	1 340	Whitten and Goodnight (1966)
Tubifex tubifex	96 hr LC$_{50}$	1 340	Whitten and Goodnight (1966)
Gammarus lacustris	96 hr LC$_{50}$	460	Sanders (1969)
Acroneuria pacifica	96 hr LC$_{50}$	24	Jensen and Gaufin (1964)
	30 day LC$_{50}$	0·2	Jensen and Gaufin (1964)
Pteronarcys californica	96 hr LC$_{50}$	39	Jensen and Gaufin (1964)
	96 hr LC$_{50}$	0·5	Sanders and Cope (1968)
	30 day LC$_{50}$	2·0	Jensen and Gaufin (1966)
Ephemerella grandis	96 hr LC$_{50}$	8·0	Gaufin *et al.* (1965)
Endrin			
Gammarus lacustris	96 hr LC$_{50}$	3·0	Sanders (1969)
Acroneuria pacifica	96 hr LC$_{50}$	0·39	Jensen and Gaufin (1964)
	30 day LC$_{50}$	0·035	Jensen and Gaufin (1966)
Pteronarcys californica	96 hr LC$_{50}$	2·4	Jensen & Gaufin (1964)
	96 hr LC$_{50}$	0·25	Sanders and Cope (1968)
	30 day LC$_{50}$	1·2	Jensen and Gaufin (1966)
Ephemerella grandis	96 hr LC$_{50}$	5·0	Gaufin *et al.* (1965)

war so its use extended in agriculture to crop protection and, in aquatic environments, to the control of 'nuisance' midges. DDT also entered water when aerial applications were made to control forest pests. Published data on the toxicity of DDT to fish and invertebrates differ widely. Some differences, especially between early data and more recent determinations, may be the result of widespread use of static tests in the earlier investigations. Aeration may cause material to be lost from the water surface, adsorption onto the tank surfaces or uptake by bacterial films may also occur, all of which change the constitution of the test solution (Holden, 1973). These problems are avoided in continuous flow-through tests. The toxicity of DDT to fish is quite high, 96 hr LC_{50} values ranging from about 1 to 30 µg litre^{-1} (Table 7.30). Data for toxicity to invertebrates are more variable (Table 7.31) but of a similar order of magnitude. The EPA criterion for DDT to protect freshwater aquatic life, having regard for the high bioaccumulation potential, is 0·001 µg litre^{-1} (Train, 1979).

Fish are able to detect DDT: non-resistant mosquito fish (*Gambusia affinis*) avoid DDT (Kynard, 1974) while the sheepshead minnow (*Cyprinodon variegatus*) avoided lower concentrations but preferred higher ones (Hansen, 1969). There is also a direct relationship between coughing frequency and the duration of exposure (Schaumberg *et al.*, 1967) suggesting that some fish are irritated by DDT.

Increased sensitivity to DDT exposure has been observed in the sheepshead minnow (*Cyprinodon variegatus*) in the offspring of parents previously exposed to the insecticide (Holland *et al.*, 1966), but succeeding generations can develop increased resistance, although this is not a consistent phenomenon (Holland and Coppage, 1970).

Impaired reproduction following exposure to DDT has been described for several species. For example, brook trout (*Salvelinus fontinalis*) fed sub-lethal concentrations for about five months had reduced fecundity and, in general, greater egg and alevin mortality was observed when at least one gamate came from exposed fish (Macek, 1968). Similarly, Burdick *et al.* (1972), on investigating the effect on reproduction of feeding DDT to brown and brook trout in a hatchery where no other extraneous contaminants could interfere with the experiments, found that the major loss which could be attributed to the insecticide occurred during the period after feeding began, that is during 'swim-up' when absorption of triglyceride fats begins.

Interference of DDT with normal behavioural responses is also well documented for fish. For example, in a temperature gradient Atlantic salmon parr selected a lower water temperature on exposure to DDT, but if the exposure was to relatively high levels they avoided colder water (5 °C or less) and selected a higher temperature (Ogilvie and Anderson, 1965). On entering the colder water the fish exposed to higher levels of DDT showed violent locomotory activity and, if they did not escape this section, death quickly ensued. These observations suggested that the lower lethal temperature (see Section 5.1.6.3) had been raised by exposure to DDT and this was confirmed in later work (Anderson, 1971). This change could be significant for the survival of wild fish: aerial spraying of DDT in spring to control spruce budworm on the Western Atlantic sea board was followed in the autumn (fall) by massive mortality of juvenile salmon as temperatures dropped to 5 °C or less (Elson and Kerswill, 1966; Elson, 1967). Clearly, there was a prolonged residual effect from exposure to an initially sub-lethal level of DDT which ultimately proved, indirectly, to be fatal.

Conditioned reflexes may also be affected by exposure to DDT. Anderson and Prins (1970) used the propellor reflex, in which gular stimulation causes weak propellor-like tail movements, as the unconditioned stimulus, and light as the conditioned stimulus. After exposure to sub-lethal concentration of DDT ($20 \mu g$ litre^{-1}) for 24 hours it was observed that the acquisition of a conditioned reflex was slower than in controls, but after four weeks the impairment was reduced. This work suggests that DDT affects the nervous system rather than the motor ability.

In more conventional learning experiments, using a simple visual avoidance response, Anderson and Peterson (1969) found that brook trout exposed to $20 \mu g$ DDT per litre failed to acquire a conditioned response. Later work (Jackson *et al.*, 1970) using a modified procedure established that both salmon and brook trout could acquire a conditioned avoidance response after exposure to sub-lethal levels of DDT. In a study of the ability of Atlantic salmon parr to learn and retain a simple conditioned reflex, Hatfield and Johansen (1972) discovered that when treated for 24 hours with the 96 hr LC_{50} dose, DDT caused a small improvement in learning. Similar enhancement of learning rate, following exposure to DDT, has been observed in rainbow trout (McNicholl and MacKay, 1975a). Using various sub-lethal concentrations it was found that the ability to discriminate between lights of different intensity increased with the dose but performance and retention were unaffected (McNicholl and MacKay, 1975b). In studies of the effect of sub-lethal concentrations ($10 \mu g$ litre^{-1}) of p,p'-DDT on the locomotor behaviour of the goldfish (*Carassius auratus*), Davy *et al.* (1972) found that the normal pattern of behaviour was not regained, even after the fish had been kept for almost 140 days in clean water.

Behavioural changes following exposure to DDT have also been observed in macroinvertebrates. The larva of the caddis fly, *Hydropsyche*, spins a net with which to collect its food. This net has a very regular mesh (Hynes, 1970, p. 188) but abnormal nets are made by larvae exposed to concentrations of DDT between 0·05 and 10 µg litre^{-1} and the frequency of abnormal nets increases with concentration (Décamps *et al.*, 1973).

γ-HCH, hexachlorocyclohexane (BHC or Lindane)

Like DDT, benzene hexachloride (BHC) had been synthesised in the nineteenth century but its insecticidal properties had not been discovered until the search for new insecticides was stimulated by the 1939–45 war. The gamma isomer is the effective insecticide and was named Lindane after Van der Linden who first isolated it.

Its insecticidal properties are generally similar to DDT and its acute toxicity to fish and invertebrates is of the same order, but varies widely between species. Its lower stability probably makes it less hazardous to wildlife. The EPA criterion of 0·01 µg litre^{-1} was obtained by taking 1% of the threshold median lethal concentration for a stonefly (Train, 1979).

Studies of the toxicity and accumulation of lindane in salmonid and cyprinid fish (Tooby and Durbin, 1975) have shown that the residues are readily eliminated and, therefore, short exposures to levels which are not acutely lethal are unlikely to give rise to persistent effects. In hardwater (270 mg litre^{-1} as CaCO$_3$) the acute toxicity to fry and yearlings was of the same order (96 hr LC$_{50}$ of 22 and 30 µg litre^{-1}, respectively). For roach (*Rutilus rutilus*) the median survival time at a nominal concentration of 0·6 mg litre^{-1} (0·1 mg litre^{-1} final concentration) was a little under 22 hours. Death is associated with the accumulation of a definite tissue concentration, a factor which might be useful in determining the cause of death in the event of a fish 'kill'. The minimum values of residues in fish which died in acute toxicity tests were very similar to those obtained from fish which were recovered following an incident in which lindane was suspected as the major pollutant (Table 7.29).

Abel (1980) investigated the toxicity of Lindane to the crustacean *Gammarus* and discovered that when animals were exposed to various concentrations for different periods of time (1 minute to almost 17 hours) deaths were observed for up to three weeks after the initial exposure.

Acute toxicity (96 hr LC_{50}) was found to be 0·3–0·4 mg litre^{-1}, which accorded with other published data, but the apparently delayed effect of exposure to Lindane suggests that, for this substance at least, a conventional toxicity test in which animals are continually exposed may give misleading results. In this study, for a given concentration of Lindane, the duration of exposure ultimately causing 50% mortality is much less than that indicated in conventional acute tests. Such an observation may be explained by assuming that the duration of a lethal exposure is considerably shorter than the time taken for the animal to die, so that continued exposure may not speed death.

Aldrin and dieldrin

Aldrin and dieldrin, together with endrin, belong to the cyclodiene group of insecticides, which also includes chlordane, heptachlor and endosulphan. This group has probably been responsible for the severest harm to wildlife, caused the greatest environmental concern and engendered the strongest opposition to agrochemical use by the conservation 'lobby'.

These substances are extremely persistent: dieldrin can remain in soil for more than a decade. Furthermore, some of them degrade in soil or in tissues to products which are equally or more toxic. For example, aldrin can be transformed both in living tissues and in soil to dieldrin. Aldrin found considerable use as a soil insecticide (although it had also a broad spectrum of uses for horticulture and agriculture) while dieldrin was particularly valuable in treating insect parasites on cattle and sheep. It was also extensively used for mothproofing.

The wildlife hazards of these insecticides and of dieldrin, in particular, became evident in Britain during the late 1950s (Mellanby, 1970). Dieldrin was used as a seed dressing and many birds dug up and ate cereal grains. Predatory birds and mammals were also affected, especially in the wheat growing areas, and their numbers declined markedly. Dieldrin in sheep dips has been blamed for the decline of the otter (*Lutra lutra*) in parts of Britain, the assumption being that the insecticide is concentrated in aquatic food chains and consequently appears in fish which form the otters' diet. Dieldrin was banned in sheep-dip from 1966 and is a 'black-list' substance in the EEC (see Table 7.2). With the phasing out of the 'drins' it is hoped that this will result in a recovery of affected populations.

Data on the toxicity of aldrin, endrin and dieldrin to fish and invertebrates are given in Tables 7.33 and 7.34. (Other cyclodiene insecticides are included in Tables 7.30 and 7.31.)

Kepone and Mirex

Kepone and Mirex are closely related compounds, their structures differing only in the presence of oxygen (Kepone) or chlorine (Mirex) at one site in the molecule. Kepone was used mainly for control of ants and cockroaches (Blattidae) while Mirex was primarily a fire-retardant before being used to control the fire ant (*Solenopis* sp.). Kepone is about 2000 times more water soluble than Mirex. Kepone is highly toxic to fish (see Table 7.32), but even more serious is the dislocation of the vertebral column, and haemorrhage associated with it, which occurred within 2 days in fish exposed to 10 µg litre^{-1} (Buckler *et al.*, 1981). By contrast, fish exposed to Mirex, even at concentrations in excess of the maximum water solubility (50–60 µg litre^{-1}), showed no mortality or any other observable effect even after 60 days. Similarly, the acute toxicity of Mirex to three taxa of invertebrates (amphipod, daphnid and dipteran) exceeded 1000 µg litre^{-1} (Sanders *et al.*, 1981), although this also exceeds 34 µg litre^{-1} for daphnids and midge larvae but was less than 2·4 µg litre^{-1} for *Gammarus pseudolimnaeus*. Adverse effects of chronic exposure to Kepone on survival and growth occur at concentrations as low as 3·1 µg litre^{-1} and reproduction may be impaired at levels as low as 0·31 µg litre^{-1}, while Mirex only impairs reproduction, if at all, at concentrations approaching the maximum solubility. The maximum acceptable toxicant concentration (MATC) of Kepone for fathead minnows is 1·2 µg litre^{-1} (Buckler *et al.*, 1981). Mirex accumulated in fathead minnows to a much greater extent than Kepone at comparable exposure concentrations. Both compounds are stable, resist degradation and accumulate in food chains (Skaar *et al.*, 1981). Bluegills (*Lepomis macrochirus*) accumulated ^{14}C-labelled Kepone and Mirex with concentration factors of 10 606 and 12 274, respectively. Mirex persisted in fish but Kepone did not. Much lower concentration factors, not exceeding 2000, were observed in a study of the effects of Kepone on fish by Roberts and Bendl (1982).

In studies of accumulation and elimination of Kepone and Mirex in invertebrates, Sanders *et al.* (1981) found the concentration factors for *Daphnia magna* to be 760 and 8025, respectively. The estimated times for elimination of 50% of the residues were 141 hours for Kepone and 12 hours for Mirex. These figures appear to be the opposite of the findings of Skaar *et al.* (1981) in which Mirex persisted in the fish while Kepone was eliminated.

7.9.3.4 *Organophosphorus insecticides*
The toxicity of organophosphorus insecticides arises from their inhibition of the enzyme acetylcholinesterase (AChE) which is essential for the

TABLE 7.35

TOXICITY OF SEVERAL ORGANOPHOSPHORUS INSECTICIDES TO FISH

Insecticide	Species	Test	Value ($\mu g\,litre^{-1}$)	Reference
Guthion®	Fathead minnow (*Pimephales promelas*)	96 hr LC_{50}	100	Henderson *et al.* (1959)
	Bluegill (*Lepomis macrochirus*)	96 hr LC_{50} 96 hr LC_{50}	1 900 5.6	Adelman *et al.* (1976) Henderson *et al.* (1959)
	Goldfish (*Carassius auratus*)	96 hr LC_{50}	2 400	Adelman *et al.* (1976)
Dursban®	Rainbow trout (*Salmo gairdneri*)	96 hr LC_{50}	8·0	Holcombe *et al.* (1982)
	Rainbow trout (*Salmo gairdneri*)	96 hr LC_{50}	7·1	Macek *et al.* (1972)
	Green sunfish (*Lepomis cyanellus*)	36 hr LC_{50}	22·0	Ferguson *et al.* (1966)
	Bluegill (*Lepomis macrochirus*)	96 hr LC_{50}	3·6	Macek *et al.* (1972)
	Golden shiner (*Notemigonus chrysoleucas*)	36 hr LC_{50}	35	Ferguson *et al.* (1966)
	Mosquito fish (*Gambusia affinis*)	36 hr LC_{50}	215	Ferguson *et al.* (1966)
	Fathead minnow (*Pimephales promelas*)	96 hr LC_{50}	203	Holcombe *et al.* (1982)

Compound	Species	Test	Value	Reference
Malathion	Fathead minnow (*Pimephales promelas*)	96 hr LC_{50}	12 500	Henderson *et al.* (1959)
	Bluegill (*Lepomis macrochirus*)	96 hr LC_{50}	20 000	Mount and Stephan (1967)
	Channel catfish (*Ictalurus punctatus*)	96 hr LC_{50}	760	Kennedy and Walsh (1970)
	Brook trout (*Salvelinus fontinalis*)	96 hr LC_{50}	120–130	Post and Schroeder (1971)
	Cut throat trout (*Salmo clarki*)	96 hr LC_{50}	150–201	Post and Schroeder (1971)
	Rainbow trout (*Salmo gairdneri*)	96 hr LC_{50}	122	Post and Schroeder (1971)
	Coho salmon (*Oncorhynchus kisutch*)	96 hr LC_{50}	265	Post and Schroeder (1971)
Parathion	Fathead minnow (*Pimephales promelas*)	96 hr LC_{50}	1 600	Henderson *et al.* (1959)
	Mosquito fish (*Gambusia affinis*)	48 hr LC_{50}	350	Chambers and Yarbrough (1974)
Methylparathion	Fathead minnow (*Pimephales promelas*)	96 hr LC_{50}	7 500	Henderson *et al.* (1959)
	Guppy (*Lebistes reticulatus*)	96 hr LC_{50}	819	Desi *et al.* (1976)
Dipterex®	Fathead minnow (*Pimephales promelas*)	96 hr LC_{50}	51 000–180 000	Henderson *et al.* (1959)
Disulfoton	Rainbow trout (*Salmo gairdneri*)	96 hr LC_{50}	3 020	Holcombe *et al.* (1982)
	Fathead minnow (*Pimephales promelas*)	96 hr LC_{50}	4 000	Holcombe *et al.* (1982)

TABLE 7.36
TOXICITY OF SEVERAL ORGANOPHOSPHORUS INSECTICIDES TO MACROINVERTEBRATES

Insecticide	Species	Test	Value (µg litre^{-1})	Reference
Fenthion	*Chaoborus* sp.	48 hr LC$_{50}$	12	Bluzat and Seuge (1979)
	Cloeon sp.	48 hr LC$_{50}$	12	Bluzat and Seuge (1979)
	Gammarus pulex	48 hr LC$_{50}$	14	Bluzat and Seuge (1979)
	Lymnaea stagnalis	48 hr LC$_{50}$	6 400	Bluzat and Seuge (1979)
Guthion®	*Pteronarcys californica*	96 hr LC$_{50}$	22	Jensen and Gaufin (1964)
	Pteronarcys californica	96 hr LC$_{50}$	1·5	Sanders and Cope (1968)
	Gammarus lacustris	96 hr LC$_{50}$	0·13	Nebeker and Gaufin (1964)
	Gammarus lacustris	96 hr LC$_{50}$	0·15	Sanders (1969)
	Gammarus fasciatus	96 hr LC$_{50}$	0·1–0·38	Sanders (1972)
	Daphnia magna	26 hr LC$_{50}$	0·18	Frear and Boyd (1967)
Dursban®	*Pteronarcys californica*	96 hr LC$_{50}$	10	Sanders and Cope (1968)
	Gammarus lacustris	96 hr LC$_{50}$	0·11	Sanders (1969)
	Gammarus fasciatus	96 hr LC$_{50}$	0·32	Sanders (1972)
Malathion	*Pteronarcys californica*	96 hr LC$_{50}$	50	Jensen and Gaufin (1964)
	Gammarus lacustris	96 hr LC$_{50}$	1·6	Nebeker and Gaufin (1964)
	Gammarus fasciatus	96 hr LC$_{50}$	0·76–0·9	Sanders (1972)
	Daphnia magna	26 hr LC$_{50}$	0·9	Frear and Boyd (1967)
Parathion	*Pteronarcys californica*	96 hr LC$_{50}$	32	Jensen and Gaufin (1964)
	Pteronarcys californica	96 hr LC$_{50}$	5·4	Sanders and Cope (1968)
	Gammarus lacustris	96 hr LC$_{50}$	12·8	Nebeker and Gaufin (1964)
	Gammarus lacustris	96 hr LC$_{50}$	3·5	Sanders (1969)
	Gammarus fasciatus	96 hr LC$_{50}$	1·3–4·5	Sanders (1972)
	Daphnia magna	26 hr LC$_{50}$	0·80	Frear and Boyd (1967)
Methyl- parathion	*Daphnia magna*	26 hr LC$_{50}$	4·8	Frear and Boyd (1967)
Trichlorophon (Dipterex^{R})	*Pteronarcys californica*	96 hr LC$_{50}$	16·5	Jensen and Gaufin (1964)
	Pteronarcys californica	96 hr LC$_{50}$	35	Sanders and Cope (1968)
	Gammarus lacustris	96 hr LC$_{50}$	50	Nebeker and Gaufin (1964)
	Gammarus lacustris	96 hr LC$_{50}$	40	Sanders (1969)

transmission of nerve impulses. When inhibition is severe, especially in brain tissue, death follows. Recovery from sub-lethal exposures is often slow. One of the earliest organophosphorus insecticides to be developed, in the 1940s, was parathion, an efficient insect killer but also very poisonous to mammals and, therefore, man. Lack of appreciation of this and failure to use adequate protective clothing, especially in tropical climates, caused a number of deaths. Fortunately, research directed towards developing effective organophosphorus insecticides with lower mammalian toxicity was successful and in 1950 malathion was introduced. This is considerably less toxic to mammals than parathion, being readily broken down in the liver, but is stable in insects. However, some individuals, who have been previously exposed to sub-lethal doses of parathion, are unable to metabolise malathion and may suffer when exposed to this otherwise relatively safe insecticide.

Some organophosphorus insecticides have a further advantage in being 'systemic' insecticides, that is, they are translocated within the plant. This means that it is possible to gain protection from treating only part of the plant (for example roots may be protected without applying the insecticide to the soil) and it may also mean that only pests are killed as they attempt to feed on the crop while other, non-target, insects are spared. These two features clearly contribute to reduced risk to wildlife and assist conservation.

Acute toxicities of organophosphorus insecticides to fish are given in Table 7.35. Even superficial comparison of these data with those for organochlorine insecticides (Tables 7.30 and 7.33) shows that toxicity to fish is much lower. However, comparison of organophosphorus insecticide toxicities to invertebrates (Table 7.36) with those of organochloride insecticides (Tables 7.31 and 7.34) indicates that there are virtually no differences.

In a study of the effects of fenitrothion on a stream fauna following aerial application to a Scottish pine and spruce forest for the control of larvae of a defoliating moth, Morrison and Wells (1981) found that within an hour of spraying the concentration in the forest stream rose to $18\,\mu g$ litre^{-1} but fell to $0.5\,\mu g$ litre^{-1} after 24 hours. No effects were observed on resident fish populations nor on caged fish: tissue time-concentrations profiles closely followed those of the water. Invertebrate drift increased markedly 12 to 16 hours after spraying but returned to pre-treatment levels in two days, but caged insects remained alive during five days of observation, suggesting that the invertebrates were displaced rather than killed (Fig. 3.6).

7.9.3.5 Carbamate insecticides

Carbamate insecticides are relatively new. They differ in chemical structure from the organophosphorus insecticides but have a similar mode of action in attacking the cholinesterases which are involved in the transmission of nerve impulses.

One of the most widely known carbamates is carbaryl (Sevin), 1-naphthyl-N-methyl carbamate, which has been used on a wide range of crops including cotton, fruit and vegetables.

Little is known of the fate and persistence of carbamate insecticides in aquatic systems but it appears that microbial degradation takes place in natural waters (Aly and El-Dib, 1972).

Data on the toxicity of carbaryl to fish and invertebrates are provided in Table 7.37. These show that this insecticide is much less toxic to fish and especially to aquatic invertebrates than the organochlorine or organophosphorus insecticides. It is probable that the carbamate insecticides have least potential danger for conservation of aquatic wildlife.

7.9.3.6 Mothproofing agents—Eulan® and Mitin®

Mothproofing agents are insecticides which are used to protect woollen

TABLE 7.37
TOXICITY OF CARBONATE INSECTICIDES TO FISH AND INVERTEBRATES

	Species	Test	Value (μg litre^{-1})	Reference
Carbaryl	Rainbow trout (*Salmo gairdneri*)	96 hr LC$_{50}$	1 470	Post and Schroeder (1971)
	Cut throat trout (*Salmo clarki*)	96 hr LC$_{50}$	1 500–2 169	Post and Schroeder (1971)
	Brook trout (*Salvelinus fontanalis*)	96 hr LC$_{50}$	1 070–1 450	Post and Schroeder (1971)
	Coho salmo (*Oncorhynchus kisutch*)	96 hr LC$_{50}$	1 300	Post and Schroeder (1971)
	Gammarus pulex	48 hr LC$_{50}$	290	Bluzat and Seuge (1979)
	Chaoborus sp.	48 hr LC$_{50}$	296	Bluzat and Seuge (1979)
	Cloeon sp.	48 hr LC$_{50}$	480	Bluzat and Seuge (1979)
	Lymnaea stagnalis	48 hr LC$_{50}$	21 000	Bluzat and Seuge (1979)

TABLE 7.38
ACUTE TOXICITY OF MOTHPROOFING AGENTS TO FISH

Substance	Species	Temperature (°C)	96 hr LC_{50} (mg litre^{-1})	Reference
Eulan WA New (chlorophenylid)	Rainbow trout (*Salmo gairdneri*)	4	5·4	Tooby (1981)
		14	1·5	Tooby (1981)
		15	0·5–1·0	Abram et al. (1981)
Mitin FF (sulcofenuron)	Brown trout (*Salmo trutta*)	13	3·9	Tooby (1981)
	Zebra fish (*Brachydanio rerio*)	23	1·2	Tooby (1981)
Mitin N (flucofenuron)	Rainbow trout (fry)	15	0·068	Abram et al. (1981)
	Rainbow trout (fry)	15	0·8[a]	Abram et al. (1981)
	Rainbow trout (yearlings)	15	0·13	Abram et al. (1981)
	Rainbow trout (yearlings)	14	11·20[a]	Tooby (1981)
	Brown trout	13	1·7–6·8	Tooby (1981)
	Zebra fish	23	4·0	Tooby (1981)

[a] Solvent not used to disperse formulation.

products from damage by the larvae of Lepidoptera (Tineidae and Oecophoridae) and Coleoptera (Dermestidae) which are able to digest keratin. Formerly, this protection was afforded by treating the yarn or woven fabric with the organochlorine insecticide dieldrin. However, in view of the ecological hazards of this substance, alternatives with lesser environmental risk have been sought.

Two effective compounds are Eulan (chlorophenylid or PCSD) and Mitin (sulcofenuron or flucofenuron, depending on the formulation) and, more recently, the synthetic pyrethroid permethrin has been used. This last compound is considered in the next section. Data on the toxicity to fish of these new mothproofing agents are given in Table 7.38.

Entry into the aquatic environment occurs when effluents from dye or scour-baths are discharged, along with other wastes, either directly to rivers or indirectly via sewers and effluent treatment works. Many woollen textile industries are sited well inland close to adequate supplies of clean, soft water and so these effluents have the potential to affect extensive lengths of rivers. In addition, the use of these materials in more or less continual industrial process means that the potential for bioaccumulation is considerable and their entry into the environment in relatively low concentration could pose a serious threat to wildlife if these materials are concentrated in food chains. Abram *et al.* (1981) studied the toxicity and accumulation of Mitin and Eulan to rainbow trout. They found that the lethal threshold concentration for Mitin was about 8 μg litre^{-1} (active ingredient) but no threshold was observed for Eulan, although it must be below 0·02 μg litre^{-1}. The toxicity of Mitin varies according to the way in which the test solution is prepared; when dissolved in alcohol the 48 hr LC_{50} was found to be 76 μg litre^{-1} but in the commercial formulation the 48 hr LC_{50} exceeded 1000 μg litre^{-1}. The low water solubility of Mitin means that the risk to fish from accidental spillages is minimal as it would not attain a high enough concentration to kill fish within a day and so spillages would dissipate before lasting damage occurred.

Accumulation of Mitin was 'stepwise': the step corresponding to the lethal threshold and fish exposed to acutely toxic levels showed an accumulation plateau at about 20 mg/kg in the viscera and 10 mg/kg in muscle. In fish exposed to sub-lethal levels the accumulation was an order of magnitude lower (3 mg/kg in viscera and 1 mg/kg in muscle). Fish exposed to sub-lethal concentrations are unlikely to accumulate appreciable quantities of the substance. By contrast, Eulan showed no plateau in the accumulation curve and tissue concentrations paralleled those in the water. Doubt has been cast on the validity of these results by Tooby (1981a) who reports other data showing that accumulation and retention of residues was constant after about two weeks and on being placed in clean water the residues are eliminated with a half-life of about two weeks.

TABLE 7.39
SUGGESTED TOXICITY CRITERIA FOR MOTHPROOFING COMPOUNDS
(From Tooby, 1981)

	Tentative 'no effect' concentration (μg litre^{-1})	Accumulation factor (fish)
Chlorophenylid (Eulan)	0·1	1 000
Sulcofenuron (Mitin FF)	10	40[a]
Flucofenuron (Mitin N)	4	1 000
Permethrin	0·003	4 000

[a] Low bioaccumulation risk.

Accumulation of Mitin is rapid: a steady-state in fish exposed to 0.1 mg litre^{-1} is reached in three days and in seven days for fish exposed to 0.01 mg litre^{-1}. The accumulation factor was calculated to be about 40 for both concentrations. Elimination is also rapid, the half-life for residues being about 4 days (Tooby, 1981a).

Invertebrate toxicity reported by Tooby (1981a) indicated that the 24 hr LC$_{50}$ to *Daphnia* of Eulan was 766 mg litre^{-1}, as formulated. He noted that at concentrations less than 1.5 µg litre^{-1} a coarse (cyprinid) fishery could survive. Trout were more sensitive, 1 µg litre^{-1} was lethal to fry and a 'safe' concentration would be about 0.2 µg/litre. Suggested water quality criteria for fisheries are given in Table 7.39.

7.9.3.7 Pyrethroids—Permethrin and related compounds

Pyrethrins are natural insecticides: pyrethrum is extracted mainly from the flowers of *Chrysanthemum cinerariaefolium*. Amongst its uses is the .control of organisms in water supply mains for which it is particularly suitable because of very low toxicity to mammals. One disadvantage is its high toxicity to fish (Table 7.40), which may pose risks for fish in aquaria although wild fish are perhaps at lower risk from water mains disinfestation since pyrethrin in mains water entering streams is likely to be diluted or absorbed on solids. Recently there has been a considerable shortfall in the world supply of pyrethrum extract (Evins, 1981) which has provided a stimulus for testing alternatives for mains dosing. Permethrin, a synthetic analogue of pyrethrum, was developed to meet this deficiency and there are now several pyrethrin analogues (allethrin, decamethrin, dimethrin ethanomethrin, resmethrin, cypermethrin, fenvalerate and fenpropanate). Natural pyrethrum is highly toxic to fish, the 24 hr LC$_{50}$ to rainbow trout (*S. gairdneri*) being 56 µg litre^{-1} (Cope, 1963). Synthetic pyrethroids are also highly toxic, the 96 hr LC$_{50}$ for rainbow trout is of the order of 0.1 to 0.5 µg litre (Marking and Mauck, 1975; Mauck *et al.*, 1976) suggesting that they may be more toxic than natural pyrethroids. Similarly, cypermethin was found to have a 96 hr LC$_{50}$ to rainbow trout of 0.5 µg litre^{-1} and the equivalent value for the cyprinid fish, the rudd (*Scardinius erythrophthalmus*), was 0.4 µg litre^{-1} (Stephenson, 1982). This is not confirmed by other work; permethrin (NDRC 143) was found to be less toxic by Coats and O'Donnell-Jeffery (1979): in their study a 24 hr LC$_{50}$ of 135 µg litre^{-1} was observed. Other synthetic pyrethroids such as cypermethrin were almost twice as toxic as permethrin. In studies of pyrethroid toxicity to Atlantic salmon Zitko *et al.* (1977, 1979) also found permethrin to be less toxic (96 hr lethal

TABLE 7.40

ACUTE TOXICITY OF SYNTHETIC PYRETHROIDS TO INVERTEBRATES AND FISH

Substance	Species	Test	Value (μg litre^{-1})	Reference
Cypermethrin	*Salmo gairdneri*	96 hr LC$_{50}$	0·5	Stephenson (1982)
	Salmo trutta	96 hr LC$_{50}$	1·2	Stephenson (1982)
	Cyprinus carpio	96 hr LC$_{50}$	0·9–1·1	Stephenson (1982)
	Scardinius erythrophthalmus	96 hr LC$_{50}$	0·4	Stephenson (1982)
	Cloeon dipterum	24 hr LC$_{50}$	0·6	Stephenson (1982)
	Gammarus pulex	24 hr LC$_{50}$	0·1	Stephenson (1982)
	Asellus aquaticus	24 hr LC$_{50}$	0·2	Stephenson (1982)
	Daphnia magna	24 hr LC$_{50}$	2·0	Stephenson (1982)
Permethrin	*Ictalurus punctatus*	96 hr LC$_{50}$	1·10	Jolly *et al.* (1978)
	Micropterus salmoides	96 hr LC$_{50}$	8·50	Jolly *et al.* (1978)
	Gambusia affinis	96 hr LC$_{50}$	15·00	Jolly *et al.* (1978)
	Procambarus clarkii (juv)		0·62	Jolly *et al.* (1978)

threshold of 8·8 μg litre^{-1}) than several other analogues (0·59–1·97 μg litre^{-1}). More recently, using continuous flow tests, Kumaraguru and Beamish (1981) found the lethal toxicity for rainbow trout to be dependent upon water temperature and body weight. For 1 g trout the 96 hr LC$_{50}$ increased from 0·62 to 6·43 μg litre^{-1} between 5 and 20 °C. At 15 °C the 96 hr LC$_{50}$ values for 1 and 100 g trout were 3·17 and 314 μg litre^{-1}, respectively. Non-salmonid species have been tested by Jolly *et al.* (1978). The 96 hr LC$_{50}$ for fingerling bass (*Mictropterus salmoides*) was 8·5 μg litre^{-1} while for the mosquito fish the equivalent value was 15 μg litre^{-1}. Channel catfish (*Ictalurus punctatus*) were more susceptible; the 96 hr LC$_{50}$ was 1·1 μg litre^{-1}. Clearly, there is considerable variation in the response of different species to permethrin which, judging by the results of the study by Kumaraguru and Beamish (1981), may arise from the effects of temperature, the size of the test individuals and details of the test procedure. Permethrin is more persistent than pyrethrins and allethrin (Zitko *et al.*, 1977) and fish accumulate permethrin from water. In the field the response might be less drastic than might at first be surmised. Exposure of rainbow trout to permethin at a range of concentrations of 12·5 to 50 µg litre^{-1} for one hour followed by transfer to clean water had no permanent effect (Muirhead-Thomson, 1978). Only exposures to 100 µg/litre for one hour had any effect.

Invertebrates are also highly sensitive to permethrin. Even with concentrations as low as 5 to 10 µg litre^{-1}, over 95% of macroinvertebrate test organisms were killed, so that applications of the insecticide for control of forest pests or disease vectors (e.g. *Simulium*) would not kill fish directly, although their food organisms would clearly be at risk (Muirhead-Thomson, 1978). Jolly *et al.* (1978) observed that the 96 hr LC$_{50}$ for newly hatched and juvenile crayfish (*Procambarus clarkii*) was 0·39 and 0·62 µg litre^{-1}, respectively. The 48 hr LC$_{50}$ of permethrin for juvenile and adult *Daphnia magna* ranged from 0·2 to 0·6 µg litre^{-1} (Stratton and Corke, 1981). The insecticide also caused particulate material to adhere to the appendages and led to immobilisation and significantly earlier mortality. The effect is caused by some effect on the crustacean and not the particles; it was observed to varying degrees with other pesticides. A comprehensive study of the effect of cypermethrin on macroinvertebrates (Stephenson, 1982) showed that the 24 hr LC$_{50}$ values of susceptible species ranged from 0·05 µg litre^{-1} (Hydracarina) to 2 µg litre^{-1} for *Daphnia magna* (Table 7.40). The 24 hr LC$_{50}$ for the water main infesting target species *Asellus aquaticus* was 0·2 µg litre^{-1}, but mobility was reduced by much lower concentrations (2 hr EC$_{50}$ was 0·03 µg litre^{-1}). These results show that important fish food organisms are more susceptible to synthetic pyrethroids than some of the more sensitive fish species. In field experiments, a dosage of 100 g ha^{-1} (exceeding that likely to occur from agricultural use) was applied to a pond (Crossland, 1982). Fish survived but there was a high mortality of insects and crustaceans. A mat of filamentous algae developed after 2 to 6 weeks. Its appearance was attributed to the demise of the herbivorous anthropods. Most species had recolonised the pond after 10 weeks.

In tests which simulated more closely the kind of exposure which might be involved in field applications of the insecticide, that is, noting the mortality which occurred within 24 hours of exposure to a given concentration for one hour, Muirhead-Thomson (1978) showed that the LC$_{90-95}$ for the common invertebrates *Baetis*, *Gammarus* and *Brachycentrus* was about 1 µg litre^{-1}. *Simulium* larvae were more resistant, a value of 5 µg litre^{-1} being observed, while for *Hydropsyche* the value could not be determined within the exposure range but exceeded 1 mg litre^{-1}. Behavioural studies were also made by means of a laboratory-simulated stream, using 30 minute exposures to 0·5 and 5 µg litre^{-1}, that is, a lethal and sub-lethal exposure. During the period of exposures, *Baetis*, *Gammarus* and *Simulium* were highly active and drifted downstream, but *Brachycentrus* did not and died, *in situ*, at the lethal concentration. *Hydropsyche* showed an intermediate response, activity occurring somewhile after exposure had ceased.

7.9.4 Molluscicides

Molluscs are rarely pests in their own right but the killing of aquatic molluscs is an indirect means of controlling the causative agent of human schistosomiasis or bilharzia, a disease caused by parasitic flatworms or 'flukes' of the genus *Schistosoma*, which live in the blood vessels of the intestine or bladder depending on the species. It is a disease of warm climates and the main areas in which each form of the disease is endemic are the Far East (*S. japonicum*); Egypt and South America (*S. mansoni*); Africa and the Middle East (*S. haematobium*).

The life cycle is complex. Eggs from female worms living in abdominal blood vessels of infected humans pass into the lumen of the gut or urinary bladder and on entering freshwater (through inadequate hygeine) hatch into ciliated larvae (miracidia). The miracidia swim actively and on encountering the appropriate intermediate host (species of planorbid snails: *Oncomelania* sp. in the case of *S. japonicum*, *Biomphalaria* sp. for *S. mansoni* and *Bulinus* (*Physopsis*) for *S. haematobium*) they penetrate the soft tissues. Within the body of the snail a series of replications occurs (as sporocysts) until larval worms (cercaria) emerge into the water. These are able to penetrate exposed human skin and enter the circulatory system where they ultimately mature and are then able to continue the cycle.

Prevention of the disease is, theoretically, merely a question of education in personal hygiene coupled with public health and sanitation measures. In practice, however, social and economic constraints prevail and in the case of the Far Eastern species (*S. japonicum*) a reservoir of the disease would continue in other mammals which also act as hosts. Prophylactic treatment and mass chemotherapy are impractical: the only remaining possibility is control of the intermediate host, the mollusc, in order to break the life cycle.

The development of molluscicides has been reviewed extensively by Muirhead-Thomson (1971). One of the first compounds used as a molluscicide was copper sulphate. The toxicity of copper is affected by a number of environmental factors (see Section 7.8.8) and therefore high field doses are sometimes required. Sodium pentachlorophenate and Pentachlorophenol (PCP) are effective also against snail eggs and the cercaria while sodium pentachlorophenate has herbicidal properties so that dense vegetation in which the snails live is also controlled. The ideal concentration is 10 ppm which should be maintained for about 8 hours.

Niclosamine, the ethanolamine salt of 5,2'-dichloro-4'-nitro-salicyl-anilide ('Bayluscide', 'Bayer 73') is very toxic to snails and snail eggs at a field dose of about 4 to 8 ppm but it is toxic to fish at effective molluscicidal concentrations. Fortunately, its mammalian toxicity is low. The compound has low solubility in water and is formulated as a wettable powder which is less satisfactory than an emulsifiable concentrate.

More recently, N-tritylmorpholine ('Frescon') has been used very effectively in snail control. This also has low solubility in water but has the advantage that it appears to have little deleterious effect on other aquatic life. Control of adult snails is attained by exposure to con-centrations of 0·1 to 0·5 ppm for an hour or 0·01 to 0·05 ppm for 24 hours. Snail eggs are, however, not controlled so that repeated treat-ments or extended continuous low dosages are required. This substance seems to offer the most useful balance of high molluscicidal activity coupled with low environmental hazard.

Some herbicides such as acrolein, paraquat and diquat have been shown to control molluscs both directly and by destroying the aquatic vegetation on which they live. Organotin and organolead compounds have been found to be effective in destroying both snails and their eggs but these substances are highly toxic and, incidentally, appear on the 'black-list' of the EEC Directive (see Table 7.2).

7.9.5 Piscicides

Although certain substances have been used from antiquity to poison fish in order that they may be gathered for food (and continues today as a means of poaching), the chemical control of fishes is relatively rare in comparison with other pesticide usage. In fishery management chemicals are sometimes used to remove or control unwanted species, for example the eradication of trout in an attempt to improve salmon populations or the removal of 'coarse' or 'trash' fish from a game or sport fishery or the removal of predators or parasites such as the lamprey (*Petromyzon*). An ideal piscicide controls or eliminates the fish populations with minimal risk to man and invertebrate fish food organisms. The majority of piscicides are relatively unselective and are usually used for complete elimination of the existing fish community after which the desired species can be introduced. An important exception is the development of lampricides which have been used extensively for the selective control of the sea lamprey (*Petromyzon marinus*) in the North American Great Lakes area.

7.9.5.1 *Rotenone (Derris)*

Rotenone has been used as a fish poison from ancient times, being extracted from the plant *Derris* in Asia and *Lonchocarpus* in South America. It is also used in horticulture as an insecticide and acaricide. A review of its use in fisheries has been made by Schnick (1974). Rotenone is relatively costly and is subject to rapid breakdown (Hamilton, 1941): this is regarded as a disadvantage in that its toxicity may be lost before it has become thoroughly distributed throughout the water of the lake (Muirhead-Thomson, 1971) but one might regard the rapid degradation as a positive advantage in environmental protection. The acute toxicity of rotenone is not remarkably high, for example the 24 and 96 hr LC_{50} for rainbow trout (*S. gairdneri*) are given as 32 and 26 μg litre^{-1} and for bluegills (*Lepomis macrochirus*) as 24 and 22 μg litre^{-1}, respectively (Holden, 1973). However, these data do not indicate the minimum period of exposure necessary, at a given concentration, to ensure that mortality of the whole population will be assured and this is what is required for effective use of a piscicide. Gilderhus (1972) conducted studies of the effective contact time (ECT) for rotenone, that is, the minimum length of exposure to a given convenient concentration which will induce 100% mortality of the target animal, 'regardless of the time it takes them to die'. This last part of the definition needs to be qualified and in practice it appears that a limit of 96 hr was set. Test were conducted at 12, 17 and 22 °C. It was found that exposure time was influenced more by temperature than toxicant concentration and the effects of rotenone poisoning could sometimes be reversed on transferring fish to clean water even though they had been on their sides in the toxicant solution for 4 or 5 hours. In general, rotenone proved fatal only if fish had been incapacitated in the test solution for 1 to 5 hours. Of the seven species tested, rainbow trout and yellow perch (*Perca flavescens*) were most susceptible with ECT values of 1 to 4 hours. The common North American target species, carp (*Cyprinus carpio*) and suckers (*Catostomus commersoni*), were eliminated by 5 hours' exposure to 250 ppb (2·5 μg litre^{-1}) at 17 °C.

Several studies have been made in order to assess the affects of rotenone application on the non-target animals present and the fish-food organisms in particular. Hamilton (1941) noted that concentrations of rotenone which killed fish were too dilute to affect other animals (including metamorphosed amphibians) except for micro-crustacea which were killed in 1 to 4 hours at concentrations of $25 \,\mu g \, litre^{-1}$. He observed, however, that this would be unlikely to cause permanent changes since the eggs of micro-crustacea are often very resistant. Lindgren (1960), in a study of the effects of rotenone on the benthic fauna of a Swedish lake, concluded that even when concentrations are high enough to effect complete eradication of the fish population, sufficient invertebrates survive to ensure recolonisation. On reviewing the published literature he commented that many other workers had observed that species which were resistant to rotenone increased their numbers quite markedly following the poisoning but populations returned to their normal levels later. Such changes were noted by Cooke and Moore (1969) in a Californian stream following rotenone treatment. They cite, as an interesting example, changes in the distribution of simuliid (blackfly) larvae. These were formerly restricted to the fastest flowing water but after treatment they were widely distributed, even in pools. It was concluded that removal of predation pressure had allowed both their density to increase and their colonisation of atypical microhabitats. All the larvae were early instars which indicated recent reproduction, rather than drift from the untreated reaches upstream. Chironomids also increased their numbers, almost exclusively as early instars, within the pools.

Rotenone application may induce drift in benthic invertebrates, as Morrison (1977) observed in hill streams in Scotland. Although some insects were killed there was no significant difference in either the population density or diversity, following treatment. Even when the dose was increased tenfold (5 ppm instead of 0·5 ppm), in error, little change was observed. Morrison and Struthers (1975) showed that invertebrates survive or re-establish themselves within a relatively short time in lochs treated with $0·04–0·06 \, mg \, litre^{-1}$ rotenone.

7.9.5.2 Toxaphene

Toxaphene is a mixture of chlorinated camphenes and is used principally as an insecticide. It has been used quite extensively in the USA as a substitute for rotenone but one serious drawback is its environmental persistence: after treatment to remove undesirable fish it has been found

TABLE 7.41

TOXICITY OF TOXAPHENE TO DIFFERENT FISH SPECIES

(Data from Macek and McAllister, 1970)

		96 hr LC_{50} (μg litre^{-1})	95% confidence interval	Relative susceptibility
Coho salmon	*Oncorhynchus kisutch*	8	6–10	4·0
Brown trout	*Salmo trutta*	3	2–5	1·5
Rainbow trout	*Salmo gairdneri*	11	8–13	5·5
Carp	*Cyprinus carpio*	4	3–5	2·0
Goldfish	*Carassius auratus*	14	11–19	7·0
Minnow	*Pimephales promelas*	14	9–22	7·0
Catfish	*Ictalurus punctatus*	13	10–17	6·5
Bullhead	*Ictalurus melas*	5	4–7	2·5
Perch	*Perca flavescens*	12	9–14	6·0
Sunfish	*Lepomis microlophus*	13	8–17	6·5
Bluegill	*Lepomis macrochirus*	18	10–30	9·0
Bass	*Micropterus salmoides*	2	1–3	1·0

impossible to restock for months or even years. This substance is highly toxic to fish (Table 7.41) but has the advantage that its toxicity to man is less than rotenone and also endrin, dieldrin and aldrin (Johnson *et al.*, 1966). Used at 5 μg litre^{-1} it reduces the numbers of small fish but to effect complete control of larger fish a concentration of 0·1 mg litre^{-1} is required. Normally some 2 to 12 months must elapse before sufficient detoxification occurs to allow restocking but some lakes have proved to be toxic to fish for up to four years. There is some evidence that the amount of suspended matter present may influence the degree of detoxification by providing surfaces for adsorption. Johnson *et al.* (1966) studied the persistence of toxaphene in treated lakes and concluded that detoxification is accomplished at least in part by sorption rather than degradation. They also suggest that different components of toxaphene, which is a mixture of compounds, may degrade at different rates. One or more appear to degrade more rapidly than the bulk of the constituents and this results in a large amount of residue of lower toxicity. Evidence is accumulating that the levels recovered from water and determined in analysis may be misleading since fish have been found to survive at residue levels which one would have been thought to be toxic.

Studies of prolonged exposure (90–150 days) of fathead minnows (*Pimephales*) to toxaphene at concentrations of 55 to 1230 ng litre^{-1} have revealed reductions in collagen leading to weakening and fragility of the backbone—the so-called 'broken back' syndrome (Mehrle and Mayer,

1975a). Similar reductions in body collagen were observed in brook trout (*Salvelinus*) fry (Mehrle and Mayer, 1975b).

7.9.5.3 *Antimycin*

Antimycin is an antibiotic substance produced by a mould (*Penicillium citrinum*) and has strong fungicidal and piscicidal action. It was discovered in the mid-1950s and its piscicidal properties were exploited about a decade later. Antimycin kills rainbow trout at a concentration of 30 parts per trillion (10^{12}) or $0.03\,\mu g$ litre^{-1}, a level which cannot be measured by analytical methods (Marking and Walker, 1973). Different fish species exhibit different relative sensitivity to antimycin (Walker *et al.*, 1964) and therefore it is useful for selective control of susceptible unwanted species. Its principal use is in the selective control of 'scalefish' (carp, sunfishes, perches, etc.) to enhance the survival of channel catfish (*Ictalurus punctatus*) for food. Control is effective at 0.4–1.0 ppb (0.4–1 μg litre^{-1}) and total elimination of scalefish can be achieved at 10 ppb (10 μg litre^{-1}). The 96 hr LC$_{50}$ for channel catfish is $20.4\,\mu g$ litre^{-1} while the value for sunfish (*Lepomis cyanellus*) is $0.31\,\mu g$ litre^{-1}; the equivalent value for rainbow trout is $0.04\,\mu g$ litre^{-1} (Marking and Walker, 1973). Antimycin is most effective in soft water and low pH (Berger *et al.*, 1969). The toxic action is largely irreversible: once fish show signs of distress (loss of equilibrium), death follows even if they are subsequently transferred to clear water. It seems that there is a point at which, while they are still swimming, a lethal dose has been assimilated. The effective contact time is generally much shorter than for rotenone. For example, for carp or suckers (*Catostomus* spp.) the effective contact time for antimycin was 6 hours or less while 18 or 24 hours' exposure was needed for rotenone at normal concentrations (Gilderhus, 1972). Antimycin is a good, stable compound with high purity and which is easy to formulate. It is deactivated in water of high pH: for example the half-life at pH 8 is less than 48 hr (Marking and Walker, 1973). Morrison (1979) used antimycin in a Scottish stream at a concentration of 10 μg litre^{-1} for several hours. Wild fish and controls in cages were killed but no adverse effects on the invertebrate fauna could be observed.

7.9.5.4 *Trifluoromethylnitrophenol (TFM)*

Trifluoromethylnitrophenol has the distinction of being the most selective piscicide, although there may be a semantic question here as to whether the principal target organism, the lamprey, is a 'fish'. It was developed for the control of the marine lamprey *Petromyzon marinus* in the Great Lakes area of North America. This parasitic species was indigenous to Lake Ontario and the St Lawrence River but gradually spread westwards through Lakes Erie, Huron, Michigan and Superior, following the opening of the Welland Canal. The lamprey caused serious effects on the populations of lake trout (*Salvelinus naymacush*) and the fishing industries which depended on them.

The sea lamprey is anadromous, spawning in small rivers and streams where the larvae spend up to 5 years before undergoing metamorphosis into the adult form when they migrate downstream. Only the adult lamprey poses a direct threat to fisheries. It lives by attaching itself to bony fish with its sucking mouth and rasps the flesh, sometimes burrowing into the body cavity.

It would be almost impossible to attempt to control the adults within the vast area of the lakes but since the larvae are relatively localised in feeder streams there is a potential for control at this stage, and once the life cycle is broken the parasitic adults will disappear. After screening several thousand possible larvicides it became apparent that halogenated nitrophenols, and in particular trifluoromethylnitrophenol (TFM), showed great potential for the control of young lampreys. It was discovered that in studies of the relative toxicity of TFM to larval lampreys and rainbow trout there was no overlap in the range of concentrations at which mortality occurred in a 24 hr period at a range of normal temperatures (Applegate *et al.*, 1961). For example, 100% mortality was observed in the lampreys at 3 ppm while no mortality occurred in rainbow trout even at 7 to 8 ppm. This selectivity is not impaired by low temperatures but is actually enhanced since the toxicity of TFM to rainbow trout decreases while its effect on lampreys remains the same. The most effective seasons for application appear to be late autumn or early spring. An additional useful factor is that at 3 ppm mature adult sea-lampreys are killed when they ascend to spawn in autumn and recently metamorphosed larvae are also killed when they begin their downstream migration in spring. The toxicity of TFM is strongly influenced by hardness and pH: it is most effective as a larvicide in soft, acid waters. As pH, conductivity and alkalinity increase, the dosage has to be increased, although it appears not to be affected by increased turbidity.

The toxicity of TFM to other bony fishes was also investigated by Applegate *et al.* (1961). Brook trout (*Salvelinus fontinalis*) were not harmed by 13·0 ppm, brown trout (*Salmo trutta*) and rock bass (*Ambloplites rupestris*) were both unharmed by concentrations up to

9 ppm, while yellow perch (*Perca flavescens*) survived concentrations up to 7 ppm. Bullheads (*Ictalurus*) died when exposed to 3 ppm TFM or more but shiners (*Notropis*), chubs (*Semotilus*), dace (*Rhinichthys*) and sculpins (*Cottus*) were unaffected at concentrations less than 13 ppm. Log perch (*Percina caprodes*) were killed by 5·0 ppm.

It was observed that crayfish and caddisfly larvae were not affected by any concentrations used, nor were turtles. In field trials on the Mosquito River (Michigan), lamprey larvae were killed but all other fish survived the treatment. No effect of the larvicide on invertebrates was observed. Similar results were observed in a trial on the Silver River (Michigan) but premature 'hatching' was induced in one species of mayfly. The general conclusion from these studies was that for no species other than lampreys was mortality so severe as to prevent it from restoring its numbers.

An interesting development in lamprey control was the discovery that niclosamine ('Bayluscide'), which was originally developed as a molluscicide for the control of schistosomiasis, or bilharzia, had a synergistic effect on the toxicity of TFM. Small amounts of niclosamine were found to enhance the lampricidal activity of TFM without significantly affecting its selectivity (Howell *et al.*, 1964). A mixture of 98% TFM and 2% Bayluscide was selected as the most suitable for field treatments and permitted a halving of the estimated cost of using TFM alone.

7.10 POLYCHLORINATED BIPHENYLS—PCBs

7.10.1 Sources
Polychlorinated biphenyls (PCBs) are mixtures of chlorinated biphenyl isomers which are marketed under several names including Phenoclor, Kanechlor, Clophen and, in the United States, Aroclor (Vieth and Lee, 1970). The degree of chlorination determines their characteristics and with the Aroclor series this is indicated in the numerical suffix. For example, Aroclor 1254 contains an average of 54%, by weight, of chlorine; the first two digits indicate the molecular type (Train, 1979).

PCBs are non-ionic, non-inflammable compounds with extremely low water solubility (which tends to decrease with the degree of chlorination). They are highly lipophilic, and hence of significance to biota, with similar ecosystem cycling and potential pollution problems as the somewhat chemically similar organochlorine insecticide DDT.

PCBs have a wide range of industrial applications but are especially useful in electrical transformers or capacitators because of their high dielectric constants. They also find applications in hydraulic fluids, lubricants and plasticisers. PCBs are used in the manufacture of paints and adhesives and have many other minor industrial applications including anti-corrosion, water proofing, ice-prevention coatings and as an evaporation retardant for pesticides (Vieth and Lee, 1970).

Environmental contamination by PCBs was first identified in aquatic systems in Scandinavia (Jensen, 1966) but it is now recognised that they occur almost universally, although often in very low concentrations. They enter the aquatic environment in sewage and industrial effluents (Vieth and Lee, 1971) and through airborne transport (Eisenreich *et al.*, 1979). A brief review of sources and means of dissipation has been provided by Strek and Weber (1982).

7.10.2 Persistence

Persistence of PCBs in the environment is expected to be prolonged by their non-ionic and lipophilic nature together with their low water solubility and volatility and resistance to degradation. Their hydrophobic character and readiness to adsorb onto particles or sediments also contributes to environmental persistence. The organic content of sediments seems to be more important than available surface area in the degree of adsorption of PCBs (Strek and Weber, 1982).

7.10.3 Bioaccumulation

The high lipophilic property of PCBs contributes to their pronounced bioaccumulation. Organisms take up PCBs from very low (microgram) levels in water and food and incorporate them into their tissues. In a comprehensive literature review of the processes of uptake and accumulation of PCBs in the aquatic environment, Falkner and Simonis (1982) concluded that, apart from higher vertebrates, bioconcentration in food chains was far less important than direct uptake from the water. Accumulation factors of 50 000 or more are widely reported for fishes (Mayer *et al.*, 1977; Mauck *et al.*, 1978), although values for invertebrates tend to be one or more orders of magnitude lower (Table 7.42). Body accumulations of PCBs differ between species and even for the same species there are large differences between habitats: these may be attributed to differences in metabolic parameters, exposure, size and rate of growth (Jensen *et al.*, 1982). Bioconcentration factors may be independent of the PCB concentration in the water (De Foe *et al.*, 1978). Total body burdens appear, as might be expected, to be related to the fat

TABLE 7.42

BIOACCUMULATION FACTORS FOR POLYCHLORINATED BIPHENYLS (PCBs) IN FISH AND INVERTEBRATES

PCB	Concentration (μg litre^{-1})	Species	Exposure period (days)	Bioaccumulation factor	Reference
(a) Fish					
Aroclor 1242	0·86	*Pimephales promelas*	260	274 000	Nebeker *et al.* (1974)
1242	0·86	*Pimephales promelas*	260	107 000	Nebeker *et al.* (1974)
1245	0·23	*Pimephales promelas*	260	235 000	Nebeker *et al.* (1974)
1245	1·8	*Pimephales promelas*	260	238 000	Nebeker *et al.* (1974)
1248	3·0	*Pimephales promelas*	250	60 000	DeFoe *et al.* (1978)
1260	2·1	*Pimephales promelas*	250	160 000	DeFoe *et al.* (1978)
1248	3·0	*Pimephales promelas*	250	120 000	DeFoe *et al.* (1978)
1260	2·1	*Pimephales promelas*	250	270 000	DeFoe *et al.* (1978)
1248	5·8	*Ictalurus punctatus*	77	56 370	Mayer *et al.* (1977)
1254	2·4	*Ictalurus punctatus*	77	61 190	Mayer *et al.* (1977)
1254	1·5	*Salvelinus fontinalis* (fry)	118	47 000	Mauck *et al.* (1978)
(b) Invertebrates					
Aroclor 1242	8·7	*Gammarus pseudolimnaeus*	60	36 000	Nebeker and Puglisi (1974)
1248	5·1	*Gammarus pseudolimnaeus*	60	108 000	Nebeker and Puglisi (1974)
1254	1·1	*Daphnia magna*	4	3 800	Mayer *et al.* (1977)
1254	1·6	*Gammarus pseudolimnaeus*	14	6 300	Mayer *et al.* (1977)
1254	2·8	*Pteronarcys dorsata*	14	750	Mayer *et al.* (1977)
1254	1·5	*Culex tarsalis*	7	3 500	Mayer *et al.* (1977)
1254	1·3	*Chaoborus punctipennis*	14	2 700	Mayer *et al.* (1977)

content of tissues (Snarski and Puglisi, 1976). Female fathead minnows (*Pimephales promelas*) accumulate twice as much of the PCBs Aroclor 1248 and 1260 as males because of their lipid content (De Foe *et al.*, 1978).

Clearance of accumulated PCB from the body may be slow. For example, in the minnow (*Phoxinus phoxinus*) after 45 days accumulation of dietary PCB to a level of about $100 \, mg \, kg^{-1}$ fresh weight, very little change was observed for 130 days and approximately a third of this original level remained after 250 days (Bengtsson, 1980). Fathead minnows (*Pimephales promelas*) exposed to PCB eliminated less than 18% of their body burden after 60 days in clean water (De Foe *et al.*, 1978) Mayer *et al.* (1977) observed that coho salmon (*Oncorhyncus kisutch*) fed PCB at dietary concentrations of 0·048, 0·48 and $4·8 \, \mu g \, g^{-1}$ of food reached equilibrium after 112 days but those fed at 48 and $480 \, \mu g \, g^{-1}$ did not reach equilibrium until 200 days had elapsed.

Spawning may afford a means of reducing PCB body burdens, especially for female fish since many eggs have a high lipid content.

7.10.4 Acute Toxicity

The short-term (4 day) acute toxicity of PCBs is relatively low and similar to that of common organochlorine insecticides (Tables 7.43 and 7.44). Aquatic macroinvertebrates appear to be more susceptible to the acute toxic effects than fish, and insects are more resistant than crustacea (Meyer *et al.*, 1977). In general, toxicity decreases as the percentage of chlorine increases. It is also evident from Tables 7.43 and 7.44 that the PCBs tested are much more toxic under continuous flow conditions where there is continual replacement by fresh PCB. In static tests almost one-third of the PCB which is added can be lost in volatilisation or adsorption on the test vessel or other materials (Nebeker and Puglisi, 1974). These data illustrate the importance of the test procedure in determining the apparent toxicity of certain substances.

7.10.5 Chronic Toxicity

The chronic toxicity of PCBs is quite high, especially to invertebrates. This is well illustrated by Aroclor 1254 in which 21 day LC_{50} values of 1·3 and $0·65 \, \mu g \, litre^{-1}$ to *Daphnia magna* and *Tanytarsus* larvae, respectively, were observed. The 30 day LC_{50} value of Aroclor 1260 for *Lepomis macrochirus* was $400 \, \mu g \, litre^{-1}$ (Mayer *et al.*, 1977) but for newly hatched *Pimephales promelas* the value was as low as $3·3 \, \mu g \, litre^{-1}$ (De Foe *et al.*, 1978). A general inverse correlation between toxicity to fish and percentage chlorination is evident although Aroclor 1248 was more

TABLE 7.43
TOXICITY OF POLYCHLORINATED BIPHENYLS (PCBs) TO FISH

PCB	Species	Test	Value (μg litre^{-1})	Reference
Aroclor 1242	Pimephales promelas	96 hr LC$_{50}$	300	Nebeker et al. (1974)
1245	Pimephales promelas	96 hr LC$_{50}$	>33	Nebeker et al. (1974)
1242	Pimephales promelas (newly hatched)	96 hr LC$_{50}$	15·0	Nebeker et al. (1974)
1254	Pimephales promelas (newly hatched)	96 hr LC$_{50}$	7·7	Nebeker et al. (1974)
1248	Pimephales promelas (newly hatched)	30 day LC$_{50}$	4·7	DeFoe et al. (1978)
1260	Pimephales promelas (newly hatched)	30 day LC$_{50}$	3·3	DeFoe et al. (1978)
1221	Salmo clarki	96 hr LC$_{50}$	1 200[a]	Mayer et al. (1977)
1232	Salmo clarki	96 hr LC$_{50}$	2 500[a]	Mayer et al. (1977)
1242	Salmo clarki	96 hr LC$_{50}$	5 400[a]	Mayer et al. (1977)
1248	Salmo clarki	96 hr LC$_{50}$	5 700[a]	Mayer et al. (1977)
1254	Salmo clarki	96 hr LC$_{50}$	42 000[a]	Mayer et al. (1977)
1260	Salmo clarki	96 hr LC$_{50}$	61 000[a]	Mayer et al. (1977)
1242	Salmo gairdneri	10 day LC$_{50}$	48	Mayer et al. (1977)
1248	Salmo gairdneri	10 day LC$_{50}$	38	Mayer et al. (1977)
1254	Salmo gairdneri	10 day LC$_{50}$	160	Mayer et al. (1977)
1260	Salmo gairdneri	10 day LC$_{50}$	326	Mayer et al. (1977)
1254	Salvelinus fontinalis, fry	96 hr LC$_{50}$	>100	Mauck et al. (1978)
1242	Ictalurus punctatus	15 day LC$_{50}$	219	Mayer et al. (1977)
1248	Ictalurus punctatus	15 day LC$_{50}$	121	Mayer et al. (1977)
1254	Ictalurus punctatus	15 day LC$_{50}$	286	Mayer et al. (1977)
1260	Ictalurus punctatus	15 day LC$_{50}$	482	Mayer et al. (1977)
1242	Lepomis macrochirus	15 day LC$_{50}$	164	Mayer et al. (1977)
1248	Lepomis macrochirus	15 day LC$_{50}$	111	Mayer et al. (1977)
1254	Lepomis macrochirus	15 day LC$_{50}$	303	Mayer et al. (1977)
1260	Lepomis macrochirus	30 day LC$_{50}$	400	Mayer et al. (1977)

[a]Static test

TABLE 7.44

TOXICITY OF POLYCHLORINATED BIPHENYLS (PCBs) TO INVERTEBRATES (DATA RELATE TO FLOW-THROUGH TESTS EXCEPT WHERE INDICATED)

PCB	Species	Test	Value (μg litre^{-1})	Reference
(a) Crustacea				
Aroclor 1221	*Daphnia magna*	21 day LC$_{50}$	189[a]	Nebeker and Puglisi (1974)
1232	*Daphnia magna*	21 day LC$_{50}$	72[a]	Nebeker and Puglisi (1974)
1242	*Daphnia magna*	21 day LC$_{50}$	67[a]	Nebeker and Puglisi (1974)
1248	*Daphnia magna*	21 day LC$_{50}$	25[a]	Nebeker and Puglisi (1974)
1254	*Daphnia magna*	21 day LC$_{50}$	31[a]	Nebeker and Puglisi (1974)
1260	*Daphnia magna*	21 day LC$_{50}$	36[a]	Nebeker and Puglisi (1974)
1262	*Daphnia magna*	21 day LC$_{50}$	43[a]	Nebeker and Puglisi (1974)
1268	*Daphnia magna*	21 day LC$_{50}$	263[a]	Nebeker and Puglisi (1974)
Aroclor 1248	*Daphnia magna*	14 day LC$_{50}$	2.6	Nebeker and Puglisi (1974)
1254	*Daphnia magna*	14 day LC$_{50}$	1·8	Nebeker and Puglisi (1974)
1254	*Daphnia magna*	21 day LC$_{50}$	1·3	Nebeker and Puglisi (1974)
Aroclor 1242	*Gammarus fasciatus*	96 hr LC$_{50}$	10	Mayer *et al.* (1977)
1248	*Gammarus fasciatus*	96 hr LC$_{50}$	52[a]	Mayer *et al.* (1977)
1254	*Gammarus fasciatus*	96 hr LC$_{50}$	2 400[a]	Mayer *et al.* (1977)
Aroclor 1242	*Gammarus pseudolimnaeus*	96 hr LC$_{50}$	73	Nebeker and Puglisi (1974)
1248	*Gammarus pseudolimnaeus*	96 hr LC$_{50}$	29	Nebeker and Puglisi (1974)
Aroclor 1254	*Palaemonetes kadiakensis*	7 day LC$_{50}$	3	Mayer *et al.* (1977)
Aroclor 1242	*Orconectes nais*	7 day LC$_{50}$	30[a]	Mayer *et al.* (1977)
1254	*Orconectes nais*	7 day LC$_{50}$	100[a]	Mayer *et al.* (1977)
(b) Insecta				
Aroclor 1242	*Ischnura verticalis*	96 hr LC$_{50}$	400	Mayer *et al.* (1977)
1254	*Ischnura verticalis*	96 hr LC$_{50}$	200	Mayer *et al.* (1977)
Aroclor 1242	*Macromia* sp.	7 day LC$_{50}$	800[a]	Mayer *et al.* (1977)
1254	*Macromia* sp.	7 day LC$_{50}$	1 000[a]	Mayer *et al.* (1977)

TABLE 7.44—*contd.*

PCB	Species	Test	Value ($\mu g\ litre^{-1}$)	Reference
Aroclor 1254	*Tanytarsus dissimilis,* larvae	21 day LC_{50}	0·65	Nebeker and Puglisi (1974)
1254	*Tanytarsus dissimilis,* pupae	21 day LC_{50}	0·45	Nebeker and Puglisi (1974)

*a*Indicates static test.

toxic than Aroclor 1242. Mayer *et al.* (1977) noted that PCBs containing 3 or 4 chlorine atoms appear to be the most toxic to fish and these predominate in Aroclor 1248.

7.10.6 Sub-lethal Effects

7.10.6.1 Effects on growth
Growth of larval fathead minnows (*P. promelas*) was significantly reduced by exposure to 4·4 μg PCB per litre for 30 days (De Foe *et al.*, 1978). This contrasts with the enhancement of growth in minnows (*Phoxinus*) by exposure to dietary PCB which continued even after PCB was no longer fed (Bengtsson, 1980) but this could not be explained although an effect on hormone systems was postulated. Clearly there is no constant response and indeed no effect on growth could be observed in several species, including coho salmon (Halter and Johnson, 1974; Mayer *et al.*, 1977), brook trout (Snarski and Puglisi, 1976) and channel catfish (Mayer *et al.*, 1977).

Channel catfish (*Ictalurus*) treated with the PCB Aroclor 1242 grew more slowly than controls but when they were returned to a PCB-free diet they rapidly recovered. No histological differences could be observed and the regression of hepatic hypertrophy suggests that low levels of PCB are not permanently injurious to channel catfish (Hansen *et al.*, 1976). Brook trout fry exposed to concentrations of Aroclor 1254 of 0·43–13 μg litre^{-1} showed retarded growth after 48 days but fry surviving to 118 days were not significantly different from the controls (Mauck *et al.*, 1978).

7.10.6.2 Reproduction
Reproduction in the fathead minnow occurred at concentrations as high

as $1.8\,\mu g$ litre^{-1} of Aroclor 1254 and $5.4\,\mu g$ litre^{-1} of Aroclor 1242 (Nebeker *et al.*, 1974) and after 240 days' exposure to concentrations of Aroclor 1248 as high as $3\,\mu g$ litre^{-1} and Aroclor 1260 at $2.1\,\mu g$ litre^{-1}, concentrations which significantly affected the survival of fry (DeFoe *et al.*, 1978). The time to hatching, the proportion of eggs hatching and the immediate subsequent fry survival of eggs of brook trout (*Salvelinus fontinalis*) exposed to Aroclor 1254 at concentrations up to $13\,\mu g$ litre^{-1} for 10 days before hatching were not significantly different ($p < 0.05$) from controls (Mauck *et al.*, 1978).

Later survival of fry was impaired: after 48 days, fry exposed to $13\,\mu g$ litre^{-1} experienced greater mortality than controls. Similarly increased mortality was observed after 118 days among fry exposed to a concentration of $3.1\,\mu g$ litre^{-1}.

In static tests, the reproduction of *Daphnia magna*, assessed on the basis of the number of young produced, was impaired by sub-lethal concentrations of Aroclor (16–$206\,\mu g$ litre^{-1} for 16 to 50% impairment) while in continuous-flow tests, reproductive impairment occurred at or just below those concentrations which prevented survival of the adults (Nebeker and Puglisi, 1974). Reproduction of *Gammarus pseudolimnaeus* was largely unaffected by exposure to $2.8\,\mu g$ litre^{-1} of Aroclor 1242 and $2.2\,\mu g$ litre^{-1} of Aroclor 1248 but at $5.1\,\mu g$ litre^{-1} of the latter it was only at half the rate observed in the controls.

7.10.6.3 Algal toxicity

Early work on PCB inhibition of photosynthesis was undertaken using algae and this indicated some effects, although the data proved difficult to interpret (Strek and Weber, 1982). In experimental studies of the effects of PCBs on algae, Nau-Ritter *et al.* (1982) found that $8.6\,\mu g$ litre^{-1} of Arocolor 1254 desorbed from clay particles inhibited photosynthesis and reduced chlorophyll-*a* content in natural phytoplankton communities more than PCB added directly to the water. Algal growth was depressed by concentrations of PCB ranging from 11 to $111\,\mu g$ litre^{-1} (Christensen and Zielski, 1980).

CHAPTER 8

Laboratory Evaluation of Pollutants

8.1 INTRODUCTION

In the preceding sections attention has concentrated mainly on the effects of pollutants in the field as indicated by changes in populations or in communities, although some of the evidence for the mechanisms whereby pollutants exert their effects have been derived from laboratory investigations. Here, the procedures used to investigate the action of pollutants on organisms and especially those used to quantify toxicity are discussed. Some duplication of material, especially where the two aspects overlap, is inevitable and the two approaches are, of course, interrelated.

All levels of biological organisation, from molecules through cells, tissues, organs and individuals up to populations and communities, may be utilised in environmental surveillance (Hellawell, 1977a,b). Each level has particular advantages: choice is dictated by the problem under consideration. It is likely that most, if not all, effects of environmental changes, especially those caused by effluent discharges, originate at the molecular level and alter physiology, structure, behaviour or reproductive capacity but are revealed in surveys at the population or community level. Since toxic substances interfere with physiological processes it is possible to explore their effects with sub-organism preparations (isolated organs or tissues) or even at the sub-cellular level. Examples of such preparations and their use are given in Table 8.1.

In environmental management one is probably most interested in the effects of existing effluents or water-management policies at the population level, but for screening new constituents of effluents or other possible contaminants the use of physiological methods may be advantageous. Tests of the effects of substances on the biota are more valuable when the results are unequivocal, i.e. the substance is highly toxic at very

329

TABLE 8.1

EXAMPLES OF THE USE OF TISSUE CULTURE FOR STUDIES OF THE TOXICITY OF SUBSTANCES IN WATER AND EFFLUENTS
(See also 8.2.1.4)

Preparation	Culture method	Technique Measurement	Toxicants	Reference
Epithelial cells of fathead minnow (*Pimephales promelas*)	Growth in plastic tissue-culture flasks	Mitotic index	Zinc	Rachlin and Perlmutter (1968)
Bovine ovarian cells	Plated in plastic Petri dishes	Colony counts absolute plating efficiency; relative plating efficiency as function of length of exposure	Heavy metals (Cd, Cu, Zn, Cr, Se), NTA and rotenone	Malcolm *et al.* (1973)
Mouse fibroblasts, L-M strain	Suspension in Erlenmeyer Flask	Cell counts using Coulter counter determine relative growth	Metal-cresol, oil-refinery effluents	Richardson *et al.* (1977)
Mouse, subcutaneous L-cells	Suspension in culture tubes	Cell counts using haemocytometer, relative cell growth rates	Organic pollutants from surface waters extracted in organic solvents, DDT as reference toxicant	Muruoka (1978)
HeLa (human carcinoma), mouse lymphoma and green monkey	Plated in plastic tissue-culture plates	Colony counts (as % of controls)	Heavy metals (Cd, Hg, Pb), ammonia, phenol, sewage effluents	Kfir and Prozesky (1981)

low, or even analytically undetectable, concentrations or, conversely, is apparently biologically inert at almost any meaningful concentration. Such extremes are rarely encountered: many toxic substances are biodegradable, although some relatively innocuous substances could be modified within the environment and thus become potentially hazardous. Again, difficulties may arise when one attempts to 'scale up' the laboratory simulation and these can often only be overcome by limited field trials. For some substances even limited field trials may be difficult since damage could be extensive and virtually irreversible. Considerable effort has, therefore, to be expended in attempts to develop standard tests or 'bioassay' procedures, for example fish-toxicity tests or algal growth enhancement (biostimulation) tests, which may be employed for routine screening of new substances or effluents. This is a wide and rapidly developing field which can only receive a cursory survey in this chapter. It is complementary to the use of field observations of the effects of effluents or management procedures which are considered in Sections 9.2.1–9.2.7.

8.2 EXPERIMENTAL AND LABORATORY INVESTIGATIONS

Although the ultimate objectives may be somewhat different it has been thought more convenient to consider together those methods of investigation which essentially depend upon laboratory or experimental facilities. Many field techniques depend upon the availability of adequate laboratory facilities for biochemical determinations. For example, studies of changes in algal populations may be conducted by estimating biomass from chlorophyll-*a* determinations, a procedure which is less tedious than counting cells or finding dry weights of filtered residues.

For further convenience, methods for the determination of quantities of constituents of organisms or metabolic rates are gathered together under a general heading of 'Physiological Measurements' (Section 8.2.1) while those methods which utilise intact organisms are treated under Section 8.2.2 entitled 'Whole Organism Responses'. This division is admittedly somewhat arbitrary, for algae, fungi and bacteria are 'whole organisms', but the methods used in investigating their responses to environmental changes are generally similar to those employed in physiological studies of the cells or tissues of multicellular organisms and are, therefore, better considered together.

Further subdivisions of physiological measurements include investigations of metabolism and growth; the effects of introduced substances on enzyme activity, cells or tissues; and the use of chemical or biochemical methods of estimating biomass.

8.2.1 Physiological Measurements

There is no clear distinction between physiological methods of assessing the effects of polluting substances and conventional toxicological studies on whole organisms, but it is convenient to include here those methods which are related directly to the organism's metabolic processes (e.g. respiration, photosynthesis) and exclude chronic or acute toxicity test procedures and methods for investigating behavioural or reproductive effects of pollutants. General reviews of physiological methods have been made by Bick (1963) and Matulova (1969).

8.2.1.1 Biochemical methods of estimating biomass changes

Estimates of biomass usually involve lengthy sampling procedures and tedious determinations of sample weights or enumeration (e.g. algal counts). In an attempt to overcome the disadvantages of these methods several workers have chosen to use chemical determinations of constituents of organisms as indirect measurements of biomass. A well-known procedure is the use of pigments of algae (e.g. chlorophyll-*a*) but measurements of carbon, nitrogen, ATP (adenosine tri-phosphate) and DNA (deoxyribonucleic acid) and enzyme activity (e.g. dehydrogenase) have also been used. These methods suffer from two basic limitations. First, an assumption has to be made that the relationship between the constituent and total biomass is constant, not only for the organisms under investigation but also for similar organisms for which calibrations may not have been made (Vollenweider, 1969, p. 20). Second, although the substances which are determined are essentially biological in origin, it may not be possible to separate the contribution from living material (at the time of sampling) and that which is derived from the decay of dead organisms. A brief review of the more commonly used biochemical methods is provided below.

8.2.1.1.1 Dehydrogenase activity

This biochemical method was derived from a technique, originally developed as an aid to counting bacterial colonies on plates (Bucksteeg and Thiele, 1958), in which colourless TTC (2,3,4-triphenyl tetrazolium chloride) is reduced to red formazan. The formazan may be extracted in alcohol and determined photometrically against a standard derived from a known suspension of bacterial cells (Bucksteeg and Thiele, 1959; Lenhard, 1965). Jones and Prasad (1969) have drawn attention to the need for care in applying this test since even gamma-irradiated sewage still has a residual activity and in mixed cultures with complex substrates direct relationships may not hold. Kalbe (1968) has similarly stressed the necessity for caution since ferrous iron and sulphide can also reduce TTC to formazan. This problem may arise from many other reducing substances in eutrophic and hypertrophic waters but the method is generally thought to be efficient.

8.2.1.1.2 Adenosine triphosphate, ATP

Adenosine triphosphate is a fundamental component of cellular metabolism and is particularly useful in biomass determinations since it is exclusively associated with living material and may be estimated from the light emitted when luciferin is oxidised (Holm-Hansen and Booth, 1966). The luciferin–luciferase system which is responsible for light production in fireflies may be utilised *in vitro*. In suitable preparations the light emitted is directly proportional to the ATP concentration, and automated instruments are now available commercially. A further advantage of the use of ATP is that its concentration is fairly constant in active cells and the ratio of carbon to ATP is similar for a wide range of microorganisms (algae, bacteria, zooplankton). A useful review of the technique has been made by Stadelmann (1974).

Since ATP is an indicator of metabolic activity it has been used in studies of activated sludge (Patterson *et al.*, 1970; Levin *et al.*, 1975) and proposed as a method for assessing toxicity (Kennicutt, 1980).

8.2.1.1.3 Carbon

Carbon determinations are complicated by the need to distinguish between inorganic and organic sources and to exclude interference from detritus. The normal procedure for particulate carbon is to make conductometric determinations from the carbon dioxide liberated from the sample at high temperature in a stream of oxygen.

The organic content of samples may be determined from the ash-free weight of the sample. The sample is filtered, dried (either in an oven at about 100 °C or under vacuum at 60 °C or by freeze drying) and weighed. Some errors may arise from the problem of residual water. The ash content is determined by burning off the organic matter in a muffle-furnace at 500–600 °C for 4–6 hours. Volatilisation of the inorganic fraction, loss of carbon dioxide from carbonates and chemically bonded water from silicates are potential sources of error here (Crisp, 1971). The organic fraction is then found by weight differences.

The carbon content of algal samples may be estimated from the organic fraction: normally this is between 40 and 60% of the ash-free dry weight (Soeder and Talling, 1969).

8.2.1.1.4 Nitrogen

Total nitrogen may be determined by the Kjeldahl method (see Golterman, 1969) in which the sample is digested by heating in concentrated sulphuric acid with a catalyst (copper sulphate, selenium oxide or mercuric oxides). Alkali is added and the ammonia driven off by steam to a condenser and determined colorimetrically or by titration. Not all the organic nitrogen is converted to ammonia but the difference is small. Estimates of protein content may be made from the protein:nitrogen ratio, which is approximately 6:25. Further details of the method may be found in Barnes (1959) and a critical review of the methods is provided by Giese (1967).

8.2.1.1.5 Chlorophyll–a

Estimates of algal biomass from chlorophyll-*a* content may be made by spectrophotometry or by measurements of fluorescence. This latter technique is particularly useful in field situations where continuous flow measurements may be made from a boat by pumping samples from known depths through the fluorimeter (Lorenzen, 1966). Chlorophylls are excited by blue light and the red fluorescence is measured (Loftus and Carpenter, 1971; Holm-Hansen *et al.*, 1965). Calibration is attained by extraction and spectrophotometric determination.

In determining the chlorophyll-*a* the sample is concentrated by centrifuging or by filtration through a membrane or glass-fibre filter under mild pressure. A small quantity of magnesium carbonate should be added to prevent acidification, which may lead to the formation of phaeophytin, the main chlorophyll degradation product. The chlorophyll may be extracted in acetone or methanol. Other solvents have been used but the equations for calculating chlorophyll-*a* concentrations have been fully worked out only for methanol and acetone. Acetone (90–100% aqueous solution) has been found to work well with marine algae but is less suitable for many freshwater forms (Marker, 1972). Ultrasonic vibration (Talling, 1969) or the use of tissue grinders (Yentsch and

Menzel, 1963) have been recommended for increasing the efficiency and speed of extraction. Acetone extraction is achieved by storage for 24 hours in a refrigerator in darkness but, even so, some pigment changes may occur (Talling, 1969). Methanol (95% usually, though not critical) may sometimes lead to some pigment precipitation (Marker, 1972) but is more efficient than acetone and quicker; boiling methanol is used for up to 30 seconds and then the mixture is allowed to stand from 30 minutes (Talling, 1969) to two hours. The filtered or centrifuged supernatant is then used for spectrophotometric determination to obtain the optical density. For freshwater algae 1 or 4 cm cuvettes are generally used. Readings of optical densities may be taken at one wavelength, usually 663–665 nm (where chlorophyll-*a* is dominant compared with its degradation products), or at three (Strickland and Parsons, 1963) though Talling (1969) questions the value of this trichromatic method. Readings must be corrected for turbidity by subtraction of a 750 nm reading, at which value chlorophyll and phaeophytin absorption is insignificant. Marker (1972) showed that greater precision could be obtained by taking readings in the 410–440 nm range but variations in pH values caused anomalous results.

Estimations of chlorophyll-*a* concentrations are complicated by the presence of pigments such as phaeophytin and other chlorophylls. Chlorophyll-*a* concentrations obtained by spectrophotometry in a given investigation may be compared with results obtained by other methods, e.g. fluoroscopy, and if they correspond there may be no need to correct for phaeophytin; checks should be carried out regularly in a lengthy study. Otherwise, phaeophytin may be differentiated from chlorophyll-*a* by conversion of the chlorophyll-*a* to phaeophytin by acidification, usually with hydrochloric acid, and measuring the optical densities before and after acidification. Moss (1967a,b) differs from most other workers in measuring these relative changes at the blue absorption maximun rather than the red. Marker (1972) has shown that, using methanol as solvent, spectra after acidification are affected by pH and therefore recommends neutralisation after acidification with excess magnesium carbonate or dimethyl aniline.

The necessary equations and extinction coefficients for calculating the concentration of chlorophyll-*a* may be found for acetone in Golterman (1969), Parsons and Strickland (1963), Lorenzen (1967), Talling (1969) and briefly in Biological Methods for the Assessment of Water Quality (1973). With methanol the absorption coefficients have not been fully investigated but Talling and Driver (1963) give a value for chlorophyll-*a*.

8.2.1.2 Metabolic measurements and growth of cells
Biomass cannot be estimated satisfactorily from measurements of metabolic rates, partly because of the large number of variables involved but mainly because one cannot know what the observed rate means in terms of the maximum for any given population. Metabolic measurements have been widely used for testing the effects of exposure of organisms to toxic substances or for estimating microbial autotrophic or heterotrophic growth enhancement caused by effluents or natural waters.

8.2.1.2.1 Heterotrophic growth enhancement
Heterotrophic growth enhancement is usually estimated from measurements of the rate of oxygen utilisation during respiration. The most familiar method is the biochemical oxygen demand (BOD) test in which the substance or effluent under investigation is incubated in darkness for five days at 20 °C and the dissolved oxygen is measured by difference. Serial dilutions are usually prepared with well-oxygenated water in order that some dissolved oxygen, and preferably 50% of the air saturation value, remains at the end of the test. This arbitrary test gives some indication of the relative potential for heterotrophic growth enhancement but needs to be interpreted with caution. The oxygen uptake of the microorganisms present depends on a number of factors. First, the nutritive value of the substance or effluent is important; some substances are readily assimilated and metabolised giving a high BOD in five days while other, more inert, substances may ultimately necessitate greater total oxygen consumption but show little change in the first five days (Fig. 8.1). Second, when the substance is not readily assimilated the BOD may vary quite widely depending on the numbers of microorganisms present which are capable of utilising it as a substrate. Relatively pure organic substrates may lack sufficient essential nutrients to support growth. Some effluents with a high organic content may also contain poisons or inhibitory substances. Third, some oxygen depletion may arise from chemical oxidation, for example of ferrous salts or sulphides. Further complications may arise from nitrification, depending on the presence of nitrifying bacteria and the levels of ammonia and nitrite which are present.

Better estimates of heterotrophic activity may be made using radiochemical uptake methods. The effects of effluent constituents upon heterotrophic production may be investigated by observing the rate of utilisation of carbon-14 (^{14}C), usually from labelled glucose, in culture. These are dynamic methods and complex mathematical procedures are

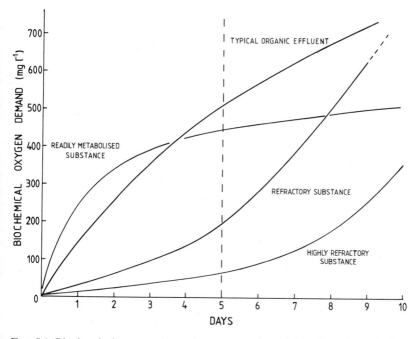

FIG. 8.1 Biochemical oxygen demand over a number of days for a hypothetical range of readily metabolised and refractory substances each having similar ultimate demands.

involved. Hobbie and Wright (1968) used bacterial uptake of ^{14}C to determine heterotrophic ability, the rate of turnover, the population size and the maximum concentration of the substrate. Andrews and Williams (1971) used radiochemical methods to estimate the utilisation of organic compounds by bacteria in seawater. Details of the methodology are given by Hobbie (1969), Vollenweider (1969) and Wetzel (1969).

8.2.1.2.2 *Autotrophic growth enhancement*

Enhancement of microbial autotrophic growth predominantly involves studies of algae although, under strongly anaerobic conditions, bacterial photosynthesis may occur. Methods of algal assay have been developed in order to estimate the potential nutrient qualities of waters or effluents, especially in connection with the problem of eutrophication. As with all assay procedures, the objectives must be clearly defined to ensure meaningful results. The objectives usually include:

(1) the determination of limiting nutrients;
(2) the evaluation of substances for their potential effects on receiving waters (e.g. a proposed new constituent of a sewage effluent);
(3) the prediction of possible effects of changes in water management (e.g. the impoundment of a lowland river water);
(4) the assessment of the efficiency of waste treatment processes by investigations of the nutrient qualities of an effluent, possibly before and after treatment;
(5) the investigation of the effects of toxic substances on algal metabolism (e.g. pesticides or herbicides).

A large number of methods have been used and proposed as 'standard' procedures in order to realise these objectives. Matulova (1969), in a survey of biological tests, identified two approaches:

(1) the exposure of natural populations in the water to be tested or in dilutions of toxic substances;
(2) the exposure of a test organism after the removal of the original population.

An example of the former approach is that of Bott and Brock (1970) who immersed slides with an existing growth of periphyton and followed the growth rate by counting, photomicrography and radioactive-uptake methods. The latter approach is exemplified by Matulova's *Chlamydomonas* technique (Matulova, 1967) and by her use of pure cultures of *Scenedesmus* and *Microcystis* (Matulova, 1969, 1970).

In addition to the choice of natural mixed test populations or isolated, pure cultures the assays may be conducted

(1) under closed, static conditions ('bottle' tests);
(2) in dynamic, continuous-flow chemostats in which a steady state of algal production is achieved in a continuously flowing water;
(3) *in-situ*, in which case the response in an isolated portion of a water body is tested.

(1) *Flask assay or batch-culture method.* In this procedure the test algae are added to the appropriate nutrient medium (control, sample of test water, medium containing a toxic substance etc.) in flasks and their growth rate is monitored over the test period. A curve of the growth rate is plotted and from this the maximum growth rate of the culture is estimated. In the standard American Algal Assay Procedure, or bottle test (AAAP, 1971), the use of one of the following algae, each having different physiological characteristics, is specified:

Selenastrum capricornutum Prinz, an autotrophic green alga, *Microcystis aeruginosa* Kutz, a non-nitrogen fixing blue-green alga, and *Anabaena flos-aquae* (Lyngb.) a nitrogen fixing blue-green alga. This allows estimates to be made of the contributions of different nutrients or choice may be made depending on the known characteristics of the test water.

(2) *Continuous-flow culture or chemostat method* (Fig. 8.2). The main advantage of this method is that the test alga is grown under steady state conditions. The nutrient medium is added continuously in order that its constituents may be maintained at the desired concentrations. Estimates can be made of the standing crop under specified conditions and also of

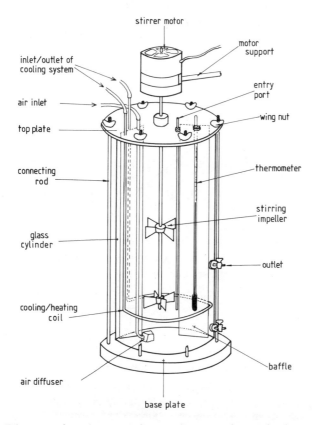

FIG. 8.2 Diagram of a chemostat for continuous culture of microorganisms. Light source used in studies of algal growth not shown.

the required concentration of a nutrient in order to produce a given increase in the rate of growth. The mathematical aspects of this method are described in the Provisional Algal Assay Procedure (1969) and by Lake Tahoe Area Council (1969). Fortunately, the steady-state conditions make the appropriate calculations relatively simple. The method has several other advantages in comparison with static cultures: for example, the effects of phosphorus absorption, a problem mentioned by Ambuhl (in Skulberg, 1966), are minimised when equilibrium is reached, as are also pH changes; gas exchange and mixing are more complete than in the bottle test. On the other hand, the continuous flow test is more difficult to set up and maintain. The methods are further compared by Porcella *et al.* (1970).

Algal cell numbers may be estimated by direct counting under a microscope (or by electronic means) of a known volume of extract or by means of absorbance measurements using a spectrophotometer, colorimeter or reflectometer. Alternatively, the chlorophyll content may be determined after extraction or by fluorimetry (see Section 8.2.1.1.5) or gravimetric determination of a centrifuged sample may be used.

(3) *Enclosed 'in situ' tests.* The principal advantage of *in situ* tests is that the physical conditions, such as light intensity (including self-shading effects) and temperature, as well as concentrations of nutrients are, initially at least, the same as those for a natural population and it is hoped, therefore, that the results can be more readily related to field situations.

Most frequently the *in situ* tests are conducted in 'light' and 'dark' bottles (or, more correctly, clear and opaque bottles) suspended in the water in lakes or in culture chambers placed on the bed of rivers so that production may be estimated from oxygen production by photosynthesis in the light and oxygen consumption by respiration in darkness. The methods suitable for studies on enclosed phytoplankton communities have been reviewed by Soeder and Talling (1969b). The use of ^{14}C techniques in connection with 'light' and 'dark' bottles is described by Goldman *et al.* (1969), and Talling and Fogg (1969) have discussed the possible limitations of these methods and provided a comparison of the oxygen and ^{14}C methods.

Glass bottles, usually flasks of 100–200 ml, are most commonly used in enclosed tests; quartz flasks have been recommended for their superior optical properties but the cost is usually prohibitive for routine work. For comparative assay purposes light absorbance may be less critical and is the same for experimental and control bottles. The 'dark' or opaque bottles are prepared from similar flasks but with their outer surfaces rendered light-proof. Black adhesive tape or paint has been used but enclosing the flasks in aluminium foil seems to be most effective. With small flasks the exposure time must be short—usually only a few hours, and rarely more than one day. In an attempt to overcome this limitation some workers have enclosed larger volumes within tubes, or bags of plastic sheet. For example, Goldman (1962) used 10 m tubes, 0·58 m and 1·16 m in diameter, with wire hoops at intervals to keep the tubes open, which were suspended from floats. Nutrients could be introduced uniformly by filling a hose with the solution, closing both ends and introducing it into the tube. On equilibrating the temperatures the lower stopper is removed and the hose slowly withdrawn when the contents are released throughout the length of the tube.

Large scale *in situ* investigations using isolated columns of lake water have been undertaken by Lund (1972). Tubes of butyl rubber sheet, 45·5 m in diameter and enclosing 16 000 m^3, were placed vertically in the lake supported by their inflatable upper rims (Fig. 8.3). The lower margin has a hollow tube containing a metal chain which assists in sinking into the mud at a depth of approximately 11 m. The tubes have been used to identify the major nutrients and other environmental factors which determine the abundance and seasonal successions of the phytoplankton (Lack and Lund, 1974).

(4) *Problems of methodology and interpretation.* Great care must be exercised in the choice and preparation of apparatus, which must be non-toxic as well as clean. Glass is usually recommended, particularly for stoppers. Every effort must also be made to ensure that there is no contamination of the test organisms by bacteria; such contamination may develop in the course of an assay. To overcome this problem in long-term tests, Braune (1971) has proposed that a short-term test should be substituted in which the productivity of the algae is measured by continuous recording of the oxygen output and consumption and which gives results within a few hours. Gloyna and Espino (in Skulberg, 1966) noted contamination by bacteria but chose to regard these as one of their experimental 'toxins'. Samples are usually sterilised by heat treatment before assay but even then some bacteria may be present and Skulberg (1966) considered that sterilisation of the samples may significantly affect algal growth through the release of some nutrients. The water sample may be filtered before assay, as in the standard American Algal Assay Procedure (AAAP, 1971), but this has been found to remove some nutrients (Porcella *et al.*, 1970) as well as the indigenous algae and is,

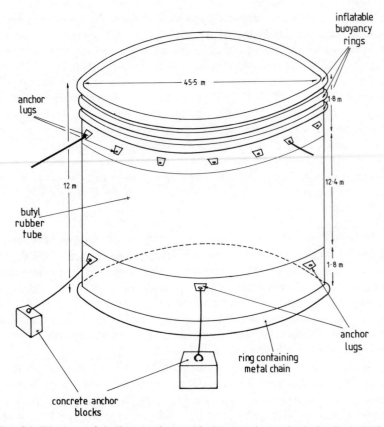

FIG. 8.3 Diagram of the Lund tube used in large-scale manipulative limnological studies. After Lack and Lund (1974).

therefore, a problematical procedure. However, McPhee (1961) has drawn attention to the problem of salt precipitation caused by autoclaving samples and preferred to use vacuum 'Millipore' filtration. Clasen and Bernhardt (1974) chose to use unfiltered samples after releasing the nutrients from the indigenous algae by sterilisation, so that these could be utilised by the test algae, in the hope that underestimations of the nutrient levels could be prevented. Unfiltered, diluted samples were used by Forsberg and Hökervall (1971).

In most algal assays unialgal cultures are used, preferably in the logarithmic phase of growth when they are in good physiological condition (Lake Tahoe Area Council, 1969). Berland *et al.* (1973) have recommended the use of starved algal cultures. Their experiments showed that starved cells were more sensitive when inoculated with seawater and were also more susceptible to toxic substances. Hannan and Patouillet (1970) found algae were generally more sensitive to lead, DDT and an anionic detergent when grown in diluted Burk's culture medium than when grown in the standard medium, and that cells suffering from nutrient deficiencies were less able to adapt to the presence of toxicants. The dimensions of the culture vessels were also found to be significant; with surface-active agents (i.e. the anionic detergent) the toxicity was more pronounced with lower surface/volume ratios, probably as a result of lower adsorption on the walls of the vessels. In the absence of an organic detergent *Chlorella* was found to grow more successfully in very dilute nutrient medium in a larger vessel.

Skulberg (1964) recognised the difficulty of relating laboratory tests to natural conditions and also pointed out that if algae naturally present in the test water are used for assays the results cannot be applied generally. If unialgal cultures are used, however, the interspecific competition which occurs in natural conditions is not reproduced in the laboratory (Oglesby, in Skulberg, 1966). In attempting to overcome this last problem Mitchell (1973) has developed a Microcosm Algal Assay Procedure (MAAP) in which laboratory microcosms were prepared from bottom mud and lake water. The planktonic algal populations were identified and enumerated and the diversity indices (see Section 9.4.1.1) were calculated. The effects on the microcosm of sodium nitriloacetate (NTA) at three different concentrations was observed and the results were compared with those obtained from PAAP (bottle) tests using *Selenastrum capricornutum*. NTA at concentrations up to 200 mg litre^{-1} in no case stimulated or inhibited growth in the bottle test. In the MAAP, NTA had no effect up to concentrations of 80 mg litre^{-1} (a level much higher than the expected environmental ones) but at 200 mg litre^{-1} algal growth was stimulated and the diversity index dropped to near zero. Mitchell concluded that here the effect of interaction between the NTA and the sediment is shown. It is a difficult procedure to employ since skill and subjective judgements are needed in interpretation of the results.

No uniformity exists with regard to the duration of algal assay tests. Shorter times are invaluable for routine survey work both for securing results rapidly and for reducing demands on laboratory facilities. The very short time required for Braune's (1971) test has already been mentioned. Gloyna and Espino (in Skulberg, 1966) conducted five day tests and MacPhee's (1961) tests took up to 38 days, but the most usual time appears to be about 21 days (e.g. Matulova, 1967; Morgan, 1972). Matulova (1967) has drawn attention to the importance of constructing growth curves rather than simply recording the final biomass in order to determine whether the growth curve is merely displaced in time (see Morgan, 1972).

(5) *Examples of the use of autotrophic assays*. Autotrophic assay procedures have three principal uses, namely, limiting nutrient studies, evaluation of toxic materials and the assessment of waste treatment processes.

(i) *Limiting nutrient studies*

It would be out of place to attempt to summarise the immense amount of work on the problems of eutrophication, reviewed by Rohlich (1969), Middlebrooks *et al.* (1969) and Milway (1970). Lund (1972) has summarised the problem with special reference to British waters. He notes that the chemical element most commonly considered to be the major cause of eutrophication is phosphorus though it is obviously an over-simplification to consider a single element without reference to other environmental factors. The other main elements are thought to be carbon and nitrogen.

Cullimore and McCann (1972) determined algal growth rates in their investigations of the effects of carbon, nitrogen and phosphorus. They compared the growth curves of several different species before and after the addition of test substances to the nutrient medium. When, for example, phosphorus was added to the test water and caused no increase in growth the authors were able to conclude that there was already an adequate amount for growth at the sampling site. They believe that their method could be used to determine the effects of such micronutrients as magnesium, iron and molybdenum on algal growth. Samsel *et al.* (1972), in an investigation of the nutrient factors limiting productivity in two ponds in Virginia, were able to set up laboratory microecosystems and from these, and from field experiments, suggested that the ammonium ion was the major single nutrient limiting productivity. McPhee (1961) showed in his study of water from Bear Lake, Idaho, using *Chlorella vulgaris* as an assay organism, that algal growth was limited by nutrient deficiency rather than by the chemical characteristics of the water.

These few examples illustrate some of the variations in limiting nutrients in differing water bodies; they emphasise the necessity of conducting assays for each specific problem, the need for cautious interpretation of experimental results, and the dangers of attempting to generalise from conclusions reached.

(ii) *Evaluation of toxic materials*

Many compounds present in consumer products such as paints, detergents and garden chemicals or commercial agricultural and horticultural pesticides eventually enter watercourses. Their possible effects on primary production may be predicted from algal assays. A very limited selection of examples of some uses of algal assays is given here.

(a) *Detergents*

Much recent concern about the role of phosphorus in promoting eutrophication has stimulated research into possible alternative detergent builders. One possible substitute is sodium citrate, which was investigated by Glooschenko and Moore (1973). They present a review of the literature and the results of assays of different concentrations of citrate, EDTA (a chelating substance) and combination of citrate, EDTA and heavy metals. The interactions were complex and the results dependent upon the chemical characteristics of the test water. These authors emphasised that assay is the only method which will give some indication of the likely effects of discharging citrate into a given water body. An alternative to citrate is NTA (nitrilo-triacetate), a substance tested by Sturm and Payne (1973). Using the AAAP bottle test (AAAP, 1971), they concluded that the stimulating effects of NTA were negligible.

(b) *Pesticides, including insecticides, herbicides and algicides*

Pal and Gopalakrishnan (1968) tested the effects of Algistat, an algicide based on 2,3 dichloro-1,4-naphthoquinone. This was found to kill *Spirogyra* at a concentration of 0·8 ppm, had adverse effects on blue-green algae at 0·5 ppm, and was toxic to fish at 0·7 ppm. It had no effect on *Euglena* even at 1 ppm. Five herbicides (Simazine; Dacthal; Amitrol-T; 2,4-D and 2,4,5-T) were used by Vance and Smith (1969) in assays to determine their toxicity to unialgal cultures of *Scenedesmus quadricauda*, *Chlamydomonas eugametos* and *Chlorella pyrenoidosa*. These showed that Dacthal, Simazine, 2,4-D and 2,4,5-T had no toxic effects up to concentrations of $200 \, \mu g \, ml^{-1}$ ($0·2 \, g \, litre^{-1}$) and Simazine at these concentrations actually increased the growth of *Chlamydomonas*. Amitrol inhibited all the algae at concentrations higher than 150 μg ml^{-1} (1·5 g litre^{-1}). Vance and Drummond (1969) investigated the effects of the pesticides aldrin, dieldrin, endrin and DDT on cultures of *Microcystis aeruginosa*, *Anabaena cylindrica*, *Scenedesmus quadricauda* and *Oedogonium* sp. The algae were highly resistant except for *Microcystis*, which was susceptible to dieldrin, aldrin and endrin. The algae concentrated the pesticides more than 100-fold but Warburg analyses indicated that there were no adverse effects on the respiration rates. Rose and McIntire (1970) demonstrated the accumulation of dieldrin by benthic algae which, after several months, showed increased concentrations up to 30 000 times that of the water. The effects of Aroclor (a polychlorinated biphenyl: PCB), and DDT upon the growth of cultures of *Chlamydomonas reinhardtii* was investigated by Morgan (1972). With Aroclor an initial reduction in growth rate, compared with controls, was observed but by day 22 the growth was parallel. Radiocarbon uptake was reduced and gas chromatographic analysis led Morgan to conclude that Aroclor was concentrated by the *Chlamydomonas*. The relative herbicidal efficiencies of MCPA (4-chloro-2-methylphenoxy acetic acid) and MCPB (4-(4-chloro-2-methylphenoxy) butyric acid) were tested on *Chlamydomonas globosa*, *Chlorella pyrenoidosa* and *Stichococcus bacillaris* by Kirkwood and Fletcher (1970). It was found that MCPB had a greater inhibitory effect on growth, respiration and phosphorus uptake than MCPA and was toxic at a minimal concentration of 0·0025 moles. Finally, attention may be drawn to the description by Gloyna and Espino (in Skulberg, 1966) of detailed studies on the toxicity of numerous compounds to *Chlorella pyrenoidosa*, in which they hoped to find some relationship between molecular structure and toxicity.

(iii) *Assessment of waste treatment*
Waste treatment efficiency may be measured by bioassay. Forsberg and Hökervall (1971, 1972a,b) used an algal growth potential test (AGP test) to monitor sewage effluent at two sites in Sweden. They used five species of algae and took as a measure of AGP the difference between the maximum algal growth in a standard medium with and without the addition of diluted sewage effluent. Algal growth was found to be highest in the mid-week period and there were very low values during the factory holiday period when the phosphate content

of the effluent was very low, probably due to comparatively high aluminium sulphate levels. From this work they were able to suggest that efforts should be made to decrease the fertilising capacity of the mid-week effluent and that a reduction in the vacation period dosage of aluminium sulphate should be made. They also found relationships between total nitrogen, total phosphorus, BOD_5 and algal volume.

Clasen and Bernhardt (1974) used continuous flow cultures of five species of algae to measure algal growth in water taken from different sampling points of the Wahnbach River/Reservoir system. Their pilot study of a process to eliminate phosphates and suspended solids from the river, which is the main tributary of the reservoir, was intended to determine whether this process would affect algal growth in the reservoir. The workers were able to show that algal growth decreased after treatment of the water in the pilot plant. Water from the reservoir supported less algal growth than untreated river water but nevertheless they were able to estimate a possible 20% decrease in algal growth in the reservoir if all the waters were treated. They showed the main decrease to be due to phosphate removal but iron removal was also significant, more so with some species than others.

Algal assays of polluted river water were made by Cain *et al.* (1979) in order to compare the results with those from routine chemical analyses. Their studies showed that algal assays were as efficient as the chemical methods in differentiating between sites; strongest correlation was observed with ortho-phosphate levels. These authors draw attention to the advantages of direct assay by pointing out that it is still necessary to interpret the significance of chemical analyses when attempting to predict the growth enhancement potential of a water. Often this may present difficulties when, for example, phosphates increase but nitrate levels do not. It is then necessary to measure the actual growth potential by algal assay. They also noted that enhanced algal growth, as indicated by assays, does not actually identify the cause. One must, therefore, have recourse to chemical analysis in order to identify the remedial action which may be required.

8.2.1.2.3 Heterotrophic–autotrophic comparisons

Three tests have been developed with particular reference to pollution assays in which bacteria and algae are used as 'heterotrophic–autotrophic comparators' to assess the relative levels of organic and mineral nutrients in effluents or receiving waters.

(a) *A–Z test (assimilation-/Zehrungstest), Knöpp (1961)*
In the 'A' procedure a batch of Winkler bottles of test solution inoculated with an 'impure culture' of Chlorococcales is placed, along with a batch of uninoculated control solutions, in a light-box for 24 hours. The percentage change in oxygen production is computed from the differences between the two batches.

In the 'Z' procedure, peptone solution is added to samples in Winkler bottles and after dark incubation for 24 hours the oxygen consumption is measured to assess the effect on bacterial heterotrophic activity.

Criticisms of this test stem mainly from the use of a general inoculum (which presumably is meant to be representative of a normal flora) and it is suggested that the use of unialgal culture would give better standardisation. It would probably be a useful, rapid test for screening effluents for potential nutrient or toxic properties.

(b) *Biomass–titre procedure, Bringmann and Kühn (1958, 1962)*
The biomass–titre procedure employs the alga *Scenedesmus quadricauda* as an autotrophic indicator of eutrophication potential and *Escherichia coli* as the heterotrophic indicator. The latter is grown in a sterilised and filtered sample with the addition of glucose and inorganic nutrients to make organic nitrogen the limiting factor. In both cases the growth of the cultures is measured nephelometrically against silicon dioxide standards.

(c) *Cell-multiplication inhibition test (Bringmann and Kühn, 1980)*
This test is a development of the biomass-titre procedure but utilises three organisms, a bacterium (*Pseudomonas putida*), an alga (*Scenedesmus quadricauda*) and a flagellate (*Entosiphon sulcatum*), and the rate of growth in specified culture media is used to assess the threshold of toxicity. In practice, a toxic material will tend to inhibit cell multiplication so that after a specified period (seven days) the density of cells will be less than in the controls and the degree of inhibition in a series of dilutions of the toxicant may be compared in order to determine the threshold. The density of bacterial and algal cells is estimated by measuring the extinction of light at 436 and 578 nm, respectively. The numbers of flagellates are estimated by means of an electronic cell-counter.

8.2.1.3 Bacterial mutagenicity tests
These tests are used for relatively rapid screening of substances to detect

those which are mutagenic (mutation-causing) and, therefore, potentially carcinogenic (cancer-producing).

In the Ames test (Ames *et al.*, 1973, 1975) mutant (His −) strains of *Salmonella typhimurium*, lacking the ability to synthesise an essential nutrient, the amino acid histidine, and which can only grow in media in which histidine is provided, are used. Mutagens which cause changes in their genetic material may induce reversion to the normal (His +) condition and thus facilitate growth in a histidine-deficient medium whereupon the change can be readily detected. Spontaneous reversion mutations occur but the extent of this is low and can be determined in experimental controls.

Some substances are not intrinsically mutagenic or carcinogenic but only become so when modified by metabolic processes. This may be simulated in the Ames test by adding mammalian liver homogenate (rat, S-9) to the preparation. Further refinements of the test include the use of bacterial strains in which the ability to repair the genetic material (DNA) is impaired, thereby increasing the probability that mutation will occur, and also by using strains with cell-wall defects which allow easier penetration by large molecules. Strains of histidine-deficient *S. typhimurium* have been selected in which the actual genetic structural fault is known and thus the mode of action of the mutagen can be determined.

Other bacteria may be used in screening tests (for example, *Escherichia coli* and *Bacillus subtilis*). Tests of aquatic pollutants (effluents, for example) may be conducted in liquid media in the 'fluctuation test' in which many replicates (usually multiples of 50) are incubated in tubes or wells in a plastic tray. By adding a pH indicator, metabolic activity may be readily seen as a colour change and the number of His + tubes or cells counted.

Recently, fish liver homogenates, prepared from bream (*Abramis brama*) naturally exposed to potential mutagens, have been used instead of the standard rat liver preparation (Slooff and Van Kreijl, 1982). Liver homogenates prepared from fish from the River Rhine were found to be equally active to those from the rats, but lake fish homogenates showed no metabolic activation.

8.2.1.4 Tissue-culture assays

Tissue-culture methods have proved invaluable in studies of cell nutrition, genetics, virology and radiobiology and their successful application to a wide range of problems suggests that they may be potentially useful for routine screening of possibly toxic substances. The measurement of the rates of growth of cultured cells could conceivably provide a simple and rapid means of assessing the likely effects of pollutants if consistent relationships between the experimental observations and environmental consequences could be demonstrated. The responses of an isolated tissue, which may lack the homeostatic capabilities of the whole organism, could be misleading and need to be interpreted with great care.

Malcolm *et al.* (1973) used an established bovine ovarian cell line to explore the usefulness of single-cell plating techniques to assess the toxicity of a number of substances including cadmium, copper, zinc, chromium, selenium, nitrilotriacetate (NTA) and rotenone. These authors concluded that the method was sufficiently sensitive to measure differences in toxicity with small changes in concentration and also between chemical species. In addition, sufficient precision and sensitivity were obtained to permit a quantitative assessment of the kinetics of toxic action. In the case of ionic cadmium, no distinct concentration threshold was found for the killing of cells in culture.

Li and Jordan (1969) measured the sensitivity of mouse fibroblast L-cells (spinner culture) to several pesticides and a herbicide. The effects of these substances were measured by cell counts made with an electronic cell-counter; inhibition was marked and concentration-dependent in the case of DDT. Li and Traxler (1972) used a similar method with mercuric chloride which induced cytological changes including marked increase in size, incomplete division and some multi-nucleate cells. Cell respiration was also measured but this was found to be less sensitive to mercuric chloride. The authors concluded that cells used for tests should be in the stationary phase of growth with an initial density of about $2 \times 10^5 \text{ ml}^{-1}$ and without foetal calf serum in the medium. The method might be useful for immediate screening of pollutants as it is simple and gives results in hours rather than days.

Commercially obtainable cells of the fathead minnow (*Pimephales promelas* Raf.) were used by Rachlin and Perlmutter (1968) in tests on the effects of zinc sulphate upon growth and the results were compared with conventionally obtained TL_m values. Cytological changes were difficult to distinguish and so the mitotic index (the ratio of cells in mitosis to the entire population of cells present) was measured after a set exposure time. The 'safe' concentration derived from whole animal studies reduced the mitotic index by half; only when a concentration as low as one-tenth of the TL_M 96 hr concentration for the whole animal was used could no effect on the mitotic index be observed. These workers suggested, therefore, that it might be better to base biologically safe concentration values on mitotic index measurements. However, in later work, Rachlin and Perlmutter (1969) found the opposite effect in a similar investigation using rainbow trout gonad cells in which significant reductions in the mitotic index were observed only at 18 ppm zinc sulphate, whereas the lethal concentration to trout is between 0·5 and 6 ppm. Clearly, further characterisation of the responses of different cell lines to a wide range of toxicants is necessary before the merits of this approach can be evaluated.

8.2.1.5 Haematological methods

Blaxhall and Daisley (1973) described routine haematological methods suitable for fishes including haemoglobin and haematocrit estimations, erythrocyte counts, erythrocyte sedimentation rate, total and differential leucocyte counts and cytochemical staining, and give the ranges and mean values for brown trout (*Salmo trutta* L.). These methods are suggested as a possible means of assessing fish health but it will be necessary first to establish values in health, disease and various stress conditions before the effectiveness of the approach can be ascertained. In addition it will be necessary to relate the information to the experimental conditions; for example, species, age, source of fish, diet, temperature and the sampling and testing procedures.

Haematological investigations are potentially useful for laboratory investigations on sub-lethal effects of toxic substances (Blaxhall, 1972) and field or post-mortem examinations of exposure to critical conditions. For example, Bouck and Ball (1965) have utilised elevated serum protein level, which in mammals is indicative of stress, as a test for exposure to low oxygen concentrations, while Eisler and Edmunds (1966) have used ionic and cholesterol blood levels. Haematocrit measurements may be used to measure stress (Sniesko, 1960) but may be less sensitive than other methods and show a variable response (Sprague, 1971). Schniffman and Fromm (1959) measured haematocrits of both splenecto-mised and intact rainbow trout exposed to 20 mg litre^{-1} of chromium and found them to be significantly higher than the controls. Olson *et al.* (1973) have demonstrated a high affinity for methyl mercury by eryth-rocytes of rainbow trout *in vitro*. Finally, Morgan *et al.* (1973), studying the effects of Baltimore Harbour water on two fish species, found no significant trends in erythrocyte counts or serum protein concentrations. There were significant changes in leucocytes but the differential counts which were required to assess these changes were regarded as too difficult and time-consuming for general use.

8.2.1.6 Enzyme analysis
Some toxicants act at subcellular levels through the inhibition of
enzymes, and the experimental determination of enzyme activity is,
therefore, a potentially useful bioassay procedure. Considerable effort has
been devoted to studying the effects of pesticides on acetylcholinesterase
activity and this is considered separately in Section 8.2.1.7.

Jackim *et al.* (1970) investigated the effects of the metals beryllium,
cadmium, copper, lead, mercury and silver on five liver enzymes (liver
alkaline phosphatase, xanthine oxidase, catalase, ribonuclease and acid
phosphatase) in the killifish *Fundulus heteroclitus* in an attempt to study
the feasibility of biochemical autopsy for the diagnosis of sub-lethal
metal poisoning and also to understand the mechanisms involved.
Known concentrations of the metal salts were added to jars containing
the fish and after 96 hours the surviving fish were removed, decapitated
and their livers removed for bioassay. It was concluded that catalase
activity showed the greatest change and might be useful in diagnosing
metal poisoning and also that xanthine oxidase activity might be useful
in distinguishing between poisoning by copper and by other metals.

Jackim (1973) investigated the possibility of using the measurement of
δ-aminolevulinate dehydrase (ALA-D) activity to provide an index of the
response to lead in water in the mummichog (*Fundulus heteroclitus*) and
the winter flounder (*Pseudopleuronectes americanus*). Liver homogenates
were prepared and ALA-D activity determined by the method of
Bonsignore *et al.* (1965). A comparison was first made with other heavy
metals to determine the specificity of ALA-D inhibition. It was found
that lead produced significantly greater inhibition of ALA-D than copper
while cadmium, zinc and silver caused an increase in activity. It seems,
therefore, that ALA-D depression is a characteristic response to lead
exposure although it is not specific. Unexposed fishes of the same species
should be used as controls since there may be other factors which cause
variability in enzyme activity which are not related to a given pollutant.

The effects of cadmium and copper on the oxidation of lactate by gill
tissue of rainbow trout were investigated by Bilinski and Jonas (1973).
After exposure to various concentrations of copper and cadmium the fish
were killed and the gill arches were removed. The oxidative activity of
the gill tissue was determined by measurements of the formation of
$^{14}CO_2$ from sodium-lactate-3-^{14}C. It was concluded that this test is only
valid at very high exposure levels of copper and cadmium. Often no
measurable change was observed at lethal concentrations of the metals
but the method might be applicable to other toxicants.

In tests on the effects of Baltimore Harbour water on two fish species, Morgan *et al.* (1973) found that serum lactate dehydrogenase activity increased after prolonged exposure, liver ribonuclease activity showed no significant trends, and liver catalase activity decreased.

8.2.1.7 Acetylcholinesterase (AChE) inhibition in fish brains
Weiss (1965) suggested that the inhibition of brain acetylcholinesterase might be used to evaluate the effects of certain pesticides in fish. Organophosphorus pesticides are known to be concentrated in fish brain tissue (possibly bound by sphingolipids) and they inhibit the action of AChE. All the current methods involve calculation of the percentage inhibition of the rate at which AChE in brain tissue homogenates hydrolyses specific substrates when compared with controls which have not been exposed to the pollutant or test substance. The technique has been used both in the detection of past exposure and also as a screening procedure.

(a) Past exposure:
Work by Holland *et al.* (1967), Williams and Sova (1966) and Gibson *et al.* (1969) gave anomalous results. Many workers, however, seemed to think that the method was worth further investigation. Hogan (1970) suggested that water temperature was a significant factor. He used bluegills (*Lepomis macrochirus*) collected over a period of 14 months in which temperatures were carefully monitored. Samples of 18 fish were prepared for each month and brain homogenates were used in estimations of AChE activity using the manometric method of Knowles and Casida (1966). The activity was found to vary directly with water temperature. He noted that his results do not agree with those of Baslow and Nigrelli (1964) who subjected killifish (*Fundulus heteroclitus*) to thermal stress, but they do agree with the results obtained by Hazel (1969). He concluded that changes in water temperature may be a source of considerable error when bluegill brain AChE is used for the biological assessment of pollution by anticholinesterase agents. Coppage (1971) believes that the anomalous results discussed by Gibson *et al.* (1969) may be due to the use of spectrophotometric methods for measuring AChE activity. Working with sheepshead minnow (*Cyprinodon variegatus*) brain homogenates he used a 'pH-stat' for the estimation of AChE activity. This apparatus combines the functions of pH control, temperature control, reagent delivery and mixing and titrant volume recording. The reasons for its

superiority over other methods are discussed in his paper. The decrease in activity of AChE was tested with one complete inhibitor and four organophosphorus pesticides. His results showed that the presence of such a pesticide may be detected from the 'pH-stat' assay but not the degree of toxicity, as the *in vitro* inhibition does not reflect the toxicity of the pesticide. This is probably because, *in vivo*, AChE inhibition will depend on factors such as the rate of pesticide penetration into the brain and the rate of its conversion into less toxic compounds. Coppage (1971) goes on to suggest how his work may be adapted for *in vivo* studies after first establishing normal AChE activity by taking fish brain samples over twelve months. In a later paper Coppage (1972) showed that death occurs in sheepshead minnows when AChE activity is reduced beyond a certain, critical level. In exposures of up to 72 hours the 'lethal AChE level' was found to be approximately 18% of normal; below this level death occurred or would occur within 24 hours. He used several different pesticides and concluded that his results would probably apply to organophosphorus pesticides in general. The AChE activity must be below 87% of normal activity to indicate exposure since large standard deviations were observed. Coppage concluded that, with the species he tested, the low levels of AChE activity caused by sub-lethal concentrations of the pesticides indicated that pollution by these substances could readily be detected by assaying brain AChE by his method. In the work of Morgan *et al.* (1973) mentioned above, inhibition of AChE activity, estimated by spectrophotometry, was observed in fishes exposed to Baltimore Harbour water for about 30 days.

(b) Screening procedure:
Cranmer and Peoples (1973) used a chromatographic method of measuring AChE activity in their procedure for measuring small amounts of anticholinesterase pesticides in air or water. They determined the percentage inhibition of AChE by samples of air or water after obtaining a standard curve using known inhibitors. Since this is basically a screening method the results may bear little relation to the effects of the pesticides on the biota within the natural environment. These workers do state, however, that from comparisons of the effectiveness of paraoxon and maroxon as inhibitors *in vitro* and the results obtained from intravenous administration of these substances, a similar order of difference in their inhibitory effectiveness was observed. Similar comparisons were made with paraoxon and carbaryl.

8.2.1.8 Tissue analysis and histology

Environmental levels of a few toxicants have been widely investigated and may be readily related to observed effects on the biota. Some substances, notably organo-chlorine pesticide residues, are known to enter organisms both directly from the environment and also through food chains. With these substances, the levels in water may not indicate the true level at which there is risk to the biota, and Lloyd (1972) doubts whether satisfactory criteria based on environmental concentrations can be established for many pesticides.

The residue levels in tissues, for example of fish that have been exposed to pesticides under experimental conditions, are likely to prove useful in assessing the significance of pesticide concentrations in wild fish. Dead fish recovered from streams following poisoning by DDT, dieldrin and BHC have been found to contain concentrations of these substances similar to those in fish which died under experimental exposures (Holden, 1972). In addition to the chemical determination of the concentration of substances within tissues, it is possible to make some assessments of the effects of some toxicants by histological investigations of important organs. Since the pathology and determination of the levels of toxicants present are likely to be investigated simultaneously it is convenient to consider them together in this section.

The two approaches have been linked together by Hinton *et al.* (1973) in a brief review of histological and histochemical methods for assessing the effects of pollutants. They identified two main approaches:

(1) Early warning of sub-lethal effects of toxicants from observations of the pathology of tissues.
(2) Post-mortem examination of tissues in order to identify the possible causative agent.

Depression of respiration is a common attribute of many poisons and these authors draw attention to the value of examinations of respiratory structures. In fish, the state of mucous glands or the histology of respiratory epithelia may indicate early stages of damage. The thickening of epithelial cells and fusion of gill lamellae on exposure to suspended solids (Herbert and Merkens, 1961), the stripping of the epithelium from secondary lamellae on exposure to phenol (Mitrovic *et al.*, 1968) in rainbow trout, and the sloughing of epithelium and gill necrosis in fish after prolonged exposure to Baltimore harbour water (Morgan *et al.*, 1973) are typical examples. Sometimes damage to nervous tissue, for example by copper, can be revealed by selective staining. Liver tissue may reveal important changes caused by exposure to pollutants by the accumulation of fat (hepatic liposis), necrosis of the cells adjacent to the central veins, increased connective tissue (hepatic fibrosis) or by presence of tumours (hepatomas). Changes in some enzyme systems (see Section 8.2.1.5) may be demonstrated by histochemical methods. Examples of some of these effects in fish exposed to methyl mercury are given by Hinton *et al.* (1973).

In a method of direct tissue analysis (Olson and Fromm, 1973) the tissue is scanned by means of an electron microprobe. This was used to detect mercury in gills where high concentrations were located in cartilage. An important advantage of this method is that little ion diffusion occurs during the preparation of the tissues and the position of elements such as sodium, potassium, chlorine and sulphur can be determined. The method is unsuitable for volatile substances but may prove useful in studies of active ion transport systems in gills and investigations of heavy metal pollutants. For example, in an investigation with rainbow trout, Olson *et al.* (1973) deduced that the gill is the primary uptake pathway for mercury in non-feeding trout, that the skin is relatively impermeable and that methyl mercury enters more easily than inorganic mercury (which does not seem to require methylation before entry). The ratio of zinc in gills and opercular bones of trout when determined at autopsy gives an indication of the dose to which they were exposed (Mount, 1964) but this technique is not, apparently, suitable for post-mortem determination of the intensity of exposure to copper (Fozzard, 1978).

An example of the value of tissue analysis is provided by the work of Silbergeld (1973) who investigated the effects of thermal stress on darters (*Etheostoma nigrum*) which had been exposed to sub-lethal levels of dieldrin. Measurements were made of growth rate, feeding rate, blood glucose levels, oxygen consumption, opercular movement rate, body lipid content and liver condition and histology. Changes in all parameters occurred during the first 15 days' exposure: oxygen consumption, growth, feeding and lipid content were all depressed during this period but adaptation occurred later. However, irreversible liver damage was observed on the fifth day of exposure and blood glucose remained elevated in exposed fish.

Most work on the analysis of toxicants and, especially, of pesticides accumulated in tissues in aquatic organisms has been undertaken on fish. A useful review has been provided by Holden (1972) and the use of pesticides in freshwater is reviewed by Muirhead-Thomson (1971). In earlier discussions of the value of bioaccumulative indicators (Chapter 3), the principles were illustrated by reference to metal uptake by plants

(Section 3.6.3.1) and to the accumulation of metals and organic compounds by fish. In the selected examples discussed below greater attention has been paid to those studies which have involved the use of macroinvertebrates since these appear to have received less attention elsewhere.

An example of early work on pesticide residues is provided in Hunt and Bischoff (1960) who reported the results of their investigation into the effects of DDD on the wildlife of Clear Lake where insecticide had been used to control gnats (*Chaoborus astictopus*). Applications of insecticide were made in 1949, 1954 and 1957, the last two resulting in a concentration of active insecticide of 1 part to 50 million parts water. Dead grebes were reported from December 1954 onwards. Collections of birds and fish were made for analysis from 1958. Fish were usually skinned and filleted and DDD in their tissues was estimated quantitatively by a modification of the Schecter-Haller method for colour development; after removal of interfering substances the presence of DDD was determined by specific colour development. Analysis of samples of nine fish species showed that the highest concentration of DDD was found in fat tissue; the amount ranged from 40 ppm in common carp (*Cyprinus carpio*) to 2500 ppm in brown bullhead (*Ictalurus nebulosus*). It was concluded that the death of grebes was caused by feeding on contaminated fish. The accumulation of DDD in fish flesh could be correlated with the food habits exhibited in normal food chain relationships. In general, smaller fish accumulated relatively less DDD than larger fish. Plankton eaters accumulated less toxic material than carnivorous fish species of the same size.

Polychlorinated biphenyls (PCBs), marketed by Monsanto under the name 'Aroclor', have been found in the environment since 1966 (Gustafson, 1970). They have relatively low acute toxicity to fish (see Table 7.43) but are now a cause for concern because of the build-up of residues, particularly in macroinvertebrates which on the whole are tolerant to relatively high concentrations of PCBs. Sanders and Chandler (1972) conducted a study to determine the rate of accumulation and biological magnification of ^{36}Cl-labelled Aroclor® 1254 from water by eight insect species. Labelled Aroclor was prepared by neutron irradiation in a nuclear reactor and purified by silicic acid chromatography. A continuous-flow bioassay system was used with a 48-hour adjustment period for the test invertebrates. Homogenates were prepared from samples and the Aroclor residue concentrations measured and computed on a whole-body dry-weight basis. Aroclor concentrations in the water were monitored radiometrically. Uptake and magnification were very rapid; for example, *Daphnia magna* exhibited total body concentration some 48 000-fold greater than water containing 1.1 ± 0.2 ppb Aroclor after 1–4 days' exposure. Late instars of the mosquito (*Culex tarsalis*) continued accumulation until pupation; there was no apparent immediate toxicity but adults often failed to emerge. Equilibrium was reached in some experiments, for example in the shrimp *Gammarus pseudolimnaeus* exposed to water containing 1.6 ± 0.1 ppb Aroclor, concentration reached a maximum after 14 days when total body residues were 'magnified' 27 500 times.

In a similar study Johnson et al. (1971) investigated the magnification and degradation of labelled DDT and aldrin. There was rapid uptake and magnification of both insecticides by all the invertebrates tested. The greatest degree of magnification was by *Daphnia magna* and the mosquito larva *Culex pipiens* which showed residue levels more than 10^5 times that of the concentration in the test water.

The accumulation of pesticides and other residues in invertebrates may have consequences for the interpretation of data on residues in fish. For example, Meeks (1968) showed that DDT levels declined after spraying but invertebrates were a source of pesticide for predators entering the area. Some degradation products may exacerbate this problem. For example, DDE, a stable metabolite of DDT, has a half-life exceeding ten years (Muirhead-Thomson, 1971) and thus macroinvertebrates may accumulate residues for long periods which are subsequently redistributed in trophic pyramids.

Exposure to combinations of pesticides was studied by Mayer et al. (1970). Rainbow trout were dosed with combinations of DDT, methoxychlor and dieldrin. After 15 days the fish were killed and the fat body analysed for lipid content and residual insecticides. This revealed insecticide interactions in which the levels of DDT and DDE increased when DDT and dieldrin were combined but dieldrin storage was reduced by the presence of DDT or methoxychlor. Methoxychlor storage was reduced by the addition of DDT. These interactions clearly have implications of data interpretation.

Examples of tissue analysis for heavy metals are to be found in the work of Mount (1964, 1967). Exposure to sub-lethal concentrations of zinc resulted in similar rates of accumulation in gill tissues and opercular bones of bluegills (*Lepomis macrochirus*), but at acutely toxic levels the ratio of zinc in gill and bone tissue increases up to 100-fold and may be used to indicate zinc-caused mortality at autopsy. Five fish species used in experimental exposures exhibited higher gill to bone ratios than field samples of some 22 species. Only carp (*Cyprinus carpio*) were considered to be unsuitable for this technique although there were no obvious reasons why this species should have such a highly variable and, generally, usually high background level of zinc in its gills (Mount, 1964). Cadmium accumulated in living bluegills to levels as high as 100 µg g^{-1} (dry weight) and in the gills of dead catfish up to 1000 µg g^{-1} were found although there was little accumulation in muscle (Mount, 1967).

Post-mortem examination of tissues does not always provide reliable evidence of past exposure levels of some toxicants. Brungs *et al.* (1973) studied the accumulation of copper by the brown bullhead (*Ictalurus*). Copper determinations were made for gill, operculum, kidney, liver, blood plasma and erythrocytes but no differences were found between dead and surviving fish so that a suitable autopsy procedure could not be derived for copper. In previous work with zinc, cadmium, endrin and parathion it had been shown that tissue concentrations of these substances in dead and surviving fish at a given external concentration of toxicant differed, so that an autopsy method was available for determining whether the cause of death in a particular fishkill was one of these toxicants.

The size of fish can influence the levels of accumulation of toxic substances. Size should always be stipulated in addition to the normal data on the toxicity characteristics, especially with pesticides when results are reported. For example, Murphy (1971) extracted and estimated DDT residues from mosquito fish (*Gambusia affinis*) and concluded that the smallest fish were most efficient in absorbing DDT from water. Thus their relative DDT content was higher than expected since, if most of the DDT normally comes from the food chain (Macek and Korn, 1970), younger fish should contain less because they occupy a lower position.

8.2.1.9 Aquatic plants as accumulators (see also Section 3.6.3.1)
Many aquatic plants are known to accumulate and concentrate pollutants and, in particular, heavy metals. This property can be used to estimate environmental levels provided that a relationship between plant tissue levels and levels in water or substrate can be established. Fortunately, many aquatic macrophytes can tolerate exposure to polluting concentrations of the metals (McLean and Jones, 1975; Benson-Evans and Williams, 1976; Say *et al.* 1981) and their tissues are convenient for metal analyses.

Deitz (1973) investigated the accumulation of heavy metals in submerged plants and calculated 'enrichment factors', that is the ratio of the concentration in the plants (wet weight) to that of the surrounding water. Although there were differences in the orders of magnitude between taxonomic groups (for example bryophytes (mosses) concentrated lead to a greater degree than flowering macrophytes), and between metals (manganese and zinc were concentrated most), there was surprisingly little variation in the concentration factors for a given metal and plant species under differing environmental levels. For this reason it was concluded that the use of analyses of water plants and knowledge of the appropriate concentration factor could be used to estimate the average concentration in the water. Confirmation of the validity of the method has been obtained by Nakada *et al.* (1979) for the macrophyte *Elodea nutallii*, exposed to cadmium, lead and copper. The method may also be applied to algal blooms (Trollope and Evans 1976) and the red alga *Lemanea* with reference to zinc, copper and lead (Harding and Whitton, 1981). Aquatic bryophytes seem particularly suitable for monitoring heavy metal contamination (Empain, 1976a,b): concentrations in actively growing shoot tips correlate significantly with ambient levels (Say *et al.*, 1981; Whitton *et al.*, 1982).

8.2.2 Whole Organism Responses

In the preceding sections methods developed for investigating the effects of pollutants at sub-cellular, cellular and tissue levels have been considered together since, although the distinction between whole-organism and physiological responses is somewhat artificial, the methodology involved tends to divide conveniently. In this section the responses of whole organisms (other than microorganisms) to pollutants will be considered. The most important responses are death, disturbed physiology, reproductive impairment and aberrant behaviour. Many of the fundamental principles relating to these investigations are, of course, much the same whatever the organism used: in order to avoid repetition they will be dealt with in Section 8.2.2.1 on fish toxicity tests.

A useful consideration of the philosophy of experimental procedures for evaluating the effects of chemical substances on aquatic life, together with outline protocols for laboratory and field assessments, has been provided by Cairns and Dickson (1978). Their bibliography contains many North American references for standard methods or recommended procedures for studying the effects of pollutants on different groups of organisms.

8.2.2.1 Toxicity tests—fish

The initial objective in measuring toxicity is to define the concentrations (and their distribution) at which a poison is capable of producing some selected harmful response (usually death since this is the most clear-cut response) in a population of organisms under controlled conditions (duration and nature) of exposure (Brown, 1973).

The effect of a toxicant is usually dependent upon its concentration and the time of exposure. These are often, but not necessarily, reciprocal. At low levels of concentration or short duration of exposure the normal homeostatic mechanisms are usually able to cope and no damage is sustained. Increasing concentration or longer exposure may initially cause reversible physiological changes and subsequently some impairment of function. Eventually a critical point is reached where significant changes are induced and even if normal conditions are restored some permanent damage may have been suffered. Ultimately, increasing concentration or period of exposure will produce disease and death (Fig. 8.4). Considerable experimentation is necessary to establish that death is hastened by long exposure to very low concentrations but it is relatively

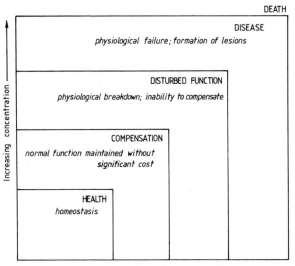

FIG. 8.4 Diagrammatic representation of the effects of increased exposure or concentration of a toxic substance upon an organism. Based on a figure in Lloyd (1972) after Hatch (1962).

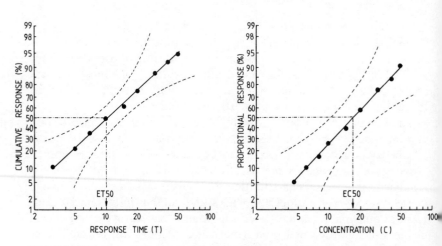

FIG. 8.5 Bioassay results: Left: Quantitative (cumulative) response of a test population to a fixed concentration. Right: Quantal response of nine test populations after a fixed period to a given concentration of the test substance. After Brown (1973).

simple to demonstrate lethal toxicity at high concentrations over short periods and for this reason most acute toxicological investigations have been short-term tests at relatively high concentrations ensuring that death occurs within hours or days.

Two approaches are used: most commonly the quantitative response of survival time is used but more appropriate and useful is the quantal response, i.e. the dosage mortality curve (Fig. 8.5). Comparisons of the toxicity of substances which are made merely on the basis of the period of elapsed time necessary for a response from a proportion (usually half) of the population are not as useful or informative as quantal data in which the responses of individuals within the test population are recorded. From these latter data one can derive the median response times and also their variances, which are essential if one wishes to make statistical assessments of the significance of differing responses. The resulting toxicity curves are shown in Fig. 8.6.

8.2.2.1.1 Terminology (see Table 7.3)

A basic premise of toxicological investigations is that there is some concentration of each toxicant below which no harmful effect on the test organisms will be observed during its normal life-span, but for practical reasons this is usually impossible to determine and normal variability

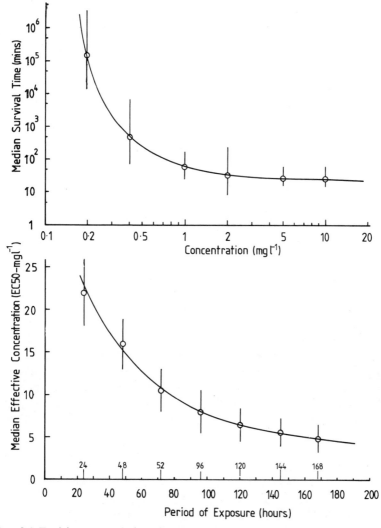

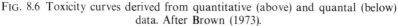

FIG. 8.6 Toxicity curves derived from quantitative (above) and quantal (below) data. After Brown (1973).

in the duration of the life-span would necessitate considerable experimentation which would increase with the subtlety of the induced change. It is also important to recognise that the effect of a poison is an interaction in which the contribution of the organisms is likely to be variable. Therefore, in quoting toxicities it is important to state not only the concentration which induces the effect within a specified time in a given proportion of the population, but also to qualify this by clearly indicating the age or size, sex and condition, where appropriate, and also the environment (temperature, oxygen tension etc.) under which the determination was made.

When the effect is lethal this term should be used since, strictly, 'acute' does not necessarily mean lethal although in common usage it has tended to acquire this meaning. The term 'chronic' has come to mean any non-lethal or 'sub-lethal' response although it should strictly only describe the duration of the effect: a chronic response might, in fact, be death.

A term commonly used in North America is the median tolerance limit (TL_m) which is equivalent to the lethal or effective dose for half the population, i.e. LD_{50} or ED_{50}. The use of the latter terminology is preferred since this has greater versatility in use as any portion of the test population can be used, for example the dose affecting 95% of the population or LD_{95}. Since 'tolerance' has other connotations, especially in the sense of acquired resistance to repeated exposures to a substance, it would seem best to avoid the use of 'median tolerance limit'. Similarly it would be preferable to avoid the terms 'threshold' or 'lethal threshold concentration' since threshold in physiological studies is the minimal stimulus necessary to elicit a response. The terms 'ultimate median tolerance limit' and 'incipient lethal level' are unsatisfactory, not least for their predictive element. The simple terms 'lethal concentration' or 'effective concentration' are self-defining and cannot be improved upon (Brown, 1973). All that is needed is to add the appropriate time factor, for example 2 hour EC_{95} or 48 hour LC_{50}.

8.2.2.1.2 Standardised toxicity tests for fish

Considerable effort has been expended in an attempt to derive a standard procedure for fishes and has resulted in the adoption of a number of routine screening methods (e.g. Doudoroff *et al.*, 1951; American Public Health Association, 1971; Weber, 1973; Alabaster and Abram, 1965; Alabaster, 1970; Lester, 1970). Standardisation of the test procedure is clearly important when comparisons of the toxicities of substances are required since this reduces the effects of variability in physical or chemical water quality, test species, method of exposure (including the rate of renewal of solutions), and perhaps even operator efficiency. A more detailed account of the ways in which standard toxicity tests are conducted is provided in Section 8.2.2.1.3.

Unfortunately, the adoption of standardised methods can induce some inflexibility in outlook, both in the approach to and the interpretation of the results. This danger is probably small for those actively engaged in this work but those who use the results may tend to lose sight of the limitations of the procedure and assume that since the method is used by, or bears the approval of, some statutory body or respected research institution the results obtained are inviolable. When only one approach is recognised the effect can be quite serious. It is evident that the more standardised or restricted the test procedure or the choice of test animal the less applicable the information obtained is likely to become (Brown, 1973). The value of adopting a standard method, or preferably several standard methods, is that some meaningful comparisons may be made on the relative toxicities of several substances under identical conditions to a given species. Lloyd (1972) has recommended that for such comparisons the lethal threshold concentration (i.e. the concentration at which the curve becomes asymptotic with the time axis, see Fig. 8.6) or, using the preferred terminology of section 8.2.2.1.1, the 'Asymptotic LC_{50}', should be used.

The choice of test species is not easy and should preferably be appropriate to the problem under investigation. In Europe, for basic, routine or general work the rainbow trout (*Salmo gairdneri*) has been widely used: the case for its use has been discussed by Herbert (1965) and stems mainly from its generally greater sensitivity than many native species so that standards derived for this species are likely to be more satisfactory for the remainder. This is not always true and more work is progressing on the responses of other fish species (Ball, 1967a,b; Alabaster, 1971). In North America a wider selection of test species has been used (see, for example, the list in Johnson, 1968).

The problem of maintaining large numbers of fish in tests is significant and is especially difficult when using rainbow trout for continuous-flow tests. Alabaster and Abram (1965) concluded that for 10 rainbow trout in a 40-litre tank some several hundred gallons of test solution would be required, and for this reason the use of the smaller exotic harlequin fish (*Rasbora heteromorpha*) was preferred. Smith (1973) has proposed the use of *Jordanella floridae* which is small, requires little space, is easily maintained and can be spawned at any time of the year. It matures in a few weeks and completes a generation in six to eight weeks.

Brown (1973) has drawn attention to the importance of recognising the need for adequate sample size in toxicity testing in view of the large variability of responses shown. For example, an EC_{50} determined from only five fish could, in a large population, affect between 0 and 100%

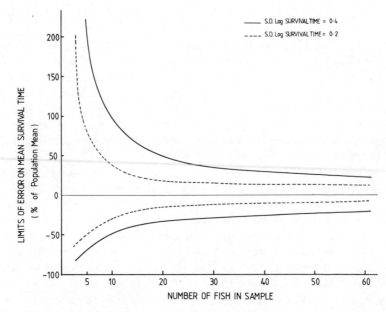

FIG. 8.7 Relationship between sample size and error of mean survival time for a population of rainbow trout. After Alabaster and Abram (1965).

while a harmful effect occurring with a probability of 0·02 (1 in 50) in a larger population would only be detectable with 95% confidence by using a test sample of 150 fishes. A useful graph showing the relationship between sample size and the error of estimates of the mean period of survival has been provided by Alabaster and Abram (1965) (see Fig. 8.7). The standard error of LC_{50} estimates and sample size has been considered theoretically and empirically by Jensen (1972). Increasing sample size significantly reduced the standard error of LC_{50} until the sample numbered about 30 fish.

In addition to the problem of numbers of fish needed, the duration of the tests, often days rather than hours, may be inconvenient. One common response of fish to poison is 'turnover', that is the loss of equilibrium and the inability to control their normal swimming position. Shaw (1979) compared median survival time, time to 10% mortality and 10% turnover time (time for 10% of the sample to turn over) and found that, for copper and cyanide, 10% turnover time (or TO 10) was shorter than the other measures at the concentrations tested. He concluded that this test may be useful in rapid detection of pollution by many hazardous chemicals.

8.2.2.1.3 *Methods used in standard toxicity tests*

Standard toxicity procedures have been established in several countries (Alabaster and Lloyd, 1980) and proposals for international standard methods are well advanced. Although these methods differ in detail there are several basic principles which are common to most standard test procedures. The underlying philosophy of the development of standard methods is to reduce sources of variability so that, hopefully, similar results will be obtained wherever the tests are performed. This aspect may be particularly important in an international context: manufacturers who wish to be sure that their products meet the required standards in other countries may prefer to conduct equivalent tests in their own laboratories.

There are several features of tests which may be standardised in order to reduce variability in results. These include the choice of test species (or even the strain of a cultivated species), the test apparatus, the dilution water, the conditions under which the test is conducted (for example, temperature and oxygen concentration) and the means by which the composition of the test solution, or suspension, is maintained.

1 *Test species*

The ideal test species should be available throughout the year, easily kept and maintained in the laboratory, and show representative sensitivity to toxic substances. This last aspect is, perhaps, unrealistic since it is unlikely that any one species will exhibit responses which are shared by other members of its taxonomic group. However, a species which is resilient and insensitive to most toxic substances is unlikely to be useful in establishing safety standards for the protection of other species. Small test species have the advantage that small volumes of test solution are required and adequate numbers (usually about 10) can be included in each test batch. This reduces the risk of spurious results from wide variability in the response of a few individuals. Some organisms, for example *Daphnia*, can be cloned; that is, all the individuals used in tests may be raised pathenogenetically from a single female and, therefore, have identical genetic constitution.

2 *Test solution*

The composition of dilution water used in tests can have considerable influence on the results. For example, it is now well established that metals are often more toxic in soft water than in hard water. As the composition of water varies widely between localities a reconstituted water may sometimes be used (Alabaster and Abram, 1965). Natural waters (springs, borehole water) have the advantage that they are unlikely to vary much in composition and may be free of contaminants. Mains water may be satisfactory provided that measures are taken to remove the chlorine which is normally present.

The test substance should be dissolved in test water and the concentration expressed gravimetrically (e.g. mg litre^{-1}) or volumetrically (ml litre^{-1}). The vague ratio 'parts per million' (ppm) is ambiguous and should be avoided. Substances which are relatively insoluble in water may first be dissolved in innocuous organic solvents such as acetone or methanol and then made up in water. It is necessary to include the organic solvent in the control to ensure that its effect can be discounted. It has been argued, quite reasonably, that if a substance is largely insoluble in water it should only be tested at concentrations up to the limit of its solubility since these are more relevant to those which might be encountered in the environment at large. Substances which are to be tested as emulsions or suspensions must be subjected to constant mixing in order to maintain them, a factor which might influence the behaviour and survival of the test organisms. Volatile test substances may be lost from solution if aeration applied for the benefit of the test organisms is excessive.

3 *Apparatus*

This will depend on the type of test and the procedure to be adopted in order to maintain the composition of the test solution. Tests may be 'static', in which the solution is not changed or only renewed at long intervals (which, depending on the frequency, might better be regarded as 'semi-static'), or 'flow through', in which the solution is replenished continually throughout the test. The rate of replacement will depend on the size and number of fish in each batch, their oxygen requirements and the characteristics of the substance under test, but 2·5 litre g^{-1} fish per day is thought to be generally adequate. By choosing an appropriate vessel size the solution will be virtually completely changed each day.

(a) *Static tests* require several tanks or vessels of adequate size for the species to be tested. At least six tanks, each of 25 litre capacity, would be needed to hold batches of five fish weighing about 10 g each. Of these, five tanks would be devoted to a range of concentrations (see below) of the substance under test and the remaining tank would serve as a control.

(b) *Flow-through tests* require more complex apparatus than static tests in order to provide for continuous renewal of the test solution at a uniform rate. The apparatus used in the United Kingdom by the Ministry of Agriculture, Fisheries and Food (Alabaster and Abram, 1965) is illustrated in Fig. 8.8. A proportional diluter, based on an earlier serial diluter (Mount and Warner, 1965), provides simultaneous dilution of a toxicant solution to give a series of test concentrations which are relatively close, up to 90% of the preceding concentration, if necessary (Mount and Brungs, 1967). However, this system is unsuitable for providing a series of

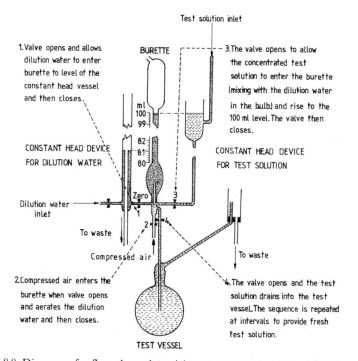

FIG. 8.8 Diagram of a flow-through toxicity test apparatus. After Alabaster and Abram (1965). The numbers indicate the sequence of operation.

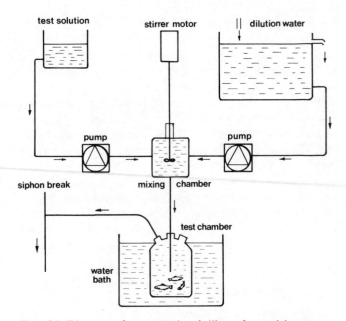

FIG. 8.9 Diagram of a proportional diluter for toxicity tests.

solutions in which the consecutive concentrations differ by a factor greater than 50%, for example in preliminary tests where one may use a series having dilutions such as 1, 0·1, 0·01 etc. A simpler arrangement is to use peristaltic or metering pumps to deliver appropriate quantities of test substance solution and dilution water (or previously prepared and mixed solutions) to the test vessel (Fig. 8.9). A simple system for serial dilutions (the 'Michigan diluter') utilises siphons and gravity feed to provide a range of duplicated concentrations from 0 to 100% of a stock solution in 8 to 10% steps (Wuerthele *et al.*, 1973). The arrangement is shown diagrammatically in Fig. 8.10.

4 *Procedure*

For substances with totally unknown toxicity it will be necessary to perform a preliminary 'sighting' test in order to establish the range of concentrations to be used in a full test. This is done in order to avoid wasting time and resources in a test having no concentrations at which a response would be observed. For example, the preliminary test may be made with concentrations of the substance of 0·1, 1, 10, 100 and

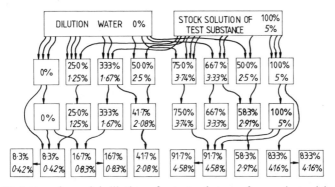

| DILUTION WATER 0% | | | | STOCK SOLUTION OF TEST SUBSTANCE | 100% 5% |

FIG. 8.10 System for serial dilution of a test substance for use in toxicity tests. After Wuerthle *et al.* (1973).

1000 mg litre^{-1}. If a rapid change in response occurs between two of these concentrations (for example, no response within a specified time at 1 mg litre^{-1}, some mortality at 10 mg litre^{-1} and total mortality at 100 mg litre^{-1}), the full test might be performed using concentrations of 1, 2, 5, 10, 20, 40, 80 and 100 mg litre^{-1}. These values span the range at which a response was observed and are logarithmically spaced (i.e. the concentration is doubled at each step). In determining the median lethal concentration by graphical methods it is preferable to have several concentrations giving responses between 0 and 100%, and usually the more there are the better will be the estimate.

The selection of animals from stock should be at random as should their allocation to each test container. It is also preferable to allocate the solution strength to be used in each vessel at random.

The test should not be considered valid if any deaths occur among the control subjects during the period of the test. When changing solutions in 'static' tests of fishes it may be easier to transfer the batch of fish to another, identical, test vessel containing freshly made solution. Containers should be inspected periodically throughout the duration of the test and dead subjects should be removed and their death noted. In 'static' tests it is also advisable to check the concentration of the test solution from time to time since some substances may be lost through evaporation, volatilisation or adsorption onto the surfaces of the test vessel. Whenever this proves impossible, the lack of verification should be declared and identified when reporting the results by expressing them as, for example, the LC(I)$_{50}$ where C(I) indicates the initial and, therefore, nominal concentration (Lloyd and Tooby, 1979).

8.2.2.1.4 Results of toxicity tests on fish

There is a vast literature on the results obtained from toxicity tests on fish and their interpretation. It would be inappropriate to consider this here, especially since a number of reviews have been produced in the last two decades. Work up to the early sixties has been reviewed by Jones (1962, 1964) and also by Herbert (1965). Alabaster has provided a review of work on suspended solids (1972), methodology (1970) and comparative sensitivities of native European species (1971). Lloyd (1972) has reviewed the development of water quality criteria which have received a critical examination in a series of documents by the European Inland Fisheries Advisory Commission (e.g. EIFAC 1964, 1968a,b, 1970 etc.) and these have been drawn together in a recent review by Alabaster and Lloyd (1980). The problems of toxicity tests using fishes and the treatment and interpretation of results together with some related subjects have been examined in a series of papers by Sprague (1969, 1970, 1971, 1973). These reviews are a useful source of background information for the interpretation of toxicity data.

8.2.2.1.5 Application of toxicity test results

All results from toxicity tests must be applied with caution in management decisions affecting field situations, especially where the conditions encountered in the field differ from those used experimentally. Generally, high temperatures decrease survival in lethal solutions but low temperatures may decrease the resistance of some species to certain poisons: for example, of rainbow trout to phenol (Brown *et al.*, 1967) and ammonia (Brown, 1968). The toxicity of ammonia also varies with pH which, in addition to temperature, controls the degree of dissociation in solution, and the toxicity depends on the concentration of undissociated ammonia (Wuhrmann and Woker, 1948; Downing and Merkens, 1955; Herbert, 1962). Low levels of dissolved oxygen usually decrease the resistance of fishes to poisons (Lloyd, 1961). Soft water generally increases the toxic effects of heavy metals (Lloyd and Herbert, 1962). There are considerable differences in the response exhibited by different species but salmonids have been generally found to be more sensitive to most common poisons than many other species.

In addition to the necessity of avoiding the application of test results outside the range of experimental conditions and the temptation to extrapolate findings, it is also important to recognise that effluents rarely consist of pure substances but are generally mixtures of poisons. The concentrations may also vary considerably so that, in practice, fishes are exposed to fluctuating levels of mixtures of poisons. Brown (1968) has described a method for calculating the toxicity of mixtures of the common industrial pollutants ammonia, phenol, zinc, copper, cadmium, lead, nickel and hydrogen cyanide in which the proportions of the 48 hr LC_{50} for each poison are summed, and has shown that this agrees with the toxicity observed in laboratory tests. Studies of fluctuating concentrations of poisons in rivers indicated that the method tended to underestimate their toxicity. In laboratory experiments fluctuating levels of ammonia, zinc and mixtures of ammonia and zinc within $\pm 50\%$ of the 48 hr LC_{50} at equal time intervals showed that results were similar to those obtained for constant levels provided that the periodicity of fluctuations did not exceed the resistance time (period of exposure for irreversible changes) for the particular poison (Brown *et al.*, 1969). Field experiments using dilutions of river waters showed a tendency to underestimate the predicted toxicity (Brown *et al.*, 1970). Some evidence has accumulated to suggest that where the concentrations of individual poisons are low and below a level to exert stress they may make no contribution to total resultant toxicity of mixtures (Sprague, 1971; Lloyd, 1972).

Studies of the cumulative frequency distributions of toxicity in several rivers have been thought to demonstrate a general empirical relationship between the estimated toxicity, expressed as the 48 hr LC_{50} to trout, of polluted water and the presence or absence of fish. The boundary condition corresponds with an annual median of about 0·28 and a 99 percentile value of 1·1 (Alabaster *et al.*, 1972).

8.2.2.1.6 *Application factors*

Application factors (Eaton, 1970, 1973) have been proposed in order to enable 'safe' concentrations to be estimated for a species for which only lethal dose data are available, from the lethal and long-term toxicity data for another species. Alternatively, application factors may be used to predict the 'safe' concentration for a species in a different type of water, provided the lethal dose in that water is known. An example of the use of application factors is given in Fig. 8.11.

8.2.2.1.7 *In situ toxicity tests*

The conventional laboratory toxicity test in which conditions may be easily maintained within desired limits is valuable for studies of the properties of specified pollutants but for many practical purposes it would be both difficult and tedious to attempt to simulate the full range

Species N
in substance S

$$96 \, \text{hr} \, LC_{50} = 100 \, \text{mg litre}^{-1}$$
$$\text{asymptotic } LC_{50} = 5 \, \text{mg litre}^{-1}$$
$$\text{Application factor} = \frac{5}{100} = 0 \cdot 05$$

Species M
in substance S

$$96 \, \text{hr} \, LC_{50} = 50 \, \text{mg litre}^{-1}$$
$$\text{Estimated asymptotic } LC_{50} = 50 \times 0 \cdot 05$$
$$= 2 \cdot 5 \, \text{mg litre}^{-1}$$

FIG. 8.11 Example of the use of application factors: estimation of asymptotic LC_{50} for a given species from acute toxicity data for that species and an application factor derived from another species.

of combinations and variations of toxicants which are likely to be encountered in the field. It is then probably more informative to observe the survival of organisms *in situ*, either in cages within the habitat under investigation, or in aquaria nearby, to which samples of the water may be pumped or carried.

Floating cages were used by Allan *et al.* (1958) for investigations on the survival of fish in a sewage effluent. Aquaria were also maintained on the banks and fed directly with water pumped from the channel under investigation and the results were found to be very similar to those obtained using cages. The responses of the fish in aquaria were much more easily observed than those in the cages but the location of aquaria was limited by the necessity of a power supply to work the pump. Cages do not have this restriction in use but they may need greater attention for cleaning and repair. These workers describe the construction of a number of different sized cages, essentially wooden boxes derived from a basic module, ballasted with building bricks, having plastic mesh screens at their up- and down-stream ends. The use of a double set of slots to hold the screens enables them to be changed for cleaning without the escape of the fishes. Similar cages were later used by Alabaster (1964) in a study of the effects of heated effluents upon fishes.

Herbert *et al.* (1965), in a study of some fishless rivers, used cages of stainless steel mesh moored in the river or, when oxygen levels were less than 35% of air saturation value, 35-litre tanks were used on the river bank and river water was pumped through these at a rate of 1 litre per minute and aerated with air from a cylinder. Between 5 and 10 one-year-old rainbow trout were used in each tank.

In studies on the relationship between water quality and the apparent absence of fish from some polluted English rivers, Brown *et al.* (1970) also held fishes in cages of stainless steel mesh or in aquaria on the bank. Approximate 48 hr median lethal concentrations were determined using serial dilutions of the river water prepared by diluting with clean water from nearby unpolluted sources. Solutions were renewed every 2 hours; in one river the test water was pumped from the river to a holding tank to provide an integrated sample while, in another, river water was pumped immediately before making the dilutions. In this way the concentration of poisons broadly followed the fluctuations occurring in the river.

Caged rainbow trout were used by Uthe *et al.* (1973) in an investigation of the uptake of mercury in the South Saskatchewan River. Approximately 75 fish were used in each cage at stations above and below a suspected source of mercury and samples of fish were removed for tissue analysis every two weeks. This work provided useful data on the rates of uptake, including seasonal differences. Accumulated levels were not as high as those found in native fishes. It was also found that the cessation of mercury discharge did not immediately halt the uptake by fish and, at some sites below the outfall, mercury was not accumulated by caged fish. These results emphasise the necessity for great care in the use of caged fish and in the interpretation of results.

In a series of papers, Cairns and his co-workers (see, for example, Cairns *et al.*, 1973a,b or Waller and Cairns, 1972) have described the activity patterns of fish on exposure to solutions of zinc and have proposed the use of in-plant monitoring of effluents by means of activity recorders linked to a computer (see 8.2.2.3.1). A number of similar units are undergoing commercial development.

A mobile laboratory for *in situ* assays has been described by Wuerthele *et al.* (1973). Pumps deliver effluent and receiving water to holding tanks to provide continuous flows of any desired mixture of waste-water discharge and receiving water. Details are also provided of the 'Michigan Diluter' in which a number of serial dilutions are automatically obtained by passing effluent and diluent through a series of test chambers (see 8.2.2.1.3).

8.2.2.2 Toxicity tests—macroinvertebrates

Tests of the toxicity of substances to macroinvertebrates may help to predict the effects of discharges of certain effluents and to interpret field observations (Murphy, 1980). They are particularly valuable for assessing the likely consequences of the application of pesticides, especially insecticides or molluscicides, which may be deleterious to species other than the 'target' organisms. In addition, macroinvertebrates are useful in experiments where the possible cumulative effects of a substance on several generations are under investigation (Arthur and Leonard, 1970). The advantages and disadvantages of macroinvertebrates in toxicity tests have been considered by Maciorowski and Clarke (1980).

8.2.2.2.1 Methodology

Simple 'static' tests involve introducing the test species into several vessels containing different concentrations of the substances under investigation, and appropriate 'controls'. The importance of adequate controls cannot be over-emphasised: the maintenance of some invertebrates in the laboratory is difficult and this may limit the range of tests which can be performed. For example, Muirhead-Thomson (1973) found that only the results of his 24-hour tests were valid as the controls did not always survive for 48 hours in the laboratory, even in continuous-flow apparatus. Warnick and Bell (1969) were able to obtain 48 hr TL_m values for various metallic salts with plecopteran and ephemeropteran nymphs and trichopteran larvae after successfully acclimatising these animals to laboratory conditions.

Static tests may be adequate for initial screening but several workers have stressed the necessity of using continuous-flow methods (in which the test solution is renewed during the experiment) to ensure that the desired levels of concentration are maintained and the effects of changes induced by the metabolic activities of the test organisms, either by breakdown of the test substance or by changes in water quality by the organism's metabolites, are reduced (Jensen and Gaufin, 1966; Arthur and Leonard, 1970; Muirhead-Thomson, 1973).

It will rarely be possible to conduct axenic or sterile tests but the influence of bacterial degradation of the test substance is minimised by the use of continuous-flow methods. The maintenance of the intended levels of exposure is also important when the investigation is designed to furnish data for the prediction of field effects since some substances are accumulated by macrophytes and muds (Wallace and Brady, 1971). Simple static tests in which the concentration is not maintained could give misleading results.

The problems of controlling the conditions under which organisms are exposed to toxicants have been attacked by many workers. For example, Mattice *et al.* (1981) have developed a system for producing short 'square-wave' changes in concentrations of a toxicant (chlorine) which they propose as a standard exposure regime. This may be contrasted with an apparatus for providing planned, varying concentrations of chemicals in order to foster more realistic studies of variations in the concentrations of toxicants as might occur naturally and to which wild populations would be exposed (Manley, 1980).

Small, weakly swimming invertebrates, such as *Daphnia*, are unable to tolerate the water currents generated by standard proportional diluters (Mount and Brungs, 1967) but a modification by Maki (1977) permits replacement of the test solution with no perceptible current-induced effects.

Jensen and Gaufin (1964, 1966) investigated the effects of several insecticides on two plecopteran species, which form an important constituent of some fish diets, and obtained TL_m values for 24, 48, 72 and 96 hours. Later they performed a 30-day continuous-flow experiment for comparison and their findings underlined the importance of long-term tests: for example, after four days in continuously renewed solutions, aldrin was found to be *less* toxic than in the static tests but from the fifth day until the tenth day individuals of both species died in solutions of lower concentration than the 96 hr TL_m static solutions. Arthur and Leonard (1970) and Arthur (1970) in two series of experiments investigated the effects of six-week exposures of amphipods (*Gammarus* sp.) and molluscs to different concentrations of copper and a detergent. The concentrations which were thought to be safe, following these investigations, were lower than those suggested by the 96 hr TL_m values.

8.2.2.2.2 Choice of test species

The choice of species depends on the objectives of the test: for investigations of the possible effects of discharges into a given water it clearly would be preferable to use key species from that habitat, but for general screening of substances the selection of the test animal may be made with respect to several general criteria. These are outlined below.

(a) Suitability

The chosen species should, preferably, be indigenous: exotic species may have many other desirable characteristics but results from tests using these are not then necessarily relevant to the local conditions. There are clear advantages in using species which have already received considerable attention and whose life histories, ecology and physiology are well documented.

(*b*) *Sensitivity*
Species which are very tenacious of life are unlikely to be useful except in that when these are affected one can be sure that the substance under test is likely to be generally undesirable. The trophic level of the species may also be important. For some substances the direct toxicity may be quite low but when accumulated in food chains the effects can be quite pronounced. For example, Morgan (1972) has shown that *Daphnia pulex* was greatly affected when fed on algae which had been cultured in a medium with a low concentration of Aroclor (a polychlorinated biphenyl, PCB).

(*c*) *Ease of culture*
Ideal test organisms must be easily cultured, requiring little space and demanding no special facilities or food. Preferably, there should be no seasonal variation in availability; short, simple life cycles are advantageous. It is probable that, in general, ease of culture and sensitivity are inversely related. That is, species which readily adapt to laboratory conditions and which make few demands for their maintenance in culture are also likely to be more resistant than others under test.

A number of different test invertebrates are documented in the literature (Murphy, 1980) but the following are listed as representative. It is interesting to observe that most of the experimental work reported in the literature was conducted in North America. European interest has tended to centre on observational studies, beginning with the classical 'saprobien system' (see Section 6.4). The protozoan *Paramecium caudatum* has been used by Pawlaczyk-Szpilowa *et al.* (1972) while Dive and Leclerc (1975, 1977) have recommended the use of *Colpidium campylum*. The oligochaete worm *Tubifex tubifex* has been used in studies of heavy metal toxicity (Brkovic-Popovic and Popovic, 1972a,b) while the tubificid oligochaetes *Tubifex tubifex* and *Limnodrilus hoffmeisteri* were used by Whitley and Sikora (1970) in investigations of the effects of heavy metals and phenol on respiration rates. Molluscs have been used in tests by Arthur and Leonard (1970). Several Crustacea, including *Gammarus* (Anderson, 1948; Arthur and Leonard, 1970; Arthur, 1970), *Daphnia* (Pawlaczyk-Szpilow *et al.*, 1972; Morgan, 1972;

Leeuwaugh, 1978) and *Artemia* (Michael *et al.*, 1956; Zilloux *et al.*, 1973) have also been used. A number of insects have been employed in specific investigations including naiads of Odonata and *Baetis*, larvae of *Simulium* (Muirhead-Thomson 1973), Plecoptera, Ephemeroptera and Trichoptera (Jensen and Gaufin, 1964, 1966; Warnick and Bell, 1969; Nagell, 1973). Gunner and Coler (1972) have used the community structure of the invertebrates (including protozoa, ostracods, nematodes and flatworms) within the rhizosphere of the duckweed (*Lemna minor*) for experimental investigation and an insecticide Diazinon. These authors emphasise the potential value of such a test system in that the microecosystem is more representative of field conditions but there is no loss of laboratory control.

From the above list it would seem that many of the more important groups appear to include suitable test animals. It is surprising that fewer tests have been conducted using molluscs since many are easily maintained in the laboratory and breed all the year round. The brine shrimp *Artemia* has many desirable features: its eggs may be purchased from aquarist suppliers and stored indefinitely, hatched when required, harvested after 24 hours and the nauplii need no feeding at this stage (Zilloux *et al.*, 1973). Adult *Artemia* were also successfully used as test organisms by Michael *et al.* (1956) for investigations of the effects of insecticides: toxicity was assessed from the degree to which they were able to maintain their normal positions at the water surface. *Artemia* is, however, well-adapted to a harsh environment and it may well be that the convenience of using this organism may be counter-balanced by an enhanced resiliance to certain substances.

A compendium of methods for the culture of invertebrates, including several freshwater groups appropriate for tests (e.g. Mollusca, Plecoptera, Odonata, Culicidae, Chironomidae and Simuliidae) has been compiled by Lutz *et al.* (1937). A bibliography of literature relating to macroinvertebrate culture has been provided by Campbell *et al.* (1973) and, more recently, the use of invertebrates in toxicity tests has been the subject of a symposium organised by the American Society for Testing and Materials (Buikema and Cairns, 1980).

Daphnia, in particular *D. magna*, is a popular subject for toxicity tests. It is readily cultured in the laboratory (Ten Berge, 1978) and is the largest and easiest species to handle (Adema, 1978). Procedures for maintaining cultures of *Daphnia magna* which are fed on the alga *Chlorella* have been outlined by Ten Berge (1978), who has stressed the importance of using filtered medium and adequate food supply

(*Chlorella*) supplemented with yeast extract. Too much food causes oxygen deficiency through algal respiration while too little is associated with a decline in the numbers of offspring accompanied by increased numbers of males. Under ideal conditions reproduction rate is between 3–9 young per *Daphnia* per day. The use of *Daphnia* cultures is not without its problems: Lee and Buikema (1979) have noted that in continuous and short-term exposures of *Daphnia pulex* to chromium, there were cyclical changes in susceptibility which were found to be related to moulting, at which time the animal is particularly sensitive. As moulting in laboratory cultures can be synchronised, the effects of variable sensitivity could introduce bias in toxicity tests. The problem can be overcome by using animals from asynchronous cultures or by using mixtures from several cultures. Results may also be influenced by the composition of the medium. Tevlin (1978) noted that some synthetic media contain a Fe-EDTA complex as a source of soluble iron for the *Daphnia* and *Chlorella*. Tevlin draws attention to the almost universal complexing ability of EDTA with metal ions and thus the risk that in tests using heavy metals, for example cadmium, as in his experiments, the metal may be chelated and thus its toxic effects may be reduced. In order to reduce this risk a ferrigluconate complex was substituted.

Sorgeloos and Persoone (1973) have developed a culture system for *Artemia*, *Daphnia* and other invertebrates with continuous separation of the larvae. The rearing and maintenance of plecopteran nymphs has been described by Kapoor (1972). Hiley (1970) has described a method of rearing Trichoptera larvae for taxonomic purposes, and White and Jennings (1973) have described a technique for rearing various aquatic Coleoptera. Cultivation of chironomid larvae was discussed by Konstantinov (1958) and an effective method for the continuous culture of *Chironomus riparius* has been developed by Credland (1973). Methods for the collection, extraction, sterilisation and low-temperature storage of eggs and for rearing *Simuliidae* in the laboratory have been described by Freeden (1959a,b). Many of these techniques could be useful in providing a supply of test organisms for routine toxicity tests. For general screening tests the species or genera listed in Table 8.2 are recommended as representatives of the major groups.

8.2.2.2.3 Interpretation of results

It is not possible to conduct tests on a few species and then generalise from these results. Patrick *et al.* (1968) have investigated the differing sensitivities of a mollusc (*Physa heterostropha*), diatom and a fish to a number of common constituents of effluents. Of twenty compounds

TABLE 8.2
MACROINVERTEBRATES RECOMMENDED FOR ROUTINE TOXICITY SCREENING TESTS

Group	Test animal	Source	Comments
Annelida	*Tubifex* sp.	Aquarist supplies	Easily maintained in the laboratory with mud in bottom of container, see La Rue in Lutz *et al.* (1937), p. 194
Mollusca	*Lymnaea* sp. or *Planorbis* sp.	Field collection	Easily maintained in laboratory in aquaria; see Lutz *et al.* (1937), pp. 520–526
Crustacea	*Gammarus* sp.	Field collection	Must be maintained in well-aerated aquaria (Mellanby, 1963); see also Arthur (1980)
	Daphnia sp.	Field collection or Aquarist supplies	Well-documented culture methods; see Lutz *et al.* (1937), pp. 207–220 or Buikema *et al.* (1980)
	Artemia sp.	Aquarist supplies	Dried eggs may be stored indefinitely and hatched when required; see Sorgeloos and Persoone (1973)
Insecta	*Chironomus* sp.	Field collection	See Credland (1973) and Anderson (1980)
	Simulium sp.	Field collection	See Freeden (1959a,b)
	Hexagenia sp.	Field collection	See Fremling and Mauck (1980)

tested, eight were used on all three organisms and of these the mollusc was most sensitive to two compounds (potassium chloride and zinc). Stewart *et al.* (1967) found the insecticide Sevin® to be more toxic to crustaceans than to molluscs and fish while with 1-naphthol, the product of its hydroloysis, the situation is reversed. Muirhead-Thomson (1973) has demonstrated differing sensitivities to insecticides between naiads of Odonata, *Baetis* and *Simulium* larvae. Not only do different species exhibit differing sensitivities but there may even be intra-specific variations, especially when test animals are taken from habitats which have been exposed to the substance under investigation and some selection for tolerance has occurred (Ferguson *et al.*, 1965; Naqvi and Ferguson, 1970).

Variation in susceptibility to toxic substances at different stages in life cycles are exhibited by some species. Arthur and Leonard (1970) have shown that the growth of newly hatched *Gammarus* was impaired on exposure to levels of copper previously found to be sub-lethal to their parents. Some arthropods show greater sensitivity at the time of ecdysis. For example, *Daphnia*, a species widely used in toxicity tests, is probably most susceptible to the toxicity of metallic salts at the time of moulting (Anderson, 1948; Lee and Buikema, 1979). Such variations emphasise the importance of long-term tests.

Further variations in observed sensitivity may arise from the experimental procedures employed. In a detailed study of the effects of the dichromate ion on the survival of *Daphnia pulex*, Sherr and Armitage (1973) investigated the interactions between the concentrations of dichromate, the age, and the density of the *Daphnia* and found that the relationships were most complex. Whitley and Sikora (1970) studied the effects of lead, nickel and phenol on the rate of respiration in tubificid worms and found that the results were influenced by pH, although the rate was depressed by increased lead concentration at all pH values investigated. Generally, phenol enhanced respiration. A similar increase in respiration rate on exposure to phenol was observed by Cole and Wilhm (1973) in a study of its effects upon the larvae of *Chironomus attenuatus*. Respiration increased with increasing phenol concentrations up to 22·4 ppm, but above this the larvae died.

Differing behavioural responses between species can influence the interpretation of toxicity tests results. For example, some gastropods with opercula are able to avoid short-term exposure to potentially lethal levels of some toxic substances simply by closing the operculum tightly.

The problems of relating experimental observations to field situations were well illustrated by Nagell (1973). On investigating the relationships between oxygen consumption and dissolved oxygen concentration for nymphs of Plecoptera (*Nemoura*) and Ephemeroptera (*Cloëon*), he was able to demonstrate the existence of a critical level of dissolved oxygen below which death occurred. At concentrations just above this level the animals showed greater activity, either in increased respiratory movements (ventilation) or general mobility, apparently in an attempt to increase oxygen uptake. Nagell believed that such investigations may be useful in estimating the incidence of past critical levels from observations of the distributions of certain species but his own work shows that *Nemoura* had a higher critical value than *Cloëon*, yet appeared to be able to survive lower oxygen concentrations in the field.

8.2.2.3 Bioassays with fish

Much work has been done on the effects of non-lethal concentrations of pollutants on fish under experimental conditions. From the investigator's point of view it would be preferable to use methods which are not too time-consuming and relatively simple to set up and observe. The relationship of the results to conditions in the field, and therefore their potentially limited use in assessing water quality, needs to be fully understood. In this connection the very sensitive responses of fish when living under artificial conditions and their need for prolonged acclimatisation periods must also be taken into account. Work done by Randall and Smith (1967) comparing the effects of handling and anaesthetisation on various teleost fish (trout, tench, goldfish and salmon) suggests that the effects of handling may be quite pronounced, and that also there may be variation between species. This underlines the necessity for minimum handling, minimum use of anaesthesia and as long a time as possible for test fish to recover and/or become acclimatised to their new situation before experimentation begins.

8.2.2.3.1 Monitoring movement or activity

Jones (1962, 1964) has reviewed the development of 'choice-chambers' in which the test solution and clean water (or two test solutions) are introduced to provide either a gradient or, preferably, an abrupt change in water quality. The movements of the test fish are observed and may also be recorded on a moving chart. Tests may be repeated by reversing the flow arrangements if necessary. Checks on the pattern of flow and the presence of the gradient or discontinuity may be made by the introduction of a small quantity of a dye. Jones (1962) states that fish are unaffected by the observer but Schere and Nowark (1973) have developed an apparatus for recording the avoidance movements in which the fish are observed through a one-way mirror as they swim in a clear-acrylic tank in which the flow of water in the two halves is kept separate. Fish movements are tracked with the aid of a viewer and electrical voltage changes caused by movement of this viewer are recorded. A timer is also incorporated to measure the time spent by the fish in each half of the tank. The authors were able to record avoidance movements following the addition of mercuric chloride to one half of the tank. They suggest that the apparatus could be adapted for use in insect larvae, crustaceans and other active animals.

Some workers have developed automatic methods of recording fish activity on exposure to pollutants. For example, the movement (activity) patterns of bluegills (*Lepomis macrochirus*) have been monitored by Waller and Cairns (1972) and of goldfish (*Carassius auratus*) by Cairns *et al.* (1970) under different zinc concentrations using interruptions of light beams at three levels in the tank to record the movements. Variances were calculated for experimental and control fish and some correlation between zinc concentrations and movement abnormalities was shown. It was suggested that this method could well be applied to other toxicants provided that some effect on activity is induced.

Spoor and Drummond (1972) have used a gradient tank with four compartments, each containing a pair of stainless steel electrodes, for monitoring movement. The tracing records the location (i.e. which compartment) of the fish, its opercular rate, and its position, orientation and state of locomotor activity within the compartment.

More recently, Cripe (1979) has described an automated device (Aquatic Gradient Avoidance Response System, AGARS) in which the test animal is offered a choice between uncontaminated water and three pollutant concentration gradients. The position of the animal is monitored automatically by means of infrared sources, sensors and a microprocessor. A detailed account of the circuitry of an automatic activity monitoring system, also using infrared lights, has been provided by Morgan (1979).

Macleod and Smith (1966) used a circular metal trough to measure the swimming ability of the fathead minnow (*Pimephales promelas*). A current was created and each time a fish was involuntarily carried by it past a fixed point a 'failure' was recorded. These workers were investigating the effect of an inert suspended solid, in this case pulp wood fibre, on the swimming ability of the fish, which was impaired by the addition of the fibre.

The rheotactic response of fish to a current has been utilised in several alarm devices. Under normal conditions the fish remains more or less stationary but when poisons are present the fish flee or, if incapacitated, they are unable to maintain their position and are carried by the current into a detector system, usually employing light beams (Besch and Juhnke, 1971) or physical contact (Besch *et al.*, 1977). Fish may be deterred from entering the detector area by applying an electrical field but in one variant of this system (Fig. 8.12) incapacitated fish are carried onto a moving conveyor-belt where their presence is detected by light beams (Scharf, 1979). This arrangement is claimed to reduce the number of false alarms since only weakened fish will allow themselves to be carried up the escalator and past the detector to be deposited in a holding tank where they may recover. A further advantage of this system is that, since the fish is detected in air, there is no interference from turbidity or coloured water.

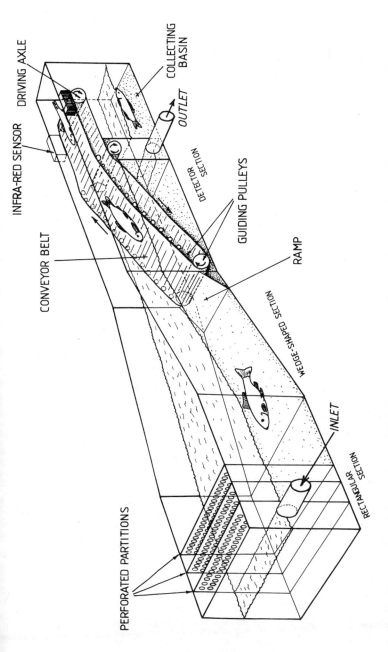

FIG. 8.12 Fish monitor (fish test alarm device) for continuous surveillance of water quality and the detection of the presence of acutely toxic substances. After Scharf (1979).

Stott and Cross (1973) used displacement from home range as a means of investigating the behavioural response to lowered dissolved oxygen. Roach (*Rutilus rutilus*) were acclimatised to the upper end of a series of connected tanks, which effectively formed a long channel, until this was accepted as their home territory. When the oxygen concentration of the water passing through the tanks was reduced, a conflict was created between the respiratory discomfort and the homing instinct so that the linear displacement indicated the intensity of the effect of a given reduction in oxygen concentration. This procedure is potentially very valuable since it seems to imitate field conditions and could be used to measure the likely behavioural effects of discharges of 'unpleasant', but non-toxic, substances.

8.2.2.3.2 Monitoring respiratory activities

Fish should be restricted as little as possible and also effectively shielded from external disturbances when respiratory activity is monitored. The least restricting method of measuring rates of breathing and 'cough' rates involves the use of free-swimming fish in tanks with electrodes, covering the maximum cross-section at each end. It should be noted that the slight electrical potential measurable between the electrodes before the introduction of the fish to the tank does not in general affect the fish, though Regnart (1931, quoted by Spoor *et al.*, 1971) found that cod (a marine species) could be affected by it. In these tanks the fish is given enough room to turn round but the tank must be narrow to encourage the fish to align towards one of the electrodes. Heath (1972) believes that the measurements obtained are related to the changes in the electrical potential of the respiratory muscles, while Spoor *et al.* (1971) believe that they are due to electrochemical changes caused by movements of the water. Spoor *et al.* (1971) have used stainless steel electrodes and shown that copper solutions affect the opercular rate. Spoor and Drummond (1972) used a gradient tank for further measurements as described above (Section 8.2.2.3.1). One measurement which cannot be ascertained accurately by this apparatus is the amplitude of breathing.

Rommel (1973) has used a tank containing carbon electrodes with salmon and eels. He found that the operculum beat was clearly distinguishable from the heart beat on the tracing and that coughs were recognisable as large spikes. If the two beats became indistinguishable he separated them by changing the low frequency filter (used to eliminate the effects of swimming movements) from 10 Hz to 1 Hz. He checked his results by the use of implanted electrodes as described by McCleave *et al.* (1971).

Variation in coughing rate is perhaps one of the easiest measurements to make. Buccal cannulation (which may be combined with opercular cannulation) and recording of buccal pressures gives an indication of the cough and breathing rates as well as having the advantage of recording the amplitude of the pressures. Using cannulation, however, the movements of the fish are restricted. Sparks *et al.* (1972) used buccal and opercular cannulation of bluegills in varying concentrations of zinc solutions. They found that the cough rate increased as the concentration of zinc increased and concluded that 'it appears that cannulation of the mouth alone would provide sufficient information to monitor toxicants'. They suggest that this is a suitable method for zinc and probably other heavy metals as the response is quick, reflects the toxicant concentration and occurs at sub-lethal concentrations.

Schaumburg *et al.* (1967) used a similar method for recording buccal pressure in evaluating the effects of kraft pulp mill wastes on juvenile coho salmon. Using varying concentrations of DDT as their reference toxicant they found a direct relationship between the frequency of coughing and the exposure time for each concentration of DDT. There was a similar relationship when mill effluent was used as the toxicant; in each case there was increased coughing with increased concentration of the toxicant. It was noted that variations in the rate of coughing and the rate of breathing were not necessarily related. Walden *et al.* (1970) observed identical aberrant cough responses in fish exposed separately to pulping and bleach effluents; some acclimatisation occurred at sub-lethal levels.

Davis (1973) also carried out an investigation of the sub-lethal effects of varying concentrations of neutralised, filtered bleached kraft pump mill effluent on sockeye salmon. In his experiments the fish were, of necessity, more restricted as measurements were made of buccal pressure, ventilation volume, breathing rate and the oxygen tension of the inspired and expired water. He also measured arterial oxygen tension by means of blood samples taken from a dorsal aortic catheter. From his very interesting results the following points may be selected as being particularly relevant. He found, like Schaumburg, an increase in coughing associated with increased concentrations of effluent, and deduced that coughing appeared in response to chemical as well as mechanical stimuli. Coughing may interfere with the normal gill flow patterns and thus may affect respiratory efficiency.

Cairns and his co-workers have carried out many investigations into the effects of zinc on fish, some of which have been mentioned above. They used the determination of bluegill breathing rates to detect zinc using an active electrode implanted into the breast of each fish with a reference electrode on the floor of the tank (Sparks *et al.*, 1972). Great care was taken to construct an isolated experimental room. Analysis of variance was used rather than simply expressing the results in terms of increased breathing rates, to ensure that only significant departures from normal were recorded. A practical application of their methods is suggested whereby 'breathing rates could be taken every minute, the variances estimated and tested every ten minutes by the means of analog-to-digital converters, mini-computers and teleprinters'. Thus, it might be possible to add this system to computers already in use in a chemical or industrial plant. The criterion for detection should be simultaneous responses by some proportion of the fish, not necessarily by all of them. In another paper Cairns (1973) suggests that if the toxic effects of a pollutant are reversible, responses by 75% of the test fish might be regarded as necessary before action is taken, but if the pollutant is fast-acting and irreversible in its effects, responses by 25% to 50% of the fish would be sufficient. He also suggests that 'in-plant' monitoring by test fish, and 'the degree of protection afforded by the in-plant ... units should be evaluated by in-stream monitoring units' (Cairns *et al.*, 1972a,b). He believes that, even if these units are not appropriate for streams with chronic exposure to pollution, they might provide a useful technique for the prevention of accidental leakage. Details of a computer-based fish monitoring system for automatically monitoring the toxicity of an industrial waste effluent stream have been provided by Van der Schalie *et al.* (1979). If a sudden increase in pollution were detected by an in-plant monitoring unit immediate action could be taken by shunting waste to a holding pond, recycling it for further treatment or reducing the operations of the plant (Cairns *et al.*, 1972c). False alarms would be reduced by exposing the test fish to higher concentrations of waste than the receiving stream. Lower concentrations would be used to test the effects on spawning in ripe bluegills, and on newly-hatched fry. In the future, Cairns and his co-workers hope to combine the monitoring of movement and breathing in one unit, using an electrode tank as described in Spoor *et al.* (1971). In this way the effects of turbidity on the light beam interruptions in the movement monitor will be minimised.

An apparatus similar to that of Spoor *et al.* (1971) was used by Carlson and Drummond (1978) to evaluate the quality of treated complex effluent using the 'cough' response of bluegill sunfish (*Lepomis macrochirus*). They found that cough rate correlated well with effluent quality and concluded that change in cough frequency is a rapid and sensitive physiological parameter and could be used in industrial, or other, effluent monitoring.

Changes in opercular rhythm in largemouth bass (*Micropterus salmoides*) were used by Morgan (1975) to investigate the toxic effects of four pesticides (carbamate, chlorodane, parathion and pentachlorophenol). Elevated breathing rates were observed on exposure to acutely lethal concentrations of these pesticides within 8 to 20 hours and well before death would have occurred and on transferring fish with elevated respiration rates to clean water none died. At $0{\cdot}1$ of the $48\,hr\,LC_{50}$ concentrations of those pesticides, an elevated respiration rate was observed within three days of exposure and continued at a high level for two weeks (the duration of the experiment) but no fish died. A definite opercular response (defined as exceedance of the maximum 99% confidence limit for normal respiration) could not be detected in fish exposed to $0{\cdot}01$ of the $48\,hr\,LC_{50}$ levels of these substances but over a five week exposure period the respiration rate did rise. It was suggested that this method might usefully be applied to a reappraisal of 'safe' concentrations of pesticides since levels which allow survival after several months' exposure may induce functional changes.

Other work on the analysis of toxic effects which should be mentioned is that of Hughes and Saunders (1970), Hughes (1973) and Hughes and Roberts (1970) who have investigated the respiratory responses of rainbow trout (*Salmo gairdneri*) to hypoxia and to temperature changes respectively. Holeton (1971) has done similar work on trout exposed to carbon monoxide. Later work by Hughes and Adeney (1977) showed that low levels of zinc affect the ability of rainbow trout to respond to the stress of low oxygen levels, and it was concluded that zinc interferes with the mechanism of oxygen uptake from the gills.

On reviewing and comparing six different methods for detecting and measuring fish respiratory movements, Heath (1972) concluded that the method which yields the most information, with minimal effect on the fish, is recording respiratory pressure. This requires cannulation of the fish but, once this has been performed, breathing frequency, depth of breathing and 'coughing' or other respiratory abnormalities can be determined simultaneously in the same fish. He does, however, suggest that for industrial or similar applications, where the effects of effluents are to be monitored, an external electrode may be preferable because it is simpler and avoids operative procedures on the fish. In such applications the fish should be of reasonable size and this excludes, for example, the mosquito fish (*Gambusia*) or the rice fish (*Oryzias*). The bluegill sunfish

(*Lepomis macrochirus*) is particularly recommended since it responds to environmental stress with pronounced changes in breathing frequency.

8.2.2.3.3 Monitoring heart rate and lateral line nerve responses

Bahr (1973) used responses from the lateral line nerve and electrocardiogram (ECG) recordings to monitor the effects of hypoxia and cyanide on trout. For the ECG recordings the electrodes were inserted under the base of the pectoral fin and pushed under the skin until they reached the mid-line, 1 cm posterior to the heart. For the nerve responses, silver wire electrodes were attached round a portion of the exposed lateral line nerve and its electrical activity was observed by means of an oscilloscope. Bahr states that the fish can be prepared in minutes and the results of subjection to varying concentrations of the toxicant (or reductions of oxygen pressure) may be obtained in a matter of hours. After establishing normal levels, ECG readings and spontaneous and evoked nerve responses of the fish in the appropriately-treated water were obtained. Marked effects were shown on nerve activity and there was reduction of the heart rate as well as changes in the pattern of the ECG. Bahr states that this 'methodology ... proved ... a useful tool for analysing neurotoxic effects of pollution-related stresses'. He does not suggest using it as a monitor of water pollution but it might be useful for screening potential effluents.

Anderson (1968) investigated the effect of sub-lethal DDT on the activity of the lateral line nerve of brook trout, *Salvelinus fontinalis*. The stimulus used was a falling drop of water hitting the water surface above the lateral line and setting up an abrupt low-frequency pressure wave. The observed neurological response was a high-frequency burst of large spikes, the duration of which was increased when DDT was present. The effect was enhanced at lower temperatures.

8.2.2.3.4 Reproduction and development

The literature relating the effects of pollutants on the reproduction and development of fish is small in comparison with that on the toxicity of substances to adult fish, but is growing. Much of this difference can almost certainly be attributed to the greater practical difficulties encountered in attempts to induce some species to mature and spawn in the laboratory.

The maintenance of a self-supporting fishery clearly depends upon the reproductive and developmental capability of the indigenous fish species, and any impairment could, eventually, lead to consequences as serious as those of the more obvious fish-kill. Some pollutants, for example suspended solids, may destroy spawning grounds or lead to asphyxiation of developing eggs. Others may cause sterility or reduced viability of fertilised eggs while some substances are most toxic to newly-hatched fry. Physical factors may also be important: much of the early work on fish reproduction has demonstrated the role of photoperiod and temperature in the initiation of breeding behaviour or in maturation but detailed studies on the thermal requirements for maturation, spawning and embryo survival have been relatively few, although the recent work of Hokanson *et al.* (1973) on *Salvelinus fontinalis* is an example of the necessary approach.

On reviewing the literature, Sprague (1971) concluded that recently hatched fish were generally more sensitive to pollutants than adults. Eggs were not especially sensitive but an exception to this generalisation has been observed for exposure to the herbicide 2,4-D (Mount and Stephan, 1967). Holland *et al.* (1966) have shown that sheepshead minnows (*Cyprinodon variegatus*) were more sensitive to exposure to DDT and endrin when their parents had been exposed to these pesticides. Later work (Holland and Coppage, 1970) showed that succeeding generations exposed to DDT and endrin can develop resistance but this is not consistent. If parents were exposed at maximal lipid metabolism during maturation, DDT was incorporated into the ova and this appears to have caused greater sensitivity. Selective mechanisms in the development of insecticide resistance have been considered by Finley *et al.* (1970).

Complete inhibition of spawning, at levels of copper for which no other changes could be demonstrated, has been described by Mount (1968). Macek (1968) studied reproduction in brook trout (*Salvelinus fontinalis*) fed sub-lethal concentrations of DDT, i.e. those causing no mortality when fed for 156 days. Those fed at the highest dosage produced fewer mature ova than the controls. In general, egg and alevin mortality was higher when at least one of the gametes came from a treated fish than in controls. Alevins usually died in the 15th week, which probably corresponded with the maximum utilisation of yolk fat. Carlson (1972) studied the effects of long-term exposure to carbaryl (Sevin®) on the fathead minnow; exposure to a concentration of 0·68 mg litre^{-1} for nine months prevented reproduction and decreased survival.

Eaton (1970) found that certain concentrations of malathion which killed or crippled bluegills after weeks or months of exposure did not stop reproduction or affect egg or early fry development. He described the symptoms of malathion poisoning, noting that some fish appeared to recover but died within a few days. Stephan and Mount (1973) described their studies of the effects of sub-lethal levels of toxic agents on seven species of fishes reared from eggs or fry younger than 20 days and continued until the offspring of those fish were at least 30 days old. Mount and Stephan (1967) found some fathead minnows were able to reproduce, apparently normally, at concentrations of malathion which killed others. Malathion is not highly persistent and it may be metabolised: such factors must always be considered in assessing results.

8.2.2.3.5 Use of conditioned reflexes in fish

The effects of sub-lethal concentrations of pollutants may be demonstrated by the failure or slowness of fish to develop conditioned responses as compared with controls. Warner *et al.* (1966) describe a conditioned avoidance response apparatus (CARA) so designed that fish placed in one end of the tank had to swim through very shallow water to reach the other end. They soon learn to do this and may then be subjected to light stimuli and to light stimuli combined with electric shock to which they become conditioned and avoid by swimming to the other end of the tank. Goldfish were shown to develop this conditioning much more slowly when the water in the tank contained sub-lethal concentrations of toxaphene (a chlorinated camphene) than they did in clean water.

Even low doses of insecticides may have profound effects on fish behaviour. If learning ability is impaired, behaviour which relies on this ability, such as territorial defence and migration, may be affected. Anderson and Prins (1970) showed the effects of sub-lethal DDT on brook trout. They made use of the fact that mild stimulation of the gular region results in a weak propellor-like movement of the tail; this stimulation was used as the unconditioned stimulus and light was the conditioned stimulus. After exposure of the test fish to a sub-lethal concentration of DDT in water for 24 hours they demonstrated that these fish became conditioned more slowly than unexposed controls. The fish were restricted during the experiment and most did not survive longer than four weeks, but those that did were retested and it was found that the DDT-treated fish had improved somewhat in their performance. Thus it seems that DDT affects the central nervous system rather than impairing motor ability.

In studies on fish treated with sub-lethal doses of DDT, Anderson and

Peterson (1969) reported failure to acquire a conditioned response. Later work, however, showed that when a different experimental procedure was employed, under appropriate conditions, fish could acquire a conditioned avoidance response even after exposure to sub-lethal doses of DDT (Jackson *et al.*, 1970).

The effect of four insecticides on the ability of Atlantic salmon parr to learn and retain a simple conditioned reflex has been investigated by Hatfield and Johansen (1972). When treated for 24 hours with 96 hr LC_{50} doses, Sumithion completely inhibited learning, Abate retarded learning, DDT mildly enhanced learning and methoxychlor had no detectable effect on learning. Learning improved in Abate and methoxychlor-treated fish from one to seven days later but DDT-treated fish showed no improvement. After a four day recovery period from Sumithion and Abate treatment, learning was still slow. When the insecticides were administered between the first and second conditioning there was little effect on learning ability. When fish were exposed to levels of insecticide one-tenth of the 96 hr LC_{50} no effect on learning ability was observed. These authors noted that the results were similar to those of Warner *et al.* (1966) and that, generally, organophosphate insecticides retarded learning while organochlorine insecticides improved it; the anomaly observed with methoxychlor (an organochlorine insecticide) was thought to be caused by its easier metabolism or degradation.

Scheier and Cairns (1968) have described an apparatus and procedure for measuring the response of bluegills to a light flicker stimulus and determining the effects of exposure to sub-lethal concentrations of toxic substances on the critical flicker frequency. Control fish responded to a critical frequency of 37 Hz, above and below which random responses were observed. Tests with parathion indicated that a toxic substance, especially one which affects the central nervous system, can change the response to critical flicker frequency.

8.2.2.3.6 Abnormal behaviour

Part of the normal behaviour patterns of fish is the ability to both detect and avoid, where possible, discharges of harmful pollutants (Sprague, 1964; Saunders and Sprague 1967; Jones 1964). A few substances are not avoided, for example, near-lethal concentrations of detergents or phenol, and chlorine has been shown to be attractive at lethal concentrations (Sprague and Drury, 1969). Some substances are now known to affect normal behaviour patterns including habitat selection, feeding, predator avoidance and orientation, when fishes are exposed to sub-lethal doses. Some examples of these effects are described below.

An important effect of some detergents is changed feeding behaviour. For example, apparently permanent changes were induced in catfish (*Ictalurus natalis*) when erosion of the taste-buds occurred and feeding stopped (Bardach *et al.*, 1965). Normal feeding behaviour was suppressed in flagfish (*Jordanella floridae*): food was taken but then rejected, presumably because the taste receptors were affected (Foster *et al.*, 1966).

A method for detecting the effects of toxicants or other stresses on predator–prey relationships has been described by Goodyear (1972). The test species are observed, after exposure of prey, predator or both, in a two-compartment aquarium, separated by a grill which allows prey species to move into the predator's compartment but not *vice versa*. Goodyear investigated the relationship between the mosquito fish, *Gambusia* (prey), and the largemouth bass, *Micropterus* (predator), in a tank which had a shallow section for the mosquito fish (Fig. 8.13) in order that they could display their normal behaviour pattern of selection of shallow water near the shore. Typical results of observations on the survival of controls, experimental fish and experimental fish retained

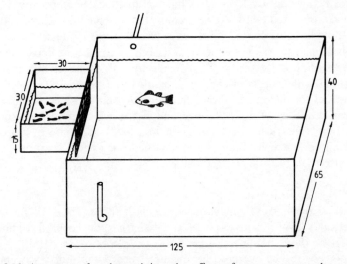

FIG. 8.13 Apparatus for determining the effect of stresses on predator–prey relationships. Dimensions in cm. Water depth is maintained by means of the inverted siphon and a constant inflow. Prey species which wander from the screened shallow-water refuge are exposed to the predator. From Goodyear (1972).

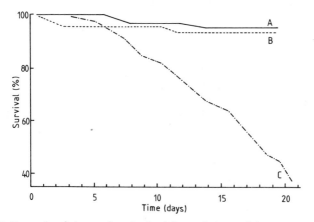

FIG. 8.14 Example of the results obtained from the use of the apparatus shown in Fig. 8.13. Group A were untreated but exposed to predation; Group B were treated but kept from predation by means of a fine screen; Group C were treated and exposed. From Goodyear (1972).

within a fine mesh screen are shown in Fig. 8.14. The apparatus could be modified to give different continuous treatments to both sections if necessary.

Increased vulnerability of Atlantic salmon parr to predation by brook trout (*Salvelinus*) after exposure to 1 ppm of the organophosphate insecticide Sumithion for 24 hours has been demonstrated by Hatfield and Anderson (1972). Sumithion at 0·1 ppm and DDT at 0·07 ppm had no noticeable effect. Predation on sockeye salmon fry by coho salmon was enhanced in the laboratory by thermal stress shocks (Sylvester, 1972a), which suggests that survival under natural conditions could be affected by sudden changes in temperature.

Habitat selection is affected by some pesticides. For example, Ogilvie and Anderson (1965) have shown that immediately after exposure to DDT, salmon selected a lower water temperature and, after more severe exposure, a higher water temperature than usual. Such changes could be important in wild fish in that they might lead to disorientation. Silbergeld (1973) found that darters (*Etheostoma nigrum*) were more susceptible to thermal stress when they had previously been exposed to sub-lethal concentrations of dieldrin. The effects of two insecticides, DDT and malathion, on salinity selection by mosquito fish was studied by Hansen (1972) who found that only DDT had any affect and caused the selection of higher salinities than the controls.

Kleerekoper *et al.* (1972) studied orientation in goldfish (*Carassius auratus*) in gradients of sub-lethal copper and found that swimming movements, including the direction and rate of turn, were affected by copper at $11-17 \mu g$ litre^{-1}. The fish appeared to be attracted towards the copper and then remained trapped within it. Behavioural effects of copper have been described from experimental and field trapping of perch and roach after exposure to low concentrations of copper (0·1–0·5 ppm) by Stott (1970), in which there was an elevenfold short-term increase in catches. Exposure to $10 \mu g$ litre^{-1} *p,p'*-DDT changed the locomotory behaviour of goldfish (Davy *et al.*, 1972) and this did not return to its normal pattern even after almost 140 days in clean water.

8.2.2.3.7 Use of growth rates

A considerable body of data has been accumulated on the growth rates of fish and other freshwater biota. It might reasonably be expected that growth rates might provide useful indications of the effects of exposure to pollutants since growth is known to be affected by many environmental factors. From a review (Sprague, 1971) of relevant literature, however, it is evident that this approach is apparently not uniformly successful. Problems arise from compensatory growth following temporary reductions in growth rates, and for most species it would be difficult to apply experimental findings to field situations since one could not adequately estimate the contributions of variations in nutrition or other environmental factors.

Recently, Haines (1973) has evaluated the use of RNA–DNA ratios for measuring long-term growth in fish populations. He concluded that this approach could reliably indicate long-term population growth in fish when the age structure is known and recruitment is controlled, and considered that the method is potentially useful for detecting responses to environmental changes before changes in growth rates become severely affected.

8.2.2.4 Bioassays with macroinvertebrates

Macroinvertebrates appear to have been used far less than fishes in bioassays to detect the effects of exposure to sub-lethal doses of pollutants. Several workers have used physiological methods such as changes in respiratory activity or oxygen consumption rates to assess the toxicity of pollutants. For example, Cole and Wilhm (1973) measured oxygen consumption of chironomid larvae (*Chironomus attenuatus*) exposed to phenol while Brkovic-Popovic and Popovic (1977b) used respirometric analyses to investigate the effects of heavy metals on oxygen consumption in the oligochaete worm. The degree of reproductive impairment of crustacean populations in culture on exposure to toxicants has been studied by several workers (Nebeker and Puglisi, 1974; Maki and Johnson, 1975; Canton *et al.*, 1975, Schober and Lampert, 1977). However, these methods hardly compare with the fish techniques described above. Some interesting and potentially useful approaches, which indicate how macroinvertebrate bioassay procedures might develop, are outlined below.

Variation in the heart-beat rate of *Caridina pristis* was used by Costa (1970) in investigations of the effects of DDT. This small shrimp was placed in an L-shaped tube, connected to aspirators containing water and insecticide, and viewed under a binocular microscope. The heart beat can easily be determined and its rate counted. Concentrations of 0·1 and 0·05 ppm DDT caused a decrease in heart rate while low concentrations of dieldrin increased the rate. Marked change in pH decreased the rate while high salinity increased it.

The efficiency of net construction by the caddis fly larva *Hydropsyche* was used by Décamps *et al.* (1973) to assess the effects of exposure to sub-lethal concentrations of dieldrin and DDT. Concentrations of insecticide from 0·05 to 10 µg litre^{-1} were used and these resulted in the formation of abnormal nets, the frequency of which increased with concentration.

A somewhat longer than usual test (taking up to 10 or more days) has been devised (Stebbing and Pomroy, 1978) which utilises the rate of asexual reproduction in *Hydra littoralis* cultured in a synthetic medium. The method is sensitive enough to detect levels of copper of 2–12 µg litre^{-1} and has been used successfully to detect cadmium.

The hatching rate of the brine shrimp *Artemia salina* (L) has been proposed (Jensen, 1975) as a laboratory bioassay. The advantage of using this animal is that the eggs are readily available and may be stored indefinitely until needed. A principal disadvantage is the necessity to use artificial seawater as the hatching medium. The toxicant is added to the medium and the percentage hatching rate (compared with the control) is noted.

CHAPTER 9

Field Assessments of Environmental Quality

9.1 INTRODUCTION

Techniques for assessing the responses of organisms to pollutants under controlled laboratory conditions were outlined in the previous chapter. These methods can provide reliable and reproducible measurements of the responses to known concentrations of toxicants or other stresses but it may sometimes prove difficult to apply the results when attempting to predict their environmental consequences.

In this chapter, methods for measuring environmental quality by examining the condition of the biological community *in situ* are described. The successful use of field observations depends, however, upon previously acquired experience from parallel studies of the distribution of biota, or community structure, and of physico-chemical parameters, together with studies of environments with known stresses—for example, downstream of an effluent outfall. Laboratory studies are invaluable in assisting in the interpretation of field observations.

Direct field observation of communities is rarely possible, with the exception, perhaps, or aquatic macrophytes, so that some indirect means have to be employed. Most commonly this is achieved by recovering a representative sample which can be examined in the field or, more usually, later in the laboratory. The actual sampling technique is governed by many factors, among which the purpose of the investigation, the community to be assessed and the nature of the habitat are paramount. Sometimes a qualitative sample (i.e. one that contains a representative range of the organisms present within the community) will suffice. Fully quantitative samples, from which the absolute abundance of organisms within a measured unit of the habitat may be estimated, are necessary in some instances. A useful compromise is the semiquantitative sample; this provides an indication of the relative abundance of the components of the community but does not enable one to relate them, in absolute terms, to a defined area or volume of the habitat. Generally, fully quantitative methods are more time-consuming but the data produced are suitable for analysis by almost any method: qualitative data naturally limit the analysis options but usually require fewer resources.

A comprehensive review of sampling methods and strategies together with data analysis techniques has been provided elsewhere (Hellawell, 1978). Here the broad outlines and key features of these methods will be presented.

9.2 SAMPLING METHODS

The relative merits of different indicator groups was considered earlier (Section 3.2). Of these, macroinvertebrates, algae, fish and macrophytes are the principal indicator communities in current use, and some methods appropriate to them will be described. The principles behind the main methods of sample collection fall conveniently under three main heads:

(1) Methods that disturb the habitat and its associated organisms which are simultaneously extracted and separated, to varying degrees, from their surroundings (e.g. nets).
(2) Methods that remove an appropriate part of the habitat, together with its associated organisms intact, from which the biota are separated or extracted subsequently (e.g. grabs).
(3) Methods that provide a habitat for colonisation by organisms which, after an appropriate period of exposure, is removed and examined (e.g. artificial substrates).

These methods are also broadly divisible into two groups, active sampling methods (1 and 2) and passive methods (3), characteristics which determine the immediacy of sample analysis and the extent to which the results may be influenced by operator skills.

The efficacy of the different methods depends on the degree to which they extract all the organisms of interest and separate them from the habitat material, and also the extent to which they may introduce bias into the results by recovering certain organisms more readily than others. The importance of these factors varies according to the purpose of the survey but most developments in sampling apparatus, and the tendency for each investigator to modify existing designs to suit his own purpose, have arisen through inadequacies, both real or perceived, in efficiency of removal or selectivity in action.

9.2.1 Qualitative Methods for Benthic Macroinvertebrates

9.2.1.1 *Hand net*
One of the most versatile implements for qualitative sampling of freshwater environments is the hand net, essentially a mesh bag attached to a rectangular or triangular metal frame which in turn is fitted to a handle or pole (Fig. 9.1a). In running water it is held vertically upon the stream bed and an area of the substratum immediately upstream of the net mouth is disturbed by hand or foot. Hand-lifting of stones is really only appropriate where stones are large enough to be handled individually and where the current is swift. More commonly 'kick-samples' are obtained; in this method the stream bed is vigorously disturbed by movements of the feet and the dislodged fauna is swept into the net by the current. Although the method is qualitative in that the area (and depth) disturbed are not fixed some operators attempt to minimise this source of variability by sampling for a definite period (say 3 minutes) or by moving forwards along a diagonal transect for a fixed distance. The method has been standardised (Department of the Environment, 1979) and calibrated (Morgan and Egglishaw, 1965; Frost *et al.*, 1971; Furse *et al.*, 1981). It is now generally recognised that the hand net can prove most effective for *in situ* studies of benthic communities.

9.2.1.2 *Shovels and dredges*
Shovels and dredges are essentially nets mounted on robust rectangular or triangular frames with strong cutting edges at the front (Fig. 9.1b,c). The main difference between these two kinds of samplers is that shovels are pushed through the substratum (and, therefore, used in relatively shallow water) while dredges are pulled (making them suitable for hauling in deeper water). The comparative performance of four patterns of dredge has been appraised by Elliott and Drake (1981b). In general, differences in their efficiencies could be related to the size of the dredges: the larger patterns required fewer replicates than the smaller models in order to obtain adequate estimates of the taxa present in the community.

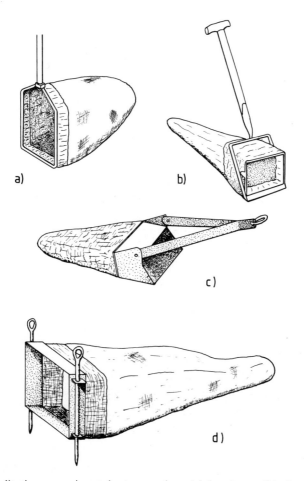

FIG. 9.1 Qualitative macroinvertebrate samplers: (a) hand net, (b) shovel, (c) dredge, (d) drift net.

9.2.1.3 Drift nets

Many macroinvertebrates move actively or drift passively in the current, a phenomenon which has been shown to exhibit a strong diurnal pattern of intensity (Waters, 1962; Elliott, 1969). Some taxonomic groups, such as insect larvae or nymphs, are more likely to be encountered in drift than others, for example molluscs or Trichoptera with their heavy shells and cases, but in spite of these limitations, Besch (1966) has suggested that, since collections of drifting invertebrates may be obtained with relatively little debris, and thus require reduced effort when sorting samples later, they may prove quite useful in food surveys. While it is possible to calibrate drift nets by estimating the volume of water which passes through them in a given period, it is not possible to relate the catches to the actual densities of organisms in the stream bed. They must, therefore, be regarded as providing qualitative samples of the macroinvertebrates and even then only of part of the community. A typical pattern of drift-net sampler is shown in Fig. 9.1d.

9.2.2 Quantitative Methods for Macroinvertebrates

9.2.2.1 Surber-type samplers
The Surber sampler combines a rectangular quadrat to delineate the area of bed to be sampled and a net into which the disturbed benthic invertebrates are swept by the current (Surber, 1937; Macan, 1958). Two triangular wings of netting, linking the lateral margins of the two frames, help to reduce the loss of material around the sides of the net (Fig. 9.2a). The apparatus has been developed to a considerable degree and many patterns now resemble rectangular boxes (Fig. 9.2b). As the area of bed to be sampled is fixed these samplers are quantitative although, of course, the depth to which the operator may disturb and dislodge the benthos may vary. Evidence that benthic invertebrates can inhabit deep gravel and that their density may increase with depth (Coleman and Hynes, 1970) means that even these methods may not provide absolute measures of population densities.

9.2.2.2 Cylinder samplers
The cylinder sampler is essentially a circular cylindrical pattern of Surber-type sampler (Fig. 9.2c) having the intrinsic advantage of being easier to insert into the substratum by means of a rotating action. The shape also means that, for a given weight of constructional material, they are stronger than box-type Surber samplers.

9.2.2.3 Grabs
Grabs are designed to remove a portion of the substrate and the animals which it contains by a biting action. The fauna within the sample is extracted subsequently, usually by sieving. Different methods are used to close the jaws of the grab as it bites into the bed. Some patterns use powerful springs which are set before the grab is lowered and are

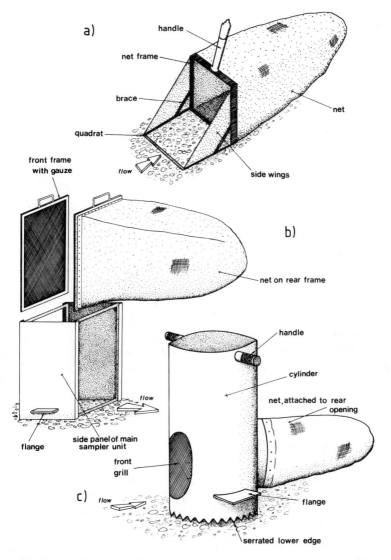

FIG. 9.2 Quantitative macroinvertebrate samplers: (a) Surber sampler, (b) rectangular box sampler, (c) cylinder sampler.

released automatically as the grab makes contact with the bottom or by a 'messenger' weight sent along the hauling cable. Other patterns use a system of levers or pulleys to close the jaws as the grab is hauled. Depth of penetration into the substrate is often dependent upon the weight of the grab. Grabs which effectively penetrate hard or compacted substrates may require quite substantial auxiliary equipment in order to cope with their great weight and may, therefore, be less suitable for freshwater environments. Many grabs were originally designed for marine use where weight is less of a disadvantage. Their complicated mechanisms were intended to reduce the risk of failure, which at sea is a major consideration since hauling may be a lengthy process and operating costs are high.

Grabs used in shallower, freshwater environments may be employed on a pole, rather than a rope or cable, which makes it easier to place them accurately and to penetrate the substrate by adding additional force through the pole. By arranging that the pole releases the jaw mechanism inadvertent premature closure is avoided and the grab only takes its sample when the operator is satisfied that it will work correctly.

A survey of the literature on the efficiencies of different patterns of grabs lead to the conclusion that in freshwater the Ponar grab was probably best for hard substrates and the Ekman grab best for soft substrates (Hellawell, 1978). In a study of the field and laboratory performance of seven grabs appropriate for use in freshwater (i.e. less than 25 kg in weight in order to make them suitable for manual operation from a small boat), Elliott and Drake (1981a) also concluded that the Ponar pattern was suitable for small stones (up to 16 mm) while the Ekman type was adequate for fine gravel (2–4 mm). Examples of some patterns of grab are shown in Fig. 9.3.

9.2.2.4 Core samplers

A core sampler (Fig. 9.4) is in essence a tube, usually not more than 150 mm in diameter, which is inserted vertically into the sediment and, when withdrawn, retains the enclosed material and the fauna within it. Single core samplers may be used or several may be combined in multiple units. They are best suited to sampling relatively soft homogeneous substrates and have found considerable use in studies of the benthos of lakes.

9.2.2.5 Air-lift samplers

Air-lift samplers (Fig. 9.5) use compressed air to scour the substrate and

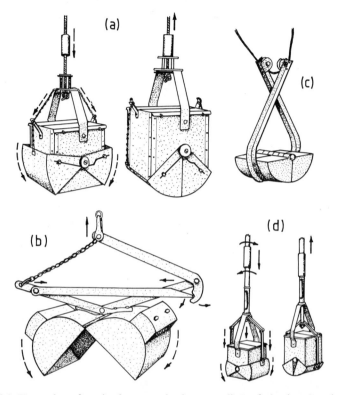

FIG. 9.3 Examples of grabs for quantitative sampling of the benthos in deep water: (a) cable operated Ekman grab, (b) Petersen grab, (c) Van Veen grab, (d) pole operated Ekman grab.

raise water, lighter substrate material and fauna as the air ascends within a delivery pipe. The material is discharged into a net where the animals and debris are retained while the water, air and very fine substrate escape. The suction power is adjusted by controlling the air pressure and the duration of the air blasts. Clearly, the air-lift works better in static or slowly moving water. The design of MacKey (1972) is simple but probably less quantitative than that of Pearson *et al.* (1973). Recently, Drake and Elliott (1982) compared the performance of both designs under field and laboratory conditions. They found that the MacKey sampler had the more powerful action, lifting stones up to 128 mm long, but it tended to 'oversample' in relation to the nominal sampling cross-sectional area of the riser pipe. The sampler of Pearson *et al.* (1973)

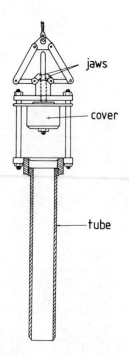

FIG. 9.4 Core sampler. After Kajak (1971).

raised stones up to 32 mm long but tended to under-sample, except at higher rates of air-flow when the results were approximately quantitative and comparable with those obtained by means of a Ponar grab (see Section 9.2.2.3).

9.2.3 Colonisation Samplers

The contagious distribution of benthic organisms in heterogeneous substrates causes considerable variation in the composition of samples taken by conventional methods. This has prompted the search for techniques which overcome this problem, and differences which may be attributed to varying abilities of operators, by providing a standard, uniform artificial substrate for colonisation by the indigenous fauna. Colonisation samplers may, however, prove very selective in attracting only those components of the fauna which find the artificial habitat to their liking. This feature is particularly evident in those samplers which are suspended in mid-water (Hester and Dendy, 1962; Mason *et al.*, 1967)

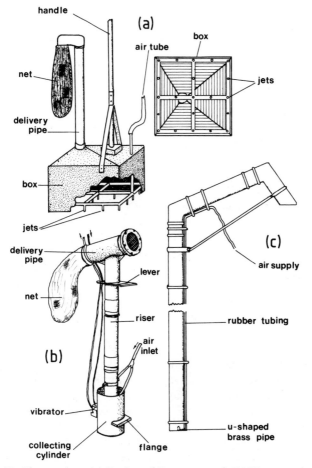

FIG. 9.5 Air lift samplers. (a) Design of Pearson *et al.* (1973); general view and view from underneath. (b) FBA design, Drake *et al.* (1983). (c) Design of MacKey (1972).

and which, therefore, can only be colonised by drifting organisms and would present an extreme case where a sampler is suspended above an inhospitable substrate, devoid of a fauna, yet is able to support a thriving community derived entirely from the drift. Yet other disadvantages of these samplers include the possibility of interference or vandalism while the artificial substrate is exposed for colonisation for a period normally extending over several weeks, the need for two visits to the site in order to set and retrieve the samplers, and, of course, the delay in obtaining results. These latter disadvantages may be less relevant in a programme of routine biological surveillance where exposed sampling units may be collected and new ones set in a single visit. Where the principal objective of the survey is to assess water quality by biological means even an 'unrepresentative' fauna may not be a disadvantage: one is simply observing that community which is able to survive in the prevailing water quality. Indeed, in rivers or streams where sampling sites are selected in relation to effluent outfalls etc., and no suitable natural substrate exists, the provision of artificial substrates may be the only means whereby biological assessments may be made. Colonisation samplers are particularly useful for routine surveillance programmes at sites which are too deep to be sampled by hand-nets or Surber-pattern samplers.

Many patterns of colonisation sampler have been described: the principal designs and their modes of operation have been reviewed elsewhere (Hellawell, 1978). Most colonisation samplers are designed to provide interstices, thus simulating the natural gravel bed, while others are intended to mimic water weeds (Macan and Kitching, 1972). The majority of colonisation samplers have been sited either on or above the natural river bed but some workers have embedded their samplers within the natural substrate (for example, Coleman and Hynes, 1970; Radford and Hartland-Rowe, 1971).

Of the various designs available, the multiple-plate sampler of Hester and Dendy (1962) and the wire-baskets of Mason *et al.* (1967) are relatively simple to make and use (Fig. 9.6). A novel colonisation-sampler, which uses plastic media originally developed for sewage-treatment percolators, has been described by Watton *et al.* (1983). There is much to be said for a simple plastic-mesh bag containing stones or for concrete or terra-cotta 'air-bricks', which are used to provide ventilation in buildings, since these are cheap, readily obtained, easily planted and recovered, virtually indestructible and vandal-proof (Hellawell, 1978). It has proved difficult to assess the relative efficiencies of colonisation samplers since their *modus operandi* is so different from the conventional samplers with which they have been compared. A review of several comparative studies suggests that they can collect large numbers of organisms and in some habitats are far more effective than other techniques (Hellawell, 1978) but one cannot escape the feeling that, in some locations, colonisation samples collect a fauna which, in the absence of the sampler, simply would not exist!

a) b)

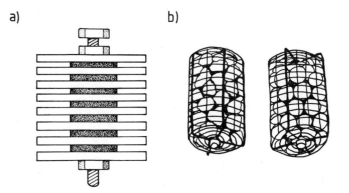

FIG. 9.6 Artificial substrate (colonisation) samplers. (a) Design of Hester and
Dendy (1962). (b) Basket of Mason *et al.* (1973).

9.2.4 Selection of Appropriate Macroinvertebrate Methods
So far, although the merits of individual patterns of sampling equipment
have been considered, no attempt has been made to give advice on
suitable methods for each habitat type. Nor has any attempt been made
to cover the planktonic macroinvertebrates of the open water of lakes,
ponds or slow rivers. Apart from the use of the hand net, plankton nets
designed for algological collections, but having different mesh apertures,
are appropriate for sampling planktonic faunas and these techniques are
considered in Section 9.2.5. Guidance on suitable methods for different
habitats will be found in Table 9.1.

9.2.5 Algal Methods
Freshwater algae are either planktonic or attached to the surfaces of
other plants or stones, sand and mud. The true planktonic species are
largely confined to lakes: river algal plankton is usually adventitious and
derived from detachment of benthic algae or from upstream lakes or
impoundments. A true algal plankton may develop in very large, slowly
moving rivers.

   Planktonic algae may be sampled simply by filling a conventional
limnological water-sampling bottle or by pumping through a hosepipe.
Both techniques can be used to sample at given depths and so obtain
some indication of the vertical distribution of planktonic algae. These
methods are appropriate for eutrophic waters where algal densities are
likely to be high enough to avoid the need for too much subsequent
concentration of the samples. In more oligotrophic waters the use of an

TABLE 9.1

RECOMMENDED METHODS FOR SAMPLING MACROINVERTEBRATE COMMUNITIES

(After Hellawell, 1978)

Water depth	Current velocity	Substrate type	Qualitative surveys	Quantitative surveys
Shallow (less than 1 m)	Swift	Boulders, large stones	Hand net—lifting stones by hand Colonisation samplers	Surber-type sampler, where this can be located properly
	Moderate	Gravel	Handnet—kicking to disturb substrate	Surber-type sampler Air-lift if depth is adequate
	Slow	Gravel-sand	Shovel sampler	Cylinder sampler Air-lift if depth is adequate
	Very slow or static	Sand mud	Shovel sampler	Core sampler (hand operated) Air-lift if depth is adequate
Moderate (1–3 m)	Slow	Gravel-sand		Ekman grab Pole-mounted corer
	Static	Sand-mud		Pole-mounted Ekman grab
Deep (more than 3 m)	Very slow	Sand-mud Compacted substrate		Ekman or Ponar grab Ponar grab
	Static	Sand-mud Compacted substrate		Air-lift sampler (diver operated) Ponar grab

Quantitative methods can also be used in qualitative surveys. Compacted substrates sometimes occur in large rivers, especially if dredged.

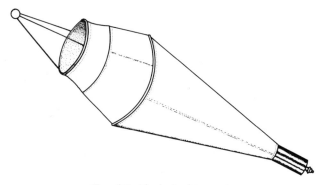

FIG. 9.7 Algal plankton net.

algal plankton net (Fig. 9.7) provides some preliminary concentration of the sample. If necessary, quantitative estimates of algal abundance may be made by calibrating the net, although for accurate work incorporation of a flow meter is necessary as in the Clarke-Bumpus (1940) plankton sampler. The composition of the catch is affected by the net mesh size; usually the smallest size has a mesh aperture of 60 μm.

Attached algae may be removed in the field by scraping (Sladeckova, 1962) or by careful brushing (Douglas, 1958) but generally the substrate and attached algae are returned to the laboratory where algae may be removed by agitation, brushing or scraping, and examined under the microscope. Dilute hydrochloric acid has been found to be of assistance in removing algae from stones. However, algae may be damaged in attempting to remove them or different species may be removed with varying success so that the community which is examined is not truly representative of that on the substrate.

In order to overcome these deficiencies many algologists have used artificial substrates which are exposed in the habitat of interest for 1 to 6 weeks after which they are returned to the laboratory for examination. Glass microscope slides have often been used as artificial substrates since these are clearly most convenient for microscopic examination, but they cannot really be used for truly quantitative work since some common species seem reluctant to grow on glass slides and the communities which do occur may be highly unstable (Butcher, 1947; Tippett, 1970). Various forms of apparatus have been devised in order to hold the microscope slides in the river or lake. Early patterns include the 'diatometer' in which slides are suspended vertically at the surface (Patrick and Hohn, 1956; Patrick et al., 1954). Sladeckova (1962) also suspended slides from

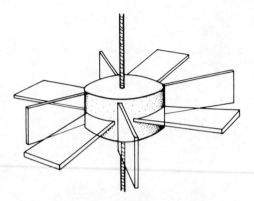

FIG. 9.8 Method of suspending glass microscope slides for colonisation by periphyton. After Sladeckova (1962).

surface floats in different orientations (Fig. 9.8). Descriptions of the design and construction of simple slide holders have been provided by Anderson and Paulson (1972) and McMahon *et al.* (1974).

9.2.6 Fish Sampling Methods

Fish are potentially valuable indicators of water quality since we know much about their environmental requirements and the effects of toxicants or other stresses upon them. Often they are of commercial value and, being vertebrates, have physiological responses which are closer to our own than, say, molluscs or insects. Their main disadvantage as indicators is the difficulty with which they may be sampled, especially by non-destructive means. Other groups, such as macroinvertebrates, algae or macrophytes, may be sampled by a single operator but this is rarely possible for fish and often many hands are needed. A considerable body of experience of fish methods has been accumulated and incorporated into the literature (Vibert, 1967; Bagenal, 1974a, 1978). Fish are rarely studied for assessments of environmental quality *per se* but more usually for fishery management purposes. However, the main methods of fish capture will be outlined here.

9.2.6.1 Netting methods

Small fish species may be taken with the hand-net used in macroinvertebrate sampling (Section 9.2.1.1) but more usually larger nets are required. Gill nets are set in still water and trawl nets are suitable

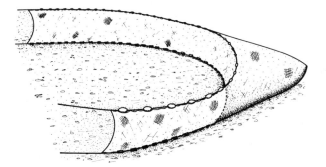

FIG. 9.9 Seine net.

only for large shallow lakes without bottom obstructions. Almost always, nets are restricted to low water velocities: even a small river can exert considerable force on a net which has accumulated weeds, leaves or other debris. A useful practical guide to the use of nets has been provided by Fort and Brayshaw (1961). For most purposes a seine net, either plain or having a bag in the centre part, will be most useful (Fig. 9.9). Mesh size governs the minimum size of fish which can be retained and usually a compromise has to be struck between the mesh size needed to catch the smallest individuals and lowest net resistance when hauling the net at a speed in order to reduce escapes. The balance between the amount of weight needed to prevent the leadline from lifting and the floats required to hold the net vertical without lifting the leadline is largely a matter of familiarity with the equipment and experience in its use. Usually a seine net has to be hauled to a beach or a shallow area in order to recover the catch, a factor which sometimes imposes limitations on the areas in which it may be deployed. Netting methods are now mainly used where electrofishing gear (see below) are less effective; for example, in deep or highly electrically-conductive waters.

Special nets have been developed for sampling fish fry (Bagenal, 1974b; Brown and Langford 1975).

9.2.6.2 *Electrofishing methods*
Electrofishing depends upon the maintenance of an electric field between (or around) electrodes by the application of alternating, direct or pulsed direct current. Alternating current (AC) stuns fish and allows them to be collected in hand-nets or in a stop-net in flowing water. A direct current (DC) induces 'galvanotaxis', that is, an involuntary swimming movement

towards the anode where again they may be recovered by means of a hand net. This property of direct current is particularly useful when fish may otherwise attempt to hide under stones, in tree roots or weed beds, or in turbid water, where stunned fish might otherwise not be seen.

Electrofishing equipment is most effective in small streams having low electrical conductivity. In highly conductive water the power demanded in order to maintain adequate voltage is considerable and not easily met in readily portable equipment. This problem may be overcome to a large extent by the use of pulsed direct current. The neurophysiological effect is equivalent to that of the same voltage of smooth DC but the power is used intermittently and so for a given maximum voltage a smaller generator is required.

Small fishes are normally less readily caught by electrofishing methods since the potential drop along their bodies is proportionally less but, fortunately, pulsed DC seems to suffer least from this defect.

9.2.6.3 Fish trapping methods

The use of traps to capture fish is very ancient. A simple pattern of fish trap (Fig. 9.10) resembling a lobster or crab 'pot' is the 'Windermere' trap (Worthington, 1980). It was developed for use in lakes but is also effective in rivers. Traps are economical on labour since it is possible to

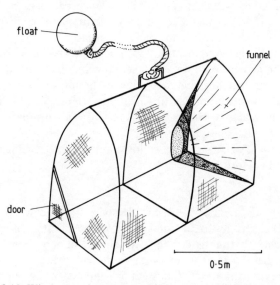

Fig. 9.10 Windermere perch trap. Redrawn after Bagenal (1972).

set and recover them single-handed, they have little adverse effect on the environment and are suitable for reaches of rivers which are too deep to be fished electrically or with seine nets. They appear to work equally well when unbaited but their overall catching power is low (Stott, 1970) and many replicates are required for reproducible results (Bagenal, 1972). This, coupled with their vulnerability to interference or vandalism, rather negates the benefits of low costs in construction and manpower requirements.

9.2.6.4 Poisons for fish capture

Fish are captured for food by means of toxic substances in some parts of the world, and two substances, antimycin and rotenone, have been used for scientific sampling. Further details of substances with piscicidal activity are to be found in Chapter 7. As a method for routine surveillance, the use of poisons has limited value although rotenone poisoning may be reversible even after prolonged exposure under natural conditions and can be detoxified by the use of potassium permanganate (up to $3\,\mathrm{mg\,litre^{-1}}$). Recovery from antimycin is not usually possible after the first signs of distress have been observed (Gilderhus, 1972).

9.2.6.5 Population estimates

The methods listed above are essentially qualitative methods which give an indication of the species present and some idea of the population size. To obtain more accurate population estimates it is necessary to employ either multiple catch or mark-recapture methods.

Multiple catch is a system whereby the population is sampled repeatedly (without replacement) and the numbers taken in successive samplings are used to estimate the total population. If the number of fish caught in the first, second and subsequent samplings is plotted on the ordinate and the cumulative catch from previous fishings is plotted on the abscissa, a line fitted through the resulting points (Fig. 9.11) will intersect at the estimate of the total population (De Lury, 1947). Recently, Cross and Stott (1975) and shown that the method may require correction for decreased catchability after repeated electric fishings.

A rapid method of estimating the total population is the two-catch methods of Seber and Le Cren (1967). This is given by the equation:

$$N = \frac{C_1^2}{C_1 - C_2}$$

where N = the estimate of the total population, C_1 = the number taken in the first catch and C_2 = the number taken in the second catch.

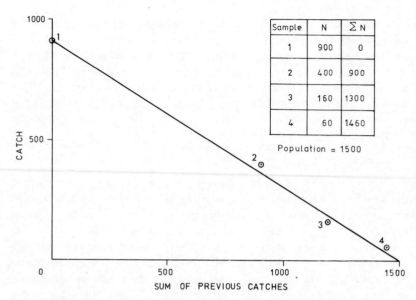

Sample	N	Σ N
1	900	0
2	400	900
3	160	1300
4	60	1460

Population = 1500

FIG. 9.11 Graphical or De Lury method of estimating fish populations by repeated sampling using identical effort.

In mark-recapture methods a sample is taken, marked distinctively (for example, by means of a tag, branding or a dye) and released again. After an interval, during which the marked fish distribute themselves through the population, a second sample is taken, using the same fishing effort. This catch is examined and the proportion of marked and unmarked fish is noted. The total population (N) is then estimated by the formula:

$$N = \frac{ms}{r}$$

where m = the number of fish marked in the first sample, s = the total number of fish in the second sample and r = the number of recaptures, that is, marked fish in the second sample.

Several modifications and improvements to both multiple catch and mark-recapture methods have been proposed; a recent comprehensive review has been provided by Cowx (1983).

9.2.7 Macrophyte Sampling Methods

Of the major taxa employed as indicators, macrophytes are somewhat unusual in that they may be examined *in situ* fairly easily. Indeed, this is claimed to be one of their principal advantages as aquatic environmental indicators.

In qualitative surveys macrophytes may be identified from the bank, a boat or bridges (Haslam, 1982b). Where visibility is impaired by turbidity or depth, samples may be recovered by means of a grapnel. Simple estimates of percentage cover or presence or absence within a rectangular grid or along transects provides semi-quantitative data on the relative abundance of species.

For fully quantitative studies biomass may be measured by direct cropping and removal. In deeper water this is best achieved by Scuba diving. The biomass may be measured easily and conveniently as wet weight after spin-drying (Edwards and Owens, 1960). Indirect estimates of biomass may be made in the field by measuring the attenuation of light in beds of weed although it is necessary, of course, to calibrate the measurements against known quantities of weed (Westlake, 1964; Owens *et al.*, 1967).

9.3 SAMPLING STRATEGIES AND PROGRAMMES

Experience suggests that sample collection is often the weakest part of a programme of biological surveillance by means of *in situ* community analysis. Now that quite powerful computing facilities are widely available it is possible to subject data to the most searching and complicated analyses but no matter how sophisticated these techniques are it is virtually impossible to compensate for inadequacies in the actual sampling strategy. It is imperative, therefore, before embarking on a comprehensive survey programme, that careful, preliminary appraisal should be made of the extent to which the objectives of the exercise are likely to be met by what is planned.

There are, however, no simple solutions to the 'sampling problem' which is, in essence, that in order to detect a desired level of environmental change with a given degree of probability an inordinate number of samples will often be required. General reviews by Southwood (1966), Hynes (1970), Elliott (1971) and Hellawell (1978) provide guidance in planning sampling programmes. A review of chemical surveillance of rivers and effluents by Montgomery and Hart (1974) is also relevant.

The sampling strategy includes selection of appropriate sampling stations, determination of the periodicity (i.e. frequency) of sampling and a decision on the kind of samplers needed, that is whether quantitative samples are necessary or whether qualitative samples will suffice. All these decisions are governed by the objectives of the exercise and, more pragmatically, by the available resources, both human and scientific.

Selection of sampling points is specific to the exercise in hand but some general and proven principles may usefully be considered here. For example, in many surveys of the effects of effluent discharge, sites are selected above and below the outfall in order to assess the impact of the effluent on the receiving stream. An obvious requirement is that the two sites shall, as far as possible, be similar in their substrate material, depth, water velocity, and so on in order to minimise the influence of factors other than the effluent discharge. Such considerations may preclude the selection of sites in the immediate vicinity of the outfall. In addition one might usefully choose sites further downstream in order to determine the extent and degree of influence of the effluent while at least one, further, site might be selected upstream, thereby securing greater confidence in the estimates of the unpolluted condition of the stream. In practice, rivers often receive effluents from many sources or are subject to several managerial influences. It is usually possible, therefore, to arrange that a series of sampling stations will serve as 'controls' for their downstream neighbours and the cumulative effects of effluents and the rates of recovery of the river between outfalls can be monitored. In large rivers it is useful to select sites on both banks as some effluents may not easily mix laterally and thus will tend to exert their major influence along one bank.

Problems often arise when biological data are to be related to the findings of surveys of chemical water quality. Often, water samples are taken at road bridges since easy access means that effort is minimal and, provided the river is well mixed, such samples are as likely to be representative of the quality of that stretch as samples taken at less accessible points. Chemical quality surveys are also often based on sampling sites at, or just above, river confluences since these points provide an integrated sample of the catchment upstream. By judicious siting of stations above and below confluences the influence of each tributary can be determined with the minimum number of sampling sites. Rarely, however, are these sites suitable for biological sampling, especially macroinvertebrates, but they may be considered suitable for surveys of macrophytes provided that the sites coincide with bridges (Haslam, 1982a), which of course they often do. Surveys of fish populations at confluence sites may be difficult to interpret as the population sampled may have a home range extending both up- and down-stream and into the lowest part of the tributary. The size of sampling site is obviously governed by the group of indicators used: for algae the site may be a few square centimetres, for macroinvertebrates it would probably be less than a square metre. Macrophyte stations would be of the order of 100 m² while for fish the site may extend to 1000 m² or more.

Sampling frequency is also governed by the taxonomic group which is being used for surveillance, but the purpose of the investigation is the primary factor which determines the pattern of sample collection. Seasonal factors influence the abundance of species or, more exactly, the abundance of different life-history stages, while in temperate climates vegetation, both microscopic and macroscopic, shows marked seasonal changes. This was noted in Section 3.2.5 where the advantages and disadvantages of certain groups for routine surveillance were examined. It may help to explain why, in spite of exhortations by some workers (e.g. Whitton, 1979), algae and macrophytes do not appear to have been adopted widely for *in situ* indication of environmental quality.

For simple spatial and long-term temporal comparisons it is evident that samples should be taken at the same season, the precise season depending on the group. In temperate climates, spring is often chosen as the optimum period for annual macroinvertebrate surveys (e.g. Hynes, 1970), although the maximum number of species may be present in mid-summer (Brooker and Morris, 1980). However, summer-emerging insect species which are important for macroinvertebrate biotic indices are often well developed and easier to identify in spring while the increased contribution to total species numbers provided by other insects, such as the chironomid flies, does not materially improve the indicator value of samples since so little is known at present about the ecology of many of the species which are collected.

In surveys which are undertaken for particular purposes, for example in order to study the critical effects of low summer flows or the influence of changes in flow- or temperature-regime caused by river augumentation schemes or heated effluents, the sampling frequency strategy is largely pre-determined by the nature of the investigation. For general, routine programmes invertebrate samples are usually taken at frequencies ranging from once to twelve times a year with median frequency around two or three times a year.

Besides sampling frequency, another question which has to be

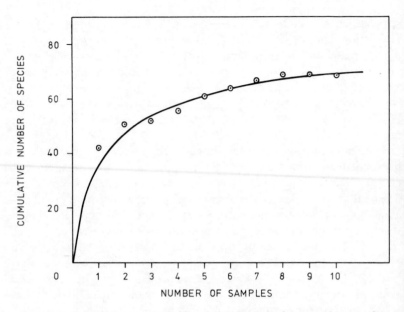

FIG. 9.12 Cumulative number of invertebrate species in successive samples.

considered is the number of sample replicates which are required. Factors which determine this are the nature of the investigation, in particular whether qualitative or quantitative data are required, the spatial distribution of the organisms, and the homogeneity of the habitat. In a qualitative survey the objective is to acquire as complete a species list as possible. Often about half of the estimated total number of macroinvertebrate species are collected in the first sample (Fig. 9.12) while subsequent samples tend to add fewer new species. The probability of catching certain groups is greater than others simply because they are more abundant and widely distributed. Table 9.2 gives an indication of the number of samples which may be required in order to obtain at least one of the specified group. It should be emphasised, however, that this table merely furnishes examples of the variation between groups in a particular study and location and that it is impossible to derive universal rules for sampling replication. The construction of a cumulative species curve such as that shown in Fig. 9.12 will help to decide what proportion of the 'total' species list is being collected by the level of sample replication which can be accommodated in a qualitative study, usually only of the order of one to five samples.

TABLE 9.2

VARIATION IN THE NUMBER OF SAMPLES REQUIRED TO BE 95% CERTAIN THAT AT LEAST ONE REPRESENTATIVE OF A GROUP OF ORGANISMS WILL BE SECURED
(The examples are selected from data in Needham and Usinger, 1956)

Organism or group	Number of samples
Plecoptera	2
Isoperla	12–17
Ephemeroptera	1–2
Baetis	2–3
Ephemerella	2
Rhithrogena	2
Trichoptera	1
Brachycentrus	4–5
Glossosoma	2–3
Hydropsyche	6–9
Lepidostoma	3–4
Rhyacophila	9–13
Diptera	1
Simuliidae	3
Chironomidae	2

When calculating certain biotic indices (for example, the Trent Biotic Index or BMWP Score) an adequate species list is all that is required since no account is taken of the abundance of indicators present.

For estimates of relative abundance, that is, of the importance of different taxa relative to each other, more samples may be required. For example, Thorup (1970), in a study of macroinvertebrates from a uniform section of stream, concluded that about 20 sample replicates were required for reliable estimates to be made of the relative abundance of the dominant species. For many purposes a reliable estimate of the relative importance of species is quite adequate. One may wish to know whether one species is much more abundant than another or whether any species dominates the community. Several biotic indices, such as the 'relative purity' index (Knöpp, 1954, 1955), the saprobic indices of Pantle and Buck (1955) or Biotic Score of Chandler (1970), require relative abundance information for their calculation.

Fully quantitative surveys are very demanding in terms of time and resources, but they are normally neither necessary nor practicable for the

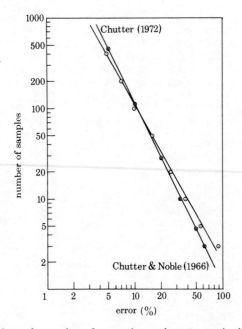

FIG. 9.13 Number of samples of macroinvertebrates required for population estimates with a given percentage error. Based on data in Chutter (1972) and Chutter and Noble (1966). From Hellawell (1977b).

conventional routine methods used for assessing environmental quality. The number of sample replicates which would be required in order to provide fully quantitative data may be inordinately high. For example, Hellawell (1978) has plotted the number of macroinvertebrate samples which would be required to obtain a population estimate with a given error at a 95% confidence level (Fig. 9.13) using published data from a study of a Californian and South African river. The similarity of the results is remarkable: approximately one hundred samples would be necessary to estimate the population with an error of 10%!

These results indicate that, for practical purposes, one may expect an increase in sample replication of an order of magnitude as one changes from qualitative to semi-quantitative to quantitative studies.

Other factors which influence the results of field investigations are the size of the sampling unit and the location of sample replicates. For quantitative studies the sampling unit should be as small as practically possible (Elliott, 1971) since this enables contagiously distributed populations to be sampled more efficiently. Small sample units do, however, have proportionably longer perimeters, and problems associated with this effect, such as fragmentation or escape losses, are increased. Handling and processing many small units may incur greater effort but in certain sampling strategies (for example, stratified sampling, explained below) the use of many small units permits more representative sampling of the microhabitats present.

The sample units may be located randomly within the habitat under investigation. Often, reported 'random' allocation of sites means that they were chosen without any particular conscious selection rather than in the strict statistical sense that all locations had an equal chance of being selected and the actual points were fixed by reference to random number tables. When truly random locations are sampled the results can be properly subjected to certain statistical analysis but it may be that the points fall out in such a way that one's instinctive reaction is that the results might be misleading, for example, in a habitat with marked variations in substrate condition. Here one may feel that a stratified random sampling procedure is more appropriate in which the samples are taken at randomly determined points but within clearly defined areas where the habitat is uniform, or the numbers of sample replicates are allocated in proportion to the areas of the habitat occupied by the various substrates.

Other methods for locating sampling points, for example, systematic sampling in which samples are taken at regular intervals, are generally less useful for sampling macroinvertebrates but may be useful for other organisms. Surveys of macrophytes can be readily undertaken by noting the species present at fixed intervals, say every half or one metre, along a transect. For rough mapping and for semi-quantitative estimates of cover the technique can be repeated along parallel transects, so forming a grid.

9.4 BIOTIC INDICES AND DATA ANALYSIS

9.4.1 Introduction

Experienced biologists are well able to make meaningful assessments of river water quality from survey results simply by inspecting the list of species found and their relative or absolute abundances (Dussart *et al.*, 1980). This is satisfactory for a single survey but when extensive surveys are undertaken over an extended period at many stations for surface water management purposes at regional or national levels, it is impossible to use or comprehend all the basic data. The problem then arises: how is one to reduce or condense the bulk of survey data and present it clearly and concisely, and, equally important, how does one convey the meaning and significance of the results to other interested parties who do not have the necessary biological expertise to comprehend them in their basic form?

A summary must be devised and this usually takes the form of an index. Although some loss of detailed information is inevitable, the overall comprehension which is provided is clearly an advantage. Often the resulting condensation or potentially more objective assessment, which an index may provide, gives new insights. Unfortunately all numerical indices are open to the possibility of abuse, especially of the kind which Elster (1960) calls 'mathematical pseudo-exactness' in which an essentially descriptive or qualitative index is given numerical values and then subjected to further statistical analysis or pressed to an accuracy totally unwarranted by the original data.

The appropriate choice of analytical method for survey data has received attention recently (Southwood, 1966; Clifford and Stephenson, 1975; Hellawell, 1977a,b, 1978). This is important because the costs of sample collection and analysis are relatively high and, as noted above, with the advent of relatively cheap and widely available computing facilities, it is essential that fullest use be made of survey results. The ideal method of analysis should be insensitive to the vagaries and problems of sampling, be indifferent to taxonomic difficulties, easy to calculate and, therefore, undemanding in comprehension. In addition it is beneficial if data analysis methods are ecologically meaningful as well as of practical significance. Few methods currently available would score highly on all those points!

One may conveniently classify the analysis methods for freshwater under three general headings: *pollution indices* which are derived from the responses of certain 'indicator' taxa to known pollution (usually organic pollution); *diversity indices*, which describe the structure of communities of organisms and, when the structure is changed by pollution or other stress, may be used to measure the extent to which a stressed community differs from an ideal structure; *comparative indices*, which may be used to compare the degree of similarity between a community subjected to stress and an unstressed community (or subjected to a known and measured stress) in either a spatial ('upstream and downstream') or temporal ('before and after') context. These three categories are considered in greater detail below.

9.4.2 Pollution Indices

Almost all pollution indices have been derived from the observation that there is a progressive loss of components of the clean water biota with increasing pollution load. From this phenomenon Kothé (1962) developed a simple index, the 'Artenfehlbetrag' or species deficit, which measures the percentage difference between the number of species occurring above and below the point of discharge of a pollutant, but the index takes no cognisance of changes in the relative abundances of individual species. Another index, proposed by Beck (1954), measures the difference between the numbers of 'tolerant' and 'intolerant' species, the distinction being made on subjective assessments of their limits of tolerance to organic pollution, but again no allowance is made for relative abundance. Other indices utilise the ratio between groups of organisms. For example, King and Ball (1964) have suggested the use of the ratio of wet weights of insects and tubificid worms but this seems both crude and naive, as does the proposal of Goodnight and Whitley (1960) in which the proportion of tubificid worms to other macroinvertebrates is used. Brinkhurst (1966) considered that a useful index of organic enrichment was provided by noting the number of tubificid worms and the proportion of *Limnodrilus hoffmeisteri* to all other species. *L. hoffmeisteri* is known to increase in abundance with organic pollution but this index may vary with seasonal abundance. Indices based on the ratios of readily recognised taxa may prove useful in detecting subtle changes in water quality when one species increasingly replaces another where their environmental tolerances overlap. Examples are few, but work by Hawkes and Davies (1971) has suggested that the ratio of *Gammarus* to *Asellus* may prove to be a useful indicator of organic enrichment.

In addition to the very simple indices described above, there are several, more complex, methods for calculating pollution indices which tend to incorporate common factors. These include an assessment of the relative abundances of each key species or group, recognition of their response to known pollution and a measure of their 'reliability' as indicators of pollution. A measure of the overall diversity of the community may be utilised in computing the index or some recognition is made of the total number of species or higher taxa present. Only qualitative data are required in order to calculate the Trent Biotic Index (Woodiwiss, 1964) and the BMWP Index (National Water Council, 1981) but quantitative or semi-quantitative (relative abundance) data are necessary for the indices of Knöpp (1954), Pantle and Buck (1955),

Dittmar (1959), Zelinka and Marvan (1961) and Chandler (1970).

An index may be derived from the product of scores for each of the factors described above and, in order to make the result less dependent upon sample size, it may be related to the total abundance of organisms by expressing the result as a quotient. The allocation of ratings for each factor is often highly subjective and the values used may be set quite arbitrarily. Identical index values may be derived from quite different combinations of factors: for example, a population of a few individuals of a sensitive pollution-intolerant species may gain the same score as that with many individuals of a ubiquitous, pollution-indifferent species. An index usually has a limited scale of possible values and may show reduced sensitivity at extreme or central values. It is, therefore, important to determine the behaviour of an index by applying it to familiar circumstances or synthetic data in order to understand its behaviour.

In using any pollution index it is imperative to note that while the basis of the index may have ecological validity it does not follow that the observation of a given index value is attributable to the kinds or intensities of the pollution from which it was originally developed. For example, a low index, suggestive of organic enrichment, may arise from adverse physical factors such as low flow, elevated temperature or even toxic waste. Indiscriminate use of pollution indices may, therefore, lead to erroneous conclusions.

9.4.1.1 Diversity indices

The use of indices of community diversity is based upon the concept that the structure of normal communities may be changed by perturbations in the environment and the degree of change in community structure may be used to assess the intensity of the environmental stress. No assumptions need be made regarding the nature of the stress, which therefore obviates the inherent weakness of most pollution indices in that they are developed from the observed responses of biota to particular pollutants, almost invariably organic wastes. It has long been recognised that in typical communities there are a few species which are abundant, several species which are less abundant and many species which are represented by very few individuals. There is no general agreement, however, as to which of several proposed models is the most adequate description of this general observation (Pielou, 1969). Examples include the logarithmic series model (Fisher *et al.*, 1943), the log-normal distribution model (Preston, 1948) and the ordered random-interval or 'broken-stick' models (e.g. MacArthur, 1957). Recently, May (1974) has suggested that these models are all cases of the log-normal distribution.

TABLE 9.3
SUMMARY OF THE PROPERTIES OF INDICES AND METHODS OF DATA ANALYSIS
(After Hellawell, 1977b)

(a) *Pollution indices*

1. Species deficit (Kothé, 1962)

$$I = \frac{S_u - S_d}{S_u} \times 100$$

S_u = no. of species upstream
S_d = no. downstream of outfall

2. Relative purity (Knöpp, 1954)

$$I = \frac{\Sigma(o + b)}{\Sigma(o + b + a + p)}$$

No. of species in each class
o = oligosaprobic
b = beta-mesosaprobic
a = alpha-mesosaprobic
p = polysaprobic

3. Saprobity index (Pantle and Buck, 1955)

$$I = \frac{\Sigma sh}{\Sigma h}$$

s = degree of saprobity (Liebmann, 1951)
oligo = 1, poly = 4
h = abundance (rare = 1, 3 = frequent,
5 = abundant)

4. Saprobic index (Zelinka and Marvan, 1961)

$$I = \frac{\Sigma ahg}{\Sigma hg}$$

a = saprobic valency in each of 5
saprobic classes (sum = 10)
h = abundance
g = indicator value (1–5; 5 = high)

5. Trent biotic index (Woodiwiss, 1964)

Derived from table provided. Highest score (10 upwards) is equivalent to lowest population. Uses qualitative data.

6. Biotic score (Chandler, 1970)

Sum of scores derived from table provided. Score of clean water rarely exceeds 2500. Uses quantitative data.

7. Pollution index (Beck, 1954)

$$I = 2S_i - S_t$$

S_i = no. of species, intolerant
S_t = tolerant to pollution
rarely exceeds 10

(b) *Diversity indices*

8. Williams' alpha index (Fisher *et al.*, 1943)

$$S \approx \log_e N/\alpha$$

S = no. of spp.
N = no. individuals
α = index of diversity

Derived from nomogram.

TABLE 9.3—*contd.*

9. Diversity index (Menhinick, 1964)

$I = S/\sqrt{N}$ Symbols as above

10. Diversity index (Margalef, 1951)

$I = S - 1/\log_e N$ Symbols as above

11. Diversity index (Simpson, 1949 or Pielou, 1969; Duffey, 1968)

$$I = -\sum \frac{n_i(n_i - 1)}{N(N - 1)}$$ n_i = no. of individuals in ith spp., otherwise as above. (values inverted)

12. Information theory diversity index (Shannon, 1948)

$$I = \sum_{r=1}^{s} p_r \log_2 p_r$$ p_r = proportion of individuals in rth spp.

13. Diversity index (McIntosh, 1967)

$$I = \sqrt{\sum_{i=1}^{s} n_i^2}$$ Symbols as above

14. Sequential comparison index (Cairns *et al.*, 1968)

$I = R/N$ R = no. of changes in species per scan
N = total number scanned

(c) *Comparative indices*

15. Coefficient of similarity (Jaccard, 1912)

$$I = \frac{c}{a + b - c}$$ a = no. of spp. in community A
b = no. in **B**
c = no. common to both

16. Coefficient of similarity (Kulezynski, 1948)

$$I = \frac{c}{2}\left(\frac{1}{a} + \frac{1}{b}\right)$$ Symbols as above

17. Quotient of similarity (Sorensen, 1948)

$$I = \frac{2c}{(a + b)}$$ Symbols as above

18. Index of similarity (Mountford, 1962)

$$I = \frac{2c}{2ab - (a + b)c}$$ Symbols as above

19. Comparative measure (Raabe, 1952)

$$I = \sum \min(d, e, f, \ldots z)$$ d, e, f, etc. are minimum % values of each species common to both communities

TABLE 9.3—*contd.*

20. Comparative measure (Czekanowski, 1913)

$$I = \frac{2\Sigma \min (g, h, \ldots z)}{A + B}$$

$g, h,$ etc. = lesser measures of abundance of species common to both communities; A and B are sums of measures of abundances in each community

21. Distance measure (Sokal, 1961)

$$D_{jh} = \sqrt{\sum_{i=1}^{s} (p_j^i - p_h^i)^2}$$

D_{jh} = distance between communities
p_j^i = proportion of spp. i in community j
p_h^i = proportion of spp. i in community h

Their value in biological surveillance is that they provide a comparative basis for studies of stressed environments since it is possible to measure the degree of deviation from the expected species abundance distributions.

Simple indices which merely relate the total number of species and individuals, that is 'species richness', have been proposed by Margalef (1951) and Menhinick (1964). The distribution of the numbers of individuals per species is utilised in calculating the indices of Simpson (1949), McIntosh (1967) and in the information-theory diversity index (Shannon, 1948; Wilhm and Dorris, 1968).

The merits of these indices have been reviewed by Hairston (1959) and comparative tests have been made by Archibald (1972). Apart from problems associated with sample size, it may be difficult to interpret the results obtained from the application of diversity indices. For example, Archibald (1972) concluded that only high community diversity was related to water quality: low diversity may be observed when water quality, judged by chemical criteria, was good and arose from severe physical conditions in the habitat under investigation.

9.4.1.2 Comparative indices

Temporal and spatial changes in water quality may be assessed by comparing two or more populations or community structures. The appropriate methods were largely developed by phytosociologists in order to distinguish plant communities in space or identify probable succession in time. These methods may be used in river quality surveillance studies to identify spatial discontinuities between communities which may be attributable to environmental change or to detect and measure temporal changes between successive samples.

At the simplest level one may compare the species composition of the communities. Several indices are available but only those which compare

TABLE 9.4

SUMMARY OF RESULTS OF COMPARISONS OF EFFECTIVENESS, TAXONOMIC DEMANDS, AND EASE OF CALCULATION OF SEVERAL INDICES AND METHODS OF DATA TREATMENT BASED ON EXPERIENCE WITH DATA DERIVED FROM A SPATIAL SURVEY OF THE POLLUTED RIVER CYNON (Learner *et al.*, 1971) AND A TEMPORAL SURVEY OF THE UNPOLLUTED RIVER DERWENT (Hynes, 1970). NUMBER OF SYMBOLS INDICATES RATING, OPEN SYMBOL INDICATES VERY LOW RATING

Index or method	Expected effectiveness on theoretical grounds	Usefulness for water management purposes	Taxonomic demand	Ease of calculation	Subjective assessment of actual performance with test data R. Cynon	Subjective assessment of actual performance with test data R. Derwent
Basic data						
Total number of individuals	•• •	••	○ •	•• ••	•• •• •	•• •• •
Numbers of individuals in given taxa	•	•	•	•• ••	•• ••	•• ••
Numbers of taxa						
(i) Species	•• •	• •	••• ••	••• ••	• •	•
(ii) Higher taxa	•• •	•	•	••• ••	• • •	—
'Species deficit' (Kothé, 1962)						—
Saprobien system						
Relative purity (Knöpp, 1954)	••• ••	•• ••	••• ••	—	—	—
Saprobic index (Pantle and Buck, 1955)	•• ••	• •	••• ••	•• ••	—	○
Saprobic index (Zelinka and Marvan, 1966)	• •	•	•	•	—	○
Pollution indices						
Trent biotic index (Woodiwiss, 1964)	• •	•• ••	•	• •	• •	•
Biotic score (Chandler, 1970)	•• •	•• ••	•	• •	• •	• •

Diversity indices

- Lognormal distribution (Preston, 1948)
- Williams α index (Fisher et al., 1943)
- Margalef (1951)
- Menhinick (1951)
- Simpson (1949)
- Information theory index (Shannon, 1948)
- McIntosh (1967)
- Sequential comparison index (Cairns et al., 1968)

Comparative indices

- Jaccard (1912)
- Kulezynski (1928)
- Sörensen (1948)
- Mountford (1962)
- Raabe (1962)
- Czekanowski (1913)
- Distance measure (Sokal, 1961)
- Coefficient of association T (Looman and Campbell, 1960)
- Rank correlation coefficient (Spearman, 1913)
- Rank correlation coefficient Tau (Kendall, 1962)

From Hellawell, J. M. (1977). Change in natural and managed ecosystems: detection, measurement and assessment. Proc. Roy. Soc. Lond. B. **197**: 31–57.
[a]When used in full matrix, otherwise poor. [b]Even better in full matrix.

joint presences (e.g. Jaccard, 1912; Kulezynski, 1928; Sorensen, 1948; Mountford, 1962) are preferred since joint absences may arise circumstantially and the close affinity indicated between samples or stations which had similar lists of 'missing' species would be spurious. Better comparisons are made with comparative indices in which the relative or absolute abundances of species are utilised, for example those of Czekanowski (1913) and Raabe (1952).

In using these indices one may compare successive pairs of stations or samples but comparisons of all samples or stations by means of a matrix are likely to prove more informative.

Another useful method for comparing communities is the 'distance measure' (Sokal, 1961; Sokal and Sneath, 1963) in which the abundances of a species within the two communities are represented in an *n*-dimensional hyperspace and the spatial separation or 'distance' between these communities provides a measure of their affinity.

Finally, communities may be compared by means of ranking methods in which the relative importance of species in each community are ranked and then compared (Spearman, 1913; Kendall, 1962). This approach has the advantage that only the relative importance of species need be measured and sampling difficulties are thereby reduced. It is, however, possible to have identical rankings in communities with very dissimilar structures.

9.4.1.3 Performance of indices

It is not possible to provide assessments of the relative performances and suitabilities of all the indices mentioned above. Some basic properties of selected indices are given in Table 9.3.

Comparative tests of the performance of certain indices have been made and published elsewhere (Hellawell, 1978). In this section the results of these comparisons will be provided.

Index performance was assessed by comparing the direction and relative magnitude of changes with those obtained from detailed preliminary consideration of the basic data and also in relation to the overall pattern which emerged when a subjective synthesis of all analyses was prepared. This strategy was considered to be quite valid since the main purpose of using indices is to provide a rapid summary of data which could only be otherwise obtained by lengthy subjective analysis. If, in addition, a consistent pattern emerged from the use of several different methods one might reasonably be suspicious of any technique which did not produce similar results, although it is, of course, possible that the consensus is unreliable or misleading.

TABLE 9.5
COMPARISON OF METHODS OF DATA ANALYSIS

(a) *Basic data*
  1. Abundance of individuals of given taxa (86)
  2. Total numbers of individuals (without identification) (66)
  3. Number of species present ('species richness') (52)
(b) *Pollution indices*
  1. Biotic score (Chandler, 1970) (93)
  2. Saprobic Index (Zelinka and Marvan, 1966) (69)
  3. Trent Biotic Index (Woodiwiss, 1964) (62)
(c) *Diversity indices*
  1. Sequential comparison index (Cairns *et al.*, 1968) (86)
  2. Simpson's (1949) index (79) and McIntosh's (1967) index (79)
  3. Information theory index (Shannon, 1948) (66)
(d) *Comparative indices*
  1. Distance measure (Sokal, 1961) (80)
  2. Czekanowski's (1913) coefficient (72)
  3. Raabe's (1952) coefficient (67)
  4. Rank correlation coefficient (Spearman, 1913) (62)

Ranking of methods on the basis of attributes listed in Table 9.4. Figures in parentheses indicate the percentage of maximum score.

For each index or method of data analysis used in the comparative tests, points were allocated for its theoretical conceptual basis, for its usefulness as a management tool, the taxonomic demand placed upon the user, and the ease with which it was calculated. In addition, points were given for its actual performance on test data (Hellawell, 1977a,b). Although the allocation of points was subjective under each heading it was thought that the overall score was a useful assessment of the relative merits of the different indices used. The results are shown in Table 9.4. A summary of the preferred methods, in order of preference, is given in Table 9.5.

9.4.1.4 Conclusions
The present multiplicity of indices is either an indication of the general dissatisfaction with the contributions of previous workers, or is a reflection of the wide range of differing requirements for environmental assessment. Much effort has been directed towards the quest for one index most suited to the needs of water pollution control (Balloch *et al.*, 1976; Cooke, 1976) and this phenomenon has not been confined to biologists (Brown *et al.*, 1970).

The advantage of a single, all embracing index is obvious. But it seems unlikely that any single index would satisfy all requirements. Indeed, it seems desirable that several types of index should always be used. When these are in agreement, greater confidence may be placed on the interpretation; when they differ, the discrepancy may provide valuable clues as to the nature of the environmental problem involved. Thus, one may conclude that both pollution indices and diversity indices should be determined for all sites and sampling periods. In special surveys these may be supplemented by calculating comparative indices between all stations.

.

CHAPTER 10

Biological Surveillance in Environmental Management

The preceding chapters have covered the fundamental theory and also the practical aspects of the use of biological methods in freshwater pollution control and environmental management, and reviewed the effects of different kinds of stress upon the ecology of aquatic systems. Much of what has been considered is of intrinsic interest. In this chapter an attempt will be made to fit the foregoing topics into an applied context and place them within a practical frame of reference of the work of regulatory organisations or authorities who are charged with the task of implementing the laws and regulations relating to water.

10.1 AQUATIC RESOURCE MANAGEMENT

Some resources are more evident than others. A supply of potable water or a thriving population of an edible fish species is clearly a valuable resource. Less evident, but possibly also of great value, is the effluent receiving and transporting capacity of a river. Some would regard the aesthetic landscape-enhancing value of a river as a resource, one which is often of considerable help in improving the quality of a townscape if it is unpolluted and provides a visual or actual corridor through otherwise congested areas. The same river or canal can be a strong disbenefit when it assumes the role of an open sewer. The recreational or amenity value of water is also an asset, but what of the wildlife—do these constitute a resource? Some would argue strongly that they do, others regard them merely as incidentals.

The value of water resources is largely a function of quantity and quality for these factors also govern the associated conservation, amenity, recreative or fisheries benefits. These two factors are also interrelated in their management since abstractions and discharges modify the quantity and quality of the natural flow. Abstractions increase the influence of subsequent discharges while these, in turn, augment the quantity of water though they might reduce its quality. The necessity for an integrated management policy was one of the factors which led to re-organisation of the water industry in England and Wales so that one authority was responsible for the management of whole catchments and entire river basins where previously a number of independent bodies had been responsible for water supply, drainage, effluent treatment, river regulation, flood relief and fisheries, either separately or in various combinations.

As pressures on aquatic resources increase, the necessity for policies which provide for balanced exploitation of the available resources becomes more vital. This is well illustrated by changes which have occurred in the management of some reservoirs and lakes once used exclusively for potable supply and from the environs of which all members of the public were debarred. This policy not only prevented contamination but gave cost savings since it avoided the need for anything but the minimum of treatment. Now the construction of a supply reservoir automatically means making provision for, and balancing the conflicting demands of, anglers, yachtsmen, water-skiers, bird-watchers, scuba-divers, conservationists and the sightseeing motorists. It also means monitoring the effects which these groups have on the basic asset, the water, and on ancilliary assets such as fish and wildlife. Multi-functional usage now applies to most water resources, river regulation, flood prevention, power generation schemes. Effluent treatment seems to be almost unique in that it is, so far, an activity in which this does not occur! Awareness of the complexities of interactions in biological systems gives the biologist a capacity not only to contribute to the specialised ecological aspects of scheme design and management, but also to considerations of the best, integrated, solutions of many of the problems which multifunctional usage creates.

10.2 BASIN MANAGEMENT CONCEPTS

The concept of management of whole river basins is one of those ideas which has self-evident advantages and yet has rarely been fully implemented. The creation, in the United States in 1933, of the Tennessee Valley Authority to control flooding, enhance navigation and generate hydro-electric power in the Tennessee River basin, which involves some seven states, through the ownership and operation of all the dams, navigation locks, power stations and all the associated structures and facilities, is probably the earliest attempt at basin management. Other interstate water compacts, dating back to 1922 with the Colorado River Compact, gave control of water quality and the establishment in 1948 of the Ohio River Valley Water Sanitation Commission provided extensive agreement on pollution among eight states (Okun, 1977). The re-organisation of the British water industry in 1974 formed ten regional water authorities, each based on one or more river catchment, from about 1600 independent organisations which had previously had responsibility for sewage disposal, water supply and river regulation. Although there are some aspects of basin management which are under the control of independent organisations or which are carried out by agencies, the formation of the UK water authorities is probably the nearest practical approach ever to the establishment of fully integrated management of river basins.

The advantages of this strategy are as follows:

(1) *Resolution of political conflicts.* Many small independent organisations may find that in selecting their own least-cost solutions to attain local ends they impose disbenefits on the community as a whole. Disparity in the distribution of resources between small organisations may lead to the under-utilisation of simple cheap water supplies in one area and the over-exploitation of inadequate expensive supplies elsewhere. Similarly, disposal of poor effluents by one organisation may damage the resources of another. Even when the common good is readily identified it may be inexpedient for one authority altruistically to incur costs or other penalties for the benefit of another, especially one of a different political complexion.

(2) *Better utilisation of resources.* When management is conducted in the context of whole river basins, all resources and needs can be considered together so as to provide the optimum solution for the whole system. It is likely that many more options will be available and the best solution might be to select different strategies at different times according to circumstances, thereby providing better services at lowest cost.

(3) *Integrated and realistic control.* Separation of the water supply, sewage treatment and effluent disposal functions and the vesting of

pollution control and river management powers in an independent body contributes to the possibility, if not probability, of conflict. When the control is the responsibility of one organisation this conflict is lessened since that body has to live with the consequences of its decisions. It also provides a facility for setting environmental and effluent standards which reflect the character of the basin and the objectives of the organisation rather than independent blanket emission standards or similar inflexible norms. This aspect is important, for some watercourses have greater assimilative capacity or potential for self-purification than others. Biological surveillance methods can contribute significantly towards the setting of environmentally acceptable standards in each individual case, and provide confirmation of their effectiveness.

(4) *Economies of scale.* Many construction projects are subject to economies of scale and this often extends to running costs also. For example, sewage can be treated more effectively in large works, even allowing for extended sewers to convey material from the extremities of the area.

It is possible that the only disadvantage of large basin management is in the need for greater personnel resources for liaison and communication and liability for staff not to perceive corporate identity.

In deciding strategies to be followed given the many options of water management which are possible within a river basin, or for that matter between basins, the development of mathematical models which incorporate elements covering hydrology, water quality and economic aspects has proved invaluable. These enable analyses to be performed in order to ascertain which are the sensitive or critical components of any scheme and enable optimum solutions to be found. An element of value judgement remains and some items, for example the environmental consequences, are less readily predicted. As an example of the kinds of issues which may be considered, one may take the Trent model developed for the River Trent in the British Midlands (Bowden *et al.*, 1971; Warn, 1978). Pressure on water resources in a basin where the river had multiple usage and carried a great deal of industrial and sewage effluent (Lester, 1975) led to consideration of means of abstracting river water for potable supply. This might be possible if either the effluent quality was raised or very advanced water treatment was introduced (or a mixture of both these solutions), but it was necessary to compare the costs of these options with large scale imports of water from elsewhere. It is instructive to note that the study led to the conclusion that on cost grounds alone the abstraction of potable water was an attractive alternative to increased imports. However, although water could be treated to meet the prevailing standards for potable supplies, it was not clear that the water would be 'wholesome' as required by law (Warn, 1978).

Similar problems are likely to be encountered whenever rivers receiving industrial effluents and, in particular, complex organic compounds, some of which may be known teratogenic or carcinogenic agents, are considered for use as potable supplies even when advanced treatment is available. Some of the screening tests outlined in Chapter 8 are potentially useful but in order to obtain adequate amounts of the relevant compounds it is necessary to concentrate them from very large volumes of water. Even where this is possible the results are hardly directly applicable to the problem in hand. It may be difficult to ascertain whether the process of concentrating the sample has changed its composition and, if not, it is still not clear what is the significance of drinking water containing barely perceptible quantities of such impurities. Yet other difficulties arise in determining whether certain compounds are of natural origin or were not actually present in effluents but have arisen through decomposition or microbial modification within the river.

10.3 APPLICATION AND IMPLEMENTATION OF UNIFORM STANDARDS

The need for, and application of, standards is well established and widely accepted. Those for the protection of water for human consumption (e.g. WHO, EC Directive or Department of the Environment 'Report 71') relate to the potable water, or to the raw source water before treatment, and usually the only hint of dissent on such standards is whether they are stringent enough. Standards intended to protect the environment at large tend to fall into two categories: those relating to the discharge itself ('emission standards') and those which concern the quality of the receiving environment ('environmental quality objectives').

The practical consequences of these approaches can be far-reaching.

Application of uniform emission standards, based on the reasonable expectations of current effluent treatment practice, treats all polluters equally but may penalise all of us if the assimilative capacity of the receiving water is inadequate (causing pollution) or ample (incurring unnecessary treatment costs). Environmental standards are preferable in that they afford greater protection to the common resources but may impose hardship on particular industries simply through accidents of history and topography.

Given adequate knowledge about the quantities of effluent and the assimilative capacity of the affected environment, either of these approaches is adequate—both is a luxury. It becomes a matter then of convenience. If it is easier to monitor effluents then emission standards will suffice; if it is simpler to sample the environment, an environmental quality objective determines the control which must be applied to dischargers. Differences in fundamental attitudes to standards became evident in the EEC when Britain opted for environmental control while the rest of the community preferred emission control. For some substances, especially those which need to be kept at very low levels in the environment in order to avoid serious bioaccumulation and which would require expensive chemical analyses, the control of emission levels is probably more practicable. A watching brief on levels in biota would be particularly advantageous and so biological surveillance can play an important role here.

10.4 ENVIRONMENTAL IMPACT STATEMENTS AND CONSERVATION

One consequence of the increased environmental awareness which has developed over the last few decades has been the demand that almost all new development, such as industrial, urban, mineral exploration and exploitation, transportation, power generation, irrigation and agricultural or other land-use changes, should be subjected to an environmental impact assessment and the results included in any proposals or plans which are submitted to planning or development authorities (O'Riordan and Hey, 1976). The case for such requirements has been greatly strengthened by the unfortunate experiences of certain large scale schemes in several developing countries. The classic cases in the aquatic field are the construction of the Aswan Dam in Egypt and the Kariba Dam in Tanzania, but there have been others in central and southern America of which the Bayano hydroelectric project is a typical example and illustrates the kinds of problems which can occur (Goodland, 1977). Presentation of an environmental impact assessment may not prevent some unwanted consequences but it should help to facilitate better appreciation of the implications of the proposals and could help to identify the least environmentally damaging option where permutations are possible. Often the biological consequences cannot be predicted with certainty but the development of ecological models offers the prospects of improvement. Many of the issues associated with biological environmental impact studies are considered by Ward (1978).

Recently, the recognition of the relevance of water management as an adjunct to conservation of natural wildlife resources has achieved some prominence. The 1973 Water Act which established the regional water authorities in England and Wales contained, in Section 22, a statutory requirement that, in formulating or considering any proposals relating to the discharge of any of their functions, these authorities:

> shall have regard to the desirability of preserving natural beauty, of conserving flora, fauna ... and shall take into account any effect which the proposals would have ... on any such flora, fauna. ...

This duty was later strengthened by the Wildlife and Countryside Act 1981, Section 48 of which amended Section 22 of the Water Act such that in formulating or considering any proposals, water authorities were required, so far as was consistent with their duties under the Water Act and the 1976 Land Drainage Act, to:

> so exercise their functions with respect to the proposals as to further the conservation and enhancement of natural beauty and the conservation of flora, fauna and geological or physiographical features of special interest. ...

as well as to continue to take account of the consequences of their actions.

The duty to 'further conservation' has been seen as a positive duty by the conservation lobby while the industry has stressed that the phrase must be read in the context of their duties under the Water and Land Drainage Acts. A ministerial circular issued in November 1982, providing guidance notes on the procedures to be adopted in implementing the revised statute, while noting the 'more positive' requirements of the amendment and exhorting the authorities to adopt conservation-oriented policies, nevertheless emphasised that any increase in expenditure (which, it was admitted, might be inevitable) must be consistent with their primary functions.

Similar conservation obligations are contained in the provisions of the Control of Pollution Act 1974, although these are not yet fully implemented. Section 46 of this Act requires that in the event of a consented discharge causing 'pollution injurious to fauna or flora of a stream . . .' the Authority must carry out as soon as is reasonable:

> such operations as are necessary for the purpose of restoring the fauna and flora of the stream so far as is reasonably practical to do so, to the state in which they were immediately before discharges were made. . . .

If, as seems likely, an authority would not knowingly or wilfully issue or vary a consent to discharge so as to cause such pollution then one must imagine that this would only occur in error or by an oversight. In order to prevent excessive claims regarding the damage caused and demands for ecological restoration to a condition beyond that which actually prevailed originally, it will be incumbent upon water authorities to undertake the necessary biological surveillance. This activity will become particularly important when implementation of Section 41 of the Control of Pollution Act requires registers of applications for consents, the terms under which consents were granted and the results of analyses made in support of pollution control functions to be maintained and open for public inspection. Under the Act the previous restrictions, which limited the power of prosecuting polluters to the Attorney General and the water authorities, will be removed and anyone may take proceedings against those (including water authorities) who fail to comply.

The Wildlife and Countryside Act also lays down procedures for the notification of sites of special scientific interest (SSSIs). These are sites which, in the opinion of the Nature Conservancy Council (NCC), have an intrinsic value for nature conservation purposes on account of their scientific interest and as representatives of particular habitats. Criteria for the selection of these sites have been developed (Ratcliffe, 1977), some of which are objective (such as size, location and recorded history) while others (naturalness, typicalness, intrinsic appeal) are largely subjective and depend heavily on value judgements. Under the terms of the Act the existence of the sites is notified to the owner or occupier and also the Water Authority as a matter of courtesy. This aspect can be important, for the water authorities have to ensure the furtherance of conservation in all their activities. Relevant examples include the issuing of licences to abstract water and to discharge effluents or engaging in land drainage works, all of which have conservation implications but none of which requires the authorities to own or occupy the land involved. For each SSSI there is a list of prescribed activities which must not be performed without giving three months' prior notice to the NCC except in emergencies, when post-notification will suffice. Typical examples of proscribed activities for an aquatic site might include changing of water levels, filling-in, re-alignment of banks, applications of pesticides, clearing or felling of trees, etc. Failure to comply with this requirement could involve a fine although this is not large and in many cases could prove cheaper than a three month delay. The main deterrent for public bodies such as water authorities is adverse publicity. The objective in giving advance notice of an intention to carry out one of the prescribed activities is to facilitate consultation, to prevent inadvertent destruction of the site's value and, if necessary, enable a management agreement to be worked out.

The importance of biological surveillance and the use of biological indicators of environmental quality will increase in the climate of the legislation described above. Field measurements of the pre- and post-implementation of schemes will enable a retrospective assessment to be made of the efficacy of those features of the design which were intended to 'further conservation' and of how well any anticipated consequences were 'taken into account'. Some prediction is possible on the basis of current knowledge but the art of furthering conservation will itself be promoted only if the ecological changes, if any, are fully documented. The iterative progression towards better prediction of the effects of schemes, which may represent substantial savings in costs by avoiding the need for excessive safety factors, needs to be recognised. Environmental impact analyses and ecological investigations may seem expensive (though often a minute fraction of other preparatory costs) but they constitute a wise investment which will provide dividends in the form of lower costs in future assessments. This issue has been explored more fully elsewhere (Hellawell, 1982).

Strengthening of the conservation duties of Water Authorities has led recently to the concept of the 'river corridor' as an important feature in furthering general conservation. Initially the corridor was perceived as a thoroughfare along which organisms might travel or distribute themselves since it often embraces a distinct habitat quality quite different from that through which it passes. The uses of the land through which the river flows might vary widely but the presence of the river provides habitat continuity.

Management of the river also affects the adjacent riparian habitat and has implications for habitats upstream and downstream. Thus the corridor itself becomes an important component of the landscape and it needs to be managed in its own right. Yet not all reaches of rivers have equal conservation value nor are the criteria used by NCC or others necessarily appropriate. Recent attempts to address this problem have resulted in the development of survey techniques and objective criteria with which the conservation importance of riparian sites may be gauged (Brooker, 1982). Further development is needed in order that the heavy manpower involvement may be reduced but the approach is a promising one.

10.5 FUTURE DEVELOPMENTS

Scientific prognostication is fraught with difficulties. One has also to distinguish between the particular and the general change: much future development will doubtless consist of the fuller implementation of methods which already exist but which are not yet widely appreciated. Many of the approaches outlined in this book are in their infancy; others are tried and tested by a few specialists but have not been adopted generally. Perhaps it would be too rash to say that the kinds of advances in chemical methods over the last couple of decades, from the drudgery of the burette to the almost insatiable appetite of the auto-analyser, will be reflected in the biological laboratory but it might prove equally foolish to say that they will not. The almost incredible rate of increase in power and in concomitant decrease in size of computers coupled with sophisticated means of image analysis may enable much of the tedious sample analysis of biological surveillance to be overcome. The concept of using computers to recognise biological material and even undertake difficult taxonomic identification is, in the context of the time-scale of 'microchip' technology, by no means recent (Cairns *et al.*, 1972; Almeida *et al.*, 1977). Although the commitment of human resources in biological surveillance is high the information content of the results is also high and it is almost certain that we are not yet extracting it and using it as fully as we might. Ecological modelling techniques may help to identify the critical features of community responses to environmental stress so that in field investigations the effort is directed to greatest effect.

An exciting example of the way in which the development of powerful data analysis methods can contribute to greater utility of biological surveillance method is the river communities project undertaken by the Freshwater Biological Association (FBA) in the United Kingdom (Wright *et al.*, 1984; Furse *et al.*, 1984).

Preliminary studies of river types based on hydrological characteristics other than flow frequency (for example, timing and magnitude relationships of highest and lowest flows, variation about mean flow, rates of change of flows and velocity associated with average daily flow) and comparisons with the occurrence of characteristic communities of invertebrates (Jones and Peters, 1977) showed promise and it was postulated that, given better invertebrate data (the study was based on purely qualitative data from non-standardised samples), a more robust characterisation could be developed.

This approach was greatly extended by workers at the FBA who selected 268 sites on 41 river systems from which samples of macroinvertebrates have been taken at different seasons. The rivers selected include the widest range of geological and physical characteristics in Britain, consonant with the requirement that they should be unpolluted or as little affected by human interference as possible. Data records of almost 600 species of invertebrates have been subjected to analysis using DECORANA (DEtrended CORrespondence ANAlysis) and TWINSPAN (TWo-way INdicator SPecies ANalysis) (Hill, 1979a,b). Important features which emerged from ordination of sites on the basis of species present were that much of the variation was accounted for on axis 1, which showed highest correlations with substratum condition and alkalinity. On axis 2 highest correlations were associated with distance from source and discharge. Thus sites were being distinguished between river systems and according to their location within their river system. Using TWINSPAN, it has been possible to use macroinvertebrate data to classify sites into groups, based on couplets up to 32 in total, and although initially 16 groups seemed to make a useful working level, by extending the investigation to 370 sites on 61 river systems it now appears likely that 30 TWINSPAN groups will be utilised. Given approximate environmental data, it has proved possible to assign sites to their correct TWINSPAN group with a success approaching 80%.

For the first time an objective means of assessing the effects of environmental manipulation has become a real possibility for applied hydrobiologists. Previously, potential damage had to be assessed on the basis of past experience: the collective store then being tapped with a view to making the nearest match in known circumstances. Soon it should be possible to predict the most probable type of community which will result from given changes in environmental parameters. Even more attractive is the prospect of effective sensitivity analysis in which the different variables are changed and the consequences estimated. In this way schemes may be modified in order to achieve the optimum characteristics of minimal environmental damage at lowest cost. Even if, as seems likely for a considerable period, it proves impracticable to put a cash price on the aesthetic value of the landscape or a biological community, it will at least prove possible to put a price-tag on the cost of achieving, or maintaining, a given ecological status.

Trent Biotic Index

PART 1   CLASSIFICATION OF BIOLOGICAL SAMPLES

Key indicator groups	Diversity of fauna	Total number of groups (see Part 2) present				
		0–1	2–5	6–10	11–15	16+
				Biotic index		
Plecoptera nymphs present	More than one species	—	VII	VIII	IX	X
	One species only	—	VI	VII	VIII	IX
Ephemeroptera nymphs present	More than one species[a]	—	VI	VII	VIII	IX
	One species only[a]	—	V	VI	VII	VIII
Trichoptera larvae present	More than one species[b]	—	V	VI	VII	VIII
	One species only[b]	IV	IV	V	VI	VII
Gammarus present	All above species absent	III	IV	V	VI	VII
Asellus present	All above species absent	II	III	IV	V	VI
Tubificid worms and/or Red Chironomid larvae present	All above species absent	I	II	III	IV	—
All above types absent	Some organisms such as *Eristalis tenax* not requiring dissolved oxygen may be present	0	I	II	—	—

[a] *Baetis rhodani* excluded.
[b] *Baetis rhodani* (Ephem.) is counted in this section for the purpose of classification.

447

PART 2   GROUPS

The term 'Group' here denotes the limit of identification which can be reached without resorting to lengthy techniques. Groups are as follows:

1 Each species of Platyhelminthes (flatworms)
2 Annelida (worms) excluding *Nais*
3 *Nais* (worms)
4 Each species of Hirudinea (leeches)
5 Each species of Mollusca (snails)
6 Each species of Crustacea (log-louse, shrimps)
7 Each species of Plecoptera (stone-fly)
8 Each genus of Ephemeroptera (mayfly) excluding *Baetis rhodani*
9 *Baetis rhodani* (mayfly)
10 Each family of Trichoptera (caddis-fly)
11 Each species of Neuroptera larvae (alder-fly)
12 Family Chironomidae (midge larvae) except *Chironomus thummi* (=*riparius*)
13 *Chironomus thummi* (blood worms)
14 Family Simuliidae (black-fly larvae)
15 Each species of other fly larvae
16 Each species of Coleoptera (beetles and beetle larvae)
17 Each species of Hydracarina (water mites)

APPENDIX 2
BIOTIC SCORE (Chandler, 1970)

Groups present in sample	Abundance in standard sample — Points scored				
	Present 1–2	Few 3–10	Common 11–50	Abundant 51–100	Very abundant 100+
Planaria alpina	90	94	98	99	100
Each species of Taenopterygidae, Perlidae, Perlodidae, Isoperlidae, Chloroperlidae	84	89	94	97	98
Each species of Leuctridae, Capniidae, Nemouridae (excluding *Amphinemura*)	79	84	90	94	97
Each species of Ephemeroptera (excluding *Baetis*)	75	80	86	91	94
Each species of Cased caddis, Megaloptera	70	75	82	87	91
Each species of *Ancylus*	65	70	77	83	88
— *Rhyacophila* (Trichoptera)	60	65	72	78	84
Genera *Dicranota*, *Limnophora*	56	61	67	73	75
Genus *Simulium*	51	55	61	66	72
Genera of Coleoptera, Nematoda	47	50	54	58	63
— *Amphinemura* (Plecoptera)	44	46	48	50	52
— *Baetis* (Ephemeroptera)	40	40	40	40	40
— *Gammarus*	40	40	40	40	40
Each species of Uncased caddis (excl. *Rhyacophila*)	38	36	35	33	31
Each species of Tricladida (excluding *P. alpina*)	35	33	31	29	25
Genera of Hydracarina	32	30	28	25	21
Each species of Mollusca (excluding *Ancylus*)	30	28	25	22	18
— Chironomids (excl. *C. riparius*)	28	25	21	18	15
Each species of *Glossiphonia*	26	23	20	16	13
Each species of *Asellus*	25	22	18	14	10
Each species of Leech (excl. *Glossiphonia* and *Haemopsis*)	24	20	16	12	8
— *Haemopsis*	23	19	15	10	7
— *Tubifex* sp.	22	18	13	12	9
— *Chironomus riparius*	21	17	12	7	4
— *Nais*	20	16	10	6	2
Each species of air breathing species	19	15	9	5	1
No animal life	0				

Biological Monitoring Working Party (BMWP)—Score

Families	Score
Siphlonuridae, Heptageniidae, Leptophlebiidae, Ephemerellidae, Potamanthidae, Ephemeridae Taeniopterygidae, Leuctridae, Capniidae, Perlodidae, Perlidae, Chloroperlidae Aphelocheiridae Phryganeidae, Molannidae, Beraeidae, Odontoceridae, Leptoceridae, Goeridae, Lepidostomatidae, Brachycentridae, Sericostomatidae	10
Astacidae Lestidae, Agriidae, Gomphidae, Cordulegasteridae, Aeshnidae, Corduliidae, Libellulidae Psychomyiidae, Philopotamidae	8
Caenidae Nemouridae Rhyacophilidae, Polycentropodidae, Limnephilidae	7
Neritidae, Viviparidae, Ancylidae Hydroptilidae Unionidae Corophiidae, Gammaridae Platycnemididae, Coenagriidae	6
Mesovelidae, Hydrometridae, Gerridae, Nepidae, Naucoridae, Notonectidae, Pleidae, Corixidae Haliplidae, Hygrobiidae, Dytiscidae, Gyrinidae, Hydrophilidae, Clambidae, Helodidae, Dryopidae, Eliminthidae, Chrysomelidae, Curculionidae Hydropsychidae Tipulidae, Simuliidae Planariidae, Dendrocoelidae	5

APPENDIX 3—*contd.*

Families	Score
Baetidae Sialidae Piscicolidae	4
Valvatidae, Hydrobiidae, Lymnaeidae, Physidae, Planorbidae, Sphaeriidae Glossiphoniidae, Hirudidae, Erpobdellidae Asellidae	3
Chironomidae	2
Oligochaeta (whole class)	1

References

AAAP (1971). *American Algal Assay Procedure (Bottle Test).* New York, Joint Industry/Government Task Force on Eutrophication.

Abel, P. D. (1980). Toxicity of γ-hexachlorocyclohexane (Lindane) to *Gammarus pulex*: mortality in relation to concentration and duration of exposure. *Freshwat. Biol.* **10**: 251–9.

Abrahams, H. J. (1978). The Hezekiah tunnel. *J. Amer. Watwks. Ass.* **70**: 406–10.

Abram, F. S. H., Collins, L., Hobson, J. A. and Howell, K. (1981). The toxicities of Mitin N and Eulan WA New to rainbow trout. WRC Report 156 M, 38 pp. Water Research Centre, Stevenage.

Addison, R. F., Zinck, M. E. and Leahy, J. R. (1976). Metabolism of single and combined doses of ^{14}C-aldrin and ^{3}H-*p.p'*-DDT by Atlantic salmon (*Salmo salar*) fry. *J. Fish. Res. Bd Can.* **33**: 2073–6.

Adelman, I. R., Smith, L. L. and Siesennop, G. D. (1976). Acute toxicity of sodium chloride, pentachlorophenol, Guthion®, and hexavalent chromium to fathead minnows (*Pimephales promelas*) and goldfish (*Carassius auratus*). *J. Fish. Res. Bd Can.* **33**: 203–8.

Adema, D. M. M. (1978). *Daphnia magna* as a test animal in acute and chronic toxicity test. *Hydrobiologia* **59**: 125–34.

Alabaster, J. S. (1959). The effect of a sewage effluent on the distribution of dissolved oxygen and fish in a stream. *J. Anim. Ecol.* **28**: 283–91.

Alabaster, J. S. (1964). The effect of heated effluents on fish. *Adv. Wat. Pollut. Res.* **1**: 261–92.

Alabaster, J. S. (1969a). Effects of heated discharges on freshwater fish in Britain. In: *Biological Aspects of Thermal Pollution.* (P. A. Krenkel and F. L. Parker, Eds.) pp. 354–70. Nashville, Vanderbilt University Press.

Alabaster, J. S. (1969b). Survival of fish in 164 herbicides, insecticides, fungicides, wetting agents and miscellaneous substances. *Int. Pest Control* **11**: 29–35.

Alabaster, J. S. (1970). Testing the toxicity of effluents to fish. *Chemy Indust.* 759–64.

Alabaster, J. S. (1971). The comparative sensitivity of coarse fish and trout to pollution. *Proc. 4th Br. Coarse Fish Conf.*, Liverpool, 1969.

Alabaster, J. S. (1972). Suspended solids and fisheries. *Proc R. Soc. Lond.* B **180**: 395–406.

Alabaster, J. S. and Abram, F. S. H. (1965). Development and use of a direct method of evaluating toxicity in fish. *Proc. 2nd Int. Wat. Pollut. Res. Conf.*, pp. 41–54. Tokyo, 1964. Pergamon Press, Oxford.

Alabaster, J. S. and Lloyd, R. (1980). *Water Quality Criteria for Freshwater Fish.* London, FAO and Butterworths. (Second edn, 1982.)

Alabaster, J. S., Garland, J. H. N., Hart, I. C. and Solbé, J. F. de L. G. (1972). An approach to the problem of pollution and fisheries. *Symp. Zool. Soc. Lond* **29**: 87–114.

Al-Daham, N. K. and Bhatti, M. N. (1977). Salinity tolerance of *Gambusia affinis* (Baird and Girard) and *Heteropneustes fossilis* (Bloch). *J. Fish Biol.* **11**: 309–13.

Alexander, D. G. and Clarke, R. McV. (1978). The selection and limitations of phenol as a reference toxicant to detect differences in sensitivity among groups of rainbow trout (*Salmo gairdneri*). *Wat. Res.* **12**: 1085–90.

Allan, I. R. H., Herbert, D. W. M. and Alabaster, J. S. (1958). A field and laboratory investigation of fish in a sewage effluent. *Fishery Invest. London (I)* **6**:2: 1–76.

Allanson, B. R. (1961). Investigations into the ecology of polluted inland waters in the Transvaal. *Hydrobiologia* **18**: 1–76.

Almeida, S. P., Eu, J. K. T., Lai, P. F., Cairns, J. and Dickson, K. (1977). A real-time optical processor for pattern recognition of biological specimens. In: *Applications of Holography and Optical Data Processing.* (E. Marom and A. A. Friesnem, Eds.) pp. 573–9. Oxford, Pergamon Press.

Aly, A. O. and Faust, S. D. (1964). Studies on the fate of 2,4-D and ester derivatives in natural surface waters. *J. Agric. Fd. Chem.* **12**: 541–6.

Aly, O. M. and El-Dib, M. A. (1972). Studies of the persistence of some carbamate insecticides in the aquatic environment. In: *Fate of Organic Pesticides in the Aquatic Environment.* Pp. 210–43. Advances in Chemistry, American Chemical Society.

American Public Health Association, American Water Works Association, and Water Pollution Control Federation (1971). Standard methods for the examination of water and waste water (13th edition). New York, American Public Health Association, 1971.

Ames, B. N., McCann, J., Lee, F. D. and Durston, W. E. (1973). An improved bacterial test system for the detection and classification of mutagens and carcinogens. *Proc. Nat. Acad. Sci. USA* **70**: 782–6.

Ames, B. N., McCann, J. and Yamasaki, E. (1975). Method for detecting carcinogens and mutagens with Salmonella/mammalian microsome mutagenicity test. *Mutat. Res.* **31**: 347–63.

Anderson, B. G. (1948). The apparent thresholds of toxicity to *Daphnia magna* for chlorides of various metals when added to Lake Erie water. *Trans. Am. Fish Soc.* **78**: 96–113.

Anderson, H. A. and Paulson, S. L. (1972). A simple and inexpensive wood-float periphyton sampler. *Progve Fish Cult.* **34**: 225.

Anderson, J. M. (1968). Effect of sublethal DDT on the lateral line of the brook trout, *Salvelinus fontinalis. J. Fish. Res. Bd Can.* **25**: 2677–82.

Anderson, J. M. (1971). Sublethal effects and changes in ecosystems—assessment of the effects of pollutants on physiology and behaviour. *Proc. Roy. Soc. London B* **177**: 307–20.

Anderson, J. M. and Peterson, M. R. (1969). DDT: sublethal effects on brook trout nervous system. *Science, NY* **164**:440–41.

Anderson, J. M. and Prins, H. B. (1970). Effects of sublethal DDT on a simple reflex in brook trout, *J. Fish. Res. Bd Can.* **27**: 331–4.

Anderson, R. B. and Everhart, W. H. (1966). Concentrations of DDT in landlocked salmon (*Salmo salar*) at Sebago Lake, Maine. *Trans. Am. Fish. Soc.* **95**: 160–164.

Anderson, R. B. and Fenderson, O. C. (1970). An analysis of variation of insecticide residues in landlocked Atlantic salmon (*Salmo salar*). *J. Fish. Res. Bd Can.* **27**: 1–11.

Anderson, R. L. (1980). Chironomidae toxicity tests—biological background and procedures. In: *Aquatic Invertebrate Bioassays*. (A. L. Buikema and J. Cairns, Eds.) pp. 70–80. Philadelphia, Pa. American Society for Testing and Materials.

Anderson, R. L. and DeFoe, D. L. (1980). Toxicity and bioaccumulation of endrin and methoxychlor in aquatic invertebrates and fish. *Environ. Pollut. Ser. A.* **22**: 111–21.

Andrews, P. and Williams, P. J. LeB. (1971). Heterotrophic utilization of dissolved organic compounds in the sea. III. Measurement of the oxidation rates and concentrations of glucose and amino acids in sea water. *J. mar. biol. Ass. UK* **51**: 111–25.

Applegate, V. C., Howell, J. H., Moffett, J. W., Johnson, B. G. H. and Smith, M. A. (1961). Use of 3-trifluoromethyl-4-nitrophenol as a selective sea lamprey larvicide. *Tech. Rep. Gt. Lakes Fishery Commission* **1**.

Archibald, R. E. M. (1972). Diversity in some South African diatom associations and its relation to water quality. *Wat. Res.* **6**: 1229–38.

Argyle, R. L., Williams, C. C. and Dupree, H. K. (1973). Endrin uptake and release by fingerling channel catfish (*Ictalurus punctatus*) *J. Fish. Res. Bd Can.* **30**: 1743–4.

Arthur, J. W. (1970). Chronic effects of linear alkylate sulphonate detergent on *Gammarus pseudolimnaeus, Campeloma decisum* and *Physa integra. Wat. Res.* **4**: 251–7.

Arthur, J. W. (1980). Review of freshwater bioassay procedures for selected amphipods. In: *Aquatic Invertebrate Bioassays*. (A. L. Buikema and J. Cairns, Eds) pp. 98–108. Philadelphia, Pa, American Society for Testing and Materials.

Arthur, J. W. and Leonard, E. N. (1970). Effects of copper on *Gammarus pseudolimnaeus, Physa integra* and *Campeloma decisum* in soft water. *J. Fish. Res. Bd Can.* **27**: 1227–83.

Arthur, J. W., Andrew, R. W., Mattson, V. R., Olson, D. T., Glass, G. E., Halligan, B. J. and Walbridge, C. T. (1975). Comparative toxicity of sewage-effluent disinfection to freshwater aquatic life. US Environmental Protection Agency, Duluth, Minnesota, EPA-600/3–75–012.

Asbury, C. and Coler, R. (1980). Toxicity of dissolved ozone to fish eggs and larvae. *J. Wat. Pollut. Control Fed.* **52**: 1990–96.

Aston, R. J. (1973). Field and experimental studies on the effects of a power station effluent on Tubificidae (Oligochaeta, Annelida). *Hydrobiologia* **42**: 225–42.

Aston, R. J. and Brown, D. J. A. (1975). Local and seasonal variations in populations of the leech *Erpobdella octoculata* (L) in a polluted river warmed by condenser effluents. *Hydrobiologia* **47**: 347–66.

Atkinson, K. M. (1972). Birds as transporters of algae. *Br. Phycol. J.* 7: 319–21.

Azad, H. S. (Ed.) (1976). *Industrial wastewater management handbook.* New York, McGraw-Hill.

Bagenal, T. B. (1972). The variability in numbers of perch *Perca fluviatilis* L. caught in traps. *Freshwat. Biol.* 2: 27–36.

Bagenal, T. B. (1974a). A buoyant net designed to catch freshwater fish larvae quantitatively. *Freshwat. Biol.* 4: 107–9.

Bagenal, T. B. (1974b). *Ageing of Fish.* Old Woking, Unwin Bros.

Bagenal, T. (1978). Methods for assessment of fish production in fresh waters. *IBP Handbook No. 3.* Oxford, Blackwell.

Bahr, T. G. (1973). Electrophysiological responses of trout to dissolved oxygen and cyanide. In: *Bioassay Techniques* and *Environmental Chemistry.* (G. E. Glass, Ed.) pp. 231–55. Ann Arbor.

Baker, J. H. and Bradnam, L. A. (1976). The role of bacteria in the nutrition of aquatic detritivores. *Oecologia (Berlin)* 24: 95–104.

Ball, I. R. (1967a). The relative susceptibilities of some species of fresh-water fish to poisons—I. Ammonia. *Wat. Res.* 1: 767–775.

Ball, I. R. (1967b). The relative susceptibilities of some species of freshwater fish to poisons—II. Zinc. *Wat. Res.* 1: 777–83.

Ball, I. R. (1967c). The toxicity of cadmium to rainbow trout (*Salmo gairdnerii* Richardson). *Wat. Res.* 1: 805–806.

Ball, R. C., Wojtalik, T. A. and Hooper, F. F. (1963). Upstream dispersion of radiophosphorus in a Michigan trout stream. *Pap. Mich. Acad. Sci.* 48: 47–64.

Balloch, D., Davies, C. E. and Jones, F. H. (1976). Biological assessment of water quality in three British rivers: the North East (Scotland), the Ivel (England) and the Taff (Wales). *Wat. Pollut. Control* 75: 92–110.

Bardach, J. E., Fujiya, M. and Holl, A. (1965). Detergents: effects on the chemical senses of the fish *Ictalurus natalis* (Le Sueur). *Science* 148: 1605–1607.

Barnes, H. (1959). *Apparatus and Methods of Oceanography. Part One: Chemical.* London, Allen and Unwin.

Barton, B. A. (1977). Short-term effects of highway construction on the limnology of a small stream in southern Ontario. *Freshwat. Biol.* 7: 99–108.

Bartsch, A. F. (1948). Biological aspects of stream pollution. *Sewage Works Journal* 20: 292–302.

Baslow, M. H. and Nigrelli, R. F. (1964). The effect of thermal acclimation on brain cholinesterase activity of the Killifish, *Fundulus heteroclitus. Zoologica* 49: 41–51.

Beamish, R. J. (1974). Loss of fish populations from unexploited remote lakes in Ontario, Canada, as a consequence of atmospheric fallout of acid. *Wat Res.* 8: 85–95.

Bechtel, T. J. and Copeland, B. J. (1970). Fish species diversity indices as indicators of pollution in Galveston Bay, Texas. *Contr. Mar. Sci. Univ. Texas* 15: 103–132.

Beck, W. M. (1954). Studies in stream pollution biology. I. A simplified ecological classification of organisms. *Q. J. Florida Acad. Sci.* 17: 211–27.

Beck, W. M. (1955). Suggested method for reporting biotic data. *Sewage Indust. Wastes* 27: 1193–7.

Bell, H. L. (1971). Effect of low pH on the survival and emergence of aquatic insects. *Wat. Res.* **5**: 313–19.

Bell, M. V., Kelly, K. F. and Sargent, J. R. (1981). The uptake from freshwater and subsequent clearance of a vanadium burden by the common eel (*Anguilla anguilla*). *Sci. Total Environment* **19**: 215–22.

Bellavere, C. and Gorbi, J. (1981). A comparative analysis of acute toxicity of chromium, copper and cadmium to *Daphnia magna, Biomphalaria glabrata* and *Brachydanio rerio. Environ. Technol. Lett.* **2**: 119–28.

Bengtsson, B.-E. (1980). Long term effects of PCB (Clophen A50) on growth, reproduction and swimming performance in the minnow *Phoxinus phoxinus. Wat. Res.* **14**: 681–7.

Benoit, D. A. (1976). Toxic effects of hexavalent chromium on brook trout (*Salvelinus fontinalis*) and rainbow trout (*Salmo gairdneri*). *Wat. Res.* **10**: 497–500.

Benoit, D. A., Leonard, E. N., Christensen, G. M. and Fiandt, J. T. (1976). Toxic effects of cadmium on three generations of brook trout (*Salvelinus fontinalis*). *Trans. Am. Fish. Soc.* **105**: 550–60.

Benson-Evans, K. and Williams, P. F. (1976). Transplanting aquatic bryophytes to assess river pollution. *J. Bryol.* **9**: 81–91.

Berger, B. L., Lennon, R. E. and Hogan, J. W. (1969). Laboratory studies on Antimycin A as a fish toxicant. US Dept. Interior, Fish and Wildlife Serv., Bureau Sports, Fish and Wildlife, Invest Fish. Control No. 26: 1–19.

Berland, B. R., Bonin, D. J., Maestrini, S. Y. and Pointier, J.-P. (1973). Etude de la fertilité des eaux marines au moyen de tests biologiques effectués avec des cultures d'algues. II. Limitation nutritionelle et viabilité de l'inoculum. *Int. Revue ges. Hydrobiol.* **58**: 203–220.

Bertram, P. E. and Hart, B. A. (1979). Longevity and reproduction of *Daphnia pulex* (de Geer) exposed to cadmium-contaminated food or water. *Environ. Pollut.* **19**: 295–305.

Besch, W. (1966). Drift netz methode und biologische Fliesswasser untersuchung. *Verh. int. Verein. theor. angew. Limnol.* **16**: 669–78.

Besch, W., Hofman, W. and Ellenberger, W. (1967). Das Makrobenthos auf polyathylene Substratum in Fliessgewassern. 1. Die Kinzig, ein Fluss der unteren Salmoniden und oberen Barbenzone. *Annls Limnol.* **3**: 331–67.

Besch, W. K. and Juhnke, H. (1971). Un nouvel appareil d'étude toxicologique utilisant des carpillons. *Annls. Limnol.* **7**: 1–6.

Besch, W. K., Ricard, M. and Cantin, R. (1972). Benthic diatoms as indicators of mining pollution in the northwest Miramichi River system, New Brunswick, Canada. *Int. Rev. ges. Hydrobiol.* **57**: 39–74.

Besch, W. K., Kemball, A., Mayer-Warden, K. and Scharf, B. (1977). A biological monitoring system employing rheotaxis of fish. *American Society for Testing and Materials STP* **607**: 56–74.

Bick, H. (1963). A review of Central European methods for the biological estimation of water pollution levels. *Bull. Wld. Hlth. Org.* **29**: 401–413.

Bick, H. (1968). Autokologische und saprobiologische Untersunchungen an Susswasserciliaten. *Hydrobiologia* **31**: 17–36.

Biesinger, K. E. and Christensen, G. M. (1972). Effects of various metals on survival, growth, reproduction and metabolism of *Daphnia magna. J. Fish. Res. Bd Can.* **29**: 1691–1700.

Bilinski, E. and Jonas, R. E. E. (1973). Effects of cadmium and copper on the oxidation of lactate by rainbow trout (*Salmo gairdneri*) gills. *J. Fish. Res. Bd Can.* **30**(10): 1553–8.

Biological Methods for the Assessment of Water Quality (1973). ASTM STP 528, American Society for Testing and Materials. (J. Cairns and K. L. Dickson, Eds.)

Blaxhall, P. C. (1972). The haematological assessment of the health of freshwater fish. A review of selected literature. *J. Fish Biol.* **4**: 593–604.

Blaxhall, P. C. and Daisley, K. W. (1973). Routine haematological methods for use with fish blood. *J. Fish. Biol.* **5**: 771–81.

Bluzat, R. and Seuge, J. (1979). Effets de trois insecticides (Lindane, Fenthion et Carbaryl): toxicité aiguë sur quatre espèces d'invertébrés limniques; toxicité chronique chez le mollusque pulmone *Lymnea. Environ. Pollut.* **18**: 51–70.

Boetius, J. (1960). Lethal action of mercuric chloride and phenylmercuric acetate on fishes. *Medd. Dan. Fisk. og Havunders. N. S.* **3**: 93–115.

Boileau, S., Baril, M. and Alary, J. G. (1979). DDT in northern pike (*Esox lucius*) from the Richelieu River, Quebec, Canada, 1974–75. *Pest. Monit J.* **13**: 109–114.

Bolas, P. M. and Lund, J. W. G. (1974). Some factors affecting the growth of *Cladophora glomerata* in the Kentish Stour. *Wat. Treat. Exam.* **23**: 25–49.

Bonsignore, D., Calissana, P. and Cartasegna, C. (1965). Un semplice metodo per la determinazione della amino-levulinico-deidratasi nel sangre. *Med. Larvaro* **56**: 199–205.

Borgmann, U., Kramar, O. and Loveridge, C. (1978). Rates of mortality, growth, and biomass production of *Lymnaea palustris* during chronic exposure to lead. *J. Fish. Res. Bd Can.* **35**: 1109–1115.

Bott, T. L. and Brock, T. D. (1970). Growth rate of *Sphaerotilus* in a thermally polluted environment. *Appl. Microbiol.* **19**: 100–102.

Bouck, G. R. and Ball, R. C. (1965). Influence of diurnal oxygen pulse on fish serum proteins. *Trans. Am. Fish. Soc.* **94**: 363–70.

Bowden, K., Green, J. A. and Newsome, D. H. (1971). A mathematical model of the Trent river system. In: *The Trent Research Programme, Proceedings of a symposium at Nottingham.* pp. 94–108. Maidstone, Institute of Water Pollution Control.

Boyd, C. E. (1970). Losses of mineral nutrients during decomposition of *Typha latifolia, Arch. Hydrobiol.* **66**: 511–17.

Braginskyi, L. P. and Shcherban, E. P. (1978). Acute toxicity of heavy metals to aquatic invertebrates at different temperatures. *Hydrobiol. J.* **14**: 78–52.

Braune, W. (1971). Zur Ermittlung der potentiellen Produktivität von Flusswasserproben im Algentest. *Int. Revue ges. Hydrobiol.* **56**: 795–810.

Brett, J. R. (1956). Some principles in the thermal requirements of fishes. *Q. Rev. Biol.* **31**: 75–87.

Brett, J. R. (1960). Thermal requirements of fish—three decades of study, 1940–1970. In: *Biological Problems in Water Pollution.* pp. 111–117. Washington DC, US Government Printing Office.

Brewin, D. J. and Hellawell, J. M. (1980). A water authority viewpoint on monitoring. *Chemistry & Industry.* August 1980 pp. 595–600. London.

Bringmann, G. and Kühn, R. (1956). Der Algen-Titer als Maßstab der Eutrophierung von Wasser und Schlamm. *Ges.-Ing.* **77**: 374–81.

Bringmann, G. and Kühn, R. (1958). Veränderung der Eutrophierung und Bio-Produktion, gemessen am Biomassentiter von Testalgen. *Ges.-Ing.* **79**: 50–4.

Bringmann, G. and Kühn, R. (1962). Biomassentiter und Saprobien—eine hydrobiologische Vergleichsanalyse an Niederrhein, Fulda und Havel. *Int. Rev. ges. Hydrobiol.* **47**: 123–45.

Bringmann, G. and Kühn, R. (1980). Comparison of the toxicity thresholds of water pollutants to bacteria, algae, and protozoa in the cell multiplication test. *Wat. Res.* **14**: 231–41.

Brinkhurst, R. O. (1963). A guide to the identification of British aquatic Oligochaeta. *Sci. Publs. Freshwat. Biol. Ass.* **22**.

Brinkhurst, R. O. (1966). The Tubificidae (Oligochaeta) of polluted waters. *Verh. int. Verein. theor. angew. Limnol.* **16**: 854–59.

Brinkhurst, R. O. and Chua, K. E. (1969). Preliminary investigation of the exploitation of some potential nutritional sources by three sympatric tubificid oligochaetes. *J. Fish. Res. Bd Can.* **26**: 2659–68.

Brinkhurst, R. O. and Cook, D. G. (1974). Aquatic earthworms (Annelida Oligochaeta) In: *Pollution Ecology of Freshwater Invertebrates* (C. W. Hart and S. L. H. Fuller, Eds.) pp. 143–156. New York, Academic Press.

Brinkhurst, R. O., Chua, K. E. and Kaushik, N. (1972). Interspecific interactions and selective feeding by tubificid oligochaetes. *Limnol. Oceanogr* **17**: 122–33.

Brkovic-Popovic, I. and Popovic, M. (1977b). Effects of heavy metals on survival and respiration rate of tubificid worms: Part I—Effects on survival. *Environ. Pollut.* **13**: 65–72.

Brkovic-Popovic, I. and Popovic, M. (1977b). Effects of heavy metals on survival and respiration rate of tubificid worms. Part II—Effects on respiration. *Environ. Pollut.* **13**: 93–8.

Broderius, S. J. and Smith, L. L. (1979). Lethal and sublethal effects of binary mixtures of cyanide and hexavalent chromium, zinc or ammonia to the fathead minnow (*Pimephales promelas*) and rainbow trout (*Salmo gairdneri*). *J. Fish. Res. Bd Can.* **36**: 164–72.

Broderius, S. J., Smith, L. L. and Lind, D. T. (1977). Relative toxicity of free cyanide and dissolved sulfide forms to the fathead minnow (*Pimephales promelas*). *J. Fish. Res. Bd Can.* **34**: 2323–32.

Brooker, M. P. (1976a). The ecological effects of the use of dalapon and 2,4-D for drainage channel management. I. Flora and chemistry. *Arch. Hydrobiol.* **78**: 396–412.

Brooker, M. P. (1976b). The ecological effects of the use of dalapon and 2,4-D for drainage channel management. II. Fauna. *Arch. Hydrobiol.* **78**: 507–525.

Brooker, M. P. (Ed.) (1982). Conservation of wildlife in river corridors, Part 1. Methods of survey and classification. Joint study by University of Wales Institute of Science and Technology, Welsh Water Authority, Royal Society for the Protection of Birds, Otter Haven Project and Nature Conservancy Council. Cardiff.

Brooker, M. P. and Edwards, R. W. (1973a). Effects of the herbicide paraquat on the ecology of a reservoir. I. Botanical and chemical aspects. *Freshwat. Biol.* **3**: 157–75.

Brooker, M. P. and Edwards, R. W. (1973b). Effects of the herbicide paraquat on the ecology of a reservoir. II. Community metabolism. *Freshwat. Biol.* **3**: 383–9.

Brooker, M. P. and Edwards, R. W. (1974). Effects of the herbicide paraquat on the ecology of a reservoir. III. Fauna and general discussion. *Freshwat. Biol.* **4**: 311–35.

Brooker, M. P. and Edwards, R. W. (1975). Aquatic herbicides and the control of water weeds. *Wat. Res.* **9**: 1–15.

Brooker, M. P. and Morris, D. L. (1980). A survey of the macro-invertebrate riffle fauna of the rivers Ystwyth and Rheidol, Wales. *Freshwat. Biol.* **10**: 459–74.

Brooker, M. P., Morris, D. L. and Hemsworth, R. J. (1977). Mass mortalities of adult salmon (*Salmo salar*) in the River Wye, 1976. *J. Appl. Ecol.* **14**: 409–417.

Brown, B. E. (1976). Observations on the tolerance of the isopod *Asellus meridianus* Rac. to copper and lead. *Wat. Res.* **10**: 555–9.

Brown, B. T. and Rattigan, B. M. (1979). Toxicity of soluble copper and other metal ions to *Elodea canadensis*. *Environ. Pollut.* **20**: 303–314.

Brown, C. B. (1960). Effects of land use and treatment on pollution. National Conference on Water Pollution, Washington DC, December, 1960.

Brown, D. A. and McLeay, D. J. (1975). Effect of nitrite on methemoglobin and total hemoglobin in juvenile rainbow trout. *Progve Fish Cult.* **37**: 36–8.

Brown, D. J. A. and Langford, T. E. (1975). An assessment of a tow net used to sample coarse fish fry in rivers. *J. Fish Biol.* **7**: 533–8.

Brown, D. J. A. and Sadler, K. (1981). The chemistry and fishery status of acid lakes in Norway and their relationship to European sulphur emissions. *J. Appl. Ecol.* **18**: 433–41.

Brown, R. M., McClelland, N. I., Deininger, R. A. and Tozer, R. G. (1970b). A water quality index—do we dare? *Water & Sewage Works*, **117**: 339–43.

Brown, V. M. (1968). The calculation of the acute toxicity of mixtures of poisons to rainbow trout. *Wat. Res.* **2**: 723–33.

Brown, V. M. (1973). Concepts and outlook in testing the toxicity of substances to fish. In: *Bioassay Techniques and Environmental Chemistry.* (G. E. Glass, Ed.) pp. 73–95. Ann Arbor.

Brown, V. M. and Dalton, R. A. (1970). The acute lethal toxicity to rainbow trout of mixtures of copper, phenol, zinc and nickel. *J. Fish Biol.* **2**: 211–16.

Brown, V. M., Jordan, D. H. M. and Tiller, B. A. (1967). The effect of temperature on the acute toxicity of phenol to rainbow trout in hard water. *Wat. Res.* **1**: 587–94.

Brown, V. M., Jordan, D. H. M. and Tiller, B. A. (1969). The acute toxicity to rainbow trout of fluctuating concentrations and mixtures of ammonia, phenol and zinc. *J. Fish Biol.* **1**: 1–9.

Brown, V. M., Shurben, D. G. and Shaw, D. (1970a). Studies on water quality and the absence of fish from some polluted English rivers. *Wat. Res.* **4**: 363–82.

Brungs, W. A. (1973). Effects of residual chlorine on aquatic life. *J. Wat. Pollut. Control Fed.* **45**: 2180–93.

Brungs, W. A., Leonard, E. N. and McKim, J. M. (1973). Acute and long-term accumulation of copper by the brown bullhead, *Ictalurus nebulosus*. *J. Fish. Res. Bd Can.* **30**: 583–6.

Bryan, K. A., Hellawell, J. M. and Harper, D. M. (1980). Environmental aspects of river augmentation by means of groundwater. *Prog. Wat. Technol.* **13**: 115–26.

Bryce, D., Caffmoor, I. M., Dale, C. R. and Jarrett, A. F. (1978). *Macroinvertebrates and the Bioassay of Water Quality: a Report Based on a Survey of the River Lee.* London, North East London Polytechnic.

Buckler, D. R., Witt, A., Mayer, F. L. and Huckins, J. N. (1981). Acute and chronic effects of Kepone and mirex on the fathead minnow. *Trans. Am. Fish. Soc.* **110**: 270–80.

Bucksteeg, W. and Thiele, H. (1959). The examination of sewage and sludge by means of TTC (2,3,5-triphenyl tetrazolium chloride). *Gas-u. Wasserfach* **100**: 916–20.

Buikema, A. L. and Cairns, J. (Eds.) (1980). *Aquatic invertebrate bioassays.* A symposium sponsored by ASTM Committee D-19 on Water. American Society for Testing and Materials Blacksburg, Va. 27–29 September 1977. ASTM Special Technical Publication 715, ASTM Publication Code Number (PCN) 04-715000-16, American Society for Testing and Materials Philadelphia, Pa 19103.

Buikema, A. L., Geiger, J. G. and Lee, D. R. (1980). *Daphnia* toxicity tests. In: *Aquatic Invertebrate Bioassays.* (A. L. Buikema and J. Cairns, Eds.) pp. 48–69. Philadelphia, Pa, American Society for Testing and Materials.

Burdick, G. E., Harris, E. J., Dean, H. J., Walker, T. M., Skea, J. and Golby, D. (1964). The accumulation of DDT in lake trout and the effect on reproduction. *Trans. Am. Fish. Soc.* **93**: 127–36.

Burrows, W. D. (1977). Aquatic aluminium: chemistry, toxicology and environmental prevalence, *CRC Crit. Rev. Environ. Control* **7**: 167–216.

Burton, M. A. S. and Peterson, P. J. (1979). Metal accumulation by aquatic bryophytes from polluted mine streams. *Environ. Pollut.* **19**: 39–46.

Butcher, R. W. (1932). Contribution to our knowledge of the ecology of sewage fungus. *Trans. Br. Mycol. Soc.* **17**: 112–24.

Butcher, R. W. (1947). Studies on the ecology of rivers. VII. The algae of organically enriched waters. *J. Ecol.* **35**: 186–91.

Butcher, R. W. (1955). Relation between the biology and the polluted condition of the Trent. *Verh. int. Verein. theor. angew. Limnol.* **12**: 823–7.

Cain, J. R., Klotz, R. L., Trainor, F. R. and Costello, R. (1979). Algal assay and chemical analysis: a comparative study of water quality assessment techniques in a polluted river. *Environ. Pollut.* **19**: 215–24.

Cairns, J. Jr (1972). Rationalization of multiple use of rivers. In: *River Ecology and Man.* (R. T. Oglesby, C. A. Carlson and J. A. McCann, Eds.) pp. 421–30. New York, Academic Press.

Cairns, J. and Dickson, K. L. (1978). Field and laboratory protocols for evaluating the effects of chemical substances on aquatic life. *J. Testing & Evaluation* **6**: 81–90.

Cairns, J. and Yongue, W. H. (1974). Protozoan colonisation rates on artificial substrates suspended at different depths. *Trans. Am. Microsc. Soc.* **93**: 206–210.

Cairns, J., Albaugh, D. W., Busey, F. and Chancy, M. D. (1968). The sequential comparison index—a simplified method for non-biologists to estimate relative differences in biological diversity in stream pollution studies. *J. Wat. Pollut. Control Fed.* **40**: 1607–13.

Cairns, J., Dickson, K., Sparks, R. E. and Waller, W. T. (1970). Preliminary report on rapid biological information systems for water pollution control. *J. Wat. Pollut. Control Fed.* **42**: 685–703.

Cairns, J., Dickson, K. A., Lanza, G. R., Almeida, S. P. and Del Balzo, D. (1972). Coherent optical spatial filtering of diatoms in water pollution monitoring. *Arch. Mikrobiol.* **83**: 141–6.

Cairns, J., Sparks, R. E. and Waller, W. T. (1973a). The design of a continuous flow biological early warning system for industrial use. Presented at 27th Purdue Industrial Waste Conference, May 2–4, 1972, Purdue University.

Cairns, J., Sparks, R. E. and Waller, W. T. (1973b). A tentative proposal for a rapid in-plant biological monitoring system. In: *Biological Methods for the Assessment of Water Quality.* (J. Cairns, and K. L. Dickson Eds.) pp. 127–47. Philadelphia, American Society for Testing and Materials. ASTM Spec. Tech. Publ. 528.

Cairns, J., Sparks, R. E. and Waller, W. T. (1973c). The use of fish as sensors in industrial waste lines to prevent fish kills. *Hydrobiologia* **41**: 151–67.

Cairns, J., Sparks, R. E. and Waller, W. T. (1973d). The relationship between continuous biological monitoring and water quality standards for chronic exposure. In: *Bioassay Techniques and Environmental Chemistry.* (G. E. Glass Ed.) pp. 383–402. Ann Arbor.

Cairns, J., Boatin, H. and Yongue, W. H. (1973e). The protozoan colonization of polyurethane foam units anchored in the benthic area of Douglas Lake, Michigan. *Trans. Am. Microsc. Soc.* **92**: 648–56.

Cairns, V. W. and Conn, K. (1979). Acute lethality of wastewater disinfection alternatives to juvenile rainbow trout (*Salmo gairdneri*). Canada-Ontario Agreement on Great Lakes Water Quality, Res. Rep. 92, Project 74-3-18. Environment Canada, Ontario.

Calamari, D. and Alabaster, J. S. (1980). An approach to theoretical models in evaluating the effects of mixtures of toxicants in the aquatic environment. *Chemosphere* **9**: 533–8.

Calamari, D., Marchetti, R. and Vailati, G. (1980). Influence of water hardness on cadmium toxicity to *Salmo gairdneri* Rich. *Wat. Res.* **14**: 1421–6.

Callely, A. G., Forster, C. F. and Stafford, D. A. (Eds.) (1977). *Treatment of Industrial Effluents.* London, Hodder & Stoughton.

Campbell, J., Flannagan, J. F. *et al.* (1973). A preliminary compilation of literature pertaining to the culture of aquatic invertebrates and macrophytes. *Tech. Rep. Fish. Res. Bd Can.* No. 227.

Canton, J. H., Greve, P. A., Sloof, W. and Van Esch, G. J. (1975). Toxicity, accumulation and elimination studies of α-hexachlorocyclohexane (α-HCH) with freshwater organisms of different trophic levels. *Wat. Res.* **9**: 1163–9.

Caponera, D. A. (1976). Earliest water law systems. Roman water law system. In: *Global Water Law Systems.* (P. D. Radosevich, V. G. Boira, D. R. Daines, G. V. Skogerboe and E. C. Vlochos, Eds.) Vol. 1, pp. 159–172 and 173–192. *Proceedings of an International Conference on Global Water Law System,* September 1975, Valencia, Spain. Colorado State University, Fort Collins, Colorado.

Carlson, A. R. (1972). Effects of long-term exposure to carbaryl (Sevin) on survival, growth and reproduction of the fathead minnow (*Pimephales promelas*). *J. Fish. Res. Bd Can.* **29**: 583–7.

Carlson, R. W. and Drummond, R. A. (1978). Fish cough response—a method for evaluating quality of treated complex effluents. *Water. Res.* **12**: 1–6.

Carpenter, K. E. (1924). A study of the fauna of rivers polluted by lead mining in the Aberystwyth district of Cardiganshire. *Ann. Appl. Biol.* **11**: 1–23.

Carpenter, K. E. (1925). On the biological factors involved in the destruction of river-fisheries by pollution due to lead-mining. *Ann. Appl. Biol.* **12**: 1–13.

Carpenter, K. E. (1926). The lead mine as an active agent in river pollution. *Ann. Appl. Biol.* **13**: 395–401.

Carpenter, K. E. (1927). The lethal action of soluble metallic salts on fishes. *J. Exp. Biol.* **4**: 378–90.

Carpenter, K. E. (1928). *Life in Inland Waters*. London, Sidgwick & Jackson.

Carrick, T. R. (1979). The effect of acid water on the hatching of salmonid eggs. *J. Fish Biol.* **14**: 165–72.

Caspars, H. and Schulz, H. (1960). Studien zur Wertung der Saprobien system. Erfahrungen an einem stadtkanal Hamburgs. *Int. Rev. ges. Hydrobiol.* **45**: 535–65.

Chambers, J. E. and Yarbrough, J. D. (1974). Parathion and methyl parathion toxicity to insecticide-resistant and susceptible mosquito fish (*Gambusia affinis*). *Bull. Environ. Contam. Toxicol.* **11**: 315–20.

Chandler, J. R. (1970). A biological approach to water quality management. *Wat. Pollut. Control Lond.* **69**: 415–22.

Chapman, G. A. (1978). Toxicities of cadmium, copper, and zinc to four juvenile stages of chinook salmon and steelhead. *Trans. Am. Fish. Soc.* **107**: 841–7.

Chapman, G. A. and Stevens, D. G. (1978). Acutely lethal levels of cadmium, copper, and zinc to adult male coho salmon and steelhead. *Trans. Am. Fish. Soc.* **107**: 837–40.

Cherry, D. S., Wasserman, C. S., Chung, M. S., Rubin, B. D. and Manning, M. (1978). Chemistry and biological hazard of a coal ash seepage stream. *J. Wat. Pollut. Control Fed.* **50**: 747–53.

Cherry, D. S., Guthrie, R. K., Sherberger, F. F. and Larrick, S. R. (1979a). The influence of coal ash and thermal discharges upon the distribution and bioaccumulation of aquatic invertebrates. *Hydrobiologia* **62**: 257–67.

Cherry, D. S., Larrick, S. R., Guthrie, R. K., Davis, E. M. and Sherberger, F. F. (1979b). Recovery of invertebrate and vertebrate populations in a coal ash stressed drainage system. *J. Fish. Res. Bd Can.* **36**: 1089–96.

Chigbo, F. E., Smith, R. W. and Shore, F. L. (1982). Uptake of arsenic, cadmium, lead and mercury from polluted waters by the water hyacinth *Eichornia crassipes*. *Environ. Pollut. Ser. A.* **27**: 31–6.

Chin, Y. N. and Sudderuddin, K. I. (1979). Effect of methamidophos on the growth rate and esterase activity of the common carp *Cyprinus carpio* L. *Environ. Pollut.* **18**: 213–20.

Christensen, E. R. and Zielski, P. A. (1980). Toxicity of arsenic and PCB to a green alga (*Chlamydemonas*). *Bull. Environ. Contam. Toxicol.* **25**: 43–8.

Chutter, F. M. (1969). The effects of silt and sand on the invertebrate fauna of streams and rivers. *Hydrobiologia* **35**: 57–76.

Chutter, F. M. (1972). An empirical biotic index of the quality of water in South African streams and rivers. *Wat. Res.* **6**: 19–30.

Clarke, G. L. and Bumpus, D. F. (1940). The plankton-sampler—an instrument for quantitative plankton investigations. *Limnol. Soc. America Spec. Publ.* **5**.

Clasen, J. and Bernhardt, H. (1974). The use of algal assays for determining the effect of iron and phosphorus compounds on the growth of various algal species. *Wat. Res.* **8**: 31–44.

Clifford, H. T. and Stephenson, W. (1975). *An Introduction to Numerical Classification.* New York, Academic Press.

Clubb, R. W., Gaufin, A. R. and Lords, J. L. (1975). Acute cadmium toxicity studies upon nine species of aquatic insects. *Environ. Res.* **9**: 332–41.

Coats, J. R. and O'Donnell-Jeffery, N. L. (1979). Toxicity of four synthetic pyrethroid insecticides to rainbow trout. *Bull. Environ. Contam. Toxicol.* **23**: 250–5.

Cocking, A. W. (1959). The effect of high temperature on roach (*Rutilus rutilus* L. 1. The effect of constant high temperatures, *J. Exp. Biol.* **36**: 203–216.

Cole, R. A. and Kelly, J. E. (1978). Zoobenthos in thermal discharge to western Lake Erie. *J. Wat. Pollut. Control Fed.* **50**: 2509–521.

Cole, S. L. and Wilhm, J. (1973). Effect of phenol on oxygen uptake rate of a laboratory population of *Chironomus attenuatus* (Walk.) *Wat. Res.* **7**: 1691–1700.

Coleman, M. J. and Hynes, H. B. N. (1970). The vertical distribution of the invertebrates in the bed of a stream. *Limnol. Oceanogr.* **15**: 31–40.

Coleman, R. L. and Cearley, J. E. (1974). Silver toxicity and accumulation in largemouth bass and bluegill. *Bull. Environ. Contam. Toxicol.* **12**: 53–61.

Conway, H. L. (1978). Sorption of arsenic and cadmium and their effects on growth, micronutrient utilization, and photosynthetic pigment composition of *Asterionella formosa. J. Fish. Res. Bd Can.* **35**: 286–94.

Cooke, S. E. K. (1976). Quest for an index of community structure sensitive to water pollution. *Environ. Pollut.* **11**: 269–88.

Cooke, S. F. and Moore, R. L. (1969). The effects of a rotenone treatment on the insect fauna of a California stream. *Trans. Am. Fish. Soc.* **98**: 539–44.

Cope, O. B. (1963). Sport fishery investigations. Pesticide–Wildlife studies, 1963. *US Fish. Wildlife Serv. Circ.* **199**: 29–43.

Coppage, D. L. (1971). Characterization of fish brain acetylcholinesterase with an automated pH stat for inhibition studies. *Bull. Environ. Contam. Toxicol.* **6**: 304–310.

Coppage, D. L. (1972). Organophosphate pesticides: specific level of brain AChE inhibition related to death in sheepshead minnows. *Trans. Am. Fish. Soc.* **101**: 534–6.

Corbett, J. R. (1974). *The Biochemical Mode of Action of Pesticides.* London, Academic Press.

Cordone, A. J. and Kelley, D. W. (1961). The influence of inorganic sediment on the aquatic life of streams. *Calif. Fish. Game* **47**: 189–228.

Costa, H. H. (1970). Effects of some common insecticides and other environmental factors on the heart beat of *Caridina pristis. Hydrobiologia* **35**: 469–80.

Coutant, C. C. (1968). Effect of temperature on the development rate of bottom organisms. In: *Biological Effects of Thermal Discharges.* pp. 11–12. Ann. Rep. Pacif. N.W. Lab. US Atomic Energy Commission. Division of Biological Medicine.

Coutant, C. C., Wasserman, C. S., Chung, M. S., Rubin, D. B. and Manning, M. (1978). Chemistry and biological hazard of a coal ash seepage stream. *J. Wat. Pollut. Control Fed.* **50**: 747–53.

Cowling, E. B. (1982). Acid precipitation in historical perspective. *Environ. Sci. Technol.* **16**: 110A–123A.

Cowx, I. G. (1983). Review of the methods for estimating fish population size from survey removal data. *Fish Mgmt* **14**: 67–82.

Craig, J. F. (1980). Growth and production of the 1955 to 1972 cohorts of perch, *Perca fluviatilis* L., in Windermere. *J. Anim. Ecol.* **49**: 291–315.

Craig, J. F., Kipling, C., Le Cren, E. D. and McCormack, J. C. (1979). Estimates of the numbers, biomass and year-class strengths of perch (*Perca fluviatilis* L.) in Windermere from 1967 to 1977 and some comparisons with earlier years. *J. Anim. Ecol.* **48**: 315–25.

Cranmer, M. and Peoples, A. (1973). Determination of trace quantities of anticholinesterase pesticides in air and water. In: *Bioassay Techniques and Environmental Chemistry.* (G. E. Glass, Ed.) pp. 367–81. Ann Arbor.

Credland, P. F. (1973). A new method for establishing a permanent laboratory culture of *Chironomus riparius* Meigen (Diptera: Chironomidae). *Freshwat. Biol.* **3**: 45–51.

Cripe, C. R. (1979). An automated device (AGARS) for studying avoidance of pollutant gradients by aquatic organisms. *J. Fish. Res. Bd. Can.* **36**: 11–16.

Crisp, D. J. (1971). Energy flow measurements. In: *Methods for the Study of Marine Benthos. IBP Handbook No. 16.* (N. A. Holme and A. D. McIntyre, Eds.) pp. 197–279, Oxford, Blackwell Scientific Publications.

Crisp, D. T. and Gledhill, T. (1970). A quantitative description of the recovery of the bottom fauna in a muddy reach in a mill stream in Southern England after draining and dredging. *Arch. Hydrobiol.* **67**: 502–41.

Crosby, D. G. and Tucker, R. K. (1966). Toxicity of aquatic herbicides to *Daphnia magna. Science, NY,* **154**: 289–91.

Cross, D. G. and Stott, B. (1975). The effect of electric fishing on the subsequent capture of fish. *J. Fish Biol.* **7**: 349–57.

Crossland, N. O. (1982). Aquatic toxicology of cypermethrin II. Fate and biological effects in pond experiments. *Aquat. Toxicol.* **2**: 205–222.

Crowther, R. A. and Hynes, H. B. N. (1977). The effect of road deicing salt on the drift of stream benthos. *Environ. Pollut.* **14**: 113–26.

Crowther, R. F. and Harkness, W. (1975). Anaerobic bacteria. In: *Ecological Aspects of Used-water Treatment, 1. The organisms and their ecology.* (C. R. Curds and H. A. Hawkes, Eds.) pp. 65–91. London, Academic Press.

Cullimore, D. R. and McCann, A. (1972). Initial studies on a method of algal assay for nutrient parameters in water. *J. Fish. Res. Bd Can.* **29**: 195–8.

Curds, C. R. and Cockburn, A. (1970). Protozoa in biological sewage-treatment processes. 1. A survey of the protozoan fauna of British percolating filters and activated-sludge plants. *Wat. Res.* **4**: 225–36.

Curtis, E. J. C. (1969). Sewage fungus: its nature and effects. *Wat. Res.* **3**: 289–311.

Curtis, E. J. C. and Curds, C. R. (1971). Sewage fungus in rivers in the United

Kingdom: the slime community and its constituent organisms. *Wat. Res.* **5**: 1147–59.

Curtis, E. J. C. and Harrington, D. W. (1971). The occurrence of sewage fungus in rivers in the United Kingdom. *Wat. Res.* **5**: 281–90.

Curtis, E. J. C., Delves-Broughton, J. and Harrington, D. W. (1971). Sewage fungus: studies of *Sphaerotilus* slimes using laboratory recirculating channels. *Wat. Res.* **5**: 267–79.

Curtis, M. W., Copeland, T. L. and Ward, C. H. (1979). Acute toxicity of 12 industrial chemicals to freshwater and saltwater organisms. *Wat. Res.* **13**: 137–41.

Czekanowski, J. (1913). *Zarys metod statystycznych.* Warsaw.

Daniels, R. E. and Allan, J. D. (1981). Life table evaluation of chronic exposure to a pesticide. *Can. J. Fish. Aquat. Sci.* **38**: 485–94.

Davies, P. H. and Everhart, W. H. (1973). Effects of chemical variations in aquatic environments: III. Lead toxicity to rainbow trout and testing application factor concept. *Ecol. Res. Ser. Rep. EPA-R3-73-011.* Washington, DC, US Environmental Protection Agency.

Davies, P. H., Goettl, J. P., Sinley, J. R. and Smith, N. F. (1976). Acute and chronic toxicity of lead to rainbow trout, *Salmo gairdneri*, in hard and soft water. *Wat. Res.* **10**: 199–206.

Davies, P. H., Goettl, J. P. and Sinley, J. R. (1978). Toxicity of silver to rainbow trout (*Salmo gairdneri*). *Wat. Res.* **12**: 113–17.

Davis, J. C. (1973). Sublethal effects of bleached kraft pulp mill effluent on respiration and circulation in sockeye salmon (*Onchorhynchus nerka*). *J. Fish. Res. Bd Can.* **30**: 369–77.

Davis, J. C. and Hoos, R. A. W. (1975). Use of sodium pentachlorophenate and dehydroabietic acid as reference toxicants for salmonid bioassays. *J. Fish. Res. Bd Can.* **32**: 411–16.

Davis, K. B. and Simco, B. A. (1976). Salinity effects on plasma electrolytes of channel catfish, *Ictalurus punctatus. J. Fish. Res. Bd Can.* **33**: 741–6.

Davy, F. B., Kleerekoper, H. and Gensler, P. (1972). Effects of exposure to sublethal DDT on the locomotor behavior of the goldfish (*Carassius auratus*). *J. Fish. Res. Bd Can.* **29**: 1333–6.

Decamps, H., Besch, K. W. and Vobis, H. (1973). Influence de produits toxiques sur la construction du filet des larves d'*Hydropsyche* (Insecta, Trichoptera). *C. R. Acad. Sci. Paris* **276**: 375–8.

DeFoe, D. L., Veith, G. D. and Carlson, R. W. (1978). Effects of Aroclor® 1248 and 1260 on the fathead minnow (*Pimephales promelas*). *J. Fish. Res. Bd Can.* **35**: 997–1002.

Deitz, F. (1973). The enrichment of heavy metals in submerged plants. In: *Advances in Water Pollution Research. Proceedings of 6th International Conference.* (S. H. Jenkins, Ed.) pp. 53–62. Oxford, Pergamon Press.

De Lury, D. B. (1947). On the estimation of biological populations. *Biometrics,* **3**: 145–67.

Department of the Environment (1972). *Water Pollution Research, 1971.* London, HMSO.

Department of the Environment (1979). *Methods of Biological Sampling. Hand-net Sampling of Aquatic Benthic Macroinvertebrates, 1978.* London, HMSO.

Desi, I., Dura, G., Gonczi, L., Kneffel, Z., Stroghmayer, A. and Szabo, A. (1976). Toxicity of malathion to mammals, aquatic organisms, and tissue culture cells. *Arch. Environ. Contam. Toxicol.* **3**: 410–25.

Devries, A. L. (1971). Freezing resistance in fishes. In: *Fish Physiology, Vol VI, Environmental Relations and Behaviour.* (W. S. Hoar and D. J. Randall, Eds.) pp. 157–90. New York, Academic Press.

Dickman, M. D. and Gochnauer, M. B. (1978). Impact of sodium chloride on the microbiota of a small stream. *Environ. Pollut.* **17**: 109–126.

D'Itri, M. (1973). Mercury in the aquatic ecosystem. In: *Bioassay Techniques and Environmental Chemistry.* (G. E. Glass, Ed.) pp. 3–70. Ann Arbor, Ann Arbor Science Publishers Inc.

Dittmar, H. (1955). Die quantitative Analyse der Fliesswasser—Benthos. Anregungen zu ihrer methodischen Anwendung und ihre praktische Bedeutung. *Arch. Hydrobiol. Suppl.* **22**: 295–300.

Dive, D. and Leclerc, H. (1975). Standardised test method using protozoa for measuring water pollutant toxicity. *Prog. Wat. Technol.* **7**: 67–72.

Dive, D. and Leclerc, H. (1977). Utilisation du protozoaire cillié *Colpidium campylum* pour la mesure de la toxicité de et de l'accumulation des micropollutants: analyse critique et applications. *Environ. Pollut.* **14**: 169–86.

Dixon, D. G. and Sprague, J. B. (1981a). Acclimation to copper by rainbow trout (*Salmo gairdneri*)—a modifying factor in toxicity. *Can. J. Fish. Aquat. Sci.* **38**: 880–88.

Dixon, D. G. and Sprague, J. B. (1981b). Acclimation-induced changes in toxicity of arsenic and cyanide to rainbow trout, *Salmo gairdnerii* Richardson. *J. Fish Biol.* **18**: 579–89.

Dixon, D. G. and Sprague, J. B. (1981c). Copper bioaccumulation and hepatoprotein synthesis during acclimation to copper by juvenile rainbow trout. *Aquat. Toxicol.* **1**: 69–81.

Dollard, G. J., Unsworth, M. H. and Harvey, M. J. (1983). Pollutant transfer in upland regions by occult precipitation. *Nature, Lond.* **302**: 241–3.

Doudoroff, P. (1956). Some experiments on the toxicity of complex cyanides to fish. *Sewage Indust. Wastes* **28**: 1020–40.

Doudoroff, P. and Katz, M. (1953). Critical review of literature on the toxicity of industrial wastes and their components to fish. 2. The metals, as salts. *Sewage, Indust. Wastes* **25**: 802–839.

Doudoroff, P., Anderson, B. G., Burdick, G. E., Galtsoff, P. S., Hart, W. B., Patrick, R., Strong, E. R., Surber, E. W. and Van Horn, W. M. (1951). Bioassay methods for the evaluation of acute toxicity of industrial wastes to fish. *Sewage Indust. Wastes* **23**: 1381–97.

Douglas, B. (1958). The ecology of the attached diatoms and other algae in a strong stream. *J. Ecol.* **46**: 295–322.

Downing, K. M. (1954). The influence of dissolved oxygen concentration on the toxicity of potassium cyanide to rainbow trout. *J. Exp. Biol.* **31**: 161–4.

Downing, K. M. and Merkens, J. C. (1955). The influence of dissolved-oxygen concentration on the toxicity of un-ionized ammonia to rainbow trout (*Salmo gairdnerii* Richardson). *Ann. Appl. Biol.* **43**: 243–6.

Duffey, E. (1968). An ecological analysis of the spider fauna of sand dunes. *J. Anim. Ecol.* **37**: 641–74.

Dussart, G. B. J., Mycock, E. R. and Scott, D. (1980). An alternative biological water quality index. *Speculations in Science and Technology* **3**: 157–65.

Duvel, W. A., Volkmar, R. D., Specht, W. L. and Johnson, F. W. (1976). Environmental impact of stream channelization. *Wat. Resour. Bull.* **12**: 799–812.

Eaton, J. G. (1970). Chronic malathion toxicity to the bluegill. *Wat. Res.* **4**: 673–84.

Eaton, J. G. (1973). Recent developments in the use of laboratory bioassays to determine 'safe' levels of toxicants for fish. In: *Bioassay Techniques and Environmental Chemistry*. (G. E. Glass, Ed.) pp. 107–115. Ann Arbor.

Edwards, C. A. (Ed.) (1973a). *Environmental Pollution by Pesticides*. London, Plenum Press.

Edwards, C. A. (1973b). Pesticide residues in soil and water. In: *Environmental Pollution by Pesticides*. (C. A. Edwards, Ed.) pp. 409–458. London, Plenum Press.

Edwards, C. A. (1973c) (2nd edn). *Persistent Pesticides in the Environment*. Cleveland, CRC Press.

Edwards, R. W. (1957). Vernal sloughing of sludge deposits in a sewage effluent channel. *Nature, Lond.* **180**: 100.

Edwards, R. W. (1972). *Pollution*. Oxford Biology readers No. 31. Oxford, Oxford Univ. Press.

Edwards, R. W. and Brown, V. M. (1967). Pollution and fisheries: a progress report. *Wat. Pollut. Control* **66**: 3–18.

Edwards, R. W. and Owens, M. (1960). The effects of plants on river conditions. IV. The oxygen balance of the chalk stream. *J. Ecol.* **50**: 207–220.

Edwards, R. W., Benson-Evans, K., Learner, M. A., Williams, P. and Williams, R. (1972). A biological survey of the River Taff. *Wat. Pollut. Control.* **71**: 144–66.

EEC (1978). Council Directive of 18 July 1978 on the quality of freshwaters needing protection or improvement in order to support fish life. *Official J. European Communities* **21** (L222): 1–10.

Effler, S. W., Litten, S., Field, S. D., Tong-Ngork, T., Hale, F., Meyer, M. and Quirk, M. (1980). Whole lake responses to low level copper sulphate treatment. *Wat. Res.* **14**: 1489–99.

Egglishaw, H. J. (1964). The distributional relationship between the bottom fauna and plant detritus in streams. *J. Anim. Ecol.* **33**: 463–76.

EIFAC (1964). Water quality criteria for European freshwater fish. Report on finely divided solids and inland fisheries. EIFAC Tech. Pap. 1 FAO, Rome.

EIFAC (1968a). Report on extreme pH values and inland fisheries. European Inland Fisheries Advisory Commission FAO, Rome. EIFAC Tech. Pap. 4.

EIFAC (1968b). Report on water temperature and inland fisheries based mainly on Slavonic literature. European Inland Fisheries Advisory Commission, FAO, Rome. EIFAC Tech. Pap. 6.

EIFAC (1969a). Water quality criteria for European freshwater fish. List of literature on the effects of water temperature on fish. EIFAC Tech. Pap. 8.

EIFAC (1969b). Water quality criteria for European freshwater fish—extreme pH values and inland fisheries. *Wat. Res.* **3**: 593–611.

EIFAC (1970). Report on ammonia and inland fisheries. European Inland Fisheries Advisory Commission, FAO, Rome. EIFAC Tech. Pap. 11.

EIFAC (1972). Report on monohydric phenols and inland fisheries. European

Inland Fisheries Advisory Commission, FAO, Rome. EIFAC Tech. Pap. 15.

EIFAC (1973a). Report on dissolved oxygen and inland fisheries. European Inland Fisheries Advisory Commission, FAO, Rome. EIFAC Tech. Pap. 19.

EIFAC (1973b). Report on chlorine and freshwater fish. European Inland Fisheries Advisory Commission, FAO, Rome. EIFAC Tech. Pap. 20.

EIFAC (1973c). Report on zinc and freshwater fish. European Inland Fisheries Advisory Commission, FAO, Rome. EIFAC Tech. Pap. 21.

EIFAC (1976). Report on copper and freshwater fish. European Inland Fisheries Advisory Commission, FAO, Rome. EIFAC Tech Pap. 27.

EIFAC (1977). Report on cadmium and freshwater fish. European Inland Fisheries Advisory Commission, FAO, Rome. EIFAC Tech. Pap. 30.

EIFAC (1980). Water quality criteria for European freshwater fish. Report on combined effects on freshwater fish and other aquatic life of mixtures of toxicants in water. EIFAC Technical Paper No. 37. European Inland Fisheries Advisory Commission, Food and Agriculture Organization of the United Nations, Rome.

Eisenreich, S. J., Hoolod, G. J. and Johnson, T. C. (1979). Accumulation of polychlorinated biphenyls (PCBs) in superficial Lake Superior sediments. Atmospheric deposition. *Environ. Sci. Technol.* **13**: 569–73.

Eisler, R. and Edmunds, P. H. (1966). Effects of endrin on blood chemistry of a marine fish. *Trans. Am. Fish. Soc.* **95**: 153–9.

Eisler, R. and Gardner, G. R. (1973). Acute toxicity to an estuarine teleost of mixtures of cadmium, copper and zinc salts. *J. Fish Biol.* **5**: 131–42.

Elliott, J. M. (1969). Diel periodicity in invertebrate drift and the effect of different sampling periods. *Oikos* **20**: 524–8.

Elliott, J. M. (1972). Effect of temperature on the time of hatching in *Baetis rhodani* (Ephemeroptera, Baetidae). *Oecologia (Berlin)* **9**: 47–51.

Elliott, J. M. (1981). Some aspects of thermal stress on freshwater teleosts. In: *Stress and Fish.* (A. D. Pickering, Ed.) pp. 209–245. London, Academic Press.

Elliott, J. M. and Drake, C. M. (1981a). A comparative study of seven grabs used for sampling benthic macroinvertebrates in rivers. *Freshwat. Biol.* **11**: 99–120.

Elliott, J. M. and Drake, C. M. (1981b). A comparative study of four dredges used for sampling benthic macroinvertebrates in rivers. *Freshwat. Biol.* **11**: 245–61.

Elliott, J. M. and Mann, K. H. (1979). A key to the British freshwater leeches with notes on their life cycles and ecology. *Sci. Publs. Freshwat. Biol. Ass.* **40**.

Ellis, M. M. (1936). Erosion silt as a factor in aquatic environments. *Ecology* **17**: 29–42.

Elson, P. F. (1967). Effects on wild young salmon of spraying DDT over New Brunswick forests. *J. Fish. Res. Bd Can.* **24**: 731–67.

Elson, P. F. and Kerswill, C. J. (1966). Impact on salmon of spraying insecticide over forest. *Adv. Wat. Pollut. Res.* **1**: 55–74.

Elster, H. J. (1966). Über die limnologischen Grundlagen der biologischen Gewässer-Beurteilung in Mitteleuropa. *Verh. int. Verein. theor. angew. Limnol.* **16**: 759–85.

Elton, C. (1942). *Voles, Mice and Lemmings: Problems in Population Dynamics.* London, Oxford Univ. Press.

Empain, A. (1976a). Estimation de la pollution par mètaux lourds dans la

Somme par l'analyse des bryophytes aquatiques. *Bull. Fr. Piscic.* **48**: 138–42.

Empain, A. (1976b). Les bryophytes aquatiques utilisès comme traceurs de la contamination en mètaux lourds des eaux douces. *Mèm. Soc. R. Bot. Belg.* **7**: 141–56.

EPA (1973). *Water Quality Criteria, 1972.* (EPA.R3.73.003) Washington, DC, Environmental Protection Agency.

Evins, C. (1975). The toxicity of chlorine to some freshwater organisms. *Tech. Rep. Wat. Res. Centre*, TR8.

Evins, C. (1981). A new substance for controlling animals in mains. *Water Services* **85**: 280, 289.

Evison, L. M. (1979). Microbial parameters of raw water quality. In: *Biological Indicators of Water Quality.* Proceedings of a symposium at the University of Newcastle-upon-Tyne, UK, September, 1978.

Extence, C. A. (1978). The effects of motorway construction on an urban stream. *Environ. Pollut.* **17**: 245–52.

Falkenmark, M. (1980). International conference on the effects of acid rain urges more research in critical areas. *Ambio* **9**: 198–9.

Falkner, R. and Simonis, W. (1982). Polychlorierte Biphenyle (PCB) in Lebensraum Wasser (Aufname und Anreicherung durch Organismen—Probleme der Weitergabe in der Nahrungspyramide). Ein Literaturbericht für den Zeitraum 1972–1979. *Archiv. Hydrobiol. Beit. Ergebn. Limnol.* **17**: 1–74.

Farmer, G. J., Ashfield, D. and Samant, H. S. (1979). Effects of zinc on juvenile Atlantic salmon *Salmo salar*: acute toxicity, food intake, growth and bioaccumulation. *Environ. Pollut.* **19**: 103–117.

Ferguson, R. G. (1958). The preferred temperature of fish and their midsummer distribution in temperate lakes and streams. *J. Fish. Res. Bd. Can.* **15**: 607–614.

Ferguson, J. F. and Gavis, J. (1972). A review of the arsenic cycle in natural waters. *Wat. Res.* **6**: 1259–74.

Ferguson, D. E., Culley, P. D. and Cotton, W. D. (1965). Tolerances of two populations of freshwater shrimp to five chlorinated hydrocarbon insecticides. *J. Miss. Acad. Sci.* **9**: 235–7.

Ferguson, D. E., Gardner, D. T. and Lindley, A. L. (1966). Toxicity of Dursban to three species of fish. *Mosquito News* **26**: 80–82.

Finley, M. T., Ferguson, D. E. and Ludke, J. L. (1970). Possible selective mechanisms in the development of insecticide-resistant fish. *Pestic. Monit. J.* **3**: 212–18.

Fisher, R. A., Corbett, A. S. and Williams, G. B. (1943). The relation between the number of species and the number of individuals in a random sample of an animal population. *J. Anim. Ecol.* **12**: 42–58.

Fjerdingstad, E. (1964). Pollution of streams estimated by benthal phytomicro-organisms. 1. A saprobic system based on communities of organisms and ecological factors. *Int. Rev. ges. Hydrobiol.* **49**: 63–131.

Fjerdingstad, E. (1965). Taxonomy and saprobic valency of benthic phytomicro-organisms. *Int. Rev. ges. Hydrobiol.* **50**: 475–604.

Fogels, A. and Sprague, J. B. (1977). Comparative short-term tolerance of zebrafish, flagfish, and rainbow trout to five poisons including potential reference toxicants. *Wat. Res.* **11**: 811–17.

Fogg, G. E. and Westlake, D. F. (1955). The importance of extracellular products of algae in freshwater. *Verh. int. Verein. theor. angew. Limnol.* **12**: 219–32.

Forsberg, C. and Hökervall, E. (1971). An algal assay procedure of value for control of sewage effluents from treatment plants in Stockholm. *Vatten* **1**: 51–7.

Forsberg, C. and Hökervall, E. (1972a). Algal growth potential test (AGP test) of sewage effluent. 1. The treatment plant at Äkeshov-Nockeby, Stockholm, February–September, 1971. e. summary p. 25. *Vatten* **5**: 17–26.

Forsberg, C. and Hökervall, E. (1972b). AGP—test av Kommunalt Avloppsvatten. 2. Eolshäll och Louddens reningsverk. *Vatten* **5**: 413–17.

Fort, R. S. and Brayshaw, A. D. (1961). *Fishery Management.* London, Faber & Faber.

Foster, N. R., Scheier, A. and Cairns, J. (1966). Effects of ABS on feeding behaviour of flagfish *Jordanella floridae. Trans. Am. Fish. Soc.* **95**: 109–110.

Fozzard, I. (1978). In discussion of paper by Hunter (1978), p. 392.

Fraser, J., Pakin, D. T. and Verspoor, E. (1978). Tolerance to lead in the freshwater isopod, *Asellus aquaticus. Wat. Res.* **12**: 637–41.

Frear, D. E. H. and Boyd, J. E. (1967). Use of *Daphnia magna* for the microbioassay of pesticides. 1. Development of standardized techniques for rearing *Daphnia* and preparation of dosage-mortality curves for pesticides. *J. Econ. Entomol.* **60**: 1228–36.

Frederick, L. L. (1975). Comparative uptake of a polychlorinated biphenyl and dieldrin by the white sucker (*Catastomus commersoni*). *J. Fish. Res. Bd Can.* **32**: 1705–1709.

Freeden, F. J. H. (1959a). Rearing black flies in the laboratory (Diptera: Simuliidae), *Can. Ent.* **91**: 73–83.

Freeden, F. J. H. (1959b). Collection, extraction, sterilization and low-temperature storage of black-fly eggs (Diptera: Simuliidae). *Can. Ent.* **91**: 450–53.

Freeman, A. M. (1978). Air and water pollution policy. In: *Current Issues in US Environmental Policy.* (P. R. Portney, Ed.) pp. 12–67. Baltimore, Johns Hopkins University Press.

Freeman, R. A. and Everhart, W. H. (1971). Toxicity of aluminium hydroxide complexes in neutral and basic media to rainbow trout. *Trans. Am. Fish. Soc.* **100**: 644–58.

Fremling, C. R. and Mauck, W. L. (1980). Methods for using nymphs of burrowing mayflies (Ephemeroptera, *Hexagenia*) as toxicity test organisms. In: *Aquatic Invertebrate Bioassays.* (A. L. Buikema and J. Cairns, Eds.) pp. 81–97. American Society for Testing and Materials ASTM STP 715.

Fromm, P. (1980). A review of some physiological and toxicological responses of freshwater fish to acid stress. *Environ. Biol. Fish.* **5**: 79–93.

Frost, W. E. and Brown, M. E. (1967). *The Trout.* London, Collins, New Naturalist Series.

Frost, S., Huni, A. and Kershaw, W. E. (1971). Evaluation of a kicking technique for sampling stream bottom fauna. *Can. J. Zool.* **49**: 167–73.

Furse, M. T., Wright, J. F., Armitage, P. D. and Moss, D. (1981). An appraisal of pond-net samples for biological monitoring of lotic macroinvertebrates. *Wat. Res.* **15**: 679–89.

Furse, M. T., Moss, D., Wright, J. F. and Armitage, P. D. (1984). The influence of seasonal and taxonomic factors in the ordination and classification of running-water sites in Great Britain and on the prediction of their macro-invertebrate communities. *Freshwat. Biol.* **14**: 257–80.

FWPCA (1968). Water quality criteria. Report of the National Technical Advisory Committee to Secretary of the Interior. Federal Water Pollution Control Administration, US Dept. of the Interior. Washington DC, Government Printing Office.

Galepp, G. (1976). Temperature as a cue for the periodicity in feeding of *Brachycentrus occidentalis* (Insecta: Trichoptera). *Anim. Behav.* **24**: 7–10.

Garside, E. T. and Tait, J. S. (1958). Preferred temperature of rainbow trout (*Salmo gairdnerii* Richardson) and its unusual relationship to acclimatization temperature. *Can. J. Zool.* **36**: 564–7.

Gaufin, A. R., Jensen, L. D., Nebeker, A. V., Nelson, T. and Teel, R. W. (1965). The toxicity of ten organic insecticides to various aquatic invertebrates. *Wat. Sewage Wks* **112**: 276–9.

Gibson, J. R., Ludke, J. L. and Ferguson, D. E. (1969). Sources of error in the use of fish-brain acetylcholinesterase activity as a monitor for pollution. *Bull. Environ. Contam. Toxicol.* **4**: 17–23.

Giese, A. C. (1967). Some methods for study of the biochemical constitution of marine invertebrates. *Oceanogr. Mar. Biol. Ann. Rev.* **5**: 159–86.

Gilderhus, P. A. (1966). Some effects of sublethal concentrations of sodium arsenite on bluegills and the aquatic environment. *Trans. Am. Fish. Soc.* **95**: 289–96.

Gilderhus, P. A. (1972). Exposure times necessary for antimycin and rotenone to eliminate certain freshwater fish. *J. Fish. Res. Bd Canada* **29**: 199–202.

Giles, M. A. and Klaverkamp, J. F. (1982). The acute toxicity of vanadium and copper to eyed eggs of rainbow trout (*Salmo gairdneri*) *Wat. Res.* **16**: 885–9.

Gipps, J. F. and Coller, B. A. W. (1980). Effect of physical and culture conditions on uptake of cadmium by *Chlorella pyrenoidosa*. *Aust. J. Mar. Freshwat. Res.* **31**: 747–55.

Gledhill, T. (1960). The Ephemeroptera, Plecoptera and Trichoptera caught by emergence traps in two streams during 1968. *Hydrobiologia* **15**: 179–88.

Glooschenko, W. A. and Moore, J. E. (1973). The effect of citrate on phytoplankton in Lake Ontario. In: (1973) *Bioassay Techniques and Environmental Chemistry.* (G. E. Glass, Ed.) pp. 321–34. Ann Arbor.

Goldman, C. R. (1962). A method of studying nutrient limiting factors *in situ*, in water columns isolated by polyethylene film. *Limnol. Oceanogr.* **7**: 99–101.

Goldman, C. R., Tunzi, M. G. and Armstrong, R. (1969). Carbon-14 uptake as a sensitive measure of the growth of algal cultures. *Proceedings of an Eutrophication-Biostimulation Assessment Workshop, University of California,* pp. 158–70, 1969.

Golterman, H. L. (Ed.) (1969). Methods for chemical analysis of freshwaters. *IBP Handbook No. 8.* Oxford, Blackwell Scientific Publications.

Goodland, R. (1977). Panamanian development and the global environment. *Oikos* **29**: 195–208.

Goodman, G. T. (1974). How do chemical substances affect the environment? *Proc. Roy. Soc. London B* **185**: 127–48.

Goodnight, C. J. and Whitley, L. S. (1960). Oligochaetes as indicators of pollution. *Proc. Am. Waste Conf. Purdue Univ.* **15**: 139–42.

Goodyear, C. P. (1972). A simple technique for detecting the effects of toxicants or other stresses on a predator–prey interaction. *Trans. Am. Fish. Soc.* **101**: 367–70.

Goreham, E. (1958a). The influence and importance of daily weather conditions in the supply of chloride, sulphate and other ions to freshwater from atmospheric precipitations. *Phil. Trans. Roy. Soc. B* **241**: 147–78.

Goreham, E. (1958b). Free acid in British soils. *Nature, Lond.* **181**: 106.

Gower, A. M. and Buckland, P. J. (1978). Water quality and the occurrence of *Chironomus riparius* Meigen (Diptera: Chironomidae) in a stream receiving sewage effluent. *Freshwat. Biol.* **8**: 153–64.

Gower, J. C. and Ross, G. J. S. (1969). Minimum spanning trees and single linkage cluster analysis. *Appl. Statist.* **18**: 54–64.

Grande, M. (1967). Effect of copper and zinc on salmonid fishes. *Adv. Wat. Pollut. Res.* **3**: 97–111.

Grant, B. F. (1976) Endrin toxicity and distribution in freshwater: A review. *Bull. Environ. Contam. Toxicol.* **15**: 283–90.

Gray, H. F. (1940). Sewerage in ancient and medieval times. *Sewage Wks J.* **12**: 939–46.

Greene, K. L. (1974). Experiments and observations on the feeding behaviour of the freshwater leech *Erpobdella octoculata* (L.) (Hirudinea: Erpobdellidae). *Arch. Hydrobiol.* **74**: 87–99.

Greenfield, J. P. and Ireland, M. P. (1978). A survey of the macrofauna of a coalwaste polluted Lancashire fluviatile system. *Environ. Pollut.* **16**: 105–122.

Gunkel, G. and Streit, B. (1980). Mechanisms of bioaccumulation of a herbicide (atrazine, s-triazine) in a freshwater mollusc (*Ancylus fluviatilis* Müll.) and a fish (*Coregonus fra* Jurnine). *Wat. Res.* **14**: 1573–84.

Gunner, H. B. and Coler, A. (1972). A microbiotic ecoassay for environmental pollutants. In: *International Symposium on Identification and Measurement of Environmental Pollutants*, Ottawa, Ontario, 14–17 June, 1971. National Research Council Canada, 1972; 314.

Gurnham, C. F. (Ed.) (1965). *Industrial Wastewater Control.* New York, Academic Press.

Gustafson, C. G. (1970). PCB's—prevalent and persistent. *Environ. Sci. Technol.* **4**: 814–19.

Hackney, C. R. and Bissonnette, G. K. (1978). Recovery of indicator bacteria in acid mine streams. *J. Wat. Pollut. Control Fed.* **50**: 775–80.

Haines, T. A. (1973). An evaluation of RNA–DNA ratio as a measure of long-term growth in fish populations. *J. Fish. Res. Bd Can.* **30**: 195–9.

Haines, T. A. (1981). Acidic precipitation and its consequences for aquatic ecosystems: a review. *Trans. Am. Fish Soc.* **110**: 669–707.

Hairston, N. G. (1959). Species abundance and community organisation. *Ecology* **40**: 404–16.

Hall, D. C., Cooper, W. E. and Werner, E. E. (1970). An experimental approach to the production dynamics and structure of freshwater animal communities. *Limnol. Oceanogr.* **15**: 839–928.

Halter, M. T. and Johnson, H. E. (1974). Acute toxicities of polychlorinated biphenyl (PCB) and DDT alone and in combination, to early life stages of coho salmon (*Oncorhynchus kisutch*). *J. Fish. Res. Bd Can.* **31**: 1543–7.

Hamelink, J. L. and Waybrant, R. C. (1976). DDE and lindane in a large-scale model lentic ecosystem. *Trans. Am. Fish. Soc.* **105**: 124–34.

Hamelink, J. L., Waybrant, R. C. and Ball, R. C. (1971). A proposal: Exchange equilibria control the degree chlorinated hydrocarbons are biologically magnified in lentic environments. *Trans. Am. Fish. Soc.* **100**: 207–214.

Hamilton, H. L. (1941). The biological action of rotenone on fresh-water animals. *Proc. Iowa Acad. Sci.* **48**: 467–79.

Hamilton, J. D. (1961). The effect of sand-pit washings on stream fauna. *Verh. int. Verin. theor. angew. Limnol.* **14**: 435–9.

Hanbury, R. G., Murphy, K. J. and Eaton, J. W. (1981). The ecological effects of 2-methylthiotriazine herbicides used for aquatic weed control in navigable canals. II. Effects on macroinvertebrate fauna, and general discussion. *Arch. Hydrobiol.* **91**: 408–426.

Hannan, P. J. and Patouillet, C. (1970). Nutrient and pollutant concentrations as determinants in algal growth rates. FAO Technical Conference on Marine Pollution and its Effects on Living Resources and Fishing. Rome, Italy.

Hansen, C. R. and Kawatski, J. A. (1976). Application of 24 hour postexposure observation to acute toxicity studies with invertebrates. *J. Fish. Res. Bd Can.* **33**: 1198–1201.

Hansen, D. J. (1969). Avoidance of pesticides by untrained sheepshead minnows. *Trans. Am. Fish. Soc.* **98**: 426–9.

Hansen, D. J. (1972). DDT and Malathion: effect on salinity selection by mosquito fish. *Trans. Am. Fish. Soc.* **101**: 346–50.

Hansen, L. G., Wiekhorst, W. B. and Simon, J. (1976). Effects of dietary Aroclor 1242 on channel catfish (*Ictalurus punctatus*) and the selective accumulation of PCB components. *J. Fish. Res. Bd Can.* **33**: 1343–52.

Hansen, P.-D. (1980). Uptake and transfer of the chlorinated hydrocarbon Lindane (BHC) in a laboratory freshwater food chain. *Environ. Pollut. Ser. A.* **21**: 97–108.

Harding, J. P. C. (1981). Macrophytes as monitors of river quality in the southern NWWA area. North West Water Authority Report No TS-BS-81-2, Warrington.

Harding, J. P. C. and Whitton, B. A. (1981). Accumulation of zinc, cadmium and lead by field populations of *Lemanea*. *Wat. Res.* **15**: 301–319.

Harrison, A. D. (1958). The effects of sulphuric acid pollution on the biology of streams in the Transvaal, South Africa. *Verh. int. Verein. theor. angew. Limnol.* **13**: 603–619.

Harrison, A. D. and Farina, T. D. W. (1965). A naturally turbid water with deleterious effects on the egg capsules of planorbid snails. *Ann. trop. Med. Parasit.* **59**: 327–330.

Hart, C. W. and Fuller, S. L. H. (1974). *Pollution Ecology of Freshwater Invertebrates*. Academic Press, New York.

Haslam, S. M. (1978). *River Plants. The Macrophytic Vegetation of Watercourses*. Cambridge, Cambridge Univ. Press.

Haslam, S. (1982a). *Vegetation in British Rivers*. Vol. 1. Text. London, Nature Conservancy Council.

Haslam, S. M. (1982b). A proposed method for monitoring river pollution using macrophytes. *Environ. Technol. Lett.* **3**: 19–34.

Hatch, T. (1962). Changing objectives in occupational health. *Am. Ind. Hyg. Ass. J.* **23**: 1–7.

Hatfield, C. T. and Anderson, J. M. (1972). Effects of two insecticides on the vulnerability of Atlantic salmon (*Salmo salar*) parr to brook trout (*Salvelinus fontinalis*) predation. *J. Fish. Res. Bd Can.* **29**: 27–9.

Hatfield, C. T. and Johansen, P. H. (1972). Effects of four insecticides on the ability of Atlantic salmon parr (*Salmo salar*) to learn and retain a simple conditioned response. *J. Fish. Res. Bd Can.* **29**: 315–21.

Hawkes, H. A. (1962a). Effects of domestic and industrial discharges on the ecology of riffles in Midland streams. Int. Conf. Wat. Pollut. Res. London, Secn. 1. No. 6.

Hawkes, H. A. (1962b) Biological aspects of river pollution In: *River Pollution 2. Causes and Effects.* (L. Klein, Ed.) pp. 311–432. London, Butterworths.

Hawkes, H. A. (1963). Effects of domestic and industrial discharges on the ecology of riffles in Midland streams. *Int. J. Air Wat. Pollut.* **7**: 565–83.

Hawkes, H. A. (1964). Effects of domestic and industrial discharges on the ecology of riffles in Midland streams. *Adv. Wat. Pollut. Res.* **1**: 293–317.

Hawkes, H. A. (1968). Ecological changes of applied significance induced by the discharge of heated waters. In: *Engineering Aspects of Thermal Pollution.* (F. L. Parker and P. A. Krenkel, Eds.) pp. 15–57. Nashville, Vanderbilt University Press.

Hawkes, H. A. (1975). River zonation and classification. In: *River Ecology. Studies in Ecology* Vol. 2. (B. A. Whitton, Ed.) pp. 312–374. Oxford, Blackwell Scientific Publications.

Hawkes, H. A. (1979). Invertebrates as indicators of river water quality (UK). In: *Biological Indicators of Water Quality.* (A. James and L. M. Evison, Eds.) pp. 2·1–2·45. Proceedings of a symposium held at University of Newcastle-upon-Tyne, 12–15 September 1978.

Hawkes, H. A. and Davies, L. J. (1971). Some effects of organic enrichment on benthic invertebrate communities in stream riffles. In: *The Scientific Management of Animal and Plant Communities for Conservation.* (E. Duffey and A. S. Watt, Eds.) pp. 271–293. Oxford, Blackwell Scientific Publications.

Heath, A. G. (1972). A critical comparison of methods for measuring fish respiratory movements. *Wat. Res.* **6**: 1–7.

Heisinger, J. F. and Green, W. (1975). Mercuric chloride uptake by eggs of the ricefish and resulting teratogenic effects. *Bull. Environ. Contam. Toxicol.* **14**: 665–73.

Hellawell, J. M. (1971). The autecology of the chub, *Squalius cephalus* (L.) of the River Lugg and Afon Llynfi. I. Age determination, population structure and growth. *Freshwat. Biol.* **1**: 29–60.

Hellawell, J. M. (1972). The growth, reproduction and food of the roach *Rutilus rutilus* (L.) of the River Lugg, Herefordshire. *J. Fish Biol.* **4**: 469–86.

Hellawell, J. M. (1974). The ecology of populations of dace, *Leuciscus leuciscus* (L.) from two tributaries of the River Wye, Herefordshire, England. *Freshwat. Biol.* **4**: 577–604.

Hellawell, J. (1977a). Biological surveillance and water quality monitoring. In: *Biological Monitoring of Inland Fisheries.* (J. S. Alabaster, Ed.) pp. 69–88. London, Applied Science Publishers.

Hellawell, J. M. (1977b). Change in natural and managed ecosystems: detection, measurement and assessment. *Proc. Roy. Soc. Lond. B.* **197**: 31–56.

Hellawell, J. M. (1978). *Biological Surveillance of Rivers.* Medmenham and Stevenage, Water Research Centre.

Hellawell, J. M. (1982). Conservation information for water authorities—What do we need and how do we get it? In: *Assessing the Conservation Value of Rivers.* (D. M. Harper, Ed.) pp. 1–20. Papers presented at a conference at Leicester University, February 1981. Vaughan Paper No. 29, Department of Adult Education, University of Leicester.

Hellawell, J. M. and Bryan, K. A. (1982). The use of herbicides for aquatic weed control in water supply catchments. *J. Inst. Wat. Eng. Sci* **36**: 221–33.

Henderson, C., Pickering, Q. H. and Tarzwell, C. M. (1959). Toxicity of organic phosphorus and chlorinated hydrocarbon insecticides to fish. In: *Biological Problems in Water Pollution,* Transactions of 1959 Seminar, Robert A. Taft Sanitary Engineering Center, Tech. Rept. W60-3: 76–92.

Hendrey, G. R. (1976). Effects of pH on the growth of periphytic algae in artificial stream channels. Sur Nedbørs virkning pa Shog og. Fisk-Project (Oslo, Norway) IR 25/72.

Herbert, D. W. M. (1962). The toxicity to rainbow trout of spent still liquors from the distillation of coal. *Ann. Appl. Biol.* **50**: 755–77.

Herbert, D. W. M. (1965). Pollution and fisheries. In: *Ecology and the Industrial Society.* pp. 173–195. Oxford, Blackwell.

Herbert, D. W. M. and Merkens, J. C. (1952). The toxicity of potassium cyanide to trout. *J. Exp. Biol.* **29**: 632–49.

Herbert, D. W. M. and Merkens, J. C. (1961). The effect of suspended mineral solids on the survival of trout. *Int. J. Air. Wat. Pollut.* **5**: 46–55.

Herbert, D. W. M. and Richards, J. M. (1963). The growth and survival of fish in some suspensions of solids of industrial origin. *Int. J. Air. Wat. Pollut.* **7**: 297–302.

Herbert, D. W. M. and Shurben, D. S. (1963). A preliminary study of the effect of physical activity on the resistance of rainbow trout (*Salmo gairdnerii* Richardson) to two poisons. *Ann. Appl. Biol.* **52**: 321–6.

Herbert, D. W. M. and Shurben, D. G. (1964). The toxicity to fish of mixtures of poisons—I. Salts of ammonium and zinc. *Ann. Appl. Biol.* **53**: 33–41.

Herbert, D. W. M. and Vandyke, J. M. (1964). The toxicity to fish of mixtures of poisons—II. Copper–ammonia and zinc–phenol mixtures. *Ann. Appl. Biol.* **53**: 415–421.

Herbert, D. W. M., Alabaster, J. S., Dart, M. C. and Lloyd, R. (1961). The effect of china-clay wastes on trout streams. *Int. J. Air Wat. Pollut.* **5**: 56–74.

Herbert, D. W. M., Jordan, D. H. M. and Lloyd, R. (1965). A study of some fishless rivers in the industrial Midlands. *J. Proc. Inst. Sew. Purif.* **1965**: 569–82.

Hester, F. E. and Dendy, J. S. (1962). A multiple plate sampler for aquatic macroinvertebrates. *Trans. Am. Fish. Soc.* **91**: 420–21.

Hewitt, L. A. and Anderson, P. D. (1978). Aspects of supra-additive interactions between cadmium and copper in fish exposed to lethal mixtures. *Pharmacology* **21**: 210 (abstract).

Hildebrand, L. D., Sullivan, D. S. and Sullivan, T. P. (1980). Effects of Roundup®

herbicide on populations of *Daphnia magna* in a forest pond. *Bull. Environ. Contam. Toxicol.* **25**: 353–7.

Hildebrand, L. D., Sullivan, D. S. and Sullivan, T. P. (1982). Experimental studies of rainbow trout populations exposed to field applications of Roundup® herbicide. *Arch. Environ. Contam. Toxicol.* **11**: 93–8.

Hiley, P. D. (1970). A method of rearing Trichoptera larvae for taxonomic purposes. *Entomologists mon. Mag.* **105**: 278–9.

Hill, M. O. (1979a). Decorana—A Fortran program for detrended correspondence analysis and reciprocal averaging. *Ecology and Systematics*. Ithaca, New York, Cornell University.

Hill, M. O. (1979b). Twinspan—A Fortran program for arranging multivariate data in an ordered two-way table by classification of the individuals and attributes. *Ecology and Systematics*, Ithaca, New York, Cornell University.

Hinton, D. E., Kendall, M. W. and Silver, B. B. (1973). Use of histologic and histochemical assessments in the prognosis of the effects of aquatic pollutants. In: *Biological Methods for the Assessment of Water Quality* (J. Cairns and K. L. Dickson, Eds.) pp. 194–208. ASTM STP 528.

Hirsch, A. (1958). Biological evaluation of organic pollution of New Zealand streams. *New Zealand J. Sci.* **1**: 500–553.

Hobble, J. E. and Wright, R. T. (1968). A new method for the study of bacteria in lakes: description and results. *Mitt. int. Verein. theor. angew. Limnol.* **14**: 64–71.

Hochachka, P. W. and Somero, G. N. (1971). Biochemical adaptation to the environment. In: *Fish Physiology Vol. VI., Environmental Relations and Behaviour.* (W. S. Hoar and D. J. Randall, Eds.) pp. 99–156. New York, Academic Press.

Hodson, P. V. (1976). δ-amino levulinic acid dehydratase activity of fish blood as an indicator of harmful exposure to lead. *J. Fish. Res. Bd Can.* **33**: 268–71.

Hodson, P. V. and Sprague, J. B. (1975). Temperature-induced changes in acute toxicity to Atlantic salmon (*Salmo salar*). *J. Fish. Res. Bd Can.* **32**: 1–10.

Hodson, P. V., Blunt, B. R. and Spry, D. J. (1978). Chronic toxicity of waterborne and dietary lead to rainbow trout (*Salmo gairdneri*) in Lake Ontario water. *Wat. Res.* **12**: 869–78.

Hodson, P. V., Spry, D. J. and Blunt, B. R. (1980). Effects on rainbow trout (*Salmo gairdneri*) of a chronic exposure to waterborne selenium. *Can. J. Fish. Aquat. Sci.* **37**: 233–40.

Hodson, P. V., Dixon, D. G., Spry, D. J., Whittle, D. M. and Sprague, J. B. (1982). Effect of growth rate and size of fish on rate of intoxication by waterborne lead. *Can. J. Fish. Aquat. Sci.* **39**: 1243–51.

Hogan, J. W. (1970). Water temperature as a source of variation in specific activity of brain acetylcholinesterase of bluegills. *Bull. Environ. Contam. Toxicol.* **5**: 347–53.

Hokanson, K. E. F., McCormick, J. H., Jones, B. R. and Tucker, J. H. (1973). Thermal requirements for maturation, spawning, and embryo survival of the brook trout, *Salvelinus fontinalis. J. Fish. Res. Bd Can.* **30**: 975–84.

Holcombe, G. W., Benoit, D. A., Leonard, E. N. and McKim, J. M. (1976). Longterm effects of lead exposure on three generations of brook trout (*Salvelinus fontinalis*). *J. Fish. Res. Bd Can.* **33**: 1731–41.

Holcombe, G. W., Benoit, D. A. and Leonard, E. N. (1979). Long-term effects of zinc exposures on brook trout (*Salvelinus fontinalis*). *Trans. Am. Fish. Soc.* **108**: 76–87.

Holcombe, G. W., Fiandt, J. T. and Phipps, G. L. (1980). Effects of pH increases and sodium chloride additions on the acute toxicity of 2,4-dichlorophenol to the fathead minnow. *Wat. Res.* **14**: 1073–77.

Holcombe, G. W., Phipps, G. L. and Tanner, D. K. (1982). The acute toxicity of kelthane, dursban, disulfoton, pydrin, and permethrin to fathead minnows *Pimephales promelas* and rainbow trout *Salmo gairdneri*. *Environ. Pollut., A.* **29**: 167–78.

Holden, A. V. (1972). The effects of pesticides on life in fresh waters. *Proc. Roy. Soc. Lond.* (*B*) **180**: 383–94.

Holden, A. V. (1973). Effects of pesticides on fish. In: *Environmental Pollution by Pesticides.* (C. A. Edwards, Ed.) pp. 213–53. London, Plenum Press.

Holdgate, M. W. (1971). The need for environmental monitoring. International Symposium on Identification and Measurement of Environmental Pollutants, Ottawa, Ontario, Canada, June 1971, pp. 1–8.

Holdway, D. A. and Sprague, J. B. (1979). Chronic toxicity of vanadium to flagfish. *Wat. Res.* **13**: 905–910.

Holeton, G. F. (1971). Oxygen uptake and transport by the rainbow trout during exposure to carbon monoxide. *J. Exp. Biol.* **54**: 239–54.

Holland, H. T. and Coppage, D. L. (1970). Sensitivity to pesticides in three generations of sheepshead minnows. *Bull. Environ. Contam. Toxicol.* **5**: 362–7.

Holland, H. T., Coppage, D. L. and Butler, P. A. (1966) Increased sensitivity to pesticides in sheepshead minnows. *Trans. Am. Fish. Soc.* **95**: 110–12.

Holland, H. T., Coppage, D. L. and Butler, P. A. (1967). Use of fish brain acetylcholinesterase to monitor pollution by organophosphorus pesticides. *Bull. Environ. Contam. Toxicol.* **2**: 156–62.

Holm-Hansen, O. and Booth, C. R. (1966). The measurement of adenosine triphosphate in the ocean and its ecological significance. *Limnol. Oceanogr.* **11**: 510–19.

Holm-Hansen, O., Lorenzen, C. J., Holmes, R. W. and Strickland, J. D. H. (1965). Fluorometric determination of chlorophyll. *J. Cons. perm. int. Explor. Mer.* **30**: 3–15.

Horoszewicz, L. (1973). Lethal and disturbing temperature in some fish species from lakes with normal and artificially elevated temperature. *J. Fish Biol.* **5**: 165–81.

Howarth, R. S. and Sprague, J. B. (1978). Copper lethality to rainbow trout in waters of various hardness and pH. *Wat. Res.* **12**: 455–62.

Howell, J. H., King, E. L., Smith, A. J. and Hanson, L. H. (1964). Synergism of 5,2'-dichloro-4', nitro-salicylanilide and 3-trifluoromethyl-4-nitrophenol in a selective lamprey larvicide. *Tech. Rep. Gt. Lakes Fishery Commission* 8.

Huet, M. (1949). Aperçu des relations entre la pente et les populations piscicoles des eaux courantes. *Schweiz. Z. Hydrol.* **11**: 333–51.

Huet, M. (1954). Biologie, profils en long et en travers des eaux courantes. *Bull. Fr. Piscic.* **175**: 41–53.

Hughes, G. M. (1973). Respiratory responses to hypoxia in fish. *Am. Zool.* **13**: 475–89.

Hughes, G. M. and Adeney, R. J. (1977). The effects of zinc on the cardiac and ventilatory rhythms of rainbow trout (*Salmo gairdneri*, Richardson) and their responses to environmental hypoxia. *Wat. Res.* **11**: 1069–77.

Hughes, G. M. and Roberts, J. L. (1970). A study of the effect of temperature changes on the respiratory pumps of the rainbow trout. *J. Exp. Biol.* **52**: 177–92.

Hughes, G. M. and Saunders, R. L. (1970). Responses of the respiratory pumps to hypoxia in the rainbow trout. *J. Exp. Biol.* **53**: 529–45.

Hughes, G. M., Perry, S. F. and Brown, V. M. (1979). A morphometric study of effects of nickel, chromium and cadmium on the secondary lamellae of rainbow trout gills. *Wat. Res.* **13**: 665–79.

Huisman, J., Ten Hoopen, M. J. G. and Fuchs, A. (1980). The effect of temperature upon the toxicity of mercuric chloride to *Scenedesmus acutus*. *Environ. Pollut. A* **22**: 133–48.

Hunt, E. G. and Bischoff, A. I. (1960). Inimical effects on wildlife of periodic DDD applications to Clear Lake. *Calif. Fish Game* **46**: 91–106.

Hunter, J. B. (1978). The role of the toxicity test in water pollution control. *Wat. Pollut. Control* **77**: 384–94.

Hunter, J. B., Ross, S. L. and Tannahill, J. (1980). Aluminium pollution and fish toxicity. *Wat. Pollut. Control* **79**: 413–20.

Hutchinson, G. E. (1957–1975). *A Treatise on Limnology*. Vol. I (1957). *Geography, Physics and Chemistry*. Vol. II (1967) *Introduction to Lake Biology and the Limnoplankton*. Vol. III (1975) *Limnological Botany*. New York, John Wiley and Sons.

Hynes, H. B. N. (1958). The effect of drought on the fauna of a small mountain stream in Wales. *Verh. int. Verein. theor. angew. Limnol.* **12**: 23–30.

Hynes, H. B. N. (1960). *The Biology of Polluted Waters*. Liverpool, Liverpool Univ. Press.

Hynes, H. B. N. (1970). *The Ecology of Running Waters*. Liverpool, Liverpool Univ. Press.

Hynes, H. B. N. and Coleman, M. V. (1968). A simple method of assessing the annual production of stream benthos. *Limnol. Oceanogr.* **13**: 569–73.

Iatomi, K., Tamura, T., Itazawa, Y., Hanyu, I. and Sugiura, S. (1958). Toxicity of endrin to fish. *Progve. Fish Cult.* **20**; 155–62.

Illies, J. (1961). Versuch einer allgemein biozönotischen Gleiderung der Fliessgewässer. *Int. Rev. Ges. Hydrobiol.* **46**: 205–213.

Ivlev, V. S. (1945). The biological productivity of waters. *Uspekhi Sovremennoi Biologii* **19**: 98–120. (Russian, Transl. Ricker, W. E., *J. Fish. Res. Bd Can.* **23**: 1727–59.)

Jaccard, P. (1912). The distribution of the flora in the alpine zone. *New Phytol.* **11**: 37–50.

Jackim, E. (1973). Influence of lead and other metals on fish δ-aminolevulinate dehydrase activity. *J. Fish. Res. Board Can.* **30**: 560–2.

Jackim, E., Hamlin, J. M. and Sonis, S. (1970). Effects of metal poisoning on five liver enzymes in the killifish (*Fundulus heteroclitus*) *J. Fish. Res. Bd Can* **27**: 383–90.

Jackson, D. A., Anderson, J. M. and Gardener, D. R. (1970). Further investigations of the effect of DDT on learning in fish. *Can. J. Zool.* **48**: 577–80.

Jarvinen, A. W., Hoffman, M. J. and Thorslund, T. W. (1977). Long-term toxic effects of DDT food and water exposure on fathead minnows (*Pimephales promelas*). *J. Fish. Res. Bd Can.* **34**: 2089–2103.

Jensen, A. L. (1972). Standard error of LC_{50} and sample size in fish bioassays. *Wat. Res.* **6**: 85–9.

Jensen, A. L., Spigarelli, S. A. and Thommes, M. M. (1982) PCB uptake by five species of fish in Lake Michigan, Green Bay of Lake Michigan, and Cayuga Lake, New York. *Can. J. Fish. Aquat. Sci.* **39**: 700–709.

Jensen, K. (1975). Hatching rate as bioassay. Proposal for a standard technique. *Bull. Environ. Contam. Toxicol.* **14**: 562–4.

Jensen, L. D. and Gaufin, A. R. (1964). Effects of ten organic insecticides on two species of stonefly naiads. *Trans. Am. Fish. Soc.* **93**: 27–34.

Jensen, L. D. and Gaufin, A. R. (1966). Acute and long-term effects of organic insecticides on two species of stonefly naiads. *J. Wat. Pollut. Control Fed.* **38**: 1273–86.

Jensen, S. (1966) Report of a new chemical hazard. *New Scientist* **32**: 612.

Jensen, S. and Jernalov, A. (1969) Biological methylation of mercury in aquatic organisms. *Nature* **223**: 753.

Jewell, W. J. (1970). Aquatic weed decay: dissolved oxygen utilisation and nitrogen and phospherus regeneration. *A. Conf. Wat. Pollut. Control Fed.* **43**: 1–27.

Johnson, B. T., Saunders, C. R., Sanders, H. O. and Campbell, R. S. (1971). Biological magnification and degradation of DDT and aldrin by freshwater invertebrates. *J. Fish. Res. Bd Canada* **28**: 705–709.

Johnson, D. W. (1968). Pesticides and fishes—a review of selected literature. *Trans. Am. Fish. Soc.* **97**: 398–424.

Johnson, D. W. (1973). Pesticide residues in fish. In *Environmental Pollution by Pesticides*. (C. A. Edwards, Ed.) pp. 181–212. London, Plenum Press.

Johnson, M. G. and Brinkhurst, R. O. (1971). Production of benthic macroinvertebrates of Bay of Quinte and Lake Ontario. *J. Fish. Res. Bd. Can.* **28**: 1699–714.

Johnson, W. D., Lee, G. F. and Spyridakis, D. (1966). Persistence of toxaphene in treated lakes. *Int. J. Air Wat. Pollut.* **10**: 555–60.

Jolly, A. L., Avault, J. W., Koonce, K. L. and Graves, J. B. (1978). Acute toxicity of permethrin to several aquatic animals. *Trans. Am. Fish. Soc.* **107**: 825–7.

Jolly, V. H. and Chapman, M. A. (1966). A preliminary biological study of the effects of pollution on Farmer's Creek and Cox's River, New South Wales. *Hydrobiologia* **27**: 160–92.

Jones, A. N. and Howells, W. R. (1969). Recovery of the River Rheidol. *Effluent Wat. Treat. J.* **9**: 605–610.

Jones, A. N. and Howells, W. R. (1975). The partial recovery of the metal polluted R. Rheidol. In: *The Ecology of Resource Degradation*. (M. J. Chadwick and G. T. Goodman, Eds.) pp. 443–59. 15th Symposium of the British Ecological Society. Oxford, Blackwell.

Jones, H. R. and Peters, J. C. (1977). Physical and biological typing of unpolluted rivers. In: *Biological Monitoring of Inland Fisheries*. (J. S. Alabaster, Ed.) pp. 39–48. London, Applied Science Publishers.

Jones, J. R. E. (1939a). Antagonism between salts of the heavy and alkaline-earth metals in their toxic action on the tadpole of the toad, *Bufo bufo bufo* (L.). *J. Exp. Biol.* **16**: 313–33.

Jones, J. R. E. (1939b). The relation between the electrolytic solution pressures of the metals and their toxicity to the stickleback (*Gasterosteus aculeatus*, L.). *J. Exp. Biol.* **16**: 425–37.

Jones, J. R. E. (1940). A study of the zinc-polluted River Ystwyth in North Cardiganshire, Wales. *Ann. Appl. Biol.* **27**: 368–78.

Jones, J. R. E. (1948). A further study of the reactions of fish to toxic solutions. *J. Exp. Biol.* **25**: 22–34.

Jones, J. R. E. (1949). An ecological study of the River Rheidol, North Cardiganshire, Wales. *J. Anim. Ecol.* **18**: 67–88.

Jones, J. R. E. (1958) A further study of the zinc-polluted River Ystwyth. *J. Anim. Ecol.* **27**: 1–14.

Jones, J. R. E. (1962). Fish and river pollution. In: *River Pollution 2: Causes and Effects.* (L. Klein, Ed.) pp. 254–310. London, Butterworth.

Jones, J. R. E. (1964). *Fish and River Pollution.* London, Butterworth.

Jones, L. H., Jones, N. V. and Radlett, A. J. (1976). Some effects of salinity on the toxicity of copper to the polychaete *Nereis diversicolor. Estuarine Coastal Mar. Sci.* **4**: 107–111.

Jones, P. H. and Prasad, D. (1969). The use of tetrazolium salts as a measure of sludge activity. *J. Wat. Pollut. Control Fed.* **41**: R441–R449.

Kaiser, K. L. E. (1980). Correlation and prediction of metal toxicity to aquatic biota. *Can. J. Fish. Aquat. Sci.* **37**: 211–218.

Kalbe, L. (1968). Investigations for the determination of sediment activity with 2,3,5-triphenyltetrazolium chloride (TTC). *Limnologica* **6**: 37–44.

Kalleberg, H. (1956). Observations in a stream tank of territoriality and competition in juvenile salmon and trout (*Salmo salar* L. and *Salmo trutta* L.). *Rep. Inst. Freshwat. Res. Drottningholm* **39**: 55–98.

Kania, H. J. and O'Hara, J. (1974). Behavioural alterations in a simple predator–prey system due to sub-lethal exposure to mercury. *Trans. Am. Fish. Soc.* **103**: 134–6.

Kapoor, N. N. (1972). Rearing and maintenance of Plecopteran nymphs. *Hydrobiologia* **40**: 51–3.

Katz, M. and Chadwick, G. G. (1961). Toxicity of endrin to some Pacific Northwest fishes. *Trans. Am. Fish. Soc.* **90**: 394–7.

Keller, W. (1956). *The Bible as History.* London, Hodder & Stoughton.

Kendall, M. G. (1962). *Rank Correlation Methods.* London, Griffin & Co Ltd.

Kennedy, H. D. and Walsh, D. F. (1970). Effects of malathion on two warmwater fishes and aquatic invertebrates in ponds. *Bur. Sport Fish. Wildl. Tech. Pap.* 55.

Kennedy, V. S. and Mihursky, J. A. (1967). *Bibliography on the Effects of Temperature in the Aquatic Environment.* University of Maryland, Natural Resources Institute, Contribution No. 326.

Kennicutt, M. C. (1980). ATP as an indicator of toxicity. *Wat. Res.* **14**: 325–8.

Kerr, S. R. and Vass, W. P. (1973). Pesticide residues in aquatic invertebrates In: *Environmental Pollution by Pesticides.* (C. A. Edwards, Ed.) pp. 134–80. London, Plenum Press.

Kfir, R. and Prozesky, O. W. (1981). Detection of toxic substances in water by means of a mammalian cell culture technique. *Wat. Res.* **15**: 553–9.

Khan, M. A. Q. (Ed.) (1977), *Pesticides in Aquatic Environments.* New York, Plenum Press.

Khoo, S. G. (1964). Studies of the biology of *Capnia bifrons* (Newman) and notes on the diapause in the nymphs of this species. *Gewäss. Abwäss* **3415**: 23–30.

King, D. L. and Ball, R. C. (1964). A quantitative biological measure of stream pollution. *J. Wat. Pollut. Control Fed.* **36**: 650.

Kirkwood, R. C. and Fletcher, W. W. (1970). Factors influencing the herbicidal efficiency of MCPA and MCPB in three species of micro-algae. *Weed Res.* **10**: 3–10.

Kleerekoper, H., Westlake, G. F., Matis, J. H. and Gensler, P. J. (1972). Orientation of goldfish (*Carassius auratus*) in response to a shallow gradient of a sublethal concentration of copper in an open field. *J. Fish. Res. Bd Can.* **29**: 45–54.

Klein, L. (1959). *River Pollution. Vol. 1. Chemical Analysis.* London, Butterworth.

Klein, L. (1962). *River Pollution. Vol. 2. Causes and Effects.* London, Butterworth.

Klotz, R. L. (1981). Algal response to copper under riverine conditions. *Environ. Pollut. Ser. A.* **24**: 1–19.

Knöpp, H. (1954). Ein neuer Weg zur Darstellung biologischer Vorfluteruntersuchungen, erlautert an einem Gütelängsschnitt des Mains. *Wasserwirtsch* **45**: 9–15.

Knöpp, H. (1961). Der A–Z Test, ein neues Verfahren zur toxologischen Prüfung von Abwässern. *Dt. Gewässerkundl. Mitt.* **5**: 66–73.

Knowles, C. O. and Casida, J. E. (1966). Mode of action of organophosphate anthelmintics. Cholinesterase inhibition of *Ascaris lumbricoides*. *J. Agr. Food Chem.* **14**: 566–72.

Kolkwitz, R. and Marsson, M. (1902). Grundsäze für die biologische Beurteilung des Wassers nach seiner Flora und Fauna. *Mitt. a.d. Kgl Prüfungsanst. f. Wasserversorg. u. Abwässerbeseitingung zu Berlin* **1**: 33–72.

Kolkwitz, R. and Marsson, M. (1908). Oekologie die pflanzlichen Saprobien. *Ber. d. Deut. Bot. Gesell.* **26**: 505–519.

Kolkwitz, R. and Marsson, M. (1909). Oekologie der tierischen Saprobien. *Int. Rev. ges Hydrobiol.* **2**: 126–52.

Konstantinov, A. S. (1958). Cultivation of chironomid larvae. Trudy saratov. Otd. vses. naucho-issled. *Inst. ozer. rech. ryb. Khoz.* **5**: 276–302. (*Trans. Ser. Fish. Res. Bd Can.* 2133.)

Koryak, M., Shapiro, M. A. and Sykora, J. L. (1972). Riffle zoobenthos in streams receiving acid mine drainage. *Wat. Res.* **6**: 1239–47.

Kothé, P. (1962). Der 'Artenfehlbetrag', ein einfaches Gütekriterium und seine Anwendung bei biologischen Vorflutersuntersuchungen. *Dt. Gewasserkundl. Mitt* **6**: 60–65.

Kovacs, T. G. and Leduc, G. (1982). Acute toxicity of cyanide to rainbow trout (*Salmo gairdneri*) acclimated to different temperatures. *Can. J. Fish. Aquat. Sc.* **39**: 1426–9.

Kramer, R. H. and Smith, L. L. (1965). Effects of suspended wood fiber on brown and rainbow trout eggs and alevins. *Trans. Am. Fish. Soc.* **94**: 252–8.

Kramer, R. H. and Smith, L. L. (1966). Survival of walleye eggs in suspended wood fibers. *Progve Fish Cult.* **31**: 79–82.

Krenkel, P. A. and Parker, F. L. (1969). Engineering aspects, sources and magnitude of thermal pollution. In: *Biological Aspects of Thermal Pollution* (P. A. Krenkel and F. L., Parker, Eds.) pp. 10–52. Nashville, Vanderbilt University Press.

Kristensen, P. (1982). Time-dependent variation of mercury in a stream sediment and the effect upon mercury content in *Gammarus pulex* (L.). *Wat. Res.* **16**: 759–64.

Kulezynski, S. (1928). Die Pflanzenassoziationen der Pieninen. *Bull. int. Acad. Pol. Sci. Lett. B Suppl.* **2**: 57–203.

Kumaraguru, A. K. and Beamish, F. W. H. (1981). Lethal toxicity of permethrin (NRDC-143) to rainbow trout, *Salmo gairdneri*, in relation to body weight and water temperature. *Wat. Res.* **15**: 503–505.

Kynard, B. (1974). Avoidance behaviour of insecticide susceptible and resistant populations of mosquitofish to four insecticides. *Trans. Am. Fish. Soc.* **103**: 557–61.

Lack, T. J. and Lund, J. W. G. (1974). Observations and experiments on the phytoplankton of Blelham Tarn, English Lake District. I. The experimental tubes. *Freshwat. Biol.* **4**: 399–415.

Ladle, M. (1971). The biology of Oligochaeta from Dorset chalk streams. *Freshwat. Biol.* **1**: 83–97.

Ladle, M. and Casey, H. (1971). Growth and nutrient relationships of *Ranunculus penicillatus* var. *calcareus* in a small chalk stream. *Proc. Eur. Weed. Res. Coun. 3rd int. Symp. aquatic Weeds.* pp. 53–64.

Lake Tahoe Area Council (1969). *Eutrophication of Surface Waters—Lake Tahoe.* Second progress report.

Langford, T. E. and Daffern, J. R. (1975). The emergence of insects from a British river warmed by power station cooling water. Part I. The use and performance of insect emergence traps in a large spate-river and the effects of various factors on total catches, upstream and downstream of the cooling water outfalls. *Hydrobiologia* **46**: 71–114.

Larkin, P. A. and Northcote, T. G. (1969). Fish as indices of eutrophication. In: *Eutrophication; Causes, Consequences, Correctives.* pp. 256–273. Proc. int. Symp. Eutrophication, Madison, Wisconsin, Nat. Acad. Sci. Washington, D.C.

Larsen, K. and Olsen, S. (1950). Ochre suffocation of fish in the River Tim Aa. Drainage from lignite pit as cause of the catastrophes af the Norre Esp trout pond farm 1941–1947. *Rep. Danish Biol. Stn.* **50**: 3–27.

Last, F. T. and Nicholson, I. A. (1982). Acid rain. *Biologist* **29**: 250–52.

Laurent, M. and Clavet, F. (1977). Utilisation des poissons comme indicateurs de qualité des eaux, comparaison avec la méthode des indices biotiques. Applications sur le Laxia: rivière du Pays Basque français et le Gabas: rivière des Landes. *Ann. Hydrobiologia.* **8**: 67–87.

Laurie, R. D. and Jones, J. R. E. (1938). The faunistic recovery of a lead polluted river in North Cardiganshire, Wales. *J. Anim. Ecol.* **7**: 272–89.

Learner, M. A. (1975). Insecta In: *Ecological Aspects of Used-water Treatment. Vol. 1. The Organisms and their Ecology.* (C. R. Curds and H. A. Hawkes, Eds.) London, Academic Press.

Learner, M. A. and Edwards, R. W. (1966). The distribution of the midge *Chironomus riparius* in a polluted river system and its environs. *Int. J. Air Wat. Pollut.* **10**: 757–68.

Learner, M. A., Williams, R., Harcup, M. and Hughes, B. D. (1971). A survey of the macro-fauna of the River Cynon, a polluted tributary of the River Taff (South Wales). *Freshwat. Biol.* **1**: 339–67.

Le Cren, E. D. (1955). Year to year variations in the year-class strength of *Perca fluviatilis*. *Verh. int. Verein. theor. angew. Limnol.* **12**: 187–92.

Leduc, G. (1978). Deleterious effects of cyanide on early life stages of Atlantic salmon (*Salmo salar*). *J. Fish. Res. Bd Can.* **35**: 166–74.

Leduc, G., Pierce, R. C. and McCracken, I. R. (1982). *The Effects of Cyanides on Aquatic Organisms with Emphasis upon Freshwater Fishes.* National Research Council of Canada, Associate Committee on Scientific Criteria for Environmental Quality, NRCC Publication No. 19246.

Lee, D. R. and Buikema, A. L. (1979). Moult-related sensitivity of *Daphnia pulex* in toxicity testing. *J. Fish. Res. Bd Canada* **36**: 1129–33.

Leeuwaugh, P. (1978). Toxicity tests with Daphnids: its application in the management of water quality. *Hydrobiologia* **59**: 145–8.

Lemly, A. D. (1982). Response of juvenile centrarchids to sublethal concentrations of waterborne selenium. I. Uptake, tissue distribution, and retention. *Aquat. Toxicol.* **2**: 235–52.

Lenhard, G. (1965). The dehydrogenase activity as a criterion of toxic effects in biological purification systems. *Hydrobiologia* **25**: 1–8.

Lester, W. F. (1975). Polluted river: River Trent, England. In: *River Ecology, Studies in Ecology Vol. 2.* (B. A. Whitton, Ed.) pp. 489–513. Oxford, Blackwell Scientific Publications.

Letterman, R. D. and Mitsch, W. J. (1978). Impact of mine drainage on a mountain stream in Pennsylvania. *Environ. Pollut.* **17**: 53–73.

Leung, T. S., Naqvi, S. M. and Leblanc, C. (1983). Toxicities of two herbicides (Basgran, diquat) and an algicide (Cutrine-Plus) to mosquito fish (*Gambusia affinis*). *Environ. Pollut. Ser. A.* **30**: 153–160.

Levin, G. V., Schrot, J. R. and Hess, W. C. (1975). Methodology for the application of adenosine triphosphate determination in waste water treatment. *Envir. Sci. Technol.* **9**: 961–5.

Li, M. F. and Jordan, C. (1969). Use of spinner culture cells to detect water pollution. *J. Fish. Res. Bd Can.* **26**: 1378–82.

Li, M. F. and Traxler, G. S. (1972). Tissue culture bioassay method for water pollution with special reference to mercuric chloride. *J. Fish. Res. Bd Can.* **29**: 501–505.

Liebmann, H. (1951; 2nd edn. 1962). *Handbuch der Frischwasser und Abwasserbiologie.* Munich.

Lindgren, P. E. (1960). About the effect of rotenone upon benthonic animals in lakes. *Rep. Inst. Freshwat. Res. Drottningholm* **41**: 172–83.

Livingstone, R. J. (1977). Review of current literature concerning the acute and chronic effects of pesticides on aquatic organisms. *CRC Crit. Rev. Environ. Control* **7**: 325–51.

Lloyd, R. (1960). The toxicity of zinc sulphate to rainbow trout (*Salmo gairdneri* Richardson). *Ann. Appl. Biol.* **48**: 84–94.

Lloyd, R. (1961a). Effect of dissolved oxygen concentrations on the toxicity of several poisons to rainbow trout (*Salmo gairdneri* Richardson). *J. Exp. Biol.* **38**: 447–55.

Lloyd, R. (1961b). The toxicity of mixtures of zinc and copper sulphates to rainbow trout (*Salmo gairdneri* Richardson). *Ann. Appl. Biol.* **49**: 535–8.

Lloyd, R. (1965). Factors that affect the tolerance of fish to heavy metal poisoning. *Biological Problems in Water Pollution*. Washington, Publ. Hlth. Serv: *99-WP-25*: 181–7.

Lloyd, R. (1972). Problems in determining water quality criteria for freshwater fisheries. *Proc. Roy. Soc. Lond. B* **180**: 439–49.

Lloyd, R. and Herbert, D. W. M. (1960). The influence of carbon dioxide on the toxicity of un-ionized ammonia to rainbow trout (*Salmo gairdnerii* Richardson). *Ann. Appl. Biol.* **48**: 399–404.

Lloyd, R. and Herbert, D. W. M. (1962). The effect of the environment on the toxicity of poisons to fish. *Instn. publ. Hlth Engrs J.* **61**: 132–45.

Lloyd, R. and Jordan, D. H. M. (1964). Some factors affecting the resistance of rainbow trout (*Salmo gairdneri*, Richardson) to acid waters. *Int. J. Air Wat. Pollut.* **8**: 393–403.

Lloyd, R. and Orr, L. D. (1969). The diuretic response by rainbow trout to sublethal concentrations of ammonia. *Wat. Res.* **3**: 335–44.

Lloyd, R. and Tooby, T. E. (1979). New terminology for short-term static fish bioassays: LC (I) 50. *Bull. Environ. Contam. Toxicol.* **22**: 1–3.

Loftus, M. E. and Carpenter, J. H. (1971). A fluorimetric method for determining chlorophylls *a*, *b* and *c*. *J. Mar. Res.* **29**: 319–38.

Looman, J. and Campbell, J. B. (1960). Adaption of Sorensen's K (1948) for estimating affinities in prairie vegetation. *Ecology* **41**: 409–416.

Lorenzen, C. J. (1966). A method for the continuous measurement of *in vivo* chlorophyll concentration. *Deep-Sea Research* **13**: 223–7.

Lorenzen, C. J. (1967). Determination of chlorophyll and pheo-pigments: spectrophotometric equations. *Limnol. Oceanogr.* **13**: 343–6.

Luedtke, R. J. and Brusven, M. A. (1976). Effects of sand sedimentation on colonization of stream insects. *J. Fish. Res. Bd Can.* **33**: 1881–6.

Lund, J. W. G. (1950). Studies on *Asterionella formosa* Hass. II. Nutrient depletion and the spring maximum. *J. Ecol.* **38**: 1–35.

Lund, J. W. G. (1964). Primary production and periodicity of phytoplankton. *Verh. int. Verein. theor. angew. Limnol.* **15**: 37–56.

Lund, J. W. G. (1972). Preliminary observations on the use of large experimental tubes in lakes. *Verh. int. Verein. theor. angew. Limnol.* **18**: 71–7.

Lutz, F. E., Welch, P. S., Galtsoff, P. S. and Needham, J. G. (1937). *Culture Methods for Invertebrate Animals*, New York, Dover Publications, Inc.

Macan, T. T. (1958). Methods of sampling the bottom fauna in stony streams. *Mitt. int. Verein. theor. angew. Limnol.* **8**: 1–21.

Macan, T. T. (1974). *Freshwater Ecology* (2nd edn). London, Longmans.

Macan, T. T. and Kitching, A. (1972). Some experiments with artificial substrata. *Verh. int. Verein. theor. angew. Limnol.* **18**: 213–20.

MacArthur, R. M. (1957). On the relative abundance of bird species. *Proc. Nat. Acad. Sci., Washington* **43**: 193–5.

Macek, K. J. (1968). Reproduction in brook trout (*Salvelinus fontinalis*) fed sublethal concentrations of DDT. *J. Fish. Res. Bd Can.* **25**: 1787–96.

Macek, K. J. and McAllister, W. A. (1970). Insecticide susceptibility of some common fish family representatives. *Trans. Am. Fish. Soc.* **99**: 20–27.

Macek, K. J. and Korn, S. (1970). Significance of the food chain in DDT accumulation by fish. *J. Fish. Res. Bd Can.* **27**: 1496–8.

Macek, K. J., Rodgers, C. R., Stalling, D. L. and Korn, S. (1970). The uptake, distribution and elimination of dietary ^{14}C-DDT and ^{14}C-dieldrin in rainbow trout. *Trans. Am. Fish. Soc.* **99**: 689–95.

Macek, K. J., Walsh, D. R., Hogan, J. W. and Holz, D. D. (1972). Toxicity of the insecticide dursban to fish and aquatic invertebrates in pond. *Trans. Am. Fish. Soc.* **101**: 420–7.

Macek, K. J., Buxton, K. S., Derr, S. K., Dean, J. W. and Sauter, S. (1976). Chronic toxicity of Lindane to selected aquatic invertebrates and fish. US Environmental Protection Agency. *Ecol. Res. Ser.* EPA-600/3-76-046.

Maciorowski, H. D. and Clarke, R. McV. (1980). Advantages and disadvantages of using invertebrates in toxicity testing. In: *Aquatic Invertebrate Bioassays.* (A. L. Buikema and J. Cairns, Eds.) pp. 36–47. Philadelphia, Pa, American Society for Testing and Materials.

MacKey, A. P. (1972). An air-lift for sampling freshwater benthos. *Oikos* **23**: 413–415.

MacLeod, J. C. and Pessah, E. (1973). Temperature effects on mercury accumulation, toxicity, and metabolic rate in rainbow trout (*Salmo gairdneri*). *J. Fish. Res. Bd Can.* **30**: 485–92.

MacLeod., J. C. and Smith, L. L. (1966). Effect of pulpwood fibre on oxygen consumption and swimming endurance of the fathead minnow. *Trans. Am. Fish. Soc.* **95**: 71–84.

MacLulich, D. A. (1937). Fluctuations in the numbers of the varying hare (*Lepus americanus*). *Univ. Toronto Studies, Biol. Ser.* No. 43.

MacPhee, C. (1961). Bioassay of algal production in chemically altered waters. *Limnol. Oceanogr.* **6**: 416–22.

Maguire, B. (1963). The passive dispersal of small aquatic organisms and their colonisation of isolated bodies of water. *Ecol. Monogr.* **33**: 161–85.

Maitland, P. S. (1978). *Biology of Freshwaters.* Glasgow, Blackie.

Maki, A. W. (1977). Modifications of continuous-flow toxicity test methods for small aquatic organisms. *Progve Fish Cult.* **39**: 172–4.

Maki, A. W. and Johnson, H. E. (1975). Effects of PCB (Aroclor 1254) and p,p'-DDT on production and survival of *Daphnia magna* Straus. *Bull. Environ. Contam. Toxicol.* **13**: 412–16.

Malcolm, A. R., Pringle, B. H. and Fisher, H. W. (1973). Chemical toxicity studies with cultured mammalian cells. In: *Bioassay Techniques and Environmental Chemistry.* (G. E. Glass, Ed.) pp. 217–230. Ann Arbor, Ann Arbor Science Publishers.

Mangi, J., Schmidt, K., Pankow, J., Gaines, L. and Turner, P. (1978). Effects of chromium on some aquatic plants. *Environ. Pollut.* **16**: 285–91.

Manley, A. R. (1980). An apparatus for the preparation of varying concentrations of chemicals for toxicity tests with aquatic organisms. *Wat. Res.* **14**: 1023–7.

Manly, R. and George, W. O. (1977). The occurrence of some heavy metals in populations of the freshwater mussel *Anodonta anatina* (L.) from the River Thames. *Environ. Pollut.* **14**: 139–54.

Mann, K. H. (1961). The oxygen requirements of leeches considered in relation to their habitats. *Ver. int. Verein. theor. angew. Limnol.* **14**: 1009–1013.

Margalef, R. (1951). Diversidad de especies en las comunidades naturales. *Publnes. Inst. Biol. apl., Barcelona,* **6**: 59–72.

Marker, A. F. H. (1972). The use of acetone and methanol in the estimation of chlorophyll in the presence of phaeophytin. *Freshwat. Biol.* **2**: 361–85.

Marking, L. L. and Bills, T. D. (1977). Chlorine: Its toxicity to fish and detoxification of antimycin. *US Fish Wildl. Serv. Invest. Fish Control* **74**.

Marking, L. L. and Mauck, W. L. (1975). Toxicity of paired mixtures of candidate forest insecticides to rainbow trout. *Bull. Environ. Contam. Toxicol.* **13**: 518–23.

Marking, L. L. and Walker, C. R. (1973). The use of fish bioassays to determine the rate of deactivation of pesticides. In: *Bioassay Techniques and Environmental Chemistry.* (G. E. Glass, Ed.) pp. 357–381. Ann Arbor, Ann Arbor Science Publishers.

Marshall, J. S. and Mellinger, D. L. (1980). Dynamics of cadmium-stressed plankton communities. *Can. J. Fish. Aquat. Sci.* **37**: 403–414.

Maruoka, S. (1978). Estimation of toxicity using cultured mammalian cells of the organic pollutants recovered from Lake Biwa. *Wat. Res.* **12**: 371–5.

Mason, C. (1981). *Biology of Freshwater Pollution.* London, Longmans.

Mason, W. T. Anderson, J. B. and Morrison, G. E. (1967). A limestone-filled, artificial substrate sampler-float unit for collecting macroinvertebrates in large streams. *Progve Fish Cult.* **29**: 74.

Matthiessen, P. and Brafield, A. E. (1977). Uptake and loss of dissolved zinc by the stickleback (*Gasterosteus aculeatus* L). *J. Fish Biol.* **10**: 399–410.

Mattice, J. S., Burch, M. B., Tsai, S. C. and Roy, W. K. (1981). A toxicity testing system for exposing small invertebrates and fish to short square-wave concentrations of chlorine. *Wat. Res.* **15**: 923–7.

Matulova, D. (1967). The application of *Chlamydomonas* cultures as bio-assay test organisms. *Hydrobiologia* **30**: 494–502.

Matulova, D. (1969). Stream pollution examination by biological tests. *Proc. 4th Int. Wat. Poll. Res. Conf.* pp. 659–69. Oxford, Pergamon Press.

Matulova, D. (1970). *Biological Assays and Water Quality in Minnesota.* Limnological Research Centre, University of Minnesota. Minneapolis. Interim Report No. 6, March, 1970.

Mauck, W. L., Olson, L. E. and Marking, L. L. (1976). Toxicity of natural pyrethrins and five pyrethroids to fish. *Arch. Environ. Contam. Toxicol.* **4**: 18–29.

Mauck, W. L., Mehrle, P. M. and Mayer, F. L. (1978). Effects of the polychlorinated biphenyl Aroclor® 1254 on growth, survival and bone development in brook trout (*Salvelinus fontinalis*). *J. Fish. Res. Bd Can.* **35**: 1084–88.

May, R. M. (1974). General introduction. In: *Ecological Stability* (Workshop Papers), (Usher, M. B. and Williamson, M. H. Eds.) London, Chapman & Hall.

Mayer, F. L., Street, J. C. and Neuhold, J. M. (1970). Organochlorine insecticide interactions affecting residue storage in rainbow trout. *Bull. Environ. Contam. Toxicol.* **5**: 300–310.

Mayer, F. L., Mehrle, P. M. and Sanders, H. O. (1977). Residue dynamics and biological effects of polychlorinated biphenyls in aquatic organisms. *Arch. Environ. Contam. Toxicol.* **5**: 501–511.

Mayhew, J. (1955). Toxicity of seven different insecticides to rainbow trout *Salmo gairdnerii* (Richardson). *Proc. Iowa Acad. Sci.* **62**: 599–606.

McCarthy, L. T. and Keighton, W. B. (1964). Quality of Delaware River water at

Trenton, New Jersey. Geological Survey Water—Supply Paper 1779-X. pp. 36–37. Washington, DC, US Government Printing Office.

McCarty, L. S., Henry, J. A. C. and Houston, A. H. (1978). Toxicity of cadmium to goldfish, *Carassius auratus*, in hard and soft water. *J. Fish. Res. Bd Can.* **35**: 35–42.

McCleave, J. D., Rommel, J. R. and Cathcart, C. L. (1971). Weak electric and magnetic fields in fish orientation *Ann. NY Acad. Sci.* **188**: 270–82.

McIntosh, R. P. (1967). An index of diversity and the relation of certain concepts to diversity. *Ecology* **48**: 392–404.

McKim, J. M. and Benoit, D. A. (1971). Effects of long-term exposures to copper on survival, growth, and reproduction of brook trout (*Salvelinus fontinalis*). *J. Fish. Res. Bd Can* **28**: 655–62.

McKnight, D. (1981). Chemical and biological processes controlling the response of a freshwater ecosystem to copper stress: a field study of the $CuSO_4$ treatment of Mill Pond Reservoir, Burlington, Massachusetts. *Limnol. Oceanogr.* **26**: 518–31.

McKnight, D. M. and Morel, F. M. M. (1979). Release of weak and strong copper-complexing agents by algae. *Limnol. Oceanogr.* **24**: 823–37.

McLean, R. O. and Benson-Evans, K. (1974). The distribution of *Stigeoclonium tenue* Kutz. in South Wales in relation to its use as an indicator of organic pollution. *Br. Phycol. J.* **9**: 83–9.

McLean, R. O. and Jones, A. K. (1975). Studies of tolerance to heavy metals in the flora of the Rivers Ystwyth and Clarach. *Freshwat. Biol.* **5**: 431–44.

McMahon, R. F., Hunter, R. D. and Russell-Hunter, W. F. (1974). Variation in *Aufwuchs* at six freshwater habitats in terms of carbon biomass and of carbon:nitrogen ratio. *Hydrobiologia* **45**: 391–404.

McNicholl, P. G. and Mackay, W. C. (1975a). Effect of DDT and M.S.222 on learning a simple conditioned response in rainbow trout (*Salmo gairdneri*). *J. Fish. Res. Bd Can.* **32**: 661–5.

McNicholl, P. G. and Mackay, W. C. (1975b). Effect of DDT on discriminating ability of rainbow trout (*Salmo gairdneri*) *J. Fish. Res. Bd Can.* **32**: 785–8.

Meeks, R. L. (1968). The accumulation of ^{36}Cl ring-labelled DDT in a freshwater marsh. *J. Wildl. Mgmt.* **32**: 376–98.

Mehrle, P. M. and Mayer, F. L. (1975a). Toxaphene effects on growth and bone composition of fathead minnows, *Pimephales promelas*. *J. Fish. Res. Bd Can.* **32**: 593–8.

Mehrle, P. M. and Mayer, F. L. (1975b). Toxaphene effects on growth and development of brook trout (*Salvelinus fontinalis*). *J. Fish. Res. Bd Can.* **32**: 609–613.

Mellanby, H. (1963). *Animal Life in Fresh Water*. London, Methuen & Co. Ltd.

Mellanby, K. (1970). *Pesticides and Pollution*. New Naturalist Series. London, Collins.

Menhinick, E. F. (1964). A comparison of some species-individuals diversity indices applied to samples of field insects. *Ecology* **45**: 859–61.

Merkens, J. C. and Downing, K. M. (1957). The effect of tension of dissolved oxygen on the toxicity of un-ionized ammonia to several species of fish. *Ann. Appl. Biol.* **45**: 521–7.

Merlini, M. and Pozzi, G. (1977). Lead and freshwater fishes: Part 1—Lead accumulation and water pH. *Environ. Pollut.* **12**: 167–72.

Metcalf, R. L., Sangha, G. K. and Kapoor, I. P. (1971). Model ecosystem for the evaluation of pesticide biodegradability and ecological magnification. *Environ. Sci. Technol.* **5**: 709–713.

Michael, A. S., Thompson, C. G. and Abramovitz, M. (1956). *Artemia salina* as a test organism for bioassay. *Science, NY* **123**: 464.

Middlebrooks, E. J., Maloney, T. E., Powers, C. F. and Kaack, L. M. (Eds.) (1969). Proceedings of the eutrophication-biostimulation assessment workshop. University of California, Berkeley and Federal Water Pollution Control Administration, Northwest Laboratory, Corvallis, Oregon.

Miller, T. G. and Mackay, W. C. (1980). The effects of hardness, alkalinity and pH of test water on the toxicity of copper to rainbow trout (*Salmo gairdneri*). *Wat. Res.* **14**: 129–33.

Milway, C. P. (Ed.) (1970). *Eutrophication in Large Lakes and Impoundments.* Uppsala Symposium, May 1968. Paris: Organisation for Economic Co-operation and Development.

Minshall, G. W. (1967). Role of allochthonous detritus in the atrophic structure of a woodland spring brook community. *Ecology* **48**: 139–49.

Mitchell, D. (1973). Algal bioassays for estimating the effect of added materials upon the planktonic algae in surface waters. In: *Bioassay Techniques and Environmental Chemistry.* (G. E. Glass, Ed.) pp. 153–8. Ann Arbor.

Mitrovic, V. V., Brown, V. M., Shurben, D. G. and Berryman, M. H. (1968). Some pathological effects of sub-acute and acute poisoning of rainbow trout by phenol in hard water. *Wat. Res.* **2**: 249–54.

Morgan, J. R. (1972). Effects of Aroclor 1242 (a polychlorinated biphenyl) and DDT on cultures of an alga, protozoan, daphnid, ostracod and guppy. *Bull. Environ. Contam. Toxicol.* **8**: 129–37.

Morgan, N. C. and Egglishaw, H. J. (1965). A survey of the bottom fauna of streams in the Scottish Highlands. Part 1. Composition of the fauna. *Hydrobiologia* **25**: 181–211.

Morgan, R. P., Fleming, R. F., Rasin, V. J. and Heinle, D. R. (1973). Sublethal effects of Baltimore harbour water on the white perch, *Morone americana*, and the hogchoker *Trinectes maculatus. Chesapeake Sci.* **14**: 17–27.

Morgan, W. S. G. (1975). Monitoring pesticides by means of changes in electrid potential caused by fish opercular rhythms. *Prog. Wat. Technol.* **7**. 33–40.

Morgan. W. S. G. (1979). Fish locomotor behaviour patterns as a monitoring tool. *J. Wat. Pollut. Control Fed.* **51**: 580–589.

Morrison, B. R. S. (1977). The effects of rotenone on the invertebrate fauna of three hill streams in Scotland. *Fish. Mgmt.* **8**: 128–39.

Morrison, B. R. S. (1979). An investigation into the effects of the piscicide Antimycin A on the fish and invertebrates of a Scottish stream. *Fish. Mgmt.* **10**: 111–22.

Morrison, B. R. S. and Struthers, G. (1975). The effects of rotenone on the invertebrate fauna of three Scottish freshwater lochs. *Fish. Mgmt.* **6**: 81–91.

Morrison, B. R. S. and Wells, D. E. (1981). The fate of fenitrothion in a stream environment and its effect on the fauna, following aerial spraying of a Scottish forest. *Sci. Total Environ.* **19**: 233–52.

Moss, B. (1967a). A note on the estimation of chlorophyll *a* in freshwater algal communities. *Limnol. Oceanogr.* **12**: 340–2.

Moss, B. (1967b). A spectrophotometric method for the estimation of percentage degradation of chlorophylls to pheo-pigments in extracts of algae. *Limnol. Oceanogr.* **12**: 335–40.

Mount, D. I. (1962). Chronic effects of endrin on bluntnose minnows and guppies. *US Fish Wildl. Serv. Res. Rep.* **58**: 1–38.

Mount, D. I. (1964). An autopsy technique for zinc-caused fish mortality. *Trans. Am. Fish. Soc.* **93**: 174–82.

Mount, D. I. (1966). The effect of total hardness and pH on the acute toxicity of zinc to fish. *Int. J. Air Wat. Pollut.* **10**: 49–56.

Mount, D. I. (1967). A method for detecting cadmium poisoning in fish. *J. Wildl. Mgmt.* **31**: 168–72.

Mount, D. I. (1968). Chronic toxicity of copper to fathead minnows (*Pimephales promelas*, Rafinesque). *Wat. Res.* **2**: 215–23.

Mount, D. I. (1973). Chronic effect of low pH on fathead minnow survival growth and reproduction. *Wat. Res.* **7**: 987–93.

Mount, D. I. and Brungs, W. A. (1967). A simplified dosing apparatus for fish toxicology studies. *Wat. Res.* **1**: 21–9.

Mount, D. I. Warner, R. E. (1965). *A Serial-Dilution Apparatus for Continuous Delivery of Various Concentrations of Materials in Water.* US Public Health Service Publ. No 999-WP-23.

Mount, D. T. and Stephan, C. E. (1967). A method for establishing acceptable limits for fish—malathion and the butoxyethanol ester of 2,4-D. *Trans. Am. Fish. Soc.* **96**: 185–95.

Mountford, M. D. (1962). An index of similarity and its application to classificatory problems. In: *Progress in Soil Zoology.* (P. W. Murphy, Ed.). London, Butterworth.

Muirhead-Thomson, R. C. (1969). A technique for establishing *Simulium* larvae in an experimental channel. *Bull. Ent. Res.* **59**: 533–6.

Muirhead-Thomson, R. C. (1971). *Pesticides and Freshwater Fauna.* London. Academic Press.

Muirhead-Thomson, R. C. (1973). Laboratory evaluation of pesticide impact on stream invertebrates. *Freshwat. Biol.* **3**: 479–98.

Muirhead-Thomson, R. C. (1978). Lethal and behavioural impact of permethrin (NRDC 143) on selected stream macroinvertebrates. *Mosquito News* **38**: 185–90.

Mullison, W. R. (1970). Effects of herbicides on water and its inhabitants. *Weed Science* **18**: 738–50.

Murphy, K. J., Hanbury R. G. & Eaton, W. J. (1981). The ecological effects of 2-methylthiotriazine herbicides used for aquatic weed control in navigable canals. I. Effects on aquatic flora and water chemistry. *Arch. Hydrobiol.* **91**: 294–331.

Murphy, P. G. (1971). The effect of size on the uptake of DDT from water by fish. *Bull. Environ. Contam. Toxicol.* **6**: 20–23.

Murphy, P. G. and Murphy. J. V. (1971). Correlations between respiration and direct uptake of DDT in the mosquito fish *Gambusia affinis. Bull. Environ. Contam. Toxicol.* **6**: 144–55.

Murphy, P. M. (1980). *A Manual for Toxicity Tests with Freshwater Macro-invertebrates and a Review of the Effects of Specific Toxicants.* University of Wales Institute of Science and Technology [Cardiff].

Murray, R. W. (1971). Temperature receptors. In: *Fish Physiology. Vol V. Sensory Systems and Electric Organs.* (W. S. Hoar and D. J. Randall, Eds.) pp. 121–32. New York, Academic Press.

Nagell, B. (1973). The oxygen consumption of mayfly (Ephemeroptera) and stonefly (Plecoptera) larvae at different oxygen concentration. *Hydrobiologia* **42**: 461–89.

Nakada, M., Fukaya, K., Takeshita, S. and Wada, Y. (1979). The accumulation of heavy metals in the submerged plant (*Elodea nuttallii*). *Bull. Environ. Contam. Toxicol.* **22**: 21–7.

Naqvi, S. M. and Ferguson, D. E. (1970). Levels of insecticide resistance in freshwater shrimp *Palaemonetes kadiakensis. Trans. Am. Fish Soc.* **99**: 696–9.

Naqvi, S. M., Lueng, T.–S. and Naqvi, N. Z. (1980). Toxicities of paraquat and diquat herbicides to freshwater copepods (*Diaptomus* sp. and *Eucyclops* sp.) *Bull. Environ. Contam. Toxicol.* **25**: 918–20.

Naqvi, S. M., Lueng, T.–S. and Naqvi, N. Z. (1981). Toxicities of paraquat and metribuzin (Senco®) herbicides to the freshwater copepods, *Eucyclops agilis* and *Diaptomus mississippiensis. Environ. Pollut. A* **26**: 275–80.

National Water Council (1981). *River Quality: The 1980 Survey and Future Outlook.* London, National Water Council.

Nau-Ritter, G. M., Wurster, C. F. and Rowland, R. G. (1982). Polychlorinated biphenyls (PCB) desorbed from clay particles inhibit photosynthesis by natural phytoplankton communities. *Environ. Pollut. Ser. A.* **28**: 177–82.

Nebeker, A. V. and Gaufin, A. R. (1964). Bioassays to determine pesticide toxicity to the amphipod crustacean *Gammarus lacustris. Proc. Utah Acad. Sci. Arts Letters* **41**: 64–7.

Nebeker, A. V. and Puglisi, F. A. (1974). Effect of polychlorinated biphenyls (PCBs) on survival and reproduction of *Daphnia, Gammarus,* and *Tanytarsus. Trans. Am. Fish. Soc.* **103**: 722–8.

Nebeker, A. V., Puglisi, F. A. and DeFoe, D. L. (1974). Effect of polychlorinated biphenyl compounds on survival and reproduction of the fathead minnow and flagfish. *Trans. Am. Fish. Soc.* **103**: 562–8.

Nebeker, A. V., Hauck, A. K. and Baker, F. D. (1979). Temperature and oxygen–nitrogen gas ratios affect fish survival in air-supersaturated water. *Wat. Res.* **13**: 299–303.

Needham, P. R. and Usinger, R. L. (1956). Variability in the macrofauna of Prosser Creek, California, as indicated by the Surber sampler. *Hilgardia* **24**: 383–409.

Nemerow, N. L. (1978). *Industrial Water Pollution. Origins, Characteristics and Treatment.* Reading, Massachusetts, Addison-Wesley.

Nesbitt, H. J. and Watson J. R. (1980a). Degradation of the herbicide 2,4-D in river water. I. Description of study area and survey of rate determining factors. *Wat. Res.* **14**: 1683–8.

Nesbitt, H. J. and Watson, J. R. (1980b). Degradation of the herbicide 2,4-D in river water. II. The role of suspended sediment, nutrients and water temperature. *Wat. Res.* **14**: 1689–94.

Neville, C. (1979). Sublethal effects of environmental acidification on rainbow trout (*Salmo gairdneri*). *J. Fish. Res. Bd. Can.* **36**: 84–7.

Newbold, C. (1974). The ecological effects of the herbicide dichlobenil within pond ecosystems. *Proc. European Weed Res. Council, 4th Int. Symp. on Aquatic Weeds.* pp. 37–52.

Newbold, C. (1975). Herbicides in aquatic systems. *Biol. Conserv.* **7**: 97–118.

Newbold, J. D., Erman, D. C. and Roby, K. B. (1980). Effects of logging on macroinvertebrates in streams with and without buffer strips. *Can. J. Fish. Aquat. Sci.* **37**: 1076–85.

Newbold, C., Purseglove, J. and Holmes, N. (1983). *Nature Conservation and River Engineering.* London, Nature Conservancy Council.

Newman, J. F. (1967). The ecological effects of paraquat and diquat when used to control aquatic weed. *Proc. 4th British Insecticide and Fungicide Conference.* pp. 45–7.

Newman, J. F. and Way, J. M. (1966). Some ecological observations on the use of paraquat and diquat as aquatic herbicides. *Proc. 8th British Weed Control Conf.* pp. 582–5.

Newson, M. D. (1979). The results of ten years experimental study on Plynlimon, Mid-Wales, and their importance for the water industry. *J. Inst. Wat. Eng. Sci.* **33**: 321–33.

Nichols, L. E. and Bulow, F. J. (1973). Effects of acid mine drainage on the stream ecosystem of the east fork of the Obey River, Tennessee. *J. Tennessee Acad. Sci.* **48**: 30–39.

Northcote, T. G. (1967). The relation of movements and migrations to production in freshwater fishes. In: *The Biological Basis of Freshwater Fish Production.* (S. D., Gerking, Ed.) pp. 315–44. Oxford, Blackwell Scientific Publications.

Nuttall, P. M. (1972). The effects of sand deposition upon the macroinvertebrate fauna of the River Camel, Cornwell. *Freshwat. Biol.* **2**: 181–6.

Nuttall, P. M. and Bielby, G. H. (1973). The effect of china-clay wastes on stream invertebrates. *Environ. Pollut.* **5**: 77–86.

Odum, E. P. (1971). *Fundamentals of Ecology.* Philadelphia, W. B. Saunders Co.

OECD (1957). *Air and water pollution. The position in Europe and in the USA.* Organisation for European Economic Co-operation and Development, Paris.

Odum, H. T. (1956). Primary production in flowing waters. *Limnol. Oceanogr.* **1**: 102–117.

Odum, H. T. (1957). Trophic structure and productivity of Silver Springs. *Ecol. Monogr.* **27**: 55–112.

Ogilvie, D. M. and Anderson, J. M. (1965). Effect of DDT on temperature selection by young Atlantic salmon, *Salmo salar. J. Fish. Res. Bd Can.* **22**: 503–12.

Okun, D. A. (1977). *Regionalization of Water Management. A Revolution England and Wales.* London, Applied Science Publishers.

Olson, K. R. and Fromm, P. O. (1973). Mercury uptake and ion distribution in gills of rainbow trout (*Salmo gairdneri*): tissue scans with an electron microprobe. *J. Fish. Res. Bd. Can.* **30**: 1575–8.

Olson, K. R., Bergman, H. L. and Fromm, P. O. (1973). Uptake of methyl mercuric chloride amd mercuric chloride by trout: a study of uptake pathways into the whole animal and uptake by erythrocytes in vitro. *J. Fish. Res. Bd Can.* **30**: 1293–9.

O'Sullivan, A. J. and Collinson, R. I. (1976). Wasted heat and aquatic ecosystems. *Effluent and Water Treatment Journal*, January 1976: 15–29.

O'Riordan, T. and Hey, R. D. (1976). *Environmental Impact Assessment.* Farnborough, Saxon House.

Owens, M., Learner, M. A. and Maris, P. J. (1967). Determination of the biomass of aquatic plants using an optical method. *J. Ecol.* **55**: 671–76.

Packer, R. (1979). Acid–base balance and gas exchange in brook trout (*Salvelinus fontinalis*) exposed to acidic environments. *J. Exp. Biol.* **79**: 127–34.

Packer, R. and Dunson, W. (1970). Effects of low environmental pH on blood pH and sodium balance of brook trout. *Exp. Zool.* **174**: 65–72.

Painter, H. (1970). A review of literature on inorganic nitrogen metabolism in micro-organisms. *Wat. Res.* **4**: 393–450.

Pal, R. N. and Gopalakrishnan, V. (1968). Preliminary trials on the toxicity of Algistat to fish and algae. *Fish. Technol.* **5**: 101–103.

Paller, M. H. and Heidinger, R. C. (1980). Mechanisms of delayed ozone toxicity to bluegill *Lepomis macrochirus* Rafinesque. *Environ. Pollut. A* **22**: 229–39.

Palmer, C. M. (1969). A composite rating of algae tolerating organic pollution. *J. Phycol.* **5**: 78–82.

Pantle, R. and Buck, H. (1955). Die biologische Überwachung der Gewässer und die Darstellung der Ergebnisse. *Gas-u. Wasserfach* **96**: 604.

Parmann, G. (1981). The acidification of Norway. *Ambio* **10**: 150–151.

Parsons, J. D. (1968). The effects of acid strip-mine effluents on the ecology of a stream. *Arch. Hydrobiol* **65**: 25–50.

Parsons, T. R. and Strickland, J. D. H. (1963). Discussion of spectrophotometric determination of marine plant pigments, with revised equations for ascertaining chlorophylls and carotenoids. *J. Mar. Res.* **21**: 155–63.

Part, P. and Svanberg, O. (1981). Uptake of cadmium in perfused rainbow trout (*Salmo gairdneri*) gills. *Can. J. Fish. Aquat. Sci.* **38**: 917–24.

Pascoe, D. and Cram, P. (1977). The effect of parasitism on the toxicity of cadmium to the three-spined stickleback, *Gasterosteus aculeatus* L. *J. Fish Biol.* **10**: 467–72.

Patrick, R. (1954) Diatoms as an indication of river change. *Proc. 9th Indust. Waste Conf. Purdue Univ. Engng Extn Ser.* **87**: 325–30.

Patrick, R. (1961). A study of the numbers and kinds of species found in rivers in eastern United States. *Proc. Acad. Nat. Sci. Philadelphia* **113**: 215–58.

Patrick, R. (1964). A discussion of the results of the Catherwood expedition to the Peruvian headwaters of the Amazon. *Verh. int. Verein. theor. angew. Limnol.* **15**: 1084–1090.

Patrick, R. and Hohn, M. H. (1956). The diatometer—a method for indicating the conditions of aquatic life. *Proc. Amer. Petroleum Inst. III* **36**: 332–9.

Patrick, R., Hohn, M. H. and Wallace, J. H. (1954). A new method for determining the pattern of the diatom flora. *Notul. Nat.* **259**: 1–12.

Patrick, R., Cairns, J. and Scheier, A. (1968). The relative sensitivity of diatoms, snails and fish to twenty common constituents of industrial wastes. *Progve Fish Cult.* **30**: 137–40.

Patterson, J. W., Brezonik, P. L. and Putnam, H. D. (1970). Measurement and significance of adenosine triphosphate in activated sludge. *Envir. Sci. Technol.* **4**: 569–75.

Pawlaczyk-Szpilowa, M., Moskal, M. and Weretelnik, J. (1972). The usefulness of biological tests for determining the toxicity of some chemical compounds in waters. *Acta Hydrobiol., Krakow.* **14**: 115–27.

Pearson, R. G., Litterick, M. R. and Jones, N. V. (1973). An air-lift for quantitative sampling of the benthos. *Freshwat. Biol.* **3**: 309–315.

Perring, F. H. and Mellanby, K. (Eds.) (1977). *Ecological Effects of Pesticides.* Linnean Society Symposium No. 5. London, Academic Press.

Perrone, S. J. and Meade, T. L. (1977). Protective effect of chloride on nitrite toxicity to coho salmon (*Oncorhynchus kisutch*). *J. Fish. Res. Bd Can.* **34**: 486–92.

Peterson, R. H., Daye, P. G. and Metcalfe, J. L. (1980) Inhibition of Atlantic salmon (*Salmo salar*) hatching at low pH. *Can. J. Fish. Aquat. Sci.* **37**: 770–4.

Phillips, D. J. H. (1977). The use of biological indicator organisms to monitor trace metal pollution in marine and estuarine environments—A review. *Environ. Pollut.* **13**: 281–317.

Phillips, D. J. H. (1978). Use of biological indicator organisms to quantitate organochlorine pollutants in aquatic environments—A review. *Environ. Pollut.* **16**: 167–229.

Phillips, D. J. H. (1980). *Quantitative Aquatic Biological Indicators. Their Use to Monitor Trace Metal and Organochlorine Pollution.* London, Applied Science Publishers.

Phillips, G. R. and Gregory, R. W. (1979). Assimilation efficiency of dietary methylmercury by northern pike (*Esox lucius*) *J. Fish. Res. Bd Can.* **36**: 1516–19.

Pickering, Q. H. and Gast, M. H. (1972). Acute and chronic toxicity of cadmium to the fathead minnow (*Pimephales promelas*) *J. Fish. Res. Bd Can.* **29**: 1099–1106.

Pickering, Q. H. and Henderson, C. (1966). The acute toxicity of some heavy metals to different species of warmwater fishes. *Int. J. Air Wat. Pollut.* **10**: 453–63.

Pickering, Q. H., Brungs, W. and Gast, M. (1977). Effect of exposure time and copper concentration on reproduction of the fathead minnow (*Pimephales promelas*). *Wat. Res.* **11**: 1079–1083.

Pielou, E. C. (1969). *An Introduction to Mathematical Ecology.* New York, Wiley-Interscience.

Pike, E. B. (1975). Aerobic bacteria. In: *Ecological Aspects of Used-water Treatment. Vol. 1. The Organisms and their Ecology.* (C. R. Curds and H. A. Hawkes, Eds.) pp. 1–63. London, Academic Press.

Pitcairn, C. E. R. and Hawkes, H. A. (1973). The role of phosphorus in the growth of *Cladophora. Wat. Res.* **7**: 159–71.

Pitt, T. K., Garside, E. T. and Hepburn, R. L. (1956). Temperature selection of the carp (*Cyprinus carpio* Linn). *Can. J. Zool.* **34**: 555–7.

Porcella, D. B., Grau, P., Huang, C. H., Radimsky, J., Toerien, D. G. and Pearson, E. A. (1970). Provisional algal assay procedures, first annual report. SERL Report No 70–8. Sanitary Engineering Research Laboratory, University of California, Berkeley.

Post, G. and Schroeder, T. R. (1971). The toxicity of four insecticides to four salmonid species. *Bull. Environ. Contam. Toxicol.* **6**: 144–55.

Potts, W. and Fryer, G. (1979). The effects of pH and salt content on sodium balance in *Daphnia magna* and *Acantholeberis curvirostris* (Crustacea: Cladocera). *J. Comp. Physiol. B. Biochem. Syst. Environ. Physiol.* **129**: 289–94.

Preston, F. W. (1948). The commonness and rarity of species. *Ecology* **29**: 254–83.

Price, D. R. H. (1977). Fish as indicators of water quality. *Wat. Pollut. Control London* **77**: 285–93.

Pringsheim, E. G. (1949a). The filamentous bacteria *Sphaerotilus, Leptothrix, Cladothrix* and their relation to iron and manganese. *Phil. Trans. Roy. Soc. B* **233**: 453–82.

Pringsheim, E. G. (1949b). Iron Bacteria. *Biol. Rev.* **24**: 200–45.

Provisional Algal Assay Procedure (1969). Joint Industry/Government Task Force on Eutrophication, New York. Chairman, Bueltman, C. G. Executive Director, Brenner, T. E.

Raabe, E. W. (1952). Über den 'Affinitätswert' in der Pflanzensoziologie. *Vegatatio, Haag* **4**: 53–68.

Rachlin, J. W. and Perlmutter, A. (1968). Fish cells in culture for study of aquatic toxicants. *Wat. Res.* **2**: 409–414.

Rachlin, J. W. and Perlmutter, A. (1969). Response of rainbow trout cells in culture to selected concentrations of zinc sulfate. *Progve Fish. Cult.* **31**: 94–8.

Radford, D. S. and Hartland-Rowe, R. (1971). Subsurface and surface sampling of benthic invertebrates in two streams. *Limnol. Oceanogr.* **16**: 114–120.

Randall, D. J. and Smith, L. S. (1967). The effect of environmental factors on circulation and respiration in teleost fish. *Hydrobiologia* **29**: 113–24.

Rao, D. A. and Saxena, A. B. (1981). Acute toxicity of mercury, zinc, lead, cadmium, manganese to the *Chironomus* sp. *Int. J. Environ. Studies* **16**: 225–6.

Rao, D. M. R. and Murty, A. S. (1982). Toxicity and metabolism of endosulfan in three freshwater catfishes. *Environ. Pollut. Ser. A.* **27**: 223–31.

Ratcliffe, D. (Ed.) (1977). *A Nature Conservation Review. The Selection of Biological Sites of National Importance to Nature Conservation in Britain.* Cambridge, Cambridge Univ. Press.

Rathore, H. S. Sanghvi, P. K. and Swarup, H. (1979). Toxicity of cadmium chloride and lead nitrate to *Chironomus tentans* larvae. *Environ. Pollut.* **18**: 173–7.

Raunkiaer, C. (1934). The area of dominance, species diversity, and formation dominants. In: *The Life Forms of Plants and Statistical Plant Geography being the Collected Papers of C. Raunkier.* Chapter XVI. Oxford, Clarendon Press.

Rehwoldt, R. E., Kelley, E. and Mahoney, M. (1977). Investigations into the acute toxicity and some chronic effects of selected herbicides and pesticides on several fresh water fish species. *Bull. Environ. Contam. Toxicol.* **18**: 361–5.

Reid, G. K. (1961). *Ecology of Inland Waters and Estuaries.* New York, Reinhold.

Reinert, R. E. and Bergman, H. L. (1974). Residues of DDT in lake trout (*Salvelinus namaycush*) and coho salmon (*Oncorhynchus kisutch*) from the Great Lakes. *J. Fish. Res. Bd Can.* **31**: 191–9.

Reinert, R. E., Stone, L. J. and Willford, W. A. (1974). Effect of temperature on accumulation of methylmercuric chloride and p,p′ DDT by rainbow trout (*Salmo gairdneri*). *J. Fish. Res. Bd Can.* **36**: 1040–48.

Remacle, J. (1981). Cadmium uptake by freshwater bacterial communities. *Wat. Res.* **15**: 67–71.

Remacle, J. and Houba, C. (1980). The influence of cadmium upon freshwater saprophytic bacteria. *Envir. Technol. Lett.* **1**: 193–200.

Reney, E. C. and Menzel, B. W. (1967). *Heated Effluents and Effects on Aquatic Life with Emphasis on Fishes: A Bibliography.* Cornell University Water Resources and Marine Science Center, Philadelphia Electric Co. and Ichthyological Associates. Bull. No. 2.

Renn, C. E. (1955). Biological properties and behaviours of cyanogenic wastes. *Sewage Indust. Wastes* **27**: 297–310.

Reynolds, F. A. and Haines, T. A. (1980). Effects of chronic exposure to hydrogen sulphide on newly hatched brown trout *Salmo trutta* L. *Environ. Pollut. A* **22**: 11–17.

Reynolds, J. H., Middlebrooks, E. J., Porcella, D. B. and Grenney, W. J. (1976). Comparison of semi-continuous and continuous flow bioassays. In: *Biostimulation and Nutrient Assessment.* (E. J. Middlebrooks, D. H. Falkenborg, T. E. Maloney, Eds.) pp. 241–65. Ann Arbor, Michigan, Ann Arbor Science.

Reynolds, W. W. (1978). The final thermal preferendum of fishes: shuttling behaviour and acclimation overshoot. *Hydrobiologia* **57**: 123–4.

Rheinheimer, G. (1974). *Aquatic Microbiology.* London, John Wiley & Sons.

Richardson, D., Dorris, T. C., Burks, S., Browne, R. H., Higgins, M. L. and Leach, F. R. (1977). Evaluation of cell culture assay for determination of water quality of oil-refinery effluents. *Bull. Environ. Contam. Toxicol* **18**: 683–90.

Roback, S. S. and Richardson, J. W. (1969). The effects of acid mine drainage on aquatic insects. *Proc. Acad. Nat. Sci. Philadelphia* **121**: 81–99.

Roberts, M. H. and Bendl, R. E. (1982). Acute toxicity of Kepone to selected freshwater fishes. *Estuaries* **5**: 158–64.

Robson, T. O. (1968). Some studies of the persistence of 2,4-D and ester derivatives in natural surface waters. *Proc. 9th British Weed Control Conf.* **1**: 404–408.

Robson, T. O. and Fearon, J. H. (Eds.) (1976). *Aquatic herbicides.* Proceedings of a symposium held at Oxford, January 1976. British Crop Protection Council Monograph No. 16.

Roch, M. and Maly, E. J. (1979). Relationship of calcium-induced hypocalcemia with mortality in rainbow trout (*Salmo gairdneri*) and the influence of temperature on toxicity. *J. Fish. Res. Bd Can.* **36**: 1297–1303.

Rohlich, G. A. (Ed.), (1969). *Eutrophication: Causes, Consequences and Correctives.* Proceedings of a Symposium at Madison, Wisconsin, 11–15 June 1967. Washington, DC, National Academy of Sciences.

Rommel, S. A. (1973). A simple method for recording fish heart and operculum beats without the use of implanted electrodes. *J. Fish. Res. Bd Can.* **30**: 693–4.

Rose, F. L. and McIntire, C. D. (1970). Accumulation of dieldrin by benthic algae in laboratory streams. *Hydrobiologia* **35**: 481–93.

Rudd, R. L. and Genelly, R. E. (1956). Pesticides: their use and toxicity in relation to wildlife. *Calif. Fish Game Bull.* **7**: 1–309.

Rudd, J. W. M., Turner, M. A. Townsend, B. E., Swick, A. and Furutani, A.

(1980) Dynamics of selenium in mercury-contaminated experimental freshwater ecosystems. *Can. J. Fish. Aquat. Sci.* **37**: 848–57.

Russo, R. C., Smith, C. E. and Thurston, R. V. (1974). Acute toxicity of nitrite to rainbow trout (*Salmo gairdneri*). *J. Fish. Res. Bd Can.* **31**: 1653–5.

Russo, R. C., Thurston, R. V. and Emerson, K. (1981). Acute toxicity of nitrite to rainbow trout (*Salmo gairdneri*): effects of pH, nitrite species, and anion species. *Can. J. Fish. Aquat. Sci.* **38**: 387–93.

Sadler, K. (1980). Effect of the warm-water discharge from a power station on fish populations in the River Trent. *J. Appl. Ecol.* **17**: 349–57.

Samioloff, M. R., Schulz, S., Jordan, Y., Denich, K. and Arnott, E. (1980). A rapid simple long-term toxicity assay for aquatic contaminants using the nematode *Panagrellus redivivus*. *Can. J. Fish. Aquat. Sci.* **37**: 1167–74.

Samsel, G. L., Reed, J. R. and Winfrey, H. J. (1972). Investigations on nutrient factors limiting phytoplankton productivity in two Central Virginia ponds. *Wat. Resources Bull.* **8**: 825–33.

Sanborn, J. R. and Yu, C. C. (1973). The fate of dieldrin in a model ecosystem. *Bull. Environ. Contam. Toxicol.* **10**: 340–6.

Sanders, H. O. (1969). Toxicity of pesticides to the crustacean *Gammarus lacustris*. US Bureau of Sport Fisheries and Wildlife. Tech. Pap. No. 25, Washington, DC, US Government Printing Office.

Sanders, H. O. (1972). The toxicities of some insecticides to four species of malacostracan crustacea. US Bureau of Sport Fisheries and Wildlife. Tech. Paper 66. Washington, DC, US Government Printing Office.

Sanders, H. O. and Chandler, J. H. (1972). Biological magnification of a polychlorinated biphenyl (Aroclor®1254) from water by aquatic invertebrates. *Bull. Environ. Contam. Toxicol.* **7**: 257–63.

Sanders, H. O. and Cope, O. B. (1966). Toxicity of several pesticides to two species of cladocerans. *Trans. Am. Fish Soc.* **95**: 165–9.

Sanders, H. O. and Cope, O. B. (1968). The relative toxicities of several pesticides to naiads of three species of stonefly. *Limnol. Oceanogr.* **13**: 112–17.

Sanders, H. O., Huckins, J., Johnson, B. T. and Skaar, D. (1981). Biological effects of Kepone and mirex in freshwater invertebrates. *Arch. Environ. Contam. Toxicol.* **10**: 531–9.

Sato, T., Ose, Y. and Sakai, T. (1980). Toxicological effect of selenium on fish. *Environ. Pollut. (A)* **21**: 217–24.

Saunders, J. H. and Smith, M. W. (1965). Changes in a stream population of trout associated with increased silt. *J. Fish. Res. Bd Can.* **22**: 395–404.

Saunders, R. L. and Henderson, E. B. (1969). Survival and growth of Atlantic salmon parr in relation to salinity. *Tech. Rep. Fish. Res. Bd Can.* No. 147.

Saunders, R. L. and Sprague, J. B. (1967). Effects of copper–zinc mining pollution on a spawning migration of Atlantic salmon. *Wat. Res.* **1**: 419–32.

Say, P. J. and Whitton, B. A. (1981). Changes in flora down a stream showing a zinc gradient. *Hydrobiologia* **76**: 255–62.

Say, P. J., Diaz, B. M. and Whitton, B. A. (1977). Influence of zinc on lotic plants. I. Tolerance of *Hormidium* species to zinc. *Freshwat. Biol.* **7**: 357–76.

Say, P. J., Harding. J. P. C. and Whitton, B. A. (1981). Aquatic mosses as monitors of heavy metal contamination in the River Etherow, Great Britain. *Environ. Pollut. Ser. B.* **2**: 295–307.

Scharf, B. W. (1979). A fish test alarm device for the continual recording of acute toxic substances in water. *Arch. Hydrobiol.* **85**: 250–56.

Schaumburg, F. D., Howard, T. E. and Walden, C. C. (1967). A method to evaluate the effects of water pollutants on fish respiration. *Wat. Res.* **1**: 731–7.

Scheier, A. and Cairns. J. (1968). An apparatus for estimating the effects of toxicants on the critical flicker frequency response of the bluegill sunfish. *J. Proc. 23rd Ind. Waste Conf.* Purdue Univ. 1968, 849–55.

Scherer, E. and Nowak, S. (1973). Apparatus for recording avoidance movements of fish. *J. Fish. Res. Bd Can.* **30**: 1594–6.

Schiffman, R. H. and Fromm, P. O. (1959). Chromium-induced changes in the blood of rainbow trout, *Salmo gairdnerii. Sewage Ind. Wastes* **31**: 205–211.

Schmeing-Engberding, F. (1953). Die Vorzugstemperaturen Knochenfische und ihre physiologische Bedeutung. *Z. Fisch. Berlin.* Band II, Neue Folge, Heft 1 and 2.

Schneider, M. J., Barraclough. S. A., Genoway, R. G. and Wolford, M. L. (1980). Effects of phenol on predation of juvenile rainbow trout *Salmo gairdneri. Environ. Pollut. A* **23**: 121–30.

Schnick, R. A. (1974). A review of the literature on the use of rotenone in fisheries. Bureau of Sport Fisheries and Wildlife, La Crosse, Wisconsin. NTIS No. PB-235 454.

Schober, U. and Lampert, W. (1977). Effects of sublethal concentrations of the herbicide Atrazin on growth and reproduction of *Daphnia pulex. Bull. Environ. Contam. Toxicol.* **17**: 269–277.

Schultz, D. P. and Harman, P. D. (1974). Residues of 2,4-D in pond waters, mud and fish, 1971. *Pest. Monit. J.* **8**: 173–9.

Scourfield, D. J. and Harding, J. P. (1966). A key to the British species of freshwater Caldocera. *Sci. Publs. Freshwat. Biol. Assn.* **5**: (3rd edn).

Scullion, J. and Edwards, R. W. (1980a). The effect of coal industry pollutants on the macroinvertebrate fauna of a small river in the South Wales coalfield. *Freshwat. Biol.* **10**: 141–62.

Scullion, J. and Edwards, R. W. (1980b). The effect of pollutants from the coal industry on the fish fauna of a small river in the South Wales coalfield. *Environ. Pollut. A* **21**: 141–53.

Seber, G. A. F. and Le Cren, E. D. (1967). Estimating population parameters from catches large relative to the population. *J. Anim. Ecol.* **36**: 631–43.

Seegeert, G. L., Brooks, A. S., Vande Castle, J. R. and Gradall, K. (1979). The effects of monochloramine on selected riverine fishes. *Trans. Am. Fish. Soc.* **108**: 88–96.

Shannon, C. E. (1948). A mathematical theory of communication. *Bell Systems Tech. J.* **27**: 379–423, 623–56.

Shaw, H. M., Saunders, R. L. and Hall, H. C. (1975). Environmental salinity: its failure to influence growth of Atlantic salmon (*Salmo salar*) parr. *J. Fish. Res. Bd Can.* **32**: 1821–4.

Shaw, T. L. (1979). A fast indicator for detecting gross pollution by hazardous chemicals. *New Zealand J. Mar. Freshwat. Res.* **13**: 393–4.

Shaw, T. L. and Brown, V. M. (1974). The toxicity of some forms of copper to rainbow trout. *Wat. Res.* **8**: 377–82.

Shell, G. L. (1976). Industrial wastewater treatment technology. In: *Industrial*

Wastewater Management Handbook. (H. S. Azad, Ed.) New York, McGraw-Hill.

Sherberger, F. F., Benfield, E. F., Dickson, K. L. and Cairns, J. (1977). Effects of thermal shocks on drifting aquatic insects: a laboratory simulation. *J. Fish. Res. Bd Can.* **34**: 529–36.

Sherr, C. A. and Armitage, K. B. (1973). Preliminary studies of the effects of dichromate ion on survival and oxygen consumption of *Daphnia pulex* (L). *Crustaceana* **25**: 51–69.

Sigmon, C. F., Kania, H. J. and Beyers, R. J. (1977). Reductions in biomass and diversity resulting from exposure to mercury in artificial streams. *J. Fish. Res. Bd Can.* **34**: 493–500.

Silbergeld, E. K. (1973). Dieldrin. Effects of chronic sublethal exposure on adaptation to thermal stress in freshwater fish. *Environ. Sci. Technol.* **7**: 846–9.

Simisan, G. V. and Chesters, G. (1976). Persistence of diquat in the aquatic environment. *Wat. Res.* **10**: 105–112.

Simpson, E. H. (1949). Measurement of diversity. *Nature (Lond.)* **163**: 688.

Sinley, J. R., Goettl, J. P. and Davies, P. H. (1974). The effects of zinc on rainbow trout (*Salmo gairdneri*) in hard and soft water. *Bull. Environ. Contam. Toxicol.* **12**: 193–201.

Skaar, D. R., Johnson, B. T., Jones, J. R. and Huckins, J. N. (1981). Fate of Kepone and mirex in a model aquatic environment: sediment, fish and diet. *Can. J. Fish. Aquat. Sci.* **38**: 931–8.

Skidmore, J. F. (1964). Toxicity of zinc compounds to aquatic animals with special reference to fish. *Q. Rev. Biol.* **39**: 227–48.

Skulberg, O. (1964). Algal problems related to the eutrophication of European water supplies, and a bio-assay method to assess fertilizing influences of pollution on inland waters. In: *Algae and Man*. (D. F. Jackson, Ed.) pp. 262–99. New York, Plenum Press.

Skulberg, O. M. (1966). Algal culture as a means to assess the fertilizing influence of pollution. *Proc. 3rd Int. Conf. Wat. Pollut. Res.* Paper 1–6.

Sladecek, V. (1964). Technische Hydrobiologie II. Tschechoslowakische Beitrage zum Saprobiensystem. *Sci. Pap. Inst. Chem. Technol. Prague, Technol. Wat.* **8**: 529–56.

Sladecek, V. (1965). The future of the saprobity system. *Hydrobiologia* **25**: 518–37.

Sladecek, V. (1966). Water quality system. *Verh. int. Verein. theor. angew. Limnol* **16**: 809–816.

Sladecek, V. (1967). The ecological and physiological trends in the saprobiology. *Hydrobiologia* **30**: 513–26.

Sladecek, V. (1971). Saprobic sequence within the genus *Vorticella*. *Wat. Res.* **5**: 1135–40.

Sladecek, V. (1973a). The reality of three British biotic indices. *Wat. Res.* **7**: 995–1002.

Sladecek, V. (1973b). System of water quality from the biological point of view. *Arch. Hydrobiol. (Ergebn. Limnol.)* **7**: 1–218.

Sladecek, V. (1981). Indicator value of the genus *Opercularia* (Ciliata). *Hydrobiologia* **79**: 229–32.

Sladeckova, A. (1962). Limnological investigation methods for the periphyton ('Aufwuchs') community. *Bot. Rev.* **28**: 286–350.

Sladeckova, A. (1968). Control of slimes and algae in cooling systems. *Verh. int. Verein. theor. angew. Limnol.* **17**: 532–8.

Slooff, W. and Van Kreijl, C. F. (1982). Monitoring the Rivers Rhine and Mease in the Netherlands for mutagenic activity using the Ames test in combination with rat or fish liver homogenates. *Aquat. Toxicol.* **2**: 89–98.

Smith, L. L. and Kramer, R. H. (1963). Survival of walleye eggs in relation to wood fibers and *Sphaerotilus natans* in Rainy River, Minnesota. *Trans. Am. Fish. Soc.* **92**: 220–34.

Smith, L. L., Kramer, R. H. and McLeod, J. C. (1965). Effects of pulpwood fibers on fathead minnows and walleye fingerlings. *J. Wat. Pollut. Control Red.* **37**: 130–140.

Smith, L., Oseid, D. M., Kimball, G. L. and El-Kandelgy, S. M. (1976). Toxicity of hydrogen sulfide to various life history stages of bluegill (*Lepomis macrochirus*). *Trans. Am. Fish. Soc.* **105**: 442–9.

Smith, R. A. (1972). *Air and Rain, the Beginnings of a Chemical Climatology.* London, Longmans Green.

Smith, W. E. (1973). A cyprinodontid fish, *Jordanella floridae*, as a laboratory animal for rapid chronic bioassays. *J. Fish Res. Bd Can.* **30**: 329–30.

Snarski, V. M. and Olson, G. F. (1982). Chronic toxicity and bioaccumulation of mercuric chloride in the fathead minnow (*Pimephales promelas*). *Aquat. Toxicol.* **2**: 143–56.

Snarski, V. M. and Puglisi, F. (1976), Effects of Aroclor 1254 on brook trout *Salvelinus fontinalis*. Ecol. Res. Ser. EPA-600/3–76–112, 1–34.

Sniesko, S. F. (1960). Microhematocrit as a tool in fishery research and management. US Fish Wildlife Serv., Spec. Sci. Rept. Fish. **341**.

Södergren, A., Svensoon, B. and Ulfstrand, S. (1972). DDT and PCB in south Swedish streams. *Environ. Pollut.* **3**: 25–36.

Soeder, C. J. and Talling, J. F. (1969). The enclosure of phytoplankton communities. In: *IBP Handbook No. 12, A Manual on Methods for Measuring Primary Production in Aquatic Environments.* pp. 62–70. Oxford, Blackwell Scientific Publications.

Sokal, R. R. (1961). Distance as a measure of taxonomic similarity *Syst. Zool.* **10**: 71–9.

Sokal, R. R. and Sneath, P. H. A. (1963). *Principles of Numerical Taxonomy.* San Francisco, Freeman.

Solbé, J. F. de L. G. (1974). The toxicity of zinc sulphate rainbow trout in very hard water. *Wat. Res.* **8**: 389–91.

Solbé, J. F. de L. G. and Flook, V. A. (1975). Studies on the toxicity of zinc sulphate and of cadmium sulphate to stone-loach *Noemacheilus barbatulus* in hard water. *J. Fish Biol.* **7**: 631–7.

Sorensen, T. (1948). A method of establishing groups of equal amplitude in plant sociology based on similarity of species content and its application to analyses of the vegetation on Danish commons. *Biol. Skr. (K. danske. vidensk. Selsk. N. S.)* **5**: 1–34.

Sorgeloos, P. and Persoone, G. (1973). A culture system for *Artemia, Daphnia*, and other invertebrates, with continuous separation of the larvae. *Arch. Hydrobiol.* **72**: 133–8.

Sparks, R. E., Cairns, J., McNabb, R. A. and Suter, G. (1972a). Monitoring zinc concentrations in water using the respiratory response of bluegills (*Lepomis macrochirus* Rafinesque). *Hydrobiologia* **40**: 361–9.

Sparks, R. E., Cairns, J. and Heath, A. G. (1972b). The use of bluegill breathing rates to detect zinc. *Wat. Res.* **6**: 895–911.

Spear, P. A. (1981). Zinc in the aquatic environment: chemistry, distribution, and toxicology. National Research Council of Canada, Associate Committee on Scientific Criteria for Environmental Quality, Ottawa. Report NRCC No. 17589.

Spearman, C. (1913). Correlations of sums and differences. *Brit. J. Psychol.* **5**: 417–26.

Spehar, R. L., Anderson, R. L. and Fiandt, J. T. (1978a). Toxicity and bioaccumulation of cadmium and lead in aquatic invertebrates. *Environ. Pollut.* **15**: 195–208.

Spehar, R. L., Leonard, E. N. and Defoe, D. L. (1978b). Chronic effects of cadmium and zinc mixtures on flagfish (*Jordanella floridae*). *Trans. Am. Fish. Soc.* **107**: 354–60.

Spehar, R. L., Fiandt, J. T. Anderson, R. L. and DeFoe, D. L. (1980). Comparative toxicity of arsenic compounds and their accumulation in invertebrates and fish. *Arch. Environ. Contam. Toxicol.* **9**: 53–63.

Spencer, D. F. and Greene, R. W. (1981). Effects of nickel on seven species of freshwater algae. *Environ. Pollut. A* **25**: 241–7.

Spoor, W. A. and Drummond, R. A. (1972). An electrode for detecting movement in gradient tanks. *Trans. Am. Fish. Soc.* **101**: 714–15.

Spoor, W. A., Neiheisel, T. W. and Drummond, R. A. (1971). An electrode chamber for recording respiratory and other movements of free-swimming animals. *Trans. Am. Fish Soc.* **100**: 22–8.

Sprague, J. B. (1964a). Lethal concentrations of copper and zinc for young Atlantic salmon. *J. Fish. Res. Bd Can.* **21**: 17–26.

Sprague, J. B. (1964b). Avoidance of copper–zinc solutions by young salmon in the laboratory. *J. Wat. Pollut. Control Fed.* **36**: 990–1004.

Sprague, J. (1966). *Iron Ore Wastes*. Pollution studies, Spec. Rep. St Andrews Biol. Stn. F–8, 15–18.

Sprague, J. B. (1968). Promising anti-pollutant: chelating agent NTA protects fish from copper and zinc. *Nature, Lond.* **200**: 1345–6.

Sprague, J. B. (1969). Measurement of pollutant toxicants to fish. I. Bioassay methods for acute toxicity. *Wat. Res.* **3**: 793–821.

Sprague, J. B. (1970). Measurement of pollutant toxicity to fish. II. Utilizing and applying bioassay results. Review paper. *Wat. Res.* **4**: 3–32.

Sprague, J. B. (1971). Measurement of pollutant toxicity to fish. III. Sublethal effects and 'safe' concentrations. *Wat. Res.* **5**: 245–66.

Sprague, J. B. (1973). The ABC's of pollutant bioassay using fish. In: *Biological Methods for the Assessment of Water Quality*. (J. Cairns and K. L. Dickson, Eds.) pp. 6–30. ASTM STP 528.

Sprague, J. B. and Drury, D. E. (1969). Avoidance reactions of salmonid fish to representative pollutants. *Proc. 4th int. Conf. Wat. Pollut. Res.*, Prague, 1969, pp. 169–79.

Stadelmann, P. (1974). Biomass estimation by measurement of adenosine triphosphate. In: *A Manual on Methods for Measuring Primary Productivity in*

Aquatic Environments, (R. A., Vollenweider, Ed.) pp. 26–30. I.B.P. Handbook No 12. 2nd edn. Oxford, Blackwells.

Stanley, R. A. (1974). Toxicity of heavy metals and salts to Eurasian water milfoil (*Myriophyllum spicatum*) *Arch. Environ. Contam. Toxicol.* **2**: 331–40.

Starmühlner, F. (1953). Die Molluskenfauna unsere Wienerwaldbächer. *Wett. Leben Sonderhg.* **2**: 184–295.

Stebbing, A. R. D. and Pomroy, A. J. (1978). A sublethal technique for assessing the effects of contaminants using *Hydra littoralis. Wat Res.* **12**: 631–5.

Stendahl, D. H. and Sprague, J. B. (1982). Effects of water hardness and pH on vanadium lethality to rainbow trout. *Wat. Res.* **16**: 1479–88.

Stephan, C. E. and Mount, D. I. (1973). Use of toxicity tests with fish in water pollution control In: *Biological Methods for the Assessment of Water Quality.* ASTM STP 528, 164–177.

Stephenson, R. R. (1982). Aquatic toxicology of cypermethrin. I. Acute toxicity to some freshwater fish and invertebrates in laboratory tests. *Aquat. Toxicol.* **2**: 175–85.

Stewart, N. E., Millemann, R. E. and Breese, W. P. (1967). Acute toxicity of the insecticide Sevin and its hydrolytic product 1-naphthol to some marine organisms. *Trans. Am. Fish. Soc.* **96**: 25–30.

Stiff, M. J. (1971a). Copper/bicarbonate equilibria in solutions of bicarbonate ion at concentrations similar to those found in natural water. *Wat. Res.* **5**: 171–6.

Stiff, M. J. (1971b). The chemical states of copper in polluted freshwater and a scheme of analysis to differentiate them. *Wat. Res.* **5**: 585–99.

Stott, B. (1967). The movements and population densities of roach (*Rutilus rutilus* (L)) and gudgeon (*Gobio gobio* (L)) in the River Mole. *J. Anim. Ecol.* **36**: 407–24.

Stott, B. (1970). Some factors affecting the catching power of unbaited traps. *J. Fish Biol.* **2**: 15–22.

Stott, B. and Cross, D. G. (1973). The reactions of roach (*Rutilus rutilus* (L)) to changes in the concentration of dissolved oxygen and free carbon dioxide in a laboratory channel. *Wat. Res.* **7**: 793–805.

Stratton, G. W. and Corke, C. T. (1981). Interaction of permethrin with *Daphnia magna* in the presence and absence of particulate material. *Environ. Pollut. A* **24**: 135–44.

Strek, H. J. and Weber, J. B. (1982). Behaviour of polychlorinated biphenyls (PCBs) in soils and plants. *Environ. Pollut. Ser. A* **28**: 291–312.

Strickland, J. D. H. and Parsons, T. R. (1963). *A Manual of Sea Water Analysis.* Fisheries Research Board of Canada, Bulletin No. 125.

Sturm, R. N. and Payne, A. G. (1973). Environmental testing of trisodium nitrilotriacetate: bioassays for aquatic safety and algal stimulation. In: *Bioassay Techniques and Environmental Chemistry* (G. E. Glass, Ed.) pp. 403–424. Ann Arbor.

Sullivan, J. F., Atchinson, G. J., Kolar, D. J. and McIntosh, A. W. (1978). Changes in the predator–prey behaviour of fathead minnows (*Pimephales promelas*) and largemouth bass (*Micropterus salmoides*) caused by cadmium. *J. Fish. Res. Bd Can.* **35**: 446–51.

Summerfelt, R. C. and Lewis, W. M. (1967). Repulsion of green sunfish by certain chemicals. *J. Wat. Pollut. Control Fed.* **39**: 2030–2038.

Surber, E. W. (1937). Rainbow trout and bottom fauna production in one mile of stream. *Trans. Am. Fish. Soc.* **66**: 193–202.

Swift, D. J. (1978). Some effects of exposing rainbow trout (*Salmo gairdnerii* Richardson) to phenol solutions. *J. Fish Biol.* **13**: 7–17.

Sykora, J. L., Smith, E. J. and Synak, M. (1972). Effect of lime neutralized iron hydroxide suspensions on juvenile brook trout (*Salvelinus fontinalis*, Mitchel). *Wat. Res.* **6**: 935–50.

Sylvester, J. R. (1972a). Effect of thermal stress on predator avoidance in sockeye salmon. *J. Fish. Res. Bd Can.* **29**: 601–603.

Sylvester, J. R. (1972b). Possible effects of thermal effluents on fish: a review. *Environ. Pollut.* **3**: 205–215.

Talling, J. F. (1969). General outline of spectrophotometric methods. In: *IBP Handbook No. 12, A Manual on Methods for Measuring Primary Production in Aquatic Environments.* pp. 22–5. Oxford, Blackwell Scientific Publications.

Talling, J. F. and Driver, D. (1963). Some problems in the estimation of chlorophyll-*a* in phytoplankton. *Proceedings, Conference of Primary Productivity Measurement, Marine and Freshwater*, Hawaii, 1961. US Atomic Energy Comm. TID-7633, 142–146.

Talling, J. F. and Fogg, G. E. (1969). Measurements (*in situ*) on isolated samples of natural communities; Possible limitations and artificial modifications. In: *IBP Handbook No. 12, A Manual on Methods for Measuring Primary Production in Aquatic Environments.* pp. 73–78. Oxford, Blackwell Scientific Publications.

Tarzwell, C. M. and Henderson, C. (1956). Toxicity of dieldrin to fish. *Trans. Am. Fish. Soc.* **86**: 245–57.

Tebo, L. B. (1955). Effects of siltation, resulting from improper logging, on the bottom fauna of a small trout stream in the Southern Appalachians. *Progve Fish Cult.* **17**: 64–70.

Teckelmann, U. (1974). The effects of temperature on growth and metabolism of flowing water cold-stenotherms. *Arch. Hydrobiol.* **74**: 479–95, 516–22.

Ten Berge, W. F. (1978). Breeding *Daphnia magna. Hydrobiologia* **59**: 121–3.

Tevlin, M. P. (1978). An improved experimental medium for freshwater toxicity studies using *Daphnia magna. Wat. Res.* **12**: 1017–1024.

Thienemann, A. (1925). Die Binnengewässer Mitteleuropas. *Die Binnengewässer* **1**: 54–83.

Tippett, R. (1970). Artificial surfaces as a method of studying populations of benthic micro-algae in fresh water. *Br. Phycol. J.* **5**: 187–99.

Tolland, M. G. (1977). Destratification/aeration in reservoirs. A literature review of the techniques used for water quality management. Technical Report TR50, Water Research Centre, Medmenham.

Tomasso, J. R., Simco, B. A. and Davis, K. B. (1979). Chloride inhibition of nitrite-induced methemoglobinemia in channel catfish (*Ictalurus punctatus*). *J. Fish. Res. Bd Can.* **36**: 1141–4.

Tonolli, L. (Ed.) (1976). Quality objectives for aquatic life in European freshwaters. Final Report to EEC of the Instituto Italiano di Idrobiologia, Pallanza, ENG/368/76E.

Tooby, T. E. (1972). Pollution studies: toxicity of dichlobenil to roach *Rutilus*

rutilus. Ann. Rep. Salm. Freshwat. Fish. Labs. Ministry of Agriculture Fisheries and Food (UK): 24–25.

Tooby, T. E. (1976). Effects of aquatic herbicides on fisheries. In: *Aquatic Herbicides.* (T. O. Robson and J. H. Fearon, Eds.) pp. 62–77. Proceedings of a symposium held in Oxford, January 1976. British Crop Protection Council Monograph 16.

Tooby, T. E. (1981a). Ecotoxicological aspects of water pollution by certain mothproofing agents with reference to Eulan®, Mitin® and Permethrin. Report prepared for the Commission of the European Communities—Environment and Consumer Protection Service, Contract No. U/80/336.

Tooby, T. E. (1981b). Predicting the direct toxic effects of aquatic herbicides to non-target organisms. In: *Aquatic Weeds and Their Control.* pp. 265–74. Proceedings of a conference held in Oxford, April 1981. Association of Applied Biologists, Wellesbourne, Warwick, UK.

Tooby, T. E. and Durbin, F. J. (1975). Lindane residue accumulation and elimination in rainbow trout (*Salmo gairdnerii*, Richardson) and roach (*Rutilus rutilus* Linnaeus). *Environ. Pollut.* **8**: 79–89.

Tooby, T. E., Hursey, P. A. and Alabaster, J. S. (1975). The acute toxicity of 102 pesticides and miscellaneous substances to fish. *Chemy Indust.* 523–6.

Train, R. E. (1979). *Quality Criteria for Water.* London, Castle House Publications.

Trollope, D. R. and Evans, B. (1976). Concentrations of copper, iron, lead, nickel and zinc in freshwater algal blooms. *Environ. Pollut.* **11**: 109–116.

Tyson, D. (1974). The fate of terbutryne in the aquatic ecosystem and its effect on non-target organisms. Technical Symposium: The use of terbutryne as an aquatic herbicide. pp. 21–32. London, Royal Commonwealth Society.

Uthe, J. F., Atton, F. M. and Royer, L. M. (1973). Uptake of mercury by caged rainbow trout (*Salmo gairdneri*) in the South Saskatchewan River. *J. Fish. Res. Bd Can.* **30**: 643–50.

Vance, B. D. and Drummond, W. (1969). Biological concentration of pesticides by algae. *J. Am. Wat. Wks Ass.* **61**: 360–362.

Vance, B. D. and Smith, D. L. (1969). Effects of five herbicides on three green algae. *Tex. J. Sci.* **20**: 329–37.

Van den Berg, C. M. G., Wong, P. T. S. and Chau, Y. K. (1979). Measurement of complexing materials excreted from algae and their ability to ameliorate copper toxicity. *J. Fish. Res. Bd Can.* **36**: 901–905.

Van der Putte, I., Lubbers, J. and Kolar, Z. (1981). Effect of pH on uptake, tissue distribution and retention of hexavalent chromium in rainbow trout (*Salmo gairdneri*). *Aquat. Toxicol.* **1**: 3–18.

Van der Schalie, W. H., Dickson, K. L., Westlake, G. F. and Cairns, J. (1979). Fish bioassay monitoring of waste effluents. *Environ. Mgmt* **3**: 217–35.

Van Hassel, J. H., Ney, J. J. and Garling, D. L. (1980). Heavy metals in a stream ecosystem at sites near highways. *Trans, Am. Fish. Soc.* **109**: 636–43.

Veith, G. D. and Lee, G. F. (1970). A review of chlorinated biphenyl contamination in natural waters. *Wat. Res.* **4**: 265–9.

Veith, G. D. and Lee, G. F. (1971). Chlorobiphenyls (PCBs) in the Milwaukee River. *Wat. Res.* **5**: 1107–1115.

Veith, G. D., DeFoe, D. L. and Bergstedt, B. V. (1979). Measuring and estimating the bioconcentration factor of chemicals in fish. *J. Fish. Res. Bd Can.* **36**: 1040–8.

Vibert, R. (Ed.) (1967). *Fishing with Electricity. Its Application to Biology and Management.* EIFAC, FAO. London, Fishing News Books.

Vollenweider, R. A. (1969). *A Manual on Methods for Measuring Primary Production in Aquatic Environments.* IBP Handbook No. 12. Oxford, Blackwell Scientific Publications.

Waiwood, K. G. and Beamish, F. W. H. (1978). Effects of copper, pH and hardness on the critical swimming performance of rainbow trout (*Salmo gairdneri* Richardson). *Wat. Res.* **12**: 611–619.

Walden, C. C., Howard, T. E. and Froud, G. C. (1970). A quantitative assay of the minimum concentrations of kraft mill effluents which affect fish respiration. *Wat. Res* **4**: 61–8.

Walker, C. R., Lennon, R. E. and Berger, B. L. (1964). Preliminary observations on the toxicity of antimycin A to fish and other aquatic animals. US Bur. Sport Fish. Wildlife, Invest, Fish Control. Circular No. **186**: 1–18.

Wallace, J. B. and Brady, U. E. (1971). Residue levels of dieldrin in aquatic invertebrates and effect of prolonged exposure on populations. *Pest. Monit. J.* **5**: 295–300.

Wallace, R. R. and Hynes, H. B. N. (1975). The catastrophic drift of stream insects after treatments with methoxychlor (1,1,1-trichloro-2,2-Bis ̓ (p-methoxyphenyl)ethane). *Environ. Pollut.* **8**: 255–68.

Wallen, I. E., Green, W. C. and Lasater, R. (1957). Toxicity to *Gambusia affinis* of certain pure chemicals in turbid waters. *Sew. Ind. Wastes* **29**: 695–711.

Waller, W. T. and Cairns, J. (1972). The use of fish movement patterns to monitor zinc in water. *Wat. Res.* **6**: 257–69.

Walsh, F. and Mitchell, R. (1972). A pH-dependent succession in iron bacteria. *Environ. Sci. Technol.* **6**: 809.

Walsh, G. E., Miller, C. W. and Heitmuller, P. L. (1971). Uptake and effects of dichlobenil in a small pond. *Bull. Environ. Contam. Toxicol.* **6**: 279–88.

Ward, D. V. (1978). *Biological Environmental Impact Studies: Theory and Methods.* New York Academic Press.

Warn, A. E. (1978). The Trent mathematical model. In: *Mathematical Models in Water Pollution Control.* (A. James, Ed.) Chichester, Wiley.

Warner, R. E., Peterson, K. K. and Bergman, L. L. (1966). Behavioural pathology in fish: a quantitative study of sublethal pesticide toxication. *J. Appl. Ecol.* **3** Supplement. Pesticides in the environment and their effects on wildlife. (N. W. Moore, Ed.) pp. 223–47.

Warner, R. W. (1971). Distribution of biota in a stream polluted by acid mine-drainage. *Ohio J. Sci.* **71**: 202–215.

Warnick, S. L. and Bell, H. L. (1969). The acute toxicity of some heavy metals to different species of aquatic insects. *J. Wat. Pollut. Control Fed.* **41**: 280–284.

Warren, C. E. (1971). *Biology and Water Pollution Control.* Philadelphia, W. B. Saunders Co.

Watanabe, T. (1962). On the biotic index of water pollution based upon the species number of Bacillariophyceae in the Tokoro River in Hokkaido. *Jap. J. Ecol.* **12**: 216–222.

Waters, T. F. (1961). Standing crop and drift of stream bottom organisms. *Ecology* **42**: 532–7.

Waters, T. F. (1962). Diurnal periodicity in the drift of stream invertebrates. *Ecology* **43**: 316–320.

Way, J. M., Newman, J. F., Moore, N. W. and Knaggs, F. W. (1971). Some ecological effects of the use of paraquat for the control of weeds in small lakes. *J. Appl. Ecol.* **8**: 509–532.

Weber, C. I. (Ed.) (1973). *Biological Field and Laboratory Methods for Measuring the Quality of Surface Waters and Effluents.* US Environmental Protection Agency. EPA 670/4–73–001.

Wedemeyer, G. A. and Yasutake, W. T. (1978). Prevention and treatment of nitrite toxicity in juvenile steelhead trout (*Salmo gairdneri*) *J. Fish. Res. Bd Can.* **35**: 822–7.

Wedemeyer, G. A., Nelson, N. C. and Yasutake, W. T. (1979). Physiological and biochemical aspects of ozone toxicity to rainbow trout (*Salmo gairdneri*) *J. Fish. Res. Bd Can.* **36**: 605–14.

Weiderholm, T. (1971). Bottom fauna and cooling water discharges in a basin of Lake Malaren. *Rep. Inst. Freshwat. Res. Drottningholm* **51**: 197–214.

Weir, P. A. and Hine, C. H. (1970). Effects of various metals on behaviour of conditioned goldfish. *Arch. Environ. Health* **20**: 45–50.

Weiss, C. M. (1965). Use of fish to detect organic insecticides in water. *J. Wat. Pollut. Control Fed.* **37**: 647–58.

Weitkamp, D. E. and Katz, M. (1980). A review of dissolved gas supersaturation literature. *Trans. Am. Fish. Soc.* **109**: 659–702.

Welch, E. B. (1980) *Ecological Effects of Waste Water.* Cambridge, Cambridge Univ. Press.

Weller, J. B. and Willetts, S. L. (1977). *Farm Wastes Management.* London, Crosby Lockwood Staples.

Westin, D. T. (1974). Nitrate and nitrite toxicity to salmonid fishes. *Progve Fish Cult.* **36**: 86–9.

Westlake, D. F. (1964). Light extinction, standing crop and photosynthesis within weed beds. *Verh. int. Verein. theor. angew. Limnol.* **15**: 415–25.

Westlake, D. F. (1968). The biology of aquatic weeds in relation to their management. *Proc. 9th British Weed Control Conf.* pp. 372–9.

Westlake, D. F. (1973). Aquatic macrophytes in rivers. A review. *Polskie Archiwum Hydrobiologii* **20**: 31–40.

Whitaker, G. A., McCuen, R. H. and Brush, J. (1979). Channel modification and macroinvertebrate community diversity in small streams. *Wat. Resour. Bull.* **15**: 875–9.

White, D. S. and Jennings, D. E. (1973). A rearing technique for various aquatic Coleoptera. *Ann. Entomol. Soc. Am.* **66**: 1174–6.

Whitley, L. S. and Sikora, R. A. (1970). The effects of three common pollutants on the respiration rate of tubificid worms. *J. Wat. Pollut. Control Fed.* **42**: (part 2): R57–R66.

Whitten, B. K. and Goodnight, C. J. (1966). Toxicity of some common insecticides to tubificids. *J. Wat. Pollut. Control Fed.* **38**: 227–35.

Whitton, B. A. (1967). Studies on the growth of riverain *Cladophora* in culture. *Arch. Mikrobiol.* **58**: 21–9.

Whitton, B. A. (1970a). Toxicity of heavy metals to freshwater algae: a review. *Phykos* **9**: 116–125.

Whitton, B. A. (1970b). Toxicity of heavy metals to Chlorophyta from flowing waters. *Arch. Mikrobiol* **72**: 353–60.

Whitton, B. A. (1970c). Biology of *Cladophora* in freshwaters. *Wat. Res.* **4**: 457–76.

Whitton, B. A. (1979). Plants as indicators of river water quality. In: *Biological Indicators of Water Quality.* Proceedings of a symposium at the University of Newcastle-upon-Tyne, UK, September 1978.

Whitton, B. A. and Say, P. J. (1975). Heavy metals. In: *River Ecology.* (B. A. Whitton Ed.) pp. 268–311. Oxford, Blackwell.

Whitton, B. A., Say, P. J. and Jupp, B. P. (1982). Accumulation of zinc, cadmium and lead by the aquatic liverwort *Scapania. Environ. Pollut. Ser. B.* **3**: 299–316.

Wilhm, J. L. and Dorris, T. C. (1968). Biological parameters for water quality criteria. *Bioscience* **18**: 477–81.

Williams, A. K. and Sova, C. R. (1966). Acetylcholinesterase levels in brains of fishes from polluted waters. *Bull. Environ. Contam. Toxicol.* **1**: 198–204.

Williams, C. B. (1953). The relative abundance of different species in a wild animal population. *J. Anim. Ecol.* **22**: 14–31.

Wilson, D. C. and Bond, C. E. (1969). The effects of the herbicides diquat and dichlobenil (Casoron) on pond invertebrates. Part I. Acute toxicity. *Trans. Am. Fish. Soc.* **98**: 438–42.

Winner, R. W. (1981). A comparison of body length, brood size and longevity as indices of chronic copper and zinc stresses in *Daphnia magna. Environ. Pollut. Ser. A* **26**: 33–7.

Winner, R. W. and Farrell, M. P. (1976). Acute and chronic toxicity of copper to four species of *Daphnia. J. Fish. Res. Bd Can.* **33**: 1685–91.

Winner, R. W., Boesel, M. W. and Farrell, M. P. (1980). Insect community structure as an index of heavy-metal pollution in lotic systems. *Can. J. Fish. Aquat. Sci.* **37**: 647–55.

Wisdom, A. S. (1956). *The Law on the Pollution of Waters.* London, Shaw and Sons.

Wisdom, A. S. (1962; 4th edn 1979). *The Law of Rivers and Watercourses.* London, Shaw & Sons.

Wisdom, A. S. (1981). *Aspects of Water Law.* Chichester, Barry Rose Publishers.

Wobeser, G. (1975). Acute toxicity of methyl mercury chloride and mercuric chloride for rainbow trout (*Salmo gairdneri*) fry and fingerlings. *J. Fish. Res. Bd Can.* **32**: 2005–13.

Wojtalik, T. A., Hall, T. F. and Hill, L. O. (1971). Monitoring ecological conditions associated with wide-scale applications of DMA 2,4-D to aquatic environments. *Pestic. Monit. J.* **4**: 184–203.

Wong, P. T. S., Chau, Y. K., Kramar, O. and Bengert, G. A. (1981). Accumulation and depuration of tetramethyllead by rainbow trout. *Wat. Res.* **15**: 621–5.

Woodiwiss, F. S. (1964). The biological system of stream classification used by the Trent River Board. *Chemy. Indust.* **11**: 443–7.

Woodward, D. F. (1976). Toxicity of herbicides dinoseb and picloram to cut throat (*Salmo clarki*) and lake trout (*Salvelinus namaycush*). *J. Fish. Res. Bd Can.* **33**: 1671–6.

Woodworth, J. and Pascoe, D. (1982). Cadmium toxicity to rainbow trout, *Salmo gairdneri* Richardson: a study of eggs and alevins. *J. Fish. Biol.* **21**: 47–57.

Worthington, E. B. (1950). An experiment with populations of fish in Windermere, 1939–48. *Proc. Zool. Soc. Lond.* **120**: 113–149.

Wright, J. F., Moss, D., Armitage, P. D. and Furse, M. T. (1984). A preliminary classification of running-water sites in Great Britain based on macro-invertebrate species and the prediction of community type using environmental data. *Freshwat. Biol.* **14**: 221–56.

Wu, Y. F. (1931). A contribution to the biology of *Simulium. Pop. Mich. Acad. Sci.* **13**: 543–99.

Wuerthele, M., Zillich, J., Newton, M. and Fetterolf, C. (1973). Descriptions of a continuous-flow bioassay laboratory trailer and the Michigan diluter. In: *Bioassay Techniques and Environmental Chemistry.* (G. E., Glass Ed.) pp. 345–54. Ann Arbor, Michigan, Ann Arbor Science Publishers.

Wuhrmann, K. (1952). Sur quelques principes de la toxicologie du poisson. *Bull. Cent. Belge Etude et Document. Eaux* **15**: 77–85.

Wuhrmann, K. and Woker, H. (1948). Experimentelle Untersuchungen über die Ammoniak- und Blausaurevergiftung, *Schweiz Z. Hydrol.* **11**: 210–44.

Wynne-Edwards, V.C. (1962). *Animal Dispersion in Relation to Social Behaviour*, New York, Hafner.

Wynne-Edwards, V. C. (1965). Self-regulating system in populations of animals. *Science, NY* **147**: 1543–8.

Yadin, Y. (1966). *Masada. Herod's Fortress and the Zealots Last Stand.* London, Weidenfeld & Nicolson.

Yentsch, C. S. and Menzel, D. W. (1963). A method for the determination of phytoplankton chlorophyll and phaeophytin by fluorescence. *Deep Sea Research* **10**: 221–31.

Yeo, R. R. (1967). Dissipation of diquat and paraquat and effects on aquatic weeds and fish. *Weeds* **15**: 42–6.

Youngs, W. F., Gutenmann, W. H. and Lisk, D. J. (1972). Residues of DDT in lake trout as a function of age. *Environ. Sci. Technol.* **6**: 451–2.

Zauke, G.-P. (1982). Cadmium in Gammaridae (Amphipoda: Crustacea) of the River Werra and Weser — II. Seasonal variation and correlation to temperature and other environmental variables. *Wat. Res.* **16**: 785–92.

Zehnder, A. J. B. (1978). Ecology of methane formation. In: *Water Pollution Microbiology Vol. 2.* (R. Mitchel, Ed.) p. 349. New York, Wiley.

Zelinka, M. and Marvan, P. (1966). Bemerkung zu neuen Methoden der sapro-biologischen Wasserbeurteilung. *Verh. int. Verein. theor. angew. Limnol,* **16**: 817–22.

Zillich, J. A. (1972). Toxicity of combined chlorine residuals to freshwater fish. *J. Wat. Pollut. Control Fed.* **44**: 212–220.

Zillioux, E. J., Foulk, H. R., Prager, J. C. and Cardin, J. A. (1973). Using *Artemia* to assay oil dispersant toxicities. *J. Wat. Pollut. Control Fed.* **45**: 2389–6.

Zitko, V. and Carson, W. G. (1977). Seasonal and developmental variation in the lethality of zinc to juvenile Atlantic salmon (*Salmo salar*). *J. Fish. Res. Bd Can.* **34**: 139–41.

Zitko, V., Carson, W. G. and Metcalfe, C. D. (1977). Toxicity of pyrethroids to juvenile Atlantic salmon. *Bull. Environ. Contam. Toxicol.* **18**: 35–41.

Zitko, V., McLeese, D. W., Metcalfe, C. D. and Carson, W. G. (1979). Toxicity of permethrin, decamethrin, and related pyrethroids to salmon and lobster. *Bull. Environ. Contam. Toxicol.* **21**: 338–43.

Zueler, M. (1908). Zur Kenntnis der biologischen Wasserbeurteilung. *Int. Rev. ges. Hydrobiol.* **1**: 439–46.

Index

Index